Fundamentals and Applications of Nano Silicon in Plasmonics and Fullerines

Micro and Nano Technologies

Fundamentals and Applications of Nano Silicon in Plasmonics and Fullerines

Current and Future Trends

Munir Nayfeh

ELSEVIER

Elsevier
Radarweg 29, PO Box 211, 1000 AE Amsterdam, Netherlands
The Boulevard, Langford Lane, Kidlington, Oxford OX5 1GB, United Kingdom
50 Hampshire Street, 5th Floor, Cambridge, MA 02139, United States

Library of Congress Cataloging-in-Publication Data
A catalog record for this book is available from the Library of Congress

British Library Cataloguing-in-Publication Data
A catalogue record for this book is available from the British Library

ISBN: 978-0-323-48057-4

Publisher: Matthew Deans
Acquisition Editor: Simon Holt
Editorial Project Manager: Sabrina Webber
Production Project Manager: Vijay Bharath. R
Designer: Miles Hitchen

Typeset by Thomson Digital

Contents

4. Seeing Atoms and Clusters on Surfaces

5. Manipulation and Patterning of Surfaces (Nanolithography)

10. Nanocarbon

Preface

Fast approaching is December 29, 2019, the 60th birthday of the vision Feynman presented in his talk at the American Physical Society meeting at Caltech "There's Plenty of Room at the Bottom" in which Feynman suggested the possibility of direct manipulation of individual atoms as a more powerful form of synthetic chemistry than those used at the time. Some considers this talk as the seed of a new nanotechnology industrial revolution despite the fact that the talk went unnoticed for a number of decades. It was rediscovered and publicized in the 1990'as a seminal event in the then emerging field of nanotechnology. President Clinton exploited it to prop his national nanotechnology initiative (NNI). In fact, so far nanotechnology, coined the technology of this century and the meeting ground of all fields, attracted billions of funding dollars, and thousands of scientists and engineers world-wide. It is exciting, novel and ventures into areas of S&T beyond nature where we never been. It captures the imagination of all, and affords us to dream to solve the problems facing the human race, such as acute disease, energy, lighting, water and food crunch. But with 60 years passing on Feynman's dream and over $20 billion invested in R&D by the US Federal government alone over the last 15 years, a question remains whether the abundancy of room at the bottom that Feynman envisioned would create room at the top for consumers. Nanotechnology has delivered some mundane products for consumer use but it yet to deliver advanced products, such as smart anticancer therapeutics that could destroy tumours, and lighter, thinner body armour that could save the lives of soldiers.

This book presents a comprehensive and smooth account of the stations, weather proposals or demonstrations that led to present day concept of nanotechnology: from the Feynman's atom by atom vision to attempts to leapfrog its realizations such as Feynman's own challenges, Drexler infinite mechanical assemblers, detection and trapping and manipulation of single atoms, imaging of individual atoms, writing and inscription with individual atoms, preassembled nanoparticles with beyond-nature exotic attributes, and the Clinton's (and others international leaders) national nanotechnology initiatives as well as Obama's initiative as the second White house initiative for nano commercialization. The book focuses on the nano structures of the trio materials: silicon, carbon, and gold, since they are emerging as the most convenient building blocks of nanotechnology, whose structures are most known and studied and make interesting, specific nanoscale configurations, while being the backbone of electronics, living systems and industrial power. Finally synergetic functionalities have been observed upon integration of nano- metal, silicon and carbon. The underlying basic science and technology for each station is taken seriously, yet simplified enough for the general readership. A bouquet of advanced and mundane applications that have been realized is presented in the same spirit as the science.

The book also reminds the readership that past generations from the 9th century through the Middle Ages may have beat our generation to nanotechnology as evidenced by use of nano metal lighting effects in ancient stained glass and lusterware pottery. Also ongoing attempts to search for nano silicon in space is presented.

The book touches on several societal implications including: consumer interest (products, commercialization, mass production, and business hurdles and what it takes to alleviate them); general public interest (safety, environment, military, international relations and ethics); and the role of the young generation in carrying the banner. Finally we define who are the nanotech pioneers and where and when are they taking us to a full development of a mature nano technology in everyday use.

Chapter 1

Atom-by-Atom Manufacturing: The Birth of Nanotechnology

1.1 FEYNMAN'S VISION

In a speech at the California Institute of Technology in December 29, 1959 by Nobel Laureate physicist Richard P. Feynman, "There's Plenty of Room at the Bottom," he discussed about the possibilities of molecular-scale engineering [1–2]. Specifically, Feynman considered the possibility of direct manipulation of individual atom as a more powerful form of synthetic chemistry than those that were used at that time. Fig. 1.1 shows the cartoon of the Feynman scheme. He also suggested the way in which it may be accomplished. He suggested that we can use big tools to make smaller tools suitable for making even smaller tools and so on, until tools suitable for directly manipulating atoms and molecules had been achieved. As to the feasibility of such a goal, he suggested that the only reason why it was not already being done was simply that no one had yet got around to do it.

In order to accomplish this goal, Feynman noted that tools are not only needed to manipulate individual atoms but also to image and record those individual atoms with enough spatial resolution and sensitivity.

Some consider this talk as the seed of a new industrial revolution, namely nanotechnology. There have been possibilities that could have startling results. In this regard, Feynman touched upon miniature writing, seeing and controlled moving of atoms, the prospect of designing molecules one atom at a time, and the challenges involved in developing miniature machines. Miniaturized devices are expected to afford a number of breakthrough applications, such as more powerful computers and increased data storage resulting from more efficient miniaturized components. Applications are also expected in many fields, such as photovoltaics, electronics, molecular electronics, and biomedicine. Many believe that once the technology is mature, it will afford the following features:

- Everything from buildings to motor cars to clothing to daily meals will be fabricated by swarms of sub-microscopic robots, which will manipulate individual atoms to build these things, or indeed.
- Components at the molecular level will be used as building blocks, as houses today are built from a pile of individual bricks.
- All this at little more cost than that of the necessary raw materials.
- Insignificant environmental impact.

Fundamentals and Applications of Nano Silicon in Plasmonics and Fullerines.
http://dx.doi.org/10.1016/B978-0-323-48057-4.00001-3

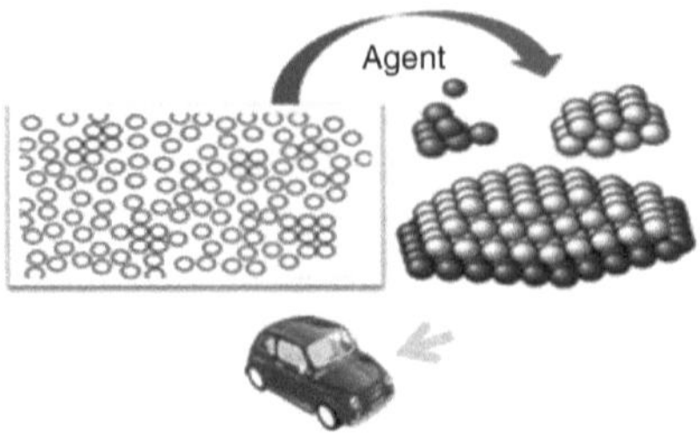

FIGURE 1.1 Cartoon of the Feynman scheme for atom-by-atom (ABA) fabrication from a single atom to a real object, such as a car.

The central tenet of nanotechnology is that any chemically stable structure, not specifically disallowed by the laws of physics, can in fact be built, and the word was originally used to describe the fabrication of materials and artifacts at molecular level by the direct manipulation of individual atoms and molecules. Today the definition has been widened to include any scientific or technical experimentation or fabrication that takes place on the nanometer scale and is often extended to include some activities at the lower end of the micrometer scale.

Richard P. Feynman

The American physicist Richard P. Feynman (1918–1988) was one of the most influential physicists of the 20th century. His series of lectures "The Feynman Lectures of Physics" is a very good introduction into the most important disciplines of modern physics. Additionally, he wrote some popular introductory science textbooks. These books are a proof of his ability to write about complex facts in a comprehensible way. Feynman was a professor of theoretical physics at the California Institute of Technology at Pasadena. His research was mainly concerned with quantum physics. In 1965, he won the Nobel Prize for his contributions to develop quantum electrodynamics.

1.2 FEYNMAN CHALLENGE FUNDING

Feynman went on to jump start research in the area of molecular-scale engineering. He decided to motive developments in two areas (see Fig. 1.2): (i) manufacturing of controlled motion devices and (ii) information storage (printing).

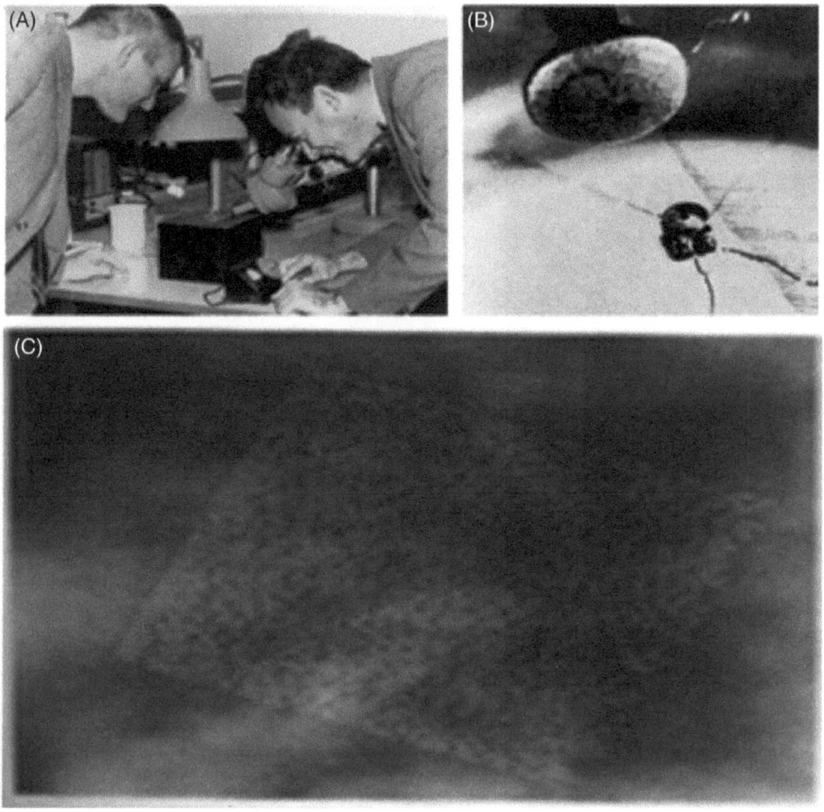

FIGURE 1.2 (A) Photograph of Richard Feynman viewing the micro motor built by William McLellan. (B) Photograph of the motor, 3.8-mm-wide, which is sitting beneath the head of a pin. (C) Enlarged photograph of the page written from *A Tale of Two Cities*, using electron beam lithography. This is 20-times smaller than the human eye can see.

To catalyze work in that direction, he offered $1,000 prizes from his personal funds to the first person to construct a working electric motor 1/64 inch or less on a side, and to the first person to produce written text at 1/25,000 scale (the size required to print the entire *Encyclopedia Britannica* on the head of a pin). In 1960, the motor prize was claimed by an engineer who constructed a very-small motor using conventional mechanical techniques. Unfortunately the size limit that Feynman set for the motor application was slightly too large to require breakthrough technology. Fig. 1.2A shows a photo of Richard Feynman viewing the micro motor, built by William McLellan and Fig. 1.2B is a photograph of the motor, which is 3.81 mm wide, sitting beneath the head of a pin.

The printing challenge on the other hand produced technological advance. In 1985, Thomas Newman, a Stanford University graduate student, used electron beam lithography to reproduce the first page of Charles Dickens' novel, *A Tale of Two Cities,* on a page measuring only 1/160 mm on a side. Fig. 1.2C shows an enlarged photo of the written page. This is 20 times smaller than the human eye can see.

This episode has been a defining moment in the history of miniaturization and nanotechnology (molecular-scale technology). Feynman call for miniaturization was heeded in a record time. Feynman assumed the process would be incremental, but miniaturization surprised all. Leaps and strides in synthesis and imaging propelled miniaturization quickly into the nanoscale. As we understand today, nanotechnology is control of matter, as well as the fabrication of devices on a scale smaller than 100 nm. At the core of this technology are the three major interests of development:

- New material
- Realization of standalone nano-device concepts
- Incorporation of nanomaterial into existing devices
- Down scaling existing devices to the limit of size

1.3 DRAWBACKS OF ATOM-BY-ATOM (ABA)

The advantage of Feynman's proposal is that Nature has provided large numbers of atoms in a discrete set. In free environments they are identical in all their electronic or molecular structure, and they have specific optical and electrical, chemical and material fingerprints. But there are several problems facing this endeavor.

1. The forgotten chemistry (Extraction factor): In general atoms do not exist in isolation. They are either found in association with others in a solid or in a molecule. Moreover, post-assembly chemical reactions and connections would potentially be limiting
2. Slow processing (Time factor): Organizing enough of them in a practical device and making enough devices for a consumer use atom-by-atom (ABA) is slow. Because of the many billions of atoms that would be needed to build even a small object, the process would be intolerably slow. At a rate of one atom per second, for example, any finished product would be many years away.
3. Visibility of atoms/molecules: Then atoms cannot be seen atoms. This requires advancement in imaging.

1.4 PARALLEL MANUFACTURING SOLUTION BY K. ERIC DREXLER

Drexler, then an undergraduate student at MIT, was motivated by Feynman's talk. He proposed a parallel robotic solution using a device called an "assembler", which might begin as a sub-microscopic robotic arm able to initiate controlled chemical reactions [3]. He went on to describe in some detail how such a device could function and be controlled and used in vast numbers. More

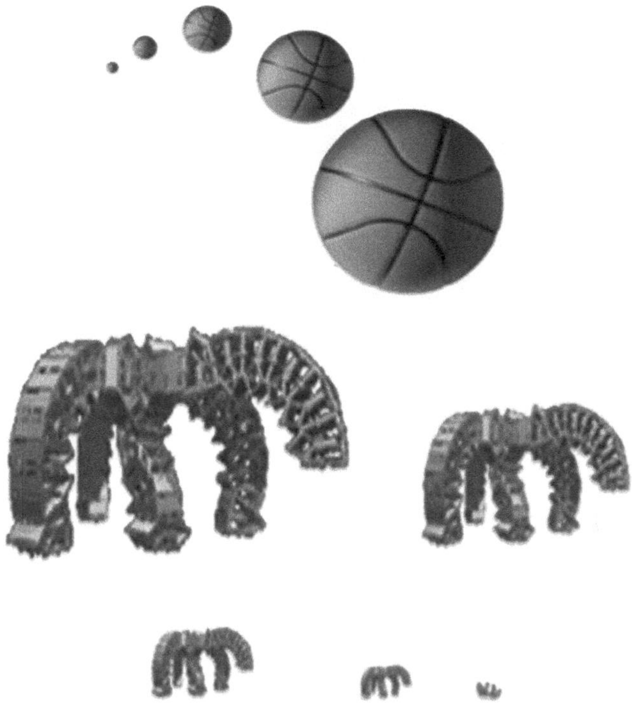

FIGURE 1.3 Cartoon of Drexler's scheme of creating assemblers in the form of nanoscale robots that can positon atoms to replicate themselves or to build larger structures one atom at a time.

importantly he suggested how it could replicate itself and be made to operate autonomously. Fig. 1.3 shows a cartoon of the Drexler's scheme of miniaturization of the robot. The concept of parallel manufacturing was proposed to alleviate the time constraint. Thus assemblers are envisioned as nanoscale robots with small mechanical arms that can positon atoms to replicate themselves or to build larger one atom structure.

But there are problems with the concept of parallel manufacturing. As most desirable products (computers, cars, etc.) are made of trillions and trillions of atoms, one nano-assembler working ABA would be rather slow. This requires not only large numbers of assemblers (trillions) but those must also need to work in parallel. In turn this requires the assembler robot arm to make copies of itself and those copies make copies and so on. Now this requires trillions of nano super-computers to control the process and those computers need to work in parallel. This is a problem of massive parallelism. The control and communication system needed is probably the holy-grail of nanotechnology. Estimates of the time needed to construct a first assembler device range from 12 years to never. Controlling millions of nanobots in parallel is a formidable task (massive parallelism).

Kim Eric Drexler

Kim Eric Drexler is an American engineer and is most known for being the driving force behind the idea of molecular nanotechnology (MNT) in the 70s and 80s while he was a student at MIT in the fields of Interdisciplinary Sciences, Astro/Aerospace Engineering, and the Media Arts and Sciences Section, School of Architecture and Planning.

REFERENCES

[1] Feynman RP. There's plenty of room at the bottom (data storage). J. Microelectromech. Syst. 1992;1(1):60–6. A reprint of the talk.

[2] Feynman R. Infinitesimal machinery. J. Microelectromech. Syst. 1993;2(1):4–14. A sequel to his first talk.

[3] Drexler K. Molecular engineering: an approach to the development of general capabilities for molecular manipulation. Proc. Natl. Acad. Sci. 1981;78(9):5275–8. ISSN 0027-8424. PMC 348724. PMID 16593078. Retrieved 6 September 2016.

FURTHER READINGS

[4] K.E. Drexler, Engines of Creation: The Coming Era of Nanotechnology, Doubleday, 1986.

[5] There's plenty of room at the bottom, Richard Feynman. Available from: http://www.nature.com/nnano/journal/v4/n12/full/nnano.2009.356.html

Chapter 2

Seeing and Detecting Atoms in a Gas Using Light

2.1 PICKING STATIONARY ATOMS WITH LASER LIGHT

The Feynman talk may have been lost in the East Coast, but it was certainly not in West Cost. The geographical (360 miles by car) and intellectual gap between Pasadena and Stanford was not big. Schawlow of Stanford re-emphasized the fact that to realize atom-by-atom (ABA) one needs first to extract individual atoms from molecules and compounds or solids then rearrange them. In the 1960s, there was absolutely no device available that would be able to see or interrogate or control them spatially whether using light photons or electrons. In fact there was no theoretical or practical clue of how that could come about. This would certainly require advanced instrumentation that was not then available in any way.

Atoms or molecules in a gas are agitated at room temperature and thermal condition, and allowed to interact with each other through collisions. The constant collisions would result in distribution of speeds, with some atoms having high speeds and others having low ones, including zero speed. The distribution is described by the Maxwell–Boltzmann distribution [1], and it depends on temperature, as shown in Fig. 2.1, for hydrogen atoms. The Maxwell–Boltzmann speed distribution function for a sample of gas at three different temperatures. At higher temperatures, the peak in the distribution function becomes lower, broader, and occurs at a higher speed.

The first attempt at Stanford was to extract atomic hydrogen gas from hydrogen gas molecules and to try to pick up the stationary atoms in the gas. Because an atom in motion radiates a frequency that directly relates to its velocity, according to the Doppler effect [2,3], measuring an ensemble of atoms having a wide distribution of velocities adds up to a similarly wide distribution of frequencies; thus masking and hiding the true structure of the atom.

According to the Doppler effect, the frequency of each photon emitted will be reduced or increased ("red"- or "blue"-shifted) depending on whether the atom is moving away or toward the observer. Thus, the higher the temperature of the gas, the wider the distribution of velocities in the gas will be, hence the wider the spectral broadening. Normally the broadening due to the motion is

Fundamentals and Applications of Nano Silicon in Plasmonics and Fullerenes.
http://dx.doi.org/10.1016/B978-0-323-48057-4.00002-5

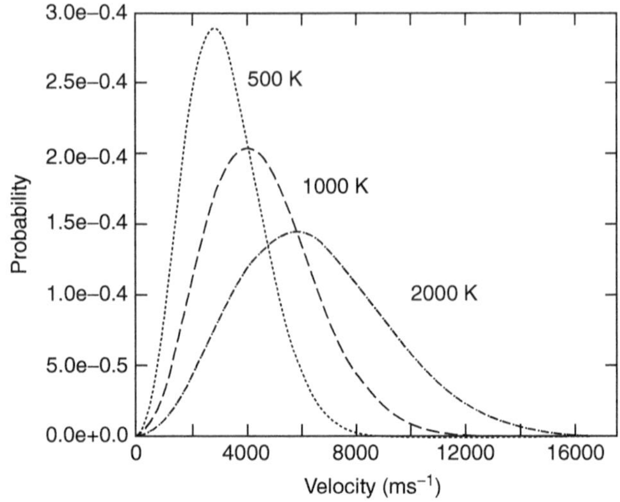

FIGURE 2.1 Maxwell–Boltzmann velocity distribution for hydrogen atoms for different temperatures.

much wider than the intrinsic one, and if broad enough it may mask several nearby natural transitions.

The hydrogen atom [4] and its isotope deuterium is the simplest of all atoms, consisting of one electron and one proton. As such, it is heaven for theorists but a menace for experimentalists for several reasons:

- Being the lightest, in a gas it has the highest thermal velocities, hence the largest Doppler masking
- The transition energy scale is one of the highest, having 13.6 eV binding and a minimum of 10.2 eV excitation energy, outside the reach of visible lasers
- It exists naturally in molecular form with ~4 eV binding, requiring an a few kV electric discharge environment to free them with small yield

In fact the development of foundation of theoretical physics is closely related to the development of our understanding of theory of the structure of hydrogen. But experimental work was lagging because of the problem of the Doppler masking and because atomic hydrogen (H) does not exist naturally as it must be extracted from molecular hydrogen (H_2), and because hydrogen has a high binding energy of −13.6 eV.

Thus it is clear, how important it is to freeze the atoms in the ensemble or at least to isolate the ones that are in motion to get at the true nature of atoms. Efforts until 1972 were focused on cooling the hydrogen atoms to reduce the speeds [5]. Samples were cooled down to 50 K or − 220°C, but unfortunately because the speed drops as a function of the square root temperature this will reduce the average temperature only by a factor less than 2.5.

George Series: The Father of the Rydberg

George William Series was born on 22 February, 1920; Lecturer, St. Edmund Hall, Oxford, 1953, Fellow 1954–69 (Emeritus); Professor of Physics, Reading University, 1968–82 (Emeritus). His early research was on optical spectroscopy of simple atoms, notably hydrogen. Hydrogen, as a testing ground for theory and a source of fundamental constants, fascinated Series, who became a world authority on the subject. In those pre-laser days, he pushed classical interferometry to its limit in studying optically the recently discovered Lamb shift, a crucial piece of evidence in the development of quantum electrodynamics (QED), the quantum theory of electromagnetic interactions. Later, he took part in a measurement of the Rydberg constant, an important combination of fundamental constants. He wrote a book, "The Spectrum of Atomic Hydrogen," in 1957, and edited a second, "The Spectrum of Atomic Hydrogen: Advances (1988)."

Theodor Hänsch and Arthur Schawlow

Theodor Hänsch (born on 30 October, 1941) is a German physicist. He received one fourth of the 2005 Nobel Prize in Physics for "contributions to the development of laser-based precision spectroscopy, including the optical frequency comb technique," sharing the prize with John L. Hall and Roy J. Glauber. Hansch invented the narrowest band tunable dye laser in 1971–72. The laser created a revolution in atomic and molecular spectroscopy, as well as precision optical spectroscopy as it enabled elimination of Doppler effects hence unmasking the true structure of atoms. It was used to precisely measure the Lamb shift and the Rydberg constant. Hänsch is presently the Director of the Max–Planck-Institut für Quantenoptik (quantum optics) and Professor of experimental physics and laser spectroscopy at the Ludwig-Maximilians University in Munich, Bavaria, Germany.

Arthur Schawlow (May 5, 1921–April 28, 1999) was an American physicist and co-inventor of the laser with Charles Townes. His central insight, which Townes overlooked, was the use of two mirrors as the resonant cavity to take MASER action to visible wavelengths. He shared the 1981 Nobel Prize in Physics with Nicolaas Bloembergen and Kai Siegbahn for his work on lasers.

To unmask the real structure of hydrogen, one needs the following:

- Extract atoms from molecules
- Stop the atoms or pick non-moving ones
- Use narrow band and tunable light source

The Stanford group consisting of Schawlow and Hansch and doctoral candidates Nayfeh, Shahin, Lee, and Curry conducted a series of experiments over a 2-year period (1972–74) on atomic hydrogen [6]. Fig. 2.2A show the energy level diagram, as well as some emission lines according to the simple Bohr Theory. When regular emission is recorded from this kind of discharge, it is found to be highly masked due to the motion of the atoms even if the gas is cooled to 50 K (−220 °C) as shown in Fig. 2.2B. It was not possible to stop atoms then, thus the Stanford group targeted the motionless hydrogen atoms in the gas sample using advances in laser technology. Instead of recording emission, the Stanford group used an ingenious procedure involving absorption from lasers. Hydrogen atoms are first extracted from the naturally existing molecular H_2 by electric discharge to form a gas. Moreover, they utilized electron excitation to target excited hydrogen atoms. Due to the electric current in the discharge, some of the atoms are already in the $n = 2$ excited state as shown in the simple energy diagram; thus putting the atoms in reach of the more convenient visible lasers. Finally, the laser they used for this demonstration was the ever narrowest band tunable laser then; it was invented by Hansch in 1972 [7].

Fig. 2.2C shows the layout of the measurement [8]. In the measurement they recorded absorption from two identical laser beams traversing the gas collinearly in opposite direction as shown in the figure. The lasers are tuned across the red Balmer line about 656 nm, coinciding excitation from $n = 2$ to $n = 3$. One of the beams is chosen to be much stronger (saturator) than the other (probe). As the Doppler effect depends on the atom's direction of motion, the two beams will be absorbed independently by different sets of moving atoms. But they will simultaneously compete for atoms that are not moving in the either one of the beam direction, namely those that are not moving. Thus if the absorption data is analyzed appropriately by, for example, modulating the strong beam with a

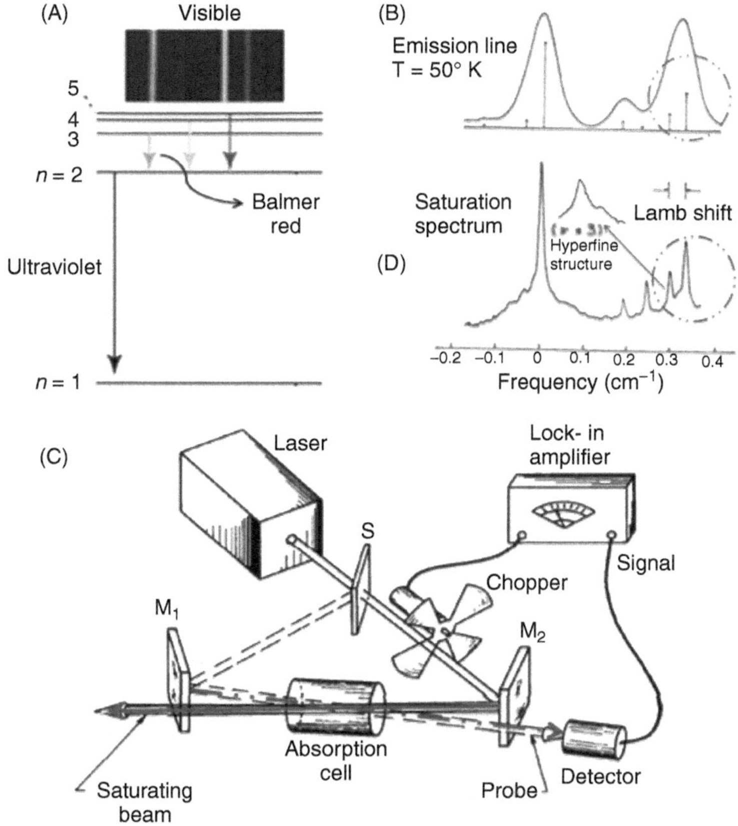

FIGURE 2.2 Spectroscopy of atomic hydrogen. (A) Energy level diagram of hydrogen atoms, as well as some emission lines according to the simple Bohr Theory. (B) Emission from a collection of hydrogen atoms in an electric discharge cooled to 50 K (−220 °C). The spectrum is highly masked due to the motion of the atoms even if the gas is cooled. (C) Layout developed at Stanford for resolving the spectrum of atomic hydrogen using Doppler-free laser spectroscopy. (D) Spectrum of hydrogen atoms obtained by the Doppler free method showing a dramatic elimination of the broadening, unmasking the true natural width of the fine structure of hydrogen, and as such is coined Doppler-free spectroscopy.

chopper at a certain chopping frequency, one can isolate the absorption of the non-moving atoms. The result given in Fig. 2.2D shows a dramatic elimination of the broadening, unmasking the true natural width of the fine structure of hydrogen, and as such is coined Doppler-free spectroscopy.

The Doppler-free technique was first introduced theoretically earlier by Russian scientists [9]. The technique allows the laser to be blind to all moving atoms in a gas sample and see only the ones that happen to be stationary along the direction of the red light beam, which unmasks the true structure of the atoms. As a doctoral student in the Stanford group, Issa Shahin was involved in using the narrow-band dye laser to study the structure of sodium and hydrogen.

As a doctoral student, under the superposition of Hansch and Schawlow Munir, Nayfeh also had the opportunity to be involved in an effort to reproduce a variation to the Jan Hall's stable (helium–neon) He–Ne light source [10] (Fig. 2.3A) and to utilize it as a wavelength standard along with the narrowband red light source to unmask and measure the hyper-fine structure and binding energy of the simplest of all atoms in nature, the hydrogen atom, whose binding energy is the corner stone of all fundamental constants in nature, namely the Rydberg constant. The study achieved then a record accuracy of 9 parts per billion, which allowed the improvements of all other fundamental constants [11]. The importance of improving the accuracy of the Rydberg constant was evident in 1995, when on the occasion of the 100 year anniversary of the top American Journal Physical Review [12], the journal listed this achievement as one of the seminal papers in the 100 year life of the Journal as shown in Fig. 2.3B. Fig. 2.3C shows a photo from a celebration at Stanford recognizing Schawlow and Hansch as California Scientist of the year for these developments and the use of the narrowband tunable laser in precise spectroscopic measurements (Figs. 2.2 and 2.4).

FIGURE 2.3 Resolution of the spectrum of atomic hydrogen. (A) Munir Nayfeh aligning a variation to the Jan Hall's stable He–Ne light source constructed at Stanford for utilization as a wavelength standard along with the narrowband red light source to unmask and measure the hyper-fine structure and binding energy of hydrogen atoms. (B) Cover of the 100 year anniversary of the American Journal Physical Review, in which the journal listed the precision measurement of the Rydberg constant as one of the seminal papers in the 100 year life of the Journal. (C) Celebration at Stanford from left to write: Arthur Schawlow, Munir Nayfeh, Stanford President, and Theodore Hansch for the choice of Schawlow and Hansch as California Scientist of the Year for the high resolution spectroscopy achievements.

FIGURE 2.4 Sketches and photos of five scientists who contributed significantly to our understanding of light effect. (Top row) Alhazen and Isaac Newton. (Bottom row) Arthur Schawlow, Theodore Haensch, and John Hall

Dispersion of Light: (From Alhazen to Newton to the Trio [13]):

Lasers were invented theoretically, in 1958, by Schawlow and Townes (who co-invented the Maser earlier) and constructed as early as 1962–1964. But not any laser will be able to selectively interrogate a certain atom in a background of other atoms. This job had to wait till the early seventies when the high-fineness, high-quality laser beams were developed. Those have a high degree of color definition of its light, which can be changed (tuned) on demand and stable to provide enough time to do the measurements. In 1015, Alhazen was concerned with understanding the different color components of light through natural phenomenon, such as propagation of light in material, reflection and refraction, and the rainbow effect. Newton, 600 years later, used manufactured glass prisms, which became available then to deliberately disperse light into its components and to introduce the seven color names red, orange, yellow, green, blue, indigo, and violet for segments of the spectrum. Nearly 300 years after Newton Theodor "Ted" Hansch and Arthur Schawlow of Stanford University set out to achieve big strides in the quest to disperse light and isolate extremely narrow components using much more advanced electronics and optics capabilities, such as prisms, gratings, interference filters, spectrographs, telescopic light expanders, etc. But such a highly pure color component would have extremely weak intensity, which would necessarily require amplification if it is to be a practical light source.

In 1972, the Stanford group dispersed the red fluorescence from a chemical dye into much finer components while simultaneously amplifying using the very advanced light–matter interaction concepts of stimulated emission (introduced by Einstein) and laser gain (Schawlow inventor) to produce a narrowband directed red laser (light amplification by stimulated emission of radiation). Pulses or flashes of red light of 8 ns duration with a record color definition (bandwidth) of 0.0004 Å (or 0.001 cm^{-1} or 30 MHz) and energy of ~1 nJ per pulse were achieved (Hansch laser). In addition to being narrow band, the laser allows for change (tuning) of wavelength with good control and precision, a pivotal property for matching (resonating with) and studying electronic structure of atoms and molecules. It is this laser that caused a revolution in high-resolution Doppler-free spectroscopy of atoms and molecules. In fact, the Hansch laser is the laser that was used in the Doppler free spectroscopy measurements that allowed the resolution of the hyperfine structure of hydrogen and to make a precision measurement of the Rydberg Constant.

John Hall on the other hand used a different approach by starting out with the red light of He–Ne gas laser. This red laser light at a given time is narrow band (very pure in color) but it is practically not as narrow over longer times because its frequency or wavelength drifts as a result of drift in ambient conditions (temperature, vibrations, etc.). Using advanced optical and electronic equipment he succeeded in making the frequency or wavelength practically stand still for long times. This high-finesse light source is extremely useful but it is not tunable in wavelength.

The two groups used these advanced devices in which the color of light has been defined much more precisely to perform high resolution laser spectroscopy measurements of unprecedented accuracy and intrinsic physical interest. The Stanford group accurately measured the Rydberg constant, the corner stone of all fundamental constant and the Lamb shift. Jan Hall used his source to measure accurately the speed of light, allowing a re-definition of the SI Meter. It should be mentioned here that the first version of the He–Ne laser was invented in 1960 by Ali Javan (Azerbajani born in Tehran, Iran) at Bell Labs [14]. However the laser produced invisible light at 1.15 μm. Only 18 months later in 1962, Javan's colleagues White and Rigden of Bell Lab constructed the visible He–Ne laser, which is more exciting, captivating, and convenient [15].

Example 2.1 Bohr Radius of Hydrogen-Like "Excitons"

http://www.umich.edu/~chem461/QMChap7.pdf.

The attraction between two opposite charges, such as the electron and proton, is given by Coulomb's law. Fig. 2.5A shows a schematic of a hydrogen atom with infinitely massive nucleus approximation; deuterium isotope (hydrogen atom with the nucleus consists of a proton and a neutron). The figure also shows hydrogen-like atom with the electron replaced by a muon charge. Highly stripped atoms in which all electrons are stripped except for one, are shown in Fig. 2.5B.

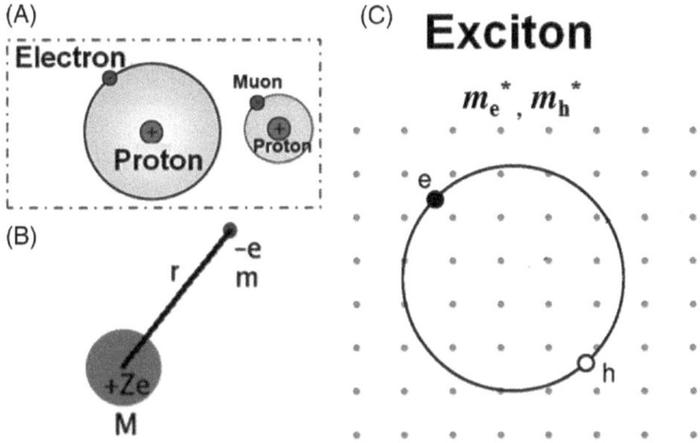

FIGURE 2.5 The bound quantum structure of two oppositely charged particles. (A) A hydrogen atom with infinitely massive nucleus approximation, the figure also shows hydrogen-like atom with the electron replaced by a muon charge. (B) Highly stripped atoms in which all electrons are stripped except for one. (C) Exciton in solid composed of a bound electron–hole pair.

Fig. 2.5C shows exciton in solids [16]. It is composed of a bound electron–hole pair where that is not localized at a crystal position. The exciton binding energy can be calculated using a simple phenomenological hydrogen-like equation.

The total mass of the system is the sum of those of the electron and the hole: $M = m_e + m_h$. The total of the wave number is also the sum of the individual ones $K = k_e + k_h$. The electron and the hole are bound together as an atom with the binding energy is $E_b = E_{b=\mu\ e4/2\varepsilon 2\ n2}$. The total energy of the exciton is the sum of three contributions of the bandgap of the crystal, the kinetic energy and the binding energy: $E_{exc} = E_G + \frac{1}{2} h^2K^2 + E_b$.

The reduced mass of the exciton is $1/\mu = 1/m_e + 1/m_h$. $E_b = \mu\ e4/2\varepsilon^2\ n^2$, where ε is the dielectric constant.

The exciton binding energy can be calculated using a simple phenomenological hydrogen-like equation, known as Wannier equation. This gives rise to a series of peaks in the absorption spectrum below the single-particle gap, the Rydberg series.

Example 2.2 Observation of the Structure of Stationary H-Atoms

Unmasking the red spectrum of hydrogen revealed for the first several features of modern quantum electrodynamics in the optical part of the spectrum. Modern quantum electrodynamics was developed by the four physicists Julian Schwinger, Richard Feynman, Ernst Stueckelberg, and Sin-ItiroTomonaga. It was not surprising to see Richard Feynman thrilled during his visits to Stanford during this period. For example, the spectrum shown in Fig. 2.2D revealed several firsts:

1. Effect of special relativity on the motion of the electron (Fine structure)
2. Interaction of the electron with vacuum (Lamb shift): It shows off as the 2p (1/2) state is slightly lower than the 2s (1/2) state resulting in a slight shift of the corresponding spectral line (the Lamb shift). The shift is only 4.372 μV or 1057 MHz or 28.37 cm. This was not predicted by the most advance equation of motion, the Dirac equation, according to which these states should have the same energy. The Lamb shift was measured in 1947 by Willis Lamb directly using microwave excitation [17]. Interaction between vacuum energy fluctuations and the hydrogen electron in these different orbitals is the cause of the Lamb shift, as was shown subsequent to its discovery. The Lamb shift has since played a significant role through its validation of vacuum energy fluctuations in the discovery of Hawking radiation from Black holes.

When the Lamb shift was experimentally determined, it provided a high precision verification of theoretical calculations made with the quantum theory of electrodynamics. These calculations predicted that electrons continually exchanged photons, this being the mechanism by which the electromagnetic force acted. The effect of the continuous emission and absorption of photons on the electron g-factor could be calculated with great precision.

The tiny Lamb shift, measured with great precision agreed too many decimal places with the calculated result from quantum electrodynamics. The measured precision gives us the electron spin g-factor as

$g = 2.002319304386$

3. Interaction of the spin of the electron with the spin of the proton (hyper-fine structure) (see Fig. 2.6). The measurements give 150 MHz, which corresponds to a wavelength of 2 m [11].
4. Effect of the mass of the nucleus hydrogen and deuterium (isotope shift). The resulting H-D isotope shift is 4.1451 (12) cm^{-1}.
5. Most precise value for the binding energy of the electron (Rydberg constant). It yielded a record accuracy of 9 parts per billion, which allowed the improvements of all other fundamental constants. The Rydberg constant is a combination of other fundamental constants including the charge of the electron, the mass of the electron, the permittivity of free space, Planck's constant and the speed of light. Because of this it is considered the cornerstone of the fundamental constants.

$$R_\infty = \frac{m_e e^4}{8\varepsilon_0^2 h^3 c}$$

The importance of improving the accuracy of the Rydberg constant was evident in 1995, when on the occasion of the 100 year anniversary of the top American

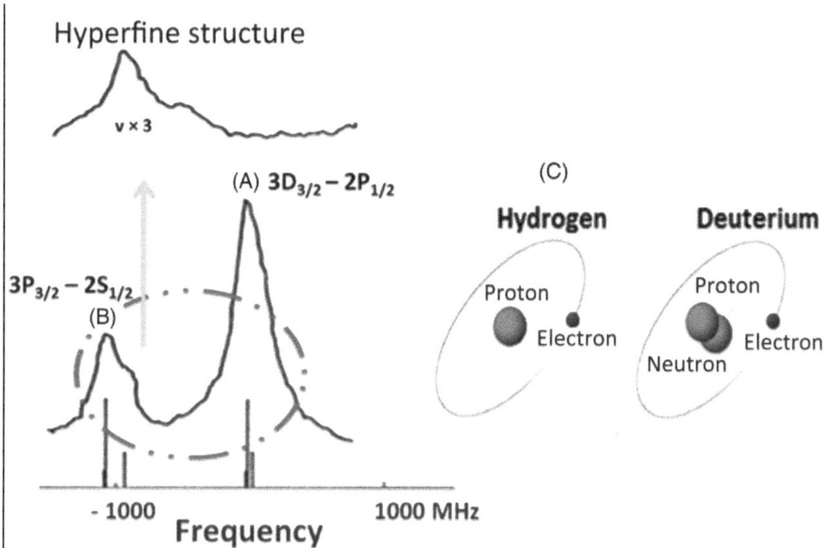

FIGURE 2.6 Hyper-fine structures due to the interaction of the spin of the electron with the spin of the proton. (A) Measurements of the structure using Doppler-free spectroscopy due to the spin interaction. (B) Isotope effect in hydrogen. (C) Sketch of the hydrogen atom and the deuterium atom.

Journal Physical Review, the journal listed this achievement as one of the seminal papers in the 100 year life of the Journal as shown in Fig. 2.3B.
http://hyperphysics.phy-astr.gsu.edu/hbase/quantum/lamb.html.
https://en.wikipedia.org/wiki/Lamb_shift.

6. The electron is an elementary particle embodied with intrinsic properties, such as charge, mass, spin, and magnetic moment with a charge and mass singularities, meaning a zero size. The last 2 decades has seen a fury of works in terms of extended models of the size. The size scale used is the Compton wavelength λc of the electron, which is the wavelength of the EM wave that results upon full conversion of the energy of the electron (mc^2) to a single frequency of EM radiation, namely $h\nu = hc/\lambda c = m_e c^2$. Recent work proposed a travelling wave electromagnetic model of an electron (TWEM). In this model the spin arise naturally as a consequence of motion of the momentum-energy of EM waves with ½ the reduced Compton wavelength $\lambda c/2\pi$ (RCW) of electron radius (see M. Asif, Physics Essays 27, 1 (2014) [18]).

2.2 COUNTING SODIUM ATOMS USING RESONANCE LIGHT (FLUORESCENCE) SPECTROSCOPY

The Feynman talk may have been lost in the East Coast, but it was certainly not in West Cost. The geographical (360 miles by car) and intellectual gap between Pasadena and Stanford was not big. In 1970, Theodor Hansch and

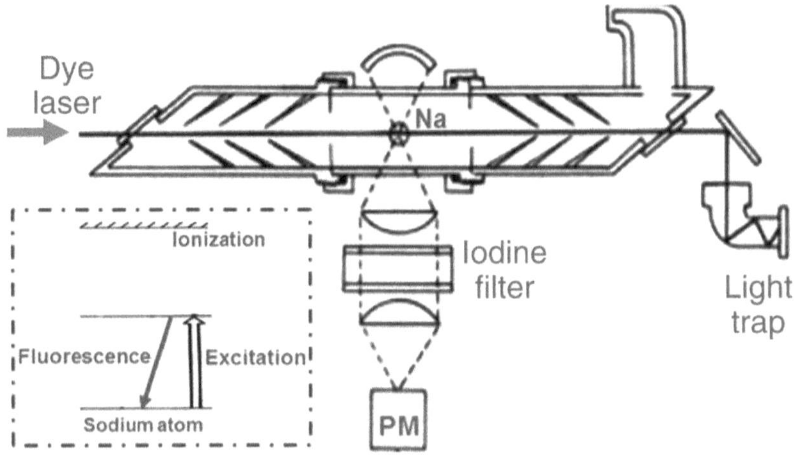

FIGURE 2.7 Setup used to measure or count number of sodium atoms using scattered photons from a laser beam (inset) partial energy-level diagram of sodium atoms showing laser excitation and scattering of photons processes.

Arthur Schawlow of Stanford University assigned Munir Nayfeh to study high-resolution spectroscopy of atomic hydrogen by picking non-moving hydrogen atoms using by narrow-band–pulsed dye laser as covered above in Section 2.1; and assigned William Fairbank Jr., a graduate research assistant to construct a tunable CW laser (dye laser) and to use it to examine the feasibility of observing and manipulating atoms of a specific material. Fig. 2.7 depicts the schematic of the experimental setup used. Sodium atoms were extracted by heating a sodium pellet. The tunable CW dye laser beam was shined through the vapor. The wavelength was tuned to resonate with one of the D lines of sodium allowing the team to measure scattered photons from atoms that happen to cross the laser beam (Fig. 2.7, inset). Due to stray light background scattering from the windows, the signal to noise ratio limited the observation to 100/cm^3 atoms within the waist of the beam [19].

2.3 DETECTION OF SINGLE ATOMS USING LIGHT-INDUCED ELECTRON RESONANCE IONIZATION

The effort of detection and counting atoms moved from Stanford to Oak Ridge National Laboratory with the joining of Munir Nayfeh the George Samuel Hurst (Saml Hurst) group. Single atom detection had become the dream of Sam Hurst, who was an expert in ionization events and its effect on health. He noted that detection of an electron is much easier and efficient than detection of a photon. Thus action related to ABA at this time spread to several sites as shown in Fig. 2.8.

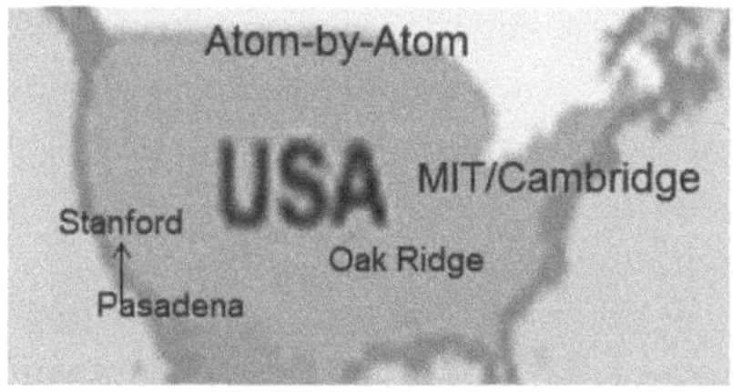

FIGURE 2.8 Four research sites in the 1970s (Pasadena, Stanford, Oakridge, and Cambridge/Boston) with activities related to Feynman's ABA.

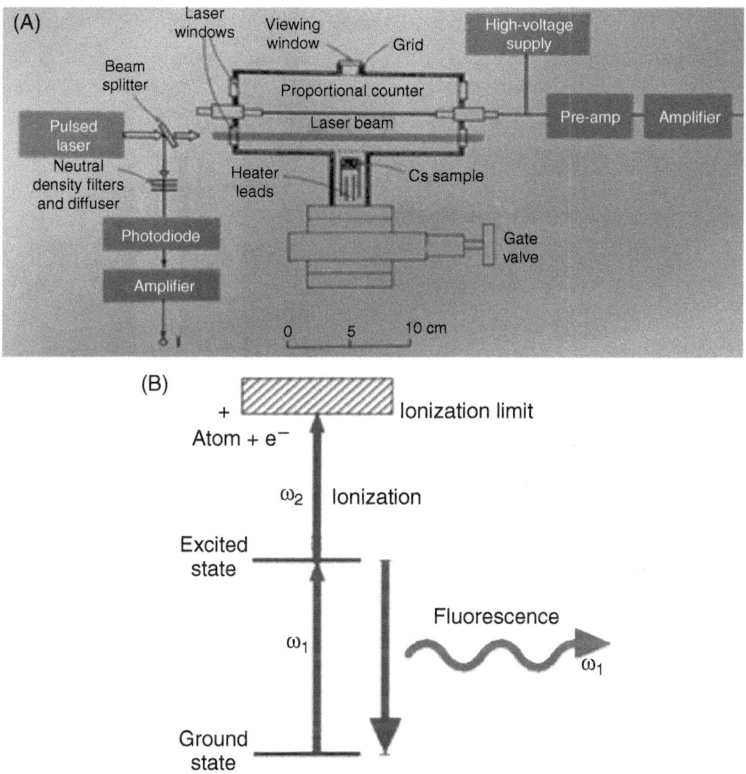

FIGURE 2.9 Detection of single atoms. (A) Experimental setup used at Oak Ridge National Laboratory, which improved the sensitivity of detection of atoms down to a single one (demonstrated first using cesium atoms). (B) Partial energy-level diagram of cesium showing resonant excitation and ionization two-photon process of the atoms to produce an electron-ion pair.

Thus the Oak Ridge National Laboratory group went on to improve the sensitivity of detection of atoms down to a single one [20–23]. The experimental setup is given in Fig. 2.9A. Cesium atoms were extracted by heating a cesium pellet and were fed into a proportional counter charge amplifier/detector which is filled by an argon gas with a small fraction of methane gas.

The detector consists of a grounded conducting cylinder with a thin tungsten wire running parallel to its axis and biased at a high voltage of ~1000 V. A laser beam enters and exits the counter via windows, such that it traverses the gas parallel to the tungsten wire. The wavelength was tuned to one of its resonance absorption lines. The laser intensity was boosted enough to further ionize the atom into an electron and ion as shown in Fig. 2.9B. As cesium atoms traverse the laser beam they get ionized and the generated electron accelerates towards the wire. The electron ionizes the gas causing a shower of electrons, which when it arrives at the wire it results in amplification. This design is the linear version of a Geiger counter used in detecting and counting charged particles for nuclear radiation research. The team monitored the generated electron, as well as the ion as a signature of the presence of the atom. By monitoring the charge produced this laser-based resonance ionization provided two orders of magnitude sensitivity over the sensitivity of resonance fluorescence. This configuration has been used in 1975–76 for identifying and counting single atoms of cesium by Hurst, Nayfeh, and Young at Oak Ridge National Laboratory.

The Oak Ridge results were covered in the search and discovery news of Physics Today as shown in Fig. 2.10A [24]. The Sciences magazine published by the New York Academy of Sciences shown in Fig. 2.10B [25]. The cover of same issue of "The Sciences" shows a 16th century painting of Islamic

FIGURE 2.10 Reports of single atom detection (A) in the search and discovery news of Physics Today Left to right: George Samule (Sam) Hurst, Jack Young, and Munir Nayfeh and (B) in the Sciences magazine published by the New York Academy of Sciences. George Samuel Hurst (1927–2010) was a health physicist. At Oak Ridge National Laboratory, he developed the omnipresent touchscreen technology and co-developed single atom detection. He was involved with neutron and alpha survey meters; Nevada Test Site radiation measurements; study of atomic bomb survivors from Hiroshima and Nagasaki; Hurst received a B.A. degree in physics and a minor in mathematics from Berea College, M.S. degree in physics from University of Kentucky, and Ph.D. in physics from the University of Tennessee.

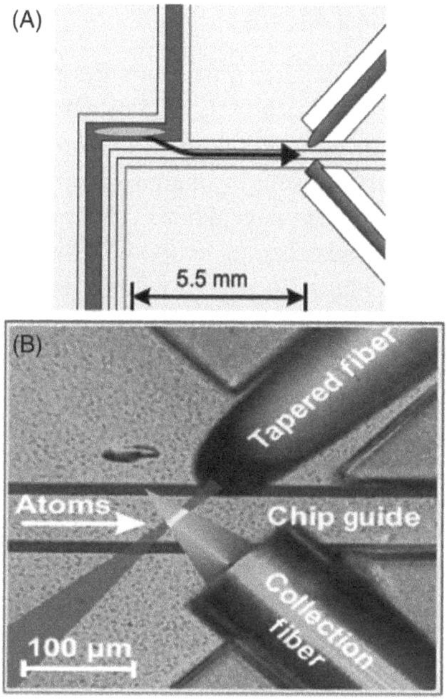

FIGURE 2.11 (A) Single atom detection by enhanced-resonance fluorescence. (B) Zooming on the fiber tips. *(From D. Heine, W. Rohringer, D. Fischer, M. Wilzbach, T. Raub, S. Loziczky, XiYuan Liu, S. Groth, B. Hessmo, J. Schmiedmayer, A single-atom detector integrated on an atom chip: fabrication, characterization and application, New J. Phys., 12 (2010) 095005 (23pp) © Deutsche Physikalische Gesellschaft. Reproduced by permission of IOP Publishing. CC BY-NC-SA)*

astronomers at work in their observatory. One of the scientists shown is Alhasan Ibn Alhaytham (Alhazen) [26]. As the process constitutes the ultimate sensitivity in analytic detection of matter, namely a single atom, the work was also covered in Britannica Yearbook of Science and the Future (1979), McGraw Hill Yearbook of Science and Technology (1979), The World Book Science Annual (1978), and more.

Enhanced-resonance fluorescence with further improvements some years later, it was also employed to detect single atoms in a configuration called atom chip. As shown in Fig. 2.11A–B, the atom chip consists of a tapered lensed single-mode fiber for precise delivery of excitation light and a multi-mode fiber to collect the fluorescence [27]. The fibers are mounted in lithographically defined holding structures on the atom chip. Neutral 87 Rb atoms propagate freely in a magnetic guide. The fluorescence detector is fully integrated into an atom chip. The variance of the fluorescence photon distribution allows determining the number of detected photons/atom; hence the atom detection efficiency is obtained. The detector allows spectrally and spatially selective detection of atoms, reaching a single-atom detection efficiency of 66%. The researches claim that one can boost the detection efficiency to >95% with simple improvements.

2.3.1 Stopping/Cooling Atoms

The concept of using cycles of laser absorption/fluorescence to cool and stop atoms was first introduced theoretically in 1975 by Theodore Hansch and Arthur Schawlow of Stanford University. In the scheme, two counter propagating narrow tunable band dye lasers are sent through a gas of moving atoms [28]. The lasers are tuned to near resonance with. Photon absorption by a thermally moving atom from two identical laser beams in opposite directions exert momentum on atom in one direction, but atom re-emits in all direction hence atom loses speed in the laser direction. This concept will be treated in detail in the following chapter (Chapter 3).

2.3.2 Atom-by-Atom Fabrication

The concept of ABA has been progressing in the 1970s at Stanford, Oak Ridge National Laboratory (ORNL) and Massachusetts Institute of Technology (MIT). At Stanford, lasers were used to pick stationary hydrogen atoms and count a small number of sodium atoms. At Oak Ridge, high-power lasers were used to detect single/individual cesium atoms. At MIT, the concept of parallel fabrication using robotic assemblers was introduced to speed up the process.

REFERENCES

[1] Mandl F. Statistical physics. second ed. Manchester Physics. Hoboken, NJ: John Wiley & Sons; 2008. ISBN 9780471915331.

[2] Landau LD, Lifshitz EM. The classical theory of fields. Course of Theoretical Physics: Volume 2 Trans. Morton Hamermesh. Oxford, UK: Elsevier Butterworth-Heinemann; 2005. (pp. 1–3, ISBN 9780750627689; Fourth revised English ed.).

[3] Nayfeh MH, Brussel MK. Electricity and Magnetism. Illinois, United States: Dover; 2016.

[4] Bethe HA, Salpeter EE. In: Flugge S, editor. The Spectrum of Atomic Hydrogen. Berlin: Springer; 1957. https://books.google.com/books?id(2UbCHAAACAAJ; Encylopedia of Physics Vol. 3QCKV.

[5] (a) Kibble BP, Howley BC, Shavryer HE, Series G. An experimental determination of the Rydberg constant. J. Phys. B: Proc. Phys. Soc. London 1973;6(1079). (b) Kessler EG. Determination of the Rydberg constant from the $n = 3$–4 He II (469-nm) line complex. Phys. Rev. A 1973;7(408). (c) Series G. The Rydberg constant. Contemp. Phys. 1974;14(49).

[6] Hänsch TW, Nayfeh MH, Lee SA, Curry SM, Shahin IS. Precision measurement of the Rydberg constant by laser saturation spectroscopy of the Balmer line in hydrogen and deuterium. Phys. Rev. Lett. 1976;32:1336.

[7] Hänsch TW. Repetitively pulsed tunable dye laser for high resolution spectroscopy. Appl. Opt. 1972;11(4):895–8.

[8] Hänsch TW, Shahin IS, Schawlow AL. High-resolution saturation spectroscopy of the sodium D lines with a pulsed tunable dye laser. Phys. Rev. Lett. 1971;27:707.

[9] Vasilenko LS, Chebotayev VP, Shishaev AV. Line shape of two-photon absorption in a standing-wave field in a gas. JETP Lett. 1970;12:113.

[10] (a) Baer T, Kowalski FV, Hall JL. Frequency stabilization of a 0.633-μm He–Ne longitudinal Zeeman laser. Appl. Opt. 1980;19:3173–7. (b) Schweitzer WGJJr, Kessler EG Jr, Deslattes RD, Layer HB, Whetstone JR. Description, performance, and wavelengths of iodine stabilized lasers. Appl. Opt. 1973;12(2927).

[11] M.H. Nayfeh, Precision Measurement of the Rydberg by Saturated Spectroscopy, PhD Thesis, Stanford University, 1974.

[12] Stroke HH. The Physical Review: The First 100 years: A Selection of Seminal Papers and Commentaries. NY: Springer and AIP Press; 1995.

[13] Nayfeh M. Optics in nanotechnology. In: Alamri M, El Gomati MM, Zubairy SM, editors. Optics in Our Time. Berlin, Germany: Springer; 2016.

[14] Javan A, Herriott D, Bennett W. Population inversion and continuous optical maser oscillation in a gas discharge containing a He–Ne mixture. Phys. Rev. Lett. 1961;6:106.

[15] White AD, Rigden JD. Continuous gas maser operation in the visible. Proc. IRE 1962;50:1697.

[16] Reynolds DC. Excitons: Their Properties and Uses. Amsterdam, Netherlands: Elsevier; 2012.

[17] Hansch TW, Shahin IS, Schawlow AL. Optical resolution of the Lamb shift in atomic hydrogen by laser saturation spectroscopy. Nature 1972;235:63–5.

[18] Asif M. de Broglie wave and electromagnetic travelling wave model of electron and other charged particles. Phys. Essays 2014;27(1):146–64.

[19] Fairbank WM Jr, Hänseh TW, Schawlow AL. Absolute measurement of very low sodium-vapor densities using laser resonance fluorescence. J. Opt. Soc. Am. 1975;65:199–204.

[20] Hurst GS, Nayfeh MH, Young JP. Demonstration of one-atom detection. Appl. Phys. Lett. 1977;30:229.

[21] Hurst GS, Nayfeh MH, Young JP. One-atom detection using resonance ionization spectroscopy. Phys. Rev. A 1977;15:2283.

[22] Hurst GS, Payne MG, Kramer SD, Young JP. Resonance ionization spectroscopy and one-atom detection. Rev. Mod. Phys. 1979;51(4):767.

[23] Nayfeh MH. Laser detection of single atoms and applications. Am. Sci. 1979;67(204):204–13.

[24] Lubkin G. Resonance electron spectroscopy detects single atoms. Phys. Today 1977;17. 17 and 20.

[25] One-atom Chemistry, in The Sciences, The New York Academy of Sciences (May/June1977) 5.

[26] The Optics of Ibn Al-Haytham, Books 1-3, Alhazen, Warburg Institute, University of London, 1989

[27] (a) Heine D, Rohringer W, Fischer D, Wilzbach M, Raub T, Loziczky S, Liu XiYuan, Groth S, Hessmo B, Schmiedmayer J. A single-atom detector integrated on an atom chip: fabrication, characterization and application. New J. Phys. 2010;12:095005. (a) Fairbank WM, Jr., She, CY Single-Atom Detection with Lasers. Optics News 5(2).

[28] Hansch TW, Schawlow AL. Cooling of gases by laser radiation. Opt. Commun. 1975;13(1):68.

Chapter 3

Trapping and Stopping/ Cooling of Atoms, Particles, and Bio-components Using Laser Light

3.1 SOLAR AND LASER SAILS

The basic principle behind the developments of laser sails is the quantum property of light. In this picture, which Einstein introduced in the 1920s, light consists of massless particles (photons). Each photon carries energy $E = hf$ and momentum E/c, where f and c are the frequency and speed of light, and h is Planck's constant. The photon can exchange those upon collisions with material objects. Therefore when the photon is absorbed, reflected, or scattered it exerts a force on the object, which can be used to manipulate the time dynamics of the motion. The direction of the force resulting from scattering or absorption is in the direction of the impinging light.

The first application is what is called the solar sail or laser sail [1,2]. In this application, light from the sun or from *powerful high intensity lasers* may be used to propel systems. The sail is just a large, highly reflective (as shown in Fig. 3.1A) and very lightweight panel. It is attached to a spaceship as shown in Fig. 3.1B. The force of the photons moves it away from the sun. The concept was first analyzed by Robert L. Forward in 1989 as a method of Interstellar travel [3].

The magnitude of the pressure depends on the photon energy and on optical properties of the sail. When the light is reflected it exerts twice the force compared to when it is totally absorbed. On the other hand, when the light is transmitted it exerts no force.

Example 3.1 Setting a Solar Sail to Take Off from Earth [1]

Consider a totally reflecting thin sail of area $A = 1\ m^2$ and mass $m = 5$ g. It is facing the sun at an angle $\theta = 45$, as shown in Fig. 3.2A. The light intensity on earth is $I = 1.36\ kW/m^2$. We want to determine the radiation pressure and determine the acceleration of the sail due to sunlight. The radiation pressure for a totally reflecting surface at an angle θ incidence is $P_r = 2I(\cos\theta)^2/c$. The force exerted

Fundamentals and Applications of Nano Silicon in Plasmonics and Fullerines.
http://dx.doi.org/10.1016/B978-0-323-48057-4.00003-7

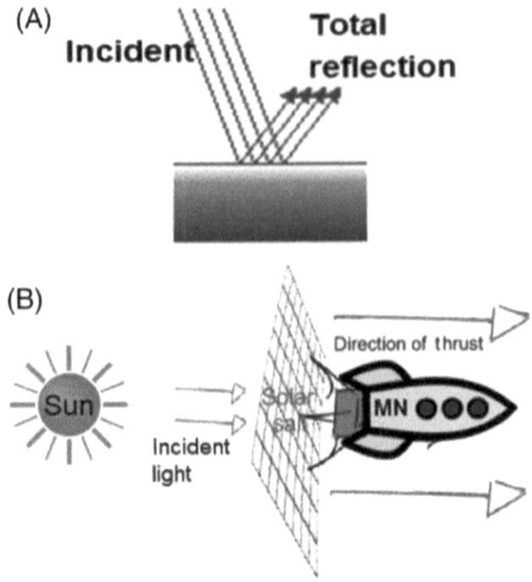

FIGURE 3.1 Solar and laser sails. (A) Layout of principle of solar sail in which a *powerful high intensity laser or sun light strikes a surface and get reflected.* (B) A solar sail consisting of a very lightweight panel attached to a spaceship. The force of the photons moves it away from the sun.

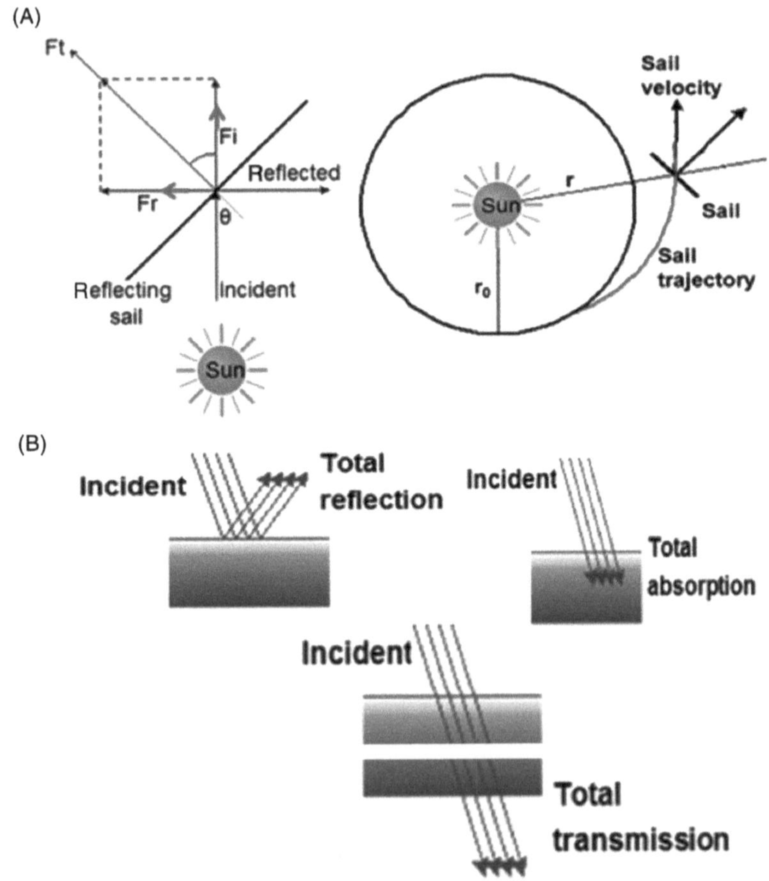

FIGURE 3.2 Calculation of solar force. (A) Specific layout of a solar sail for take-off from Earth. (B) Schematic of interaction of light with sails for three types of surfaces (totally reflecting, totally absorbing, and totally transmitting).

on the sail is $P_r A$. To determine the acceleration we apply Newton's law $F = ma$: $2\ AI\ (\cos\theta)\ ^2/c = ma$. Thus the maximum for an angle of $\theta = 0$ is $a_c = 2AI/mc = 2I/(c\ s)$ or $a_c = 9.08/s = 1.8\ \text{mm/s}^2$. This is much smaller than the Earth's gravitation acceleration of $9.8 \times 10^3\ \text{mm/s}^2$.

The radial component reduces the gravitational force on the solar sail and the transverse component affects the angular momentum of the solar sail. For a large enough reflective surface area sufficiently light material, Sun light should be able to exert enough pressure to move the spacecraft. If the sail was totally absorbing, then the force and hence the acceleration would be cut in half. And it would vanish if the sail is totally transmitting. As an exercise, find what happens to the force if the sail is totally absorbing and if it is totally transparent as shown in Fig. 3.2B.

http://ccar.colorado.edu/asen5050/projects/projects_2004/tetzlaff/.

Example 3.2 Current and Future Status of Solar Sails [4]

Current:

1. Successful solar sail propulsion was first demonstrated in Japan in June 2010. The Japanese Ikaros probe used 14-m-wide sail.
2. In November 2010, 5 months after the Japanese demonstration, NASA successfully put a solar sail in orbit (the Nano Sail-D demonstrator), operating its solar sail in January 2011; it circled the planet for eight months before burning up in the atmosphere.

Future:

3. NASA plans to launch in 2016 a sail measuring $38 \times 38\ \text{m}^2$ made from advanced material called Kapton, which is just 5 μm thick and weighs ~32 kg. It is $27 million project, called Sunjammer mission, destined to a spot about 3 million km from Earth, along the Earth–sun line.
4. Solar sail supplement: The nonprofit Planetary Society, for example, is working on a project called LightSail-1, which will launch a roughly 10-pound (4.5 kg) spacecraft to Earth orbit. The goal is to demonstrate navigational control and use sunlight pressure to increase the craft's speed, further proving out solar-sail technology, Planetary Society representatives said.
5. Solar sail to the moon: NASA plans to send a sail-equipped "CubeSat" to the moon. In addition to sail propulsion; sun light reflected by its mirror would be used as a search light down into shaded lunar craters (Lunar Flashlight).

Fiction:

6. Solar sail to escape Earth gravitation: General Astronautics and Surrey Satellite Technology Ltd. of the UK plan to construct sail areas as large as 1000–4000 m^2, which is capable of escaping from Earth and operating in interplanetary space.
7. Texas-size sails (truly enormous sail!), a hybrid solar-laser sail, is hoped to be able to send a spaceship to another star system in just a few centuries. Once it is out of range of sunlight, a powerful laser beam is used to propel the sail.

Example 3.3 – Radiation Pressure and Star Stability [5]

Radiation forces or pressure are behind the stability of stars against gravitation collapse. In fact there are three pressures involved. Fig. 3.3 shows a schematic of the three forces.

- Gravity forces are pull forces; they pull inward and try to compress the mass to the center.
- Radiation pressure is a push force; they push outward, opposing the gravitational inward pull.
- Gas (Thermal) pressure is a push force; they push outward, opposing the gravitational inward pull.

Push pressures:

- Radiation pressure for blackbody radiation: The power density radiated by a star of temperature T is $I = 4\sigma/3T^4$ where T is the temperature, s is the Stefan–Boltzmann constant and c the speed of light. Thus the radiation pressure $P_r = I/c$ is $1/3\ a\ T^4 = 4\sigma/3c\ T^4$, where a is the radiation constant $a = 4\sigma/c = 7.565 \times 10^{-16}\ \text{J m}^{-3}\ \text{K}^{-4}$
- Gas (thermal) pressure: Stars have high density of hydrogen gas that get heated and produce a large outward gas pressure according to the gas law $P_{gas} = R\rho T/\mu$, where R is the gas constant, ρ is the gas density, and μ is the degree of ionization. It is 0.5 for fully ionized and 1 for fully neutral hydrogen.

Stability condition: When pull gravitation pressure balances the total push forces stability is achieved.

$$Pg = Pr + Pgas$$

Temperature behavior: The two push pressures have different temperature dependence. For low temperature, the gas pressure may dominate, while for high temperature the radiation pressure dominates.

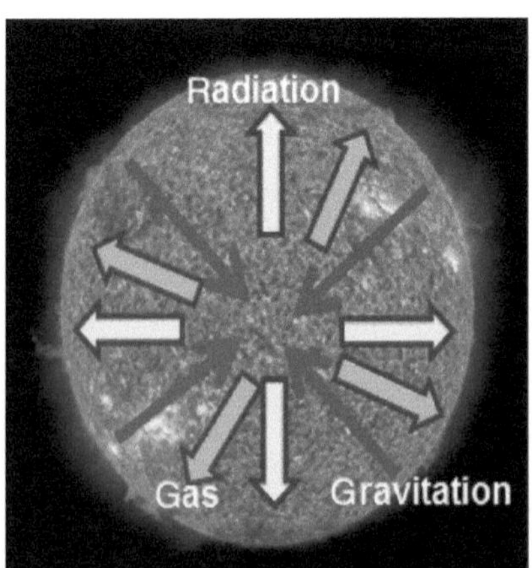

FIGURE 3.3 **Schematic of the three forces that act on a star: radiation or pressure, gas pressure, and gravitation forces.**

Small stars: The majority of stars that inhabit the main sequence have low internal temperatures (~ million degrees). Our sun is an example. The radiation pressure is not significant and contributes only a few percent while the gas pressure primarily balances gravity and hence primarily dominates, supporting stability against gravity by.

Massive stars: The temperatures of massive stars are hundreds of times hotter (~100 million degrees). Radiation pressure begins to dominate the gas pressure. In the most massive stars, the mass of the star is supported against gravity primarily by radiation pressure.

As discussed above, solar sails rely on light from the sun. Laser sails utilize powerful light laser beams [6]. The Voyager 1 spacecraft, which relies on sun light was launched in 1977. It has recently reached a speed of ~ 38,000 miles per hour or less than 0.006% of the speed of light. It took 37 years of flight to reach this speed and to get to the edge of the solar system. A great accomplishment, but with conventional propulsion technology, solar sails will never reach even the nearest stars. To have a chance, sun light must be replaced with extremely powerful lasers. But building such extremely powerful laser is a technologically challenging feat at this time.

A novel approach proposed recently by Philip Lubin at the University of California at Santa Barbara depends on development of laser arrays, as depicted in Fig. 3.4 [6]. Instead of building one extremely powerful laser the idea now is to build arrays that are made of a large number of relatively modest laser amplifiers. For this idea to have a chance, the lasers must be phased together. In other words they need to sync up to act like a single powerful laser. This strategy also eliminates the need for a single giant lens, replacing it with a phased array of smaller optics.

One prospectus laser is the ytterbium laser amplifier. Presently the laser is available with a power of 1–3 kW. It is about the size of a textbook and weighs roughly 5 kg. But it is estimated that one needs 50–70 GW to accomplish the

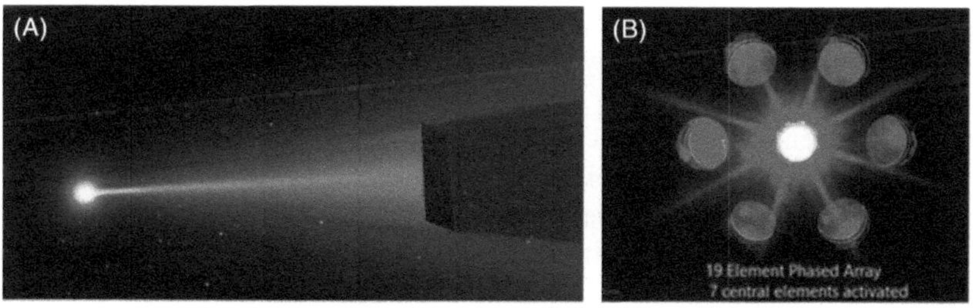

FIGURE 3.4 Laser array for sails. (A) Merged powerful laser beam (B) 19 element phased laser array of relatively modest laser amplifiers with 7 ventral elements activated.

task. This means an array of ~25 million smaller lasers is needed. It is estimated that the need array would take an area of 10 km by 10 km. If this array is placed in Earth orbit, it could propel a gram-sized wafer-like spacecraft attached to a 1-m-wide sail to more than 25% of the speed of light after about 10 min. of illumination. This could reach Mars in 30 min. and Alpha Centauri in about 20 years. Of course the scale up will have to be dome gradually, adding lasers over time.

More estimates show that the array could launch roughly 40,000 relativistic wafer-sized probes per year. Each "wafersat" would be a complete miniature spacecraft, carrying cameras, communications, power, and other systems.

As to man flights, the same array could propel a 100-ton spacecraft (about the mass of a fully loaded space shuttle, sans rockets). But it would need an 8.5-km-wide sail. Under these conditions, only 0.2% of the speed of light can be reached after ~15 years of illumination. At this speed, it would take about 2200 years to reach Alpha Centauri. Therefore, a larger array would be needed for a human interstellar trip. In addition, the braking problem must be solved.

One exciting goal aims at the realization of the laser sail. The goal is to propel a light spacecraft to Mars in 3 days. Current technology takes 4–8 months to get to Mars. It is hoped that the spacecraft would be accelerate to an unpresented 30% of the speed of light. If this target is realized, a manned shuttle could reach Mars in a month using the laser-based sail [6].

3.2 STOPPING ATOMS: ATOM TRAP (MAGNETO-OPTICAL TRAP)

The radiation pressure effect may also be used to slow down, cool, and reduce the random thermal velocities of atoms in a gas. Atoms can usually be treated as very-small classical particles.

However there are several difficulties and complications that hamper the realization of this goal [7].

- Atoms have discrete electronic structure and absorption/emission as shown in Fig. 3.5, or sodium, rubidium, and strontium atoms. Thus light absorption is strongly dependent on the frequency of the incident light. Moreover, the behavior of the atom at any time is dependent on the velocity of the atom (due to Doppler shifts).
- The frequency profile (spectrum) of the trapping beam is also very important. In general, atom trapping requires a laser beam tuned close to a transition frequency (spectral line) of the atoms in question.
- Atoms experience either attractive or repulsive gradient forces, depending on the detuning between the laser and the atomic transition. Attractive forces result from [RED/BLUE]-detuned incident light, and repulsive forces from [RED/BLUE]-detuned light.

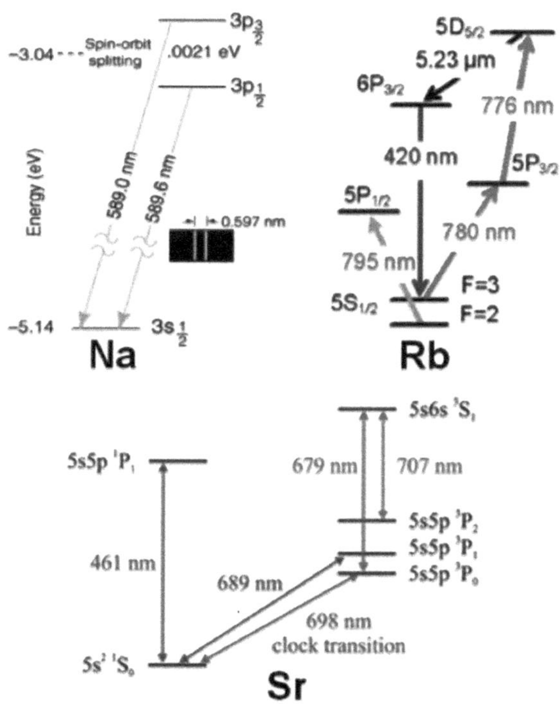

FIGURE 3.5 **Partial energy level diagrams of atoms relevant to cooling and stopping atoms of sodium and rubidium and strontium atoms.**

- Atoms experience a scattering force due to absorption (with a force in the direction of the beam) followed by spontaneous emission (in all directions, with a zero average recoil force).

The above difficulties have been alleviated and the goal has been accomplished by refining the radiation pressure effect in two ways. First, the process utilized high-quality narrow-band tunable laser radiation for which the frequency is highly defined/sharp and allows for tenability of the frequency with high control to match the sharp transitions in atoms.

Second, it exploited the problematic problem of the Doppler effect to an advantage. In this effect, the frequency of absorption/emission of an atom gets shifted by an amount proportional to the speed of the atom. More precisely, the characteristic frequency of absorption and emission of stand-still atoms ω_0 increases to ω_0 $(1 + v/c)$ if the atom moves toward a stationary observer (or opposite a laser beam) and decreases to ω_0 $(1-v/c)$ when moving away from the observer (or along a laser beam). The two cases are called blue and red shifting respectively as depicted in Fig. 3.6. It is to be noted that the frequency changes far less for atoms moving in a direction perpendicular to the line of sight between them (transverse Doppler effect).

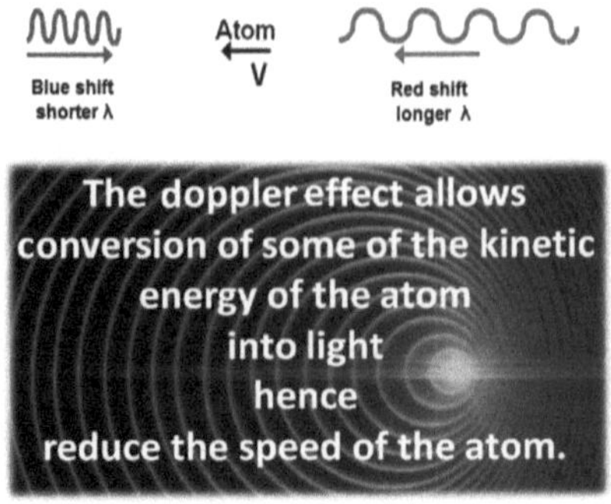

FIGURE 3.6 Doppler effect in the interaction of light with moving atoms. Two cases: atoms moving toward and along the laser light resulting in blue and red shifts, respectively.

William Philips, Claude Cohen-Tannoudji, and Steve Chu

1995 Nobel Prize "for development of methods to cool and trap atoms with laser light".

Claude Cohen-Tannoudji was born in Algeria and he is a citizen of France. He received his PhD in 1962 from the Ecole Normale Suprieure in Paris. He has been a professor at the College de France since 1973 and is a member of the Academie des Sciences. He shared the 1992 APS Lilienfeld Prize with Alan Guth. In a particularly sophisticated form of laser cooling, Cohen-Tannoudji has put helium atoms into a "dark state," whereby the coldest atoms become unable to absorb additional light and fall to temperatures even lower than previously imagined possible. The chilled atoms may also become the basis for extremely precise atomic clocks, accelerometers and gyroscopes.

Steven Chu received his PhD in physics from the University of California at Berkeley in 1976 and remained there for two more years as a postdoctoral fellow before joining the technical staff at AT&T Bell Laboratories in 1978. He headed the Quantum Electronics Research Department in 1983, leaving the company in 1987 to join the faculty of Stanford University. Chu has used his laser array to split ultracold atoms into separate waves and recombine them to form interference patterns that can provide detailed information on the atoms.

William Phillips received his PhD in 1976 from the Massachusetts Institute of Technology and completed a postdoctoral stint there before joining NIST in 1978. Phillips can create "optical lattices," crystal-like arrays of atoms held in place by light waves. He and his team studied ultra-cold trapped atoms with spin-off applications for improved accuracy in atomic clocks and in the fabrication of nanostructures.

Fig. 3.7A explains in detail how it works. First one adjusts or tunes the frequency of the light beam short of what is needed to induce absorption of the light when the atom is stationary. In other words one sets the frequency below the atomic resonance frequency ($\omega < \omega_0$). An atom moving opposite to the laser beam is shown in Fig. 3.7B then uses its own speed to supplement the frequency of the light to come into strong resonance absorption. Once this happens, the process results into a strong push force along the laser beam. Within a few nano seconds, the atom quickly and spontaneously emits a photon in a random direction. The emitted photon exerts another force on the atom in the opposite direction of its direction. Because unlike absorption, emission occurs in random directions (hence recoil) the overall process results in a net force with velocity reduction (cooling) in the direction of incident photon flux.

On the other hand for an atom moving along the direction or perpendicular to the beam the process is ineffective. In other words the atom in this case cannot use its speed to bring itself into strong absorption. This happens because the frequency gets shifted further away making the atom interact very weakly with

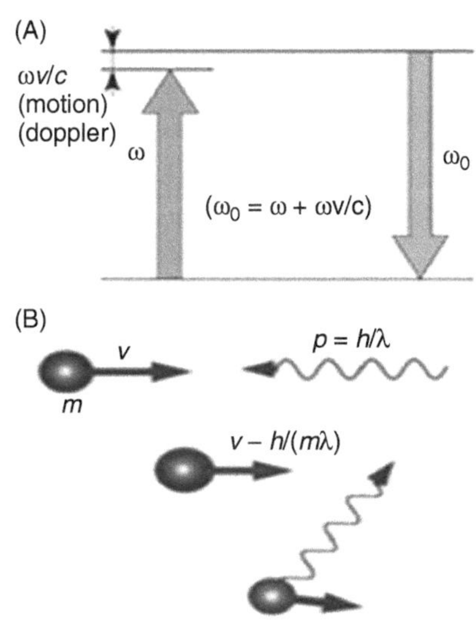

FIGURE 3.7 Detailed diagram of the cooling of atoms using the Doppler effect with one laser beam. (A) The energy level diagram and laser interaction. (B) Sketch of analysis for an atom moving opposite to the laser beam.

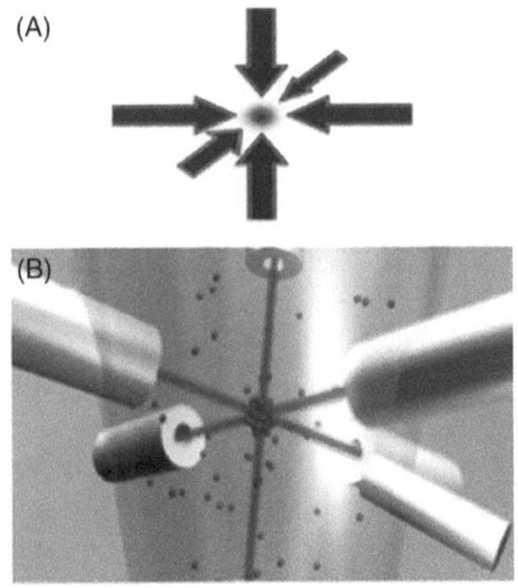

FIGURE 3.8 (A and B) Sketches of laser-based molasses or optical molasses (OM) in which one uses six identical beams (three pairs of identical counter propagating laser beams) along the three coordinate axes to ensure cooling in three dimensions.

no push force. In fact as depicted in Fig. 3.8A–B researchers shine six identical beams (three pairs of identical counter propagating laser beams) along the three coordinate axes to ensure cooling in three dimensions. This procedure is now called Doppler pressure cooling, or optical molasses (OM). It can cool the atoms down to 40 μK.

The laser-based molasses or OM arrangement provided damping forces. The damping forces may keep the atom in place only for about 0.3–0.4 s before it escapes or diffuses out. To keep it in place for long time, additional forces are needed in the form of restoring forces or trapping forces. Restoring forces may be produced using two different ways:

- *Focused laser beams* with high intensity: Focused beams of high-intensity with a large gradient provide trapping to the center of the beam (to be discussed later). But because intense lasers have high electric fields, the laser is expected to change the structure of the atom, shifting their frequencies, independent of their speed, and may interfere with the cooling process.
- *Magnetic forces*: The presence of a "focused" (inhomogeneous) magnetic field (having strong gradient) provide magnetic gradient restoring force. The field can be provided by a pair of magnetic field coils as shown in Fig. 3.9. On its own, it can cool atoms (cools sodium atoms down to only 300 μK), but this cooling ability is nearly an order of magnitude worse than the cooling ability of the six laser beam arrangement (OM) (down to 40 μK). However, its trapping ability is the attractive feature here. When combined together, the system provides good cooling and trapping for long times. As such it is called magnetic optical trap (MOT).

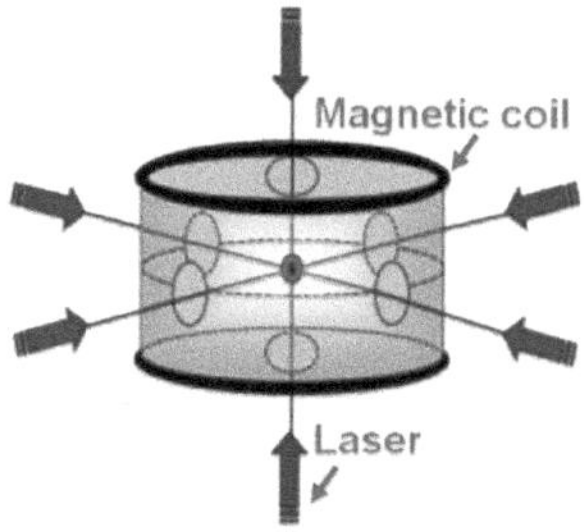

FIGURE 3.9 **Sketch of a magnetic optical trap (MOT) in which six laser beams are used along with and inhomogeneous magnetic field (having strong gradient) that provide restoring force.**

Example 3.4 Magnetic-Optical Trap

At room temperature (~300 K), a gas of sodium atoms would have a thermal kinetic energy of 20 meV. The speed range covers 0–10^4 m/s with an average of 500 m/s. These atoms have been cooled to average speed of 4.9 cm/s. The work earned Chu the 1997 Nobel Prize in Physics along with Claude Cohen-Tannoudji and William D. Phillips [8].

Example 3.5 Acceleration of a Sodium Atom by Single-Photon Absorption

Consider a laser beam of $\lambda = 586$ nm propagating along the *z*-direction toward a sodium Na atom with a mass $M = 23$ amu. As shown in Fig. 3.10, a photon in the beam is absorbed by the sodium atom. Consequently, the sodium atom emits the photon at an angle with a characteristic 16 ns lifetime.

We explore the magnitude of the force and the corresponding acceleration that the absorbed photon exerts on the sodium atom.

The magnitude of the wave vector is $k = 2\pi/\lambda = 10^7\ \text{m}^{-1}$. Using $F = \Delta p/\Delta t$ where Δp is the momentum of the photon and $\Delta t = 16$ ns gives 10^{-19} Newton for the push force. This force is minuscule but because the atom is extremely light it is enough to give the atom an incredible acceleration of $a_{max} = 10^6\ \text{m/s}^2$. By comparison the gravitational acceleration ($g = 9.81\ \text{m/s}^2$) is about 10^5 times smaller than the maximal emission force, but it last only 15 ns.

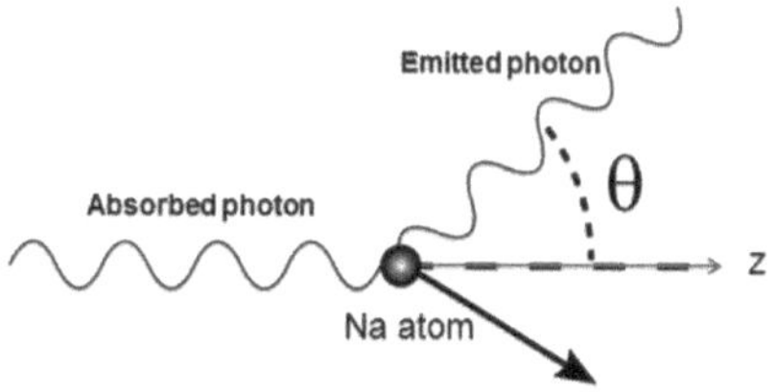

FIGURE 3.10 **Diagram of the acceleration of a sodium atom by a single-photon absorption.**

Exercise: When the photon is emitted, the atom experiences another acceleration. The photon is emitted at an angle $\theta = 30$ degree. Determine the overall acceleration. In fact it will be found that there is still a residual push acceleration. And with many absorption/emission cycles it loses a lot of speed in the direction of the laser beam.

Next we consider the time required to slow down a thermal atom: The mean thermal speed of sodium at 300 K is 500 m/s. The recoil speed of a sodium atom after absorption of a photon at 586 nm is $v_R = 3$ cm/s. Thus we need to absorb 500/0.03 = 17,000 photons, in order to completely stop a thermal sodium atom. At least two times 17,000 spontaneous lifetimes are required, which takes 300 μs.

Example 3.6 Quantum Condensation of Atoms: Bose–Einstein Condensate (BEC) [8]

There are many applications in atomic interactions that can employ cold, motionless trapped atoms. One fundamental application is testing a basic principle of the theory of quantum mechanics. Eric Cornell and Carl Wieman at the University of Colorado at Boulder NIST–JILA laboratory used the trap to bring a dilute ensemble of rubidium atoms in a gas to an ultra-cold state. The team started with 10 million rubidium atoms at thermal conditions of 300 K. They used laser radiation at a wavelength to cool the atom ensemble bringing their average temperature down to 10^{-6} (μK). Then they held them in the quadrupole-magnetic field trap. At this point the atoms are not cold enough. The temperature needs to be reduced by nearly two more orders of magnitude. Instead of cooling further, the team simply removed the hotter atoms in the ensemble leaving only the coldest ones. This has been accomplished by a cleanup "evaporation" step of the more energetic atoms in the magnetic trap, leaving behind a small number (10^4 or ~1 percent) of atoms that is colder and denser with a temperature reaching nearly 20 nK. What is left, however, is significant enough to carry out the experiments. The rubidium atoms reached a temperature of 20×10^{-9} (20 nK), the trapped atoms went into a BEC.

Unlike fermions, a spin of ½ particles, boson atoms of spin 0 are not constrained by the Pauli Exclusion Principle. A BEC, predicted by *Bose* and *Einstein* in 1924–25, is a state of matter of a dilute gas of zero spin particles (*bosons*) cooled to *temperatures* very close to *absolute zero* (that is, very near 0 K or −273.15 C) such that all of the atoms are in a single quantum state and can therefore be described by the same wavefunction. In fact, the first gaseous condensate was produced on June 5, 1995. Cornell and Wieman shared for their achievements with Ketterle of MIT, who demonstrated important BEC properties the 2001 Nobel Prize in Physics. In the intervening years there was little notice of this possibility except in the behavior of superfluid helium and in superconductivity. Fig. 3.11shows images of the BEC as it was cooled from 200 nK down to a temperature of 20 nK. The coherent BEC emerged as a peak representing a group of atoms with the same velocity, surrounded by a field of normal atoms with random velocities. The BEC represented about 2000 atoms of rubidium.

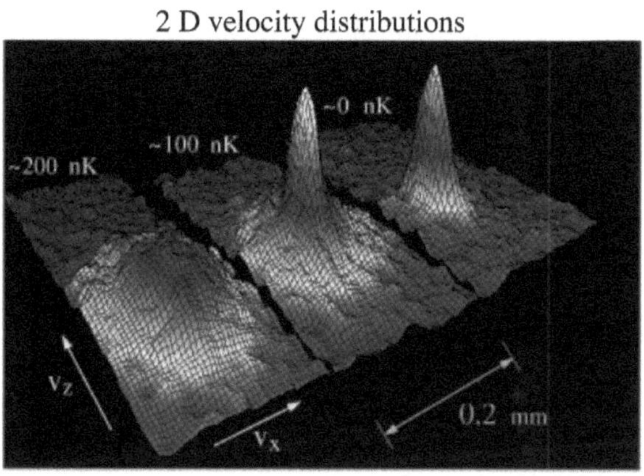

FIGURE 3.11 **Bose–Einstein condensation (BEC) of atoms about 2000 rubidium atoms cooled to a temperature of 20 nK.** *(Used with permission from Michael Matthews, University of Colorado/JILA)*

3.3 TRAPPING NANO- TO MICRO-PARTICLES (PUSH–PULL FORCES)

We have seen above how light can be used to cool and trap the smallest constituent of matter, atoms as well as how light can be used to propel macro objects as large as many meters. Here we show how we can use the interaction of light to interact and trap (optical trap) intermediate size objects ranging from several micrometers to several nanometers solid objects [9]. As in the case of atoms, optical trapping of particles is the action of immobilization of small particles in specified volumes either in vacuum or in a liquid of choice by the action of their interaction of light. This size regime is of great importance to nanoscience and technology. As to material types of nano particles, nanoparticles may include dielectric, semiconductor, or metallic nanoparticles.

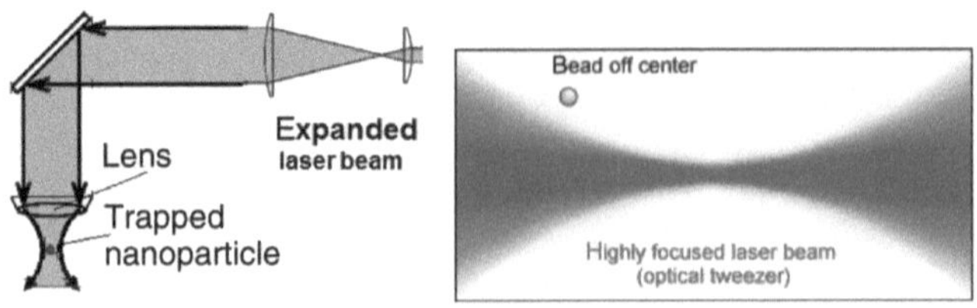

$$I(r,z) = I_0 \left(\frac{w_0}{w(z)} \right)^2 e^{-\frac{2r^2}{w^2(z)}}$$

FIGURE 3.12 Optical tweezer. (Top) (left) basic concept of the tweezer or single-beam gradient trap. (Right) sketch of the beam waist (bottom) theoretical expression describing the intensity of the laser beam I in the trap.

The basic principle of the trapping is that small particles are polarizable and hence can be trapped with gradient forces. The most common type of laser trap relies on the use of a single-beam gradient trap. This trap is also known as optical tweezers, or laser tweezers [10].

Fig. 3.12A illustrates the basic concept of the tweezer or single-beam gradient trap. A narrow laser beam is first expanded using an inverted telescope. It is then tightly focused to a very small spot size. The focusing is achieved using a high numerical aperture microscope objective. At the focus there is a strong maximum intensity. For example, a laser beam has a Gaussian profile with a waist *a* ($e{-}r^2/a^2$) when they exit the laser window [11]. When focused by a lens, the beam returns the Gaussian profile but the waist shrinks forming a much tighter waist. The intensity of the laser beam I depends on the distance along the beam z, and on the distance perpendicular to the beam r. It is shown graphically, as well as via a formula in Fig. 3.12B. In the formula, *w* is the diameter of the cross section (waist).

When a small particle of radius *a* and refractive index *n* moving in a medium of refractive index n_0 approaches the focal spot of the laser beam of wavenumber $k = 2\pi/\lambda$ (propagating along the *z*-axis) to cross it from any direction *x*, or *y*, it scatters light in all directions and as a result the laser light tends to apply a push force on particles in the beam along the direction of beam propagation (as we have seen above in the laser sail). This is effectively a de trapping scattering force. Fig. 3.13 shows the layout of the interaction.

Detailed analysis of the interaction of the particle with the beam when the particle center at a distance *r* from the center of the beam shows that the scattering force per unit area F_{scat} (pressure) is proportional to the intensity of the beam *I* and is given by the following expression:

$$F_{scat}(r) = \frac{k^4 \alpha^2}{6\pi c n_0^3 \epsilon_0^2} I(r)\hat{z} = \frac{8\pi n_0 k^4 a^6}{3c} \left(\frac{m^2-1}{m^2+2} \right)^2 I(r)\hat{z}. \quad (1)$$

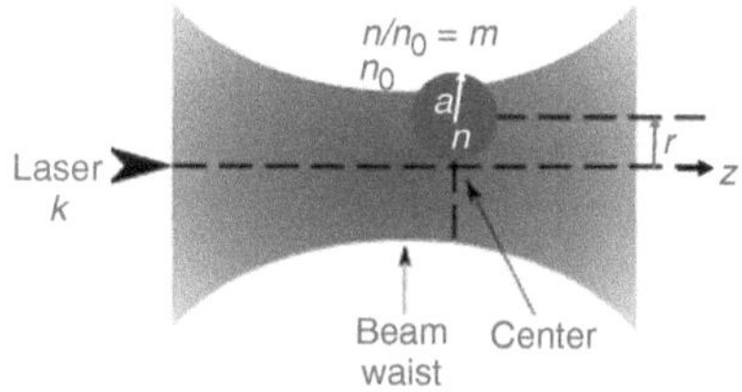

FIGURE 3.13 **The layout of the interaction a small particle moving in a medium as it approaches the focal spot of the laser beam.**

Where c and ε_0 are the speed of light and the permittivity of free space, and α and V are the polarizability and volume of the nanoparticle:

$$\alpha = 4\pi n_0^2 \varepsilon 0\, a^3\left(\left(m^2-1\right)/\left(m^2+2\right)\right) = 3n_0^2\, \varepsilon 0\left(\left(m^2-1\right)/\left(m^2+2\right)\right)V \quad (2)$$

Example 3.7 Scattering Force

Consider a light beam of intensity I propagating in the z-direction in a medium of refractive index n_0. The beam strikes a particle of radius a, and refractive index $n = mn_0$. After the particle absorbs an incident photon, it isotopically reemits the radiation. Upon absorption, the incident radiation exerts an impulse on the molecule along it propagation direction. But upon emission, the impulse will be in the opposite direction of the propagation of the emitted photon. As the photon emission can be in any direction (isotropic emission), that is, it has no preferred direction, the overall process of absorption–emission results in a resultant force in the direction of incident photon flux. This force is proportional to the magnitude of the incident intensity:

$$F = n_0 6 I / cz \quad (1)$$

where n_m is the refractive index of the surrounding medium, I is the intensity, c is the speed of light, and $\boldsymbol{\sigma}$ is the particle's cross section. For a spherical particle, the expression takes the form is given by

$$\sigma = \frac{8}{3}\pi (kr)^4 r^2 \left(\frac{n^2-1}{n^2+2}\right)^2 \quad (2)$$

with the particle radius r, the refractive index n of the particle and the wave vector k of the used light. This gives:

$$F_{scat}(r) = \frac{k^4\alpha^2}{6\pi c n_0^3 \in_0^2} I(r)\hat{z} = \frac{8\pi n_0 k^4 a^6}{3c}\left(\frac{m^2-1}{m^2+2}\right)^2 I(r)\hat{z}.$$

At the same time, when a small particle approaches the focal spot of the beam to cross the beam from any direction *x* or *y*, it experiences a pull force toward higher and higher intensity, namely to the center of the focal spot. The pull is caused by the gradient of the intensity. The gradient force's direction is toward the area of highest light intensity, that is, toward the beam axis in case of a Gaussian beam profile, and toward the focus of the laser if the beam is focused. This is a gradient force or trapping force.

$$F = \frac{1}{2}\alpha\nabla E^2 = \frac{2\pi n_0 a^3}{c}\left(\frac{m^2-1}{m^2+2}\right)\nabla I(r), \tag{2}$$

The trapping and de-trapping forces are shown graphically in Fig. 3.14A [12]. Mathematically the forces are given in Eqs. (1) and (2). In the formula

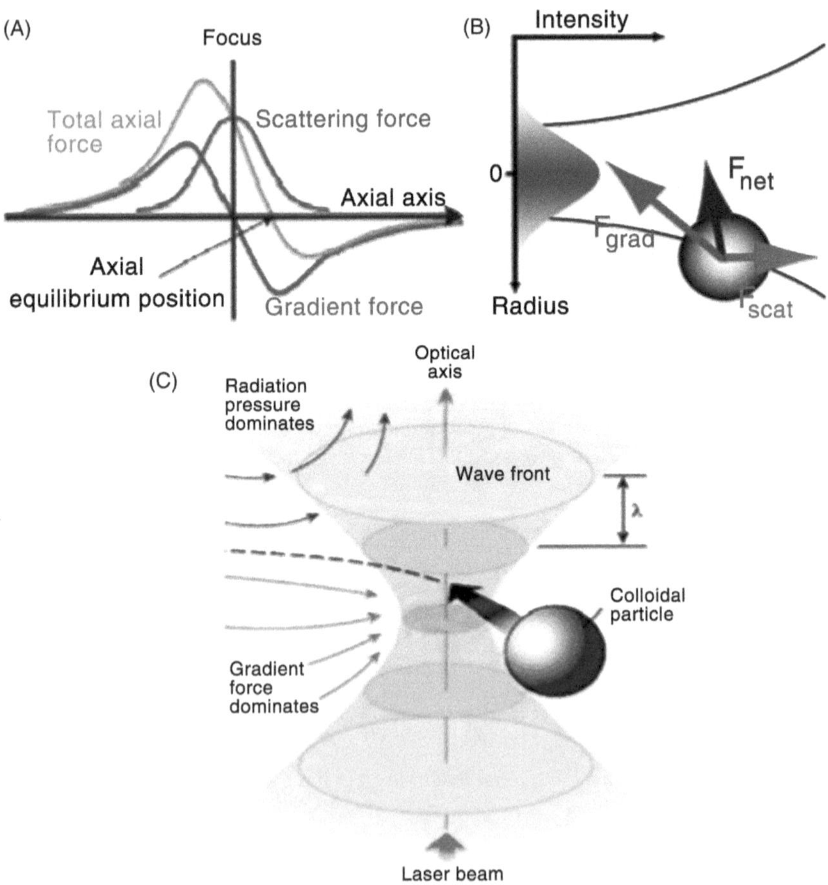

FIGURE 3.14 (A) Plot of trapping and de-trapping forces with the distance from the center of the focal spot of the laser beam. *(From http://www.isibrno.cz/omitec/index.php?action=sbt.html)* (B) Schematic of gradient trapping force and the scattering force acting on the particle. (C) Outlines the regions of dominance of scattering (pressure) and gradient (trapping) forces. *(From David G. Grier, A revolution in optical manipulation, Nature 424, 810–816 (14 August 2003))*

a is the polarizability of the object, and can be easily read if necessary. The net force acting is therefore the sum of the two as schematically shown in Fig. 3.14B. Fig. 3.14C outlines the regions of dominance of pressure and gradient forces. Fig. 3.14A shows that the trapping force acts as a restoring force toward the focus of the beam with respect to all three dimensions. In fact the force experienced by the particle is linear with respect to small displacement from the center of the trap is similar to a simple spring restoring force, which follows Hooke's law, allowing the trap to work well. Near the narrowest point of the focused beam, the beam waist, the trap force dominates and overall effect of the two processes is an attractive force Typical forces provided by the gradient effect could be as high as Pico Newton. This could accelerate a particle of 100 nm.

Example 3.8 – Trapping of Si nano Particle [13]

Porous p-silicon is prepared by chemical etching. The porous film was peeled off. The layer has approximately 50% porosity. The free standing layer was broken up into silicon nanoparticles using mechanical ball milling. Scanning electron microscopy (SEM) imaging of the powder samples shows an average size of 150 nm as shown in Fig. 3.15. The silicon nanoparticle is dissolved in water. The refractive indices of the particle is n_p ~3.7 (for Si at NIR wavelengths) and for the medium is n_m ~1.33 (for water), respectively. The optical tweezers apparatus (see the sketch in Fig. 3.1). Optical trapping is achieved by focusing a laser beam on the sample, through an oil immersion high numerical aperture (NA = 1.3) objective. Trapping is examined using 830 nm laser diode; hence the size of the trapped particle is much smaller than the trapping wavelength. The trapping beam is a Gaussian beam with a beam waist radius w_0 ~0.6 λ/ NA (diffraction-limited spot) and Rayleigh length $z_{ero}= \pi w^2/\lambda$. The transverse spring constant was measured to be about 3 pan μm^{-1}. With these parameters the electric field may be written in terms of the trapping power P as

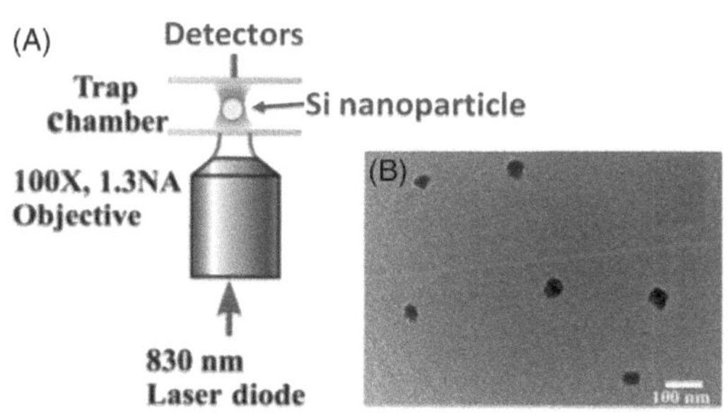

FIGURE 3.15 (A) Trapping of Si nano particle in a section of the optical tweezer apparatus. (B) Scanning electron microscopy (SEM) images of the silicon powder samples showing an average size of 150 nm. *(From Maria G. Donato, Marco A. Monaca, Giuliana Faggio, Luca De Stefano, Philip H. Jones, Pietro G. Gucciardi Onofrio M. Maragò, Optical trapping of porous silicon nanoparticles, Nanotechnology, 22, (50))*

$$E_0^2 = \frac{4P}{\pi w_0^2 n_m \varepsilon_0 c}$$

The trapping force is given by Eq. (2) is now linearized with respect to *r*, the radial direction. It is then equated to – or to get the spring constant for trapping *k*. The result is

$$k_r = \frac{\mathrm{Re}\{\alpha\}E_0^2}{w_0^2} = \frac{16P}{cn_m w_0^4}\left(\frac{d}{2}\right)^3\left(\frac{m^2-1}{m^2+2}\right)$$

Thus, by using the parameters of the experiment, a transverse spring constant of about 10 pan μm^{-1}, which is in fair agreement with the measured value of about 3 pan μm^{-1}. The discrepancy can be accounted by the fact that we have not considered any effect from the scattering force that pushes the particles away from the focal spot, hence weakening the trap.

The material can be classified according to their optical characteristics as follows:

1. Transparent particles are trapped in the beam focus if the refractive index of the particle is higher than the refractive index of the surrounding medium (*high-index particles*). Trapping requires the gradient force to be larger than the scattering and absorption forces, which will tend to push the particle along the beam axis.
2. Strongly absorbing particles: The gradient force is negligible. The particles cannot be three-dimensionally trapped by a single beam. The particle can, however, be trapped two-dimensionally in a region where the beam is converging.
3. Strongly reflective particles: The major response is similar to that of strongly absorptive particles. Because of reflection, the particle geometry becomes more important, as the direction of the momentum transfer depends on the directions of both the incident and reflected light.

Example 3.9 – Gravitation and Radiation Accelerations of Nanoparticles (Nano grains) in Vicinity of the Sun

Consider a nanoparticle at a distance *r* from the center of the sun as shown in Fig. 3.16. In Chapter 15, we will be discussing the prospect of finding nanoparticles in the proximity of the sum corona. The particle will essentially be under two

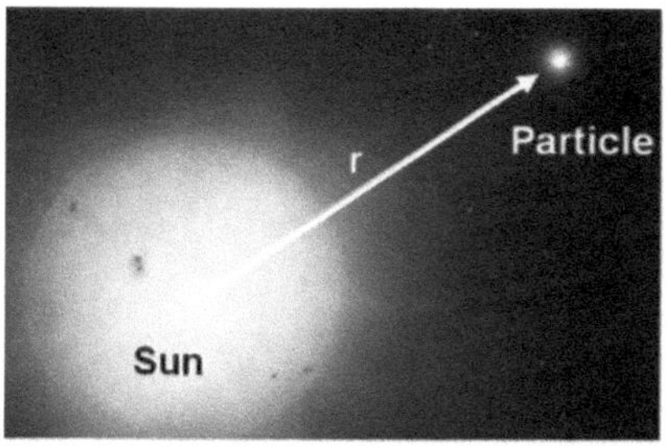

FIGURE 3.16 Diagram of nanoparticles (nano grains) in vicinity of the sun.

forces, a pull force due to the gravity of the sun and a push force due to light from the sun striking the particle. We can write down an expression for the scattering force. The scattering force due to light propagating along the z direction is given by the following expression: $F_{scat} = n_m \sigma I/cz$, where n_m is the refractive index of the surrounding medium, c is the speed of light, and σ is the particle's scattering cross section.

Micro particles: For large particles with a radius larger than the wavelength of visible light, the scattering cross section σ is simply the geometrical cross section.

The gravitational attraction of the Sun (mass M at distance r) upon a spherical particle of radius s and density ρ is $Fg = (4/3)\ \rho\pi s^3 GM/r^2$, where G is the gravitational constant. The radiation pressure force due to solar radiation is $F_r = (IA/c)$, using $I = L/4\pi r^2$for the radiation flux density at distance r; L is the solar luminosity. We can also write $I = I_o(r_o/r)^{\ 2}$, where $I_o = 1.36 \times 10^6$ ergs $cm^{-2}s^{-1}$ is the solar constant and $r_0 = 1$ AU in centimeters. We then compare the radiation force to the gravitational force, defining their ratio by the parameter $F_r/F_g = 5.7 \times 10^{-5}/\rho s$, where ρ is the density of the particle and s is the radius of the particle in cgs units. We note that the ratio is of order unity for particle radii of order 0.5 μm, or near where most of the solar energy is contained.

Nano particles:

For scattering by particles much smaller than the wavelength of the light (particles with radius a << λ) the cross section used above cannot be used. Detailed electromagnetic theory must be used. This regime is called Rayleigh scattering regime. The more general Mie theory must be used. Mie theory is a complete analytical solution of Maxwell's equations for the scattering of electromagnetic radiation by spherical particles [14]. The effect depends on the size and material type. The results, however, have to be cast numerically.

Some calculations have been carried out. Fig. 3.17 gives the ratio F_r/F_g for nanoparticles of a variety of material: ice, silica (quartz), iron, basalt, obsidian, magnetite, and ideal material. The results show that quartz and magnetite respond much more strongly in the nano regime. In fact for 100 nm, scattering dominates

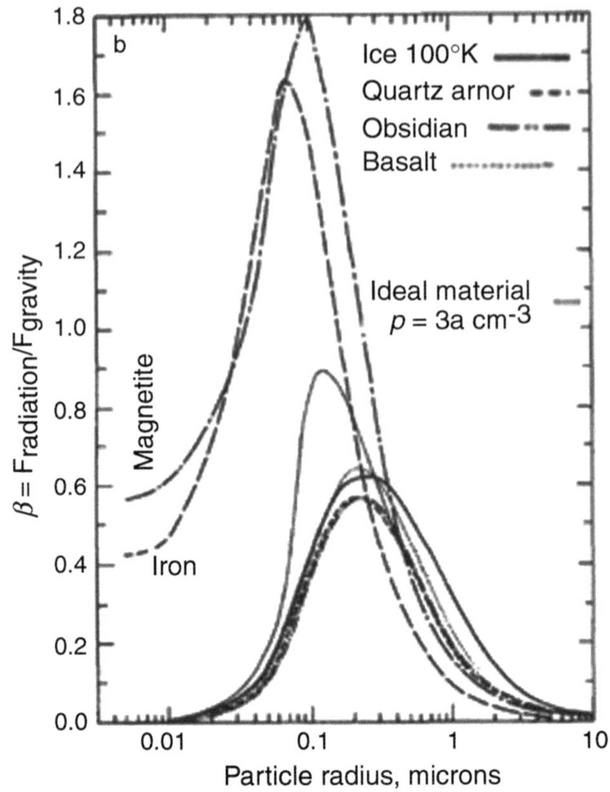

FIGURE 3.17 The ratio of the radiation and gravity forces F_r/F_g for nanoparticles of a variety of material: ice, silica (quartz), iron, basalt, obsidian, magnetite, and ideal material, as a function of their size.

gravitational forces by nearly twofold. Also note that for nanoparticles of less than 8 nm the scattering force becomes independent of size. Note that other materials show near insensitive to the type of the material in the regime for 200–500 nm, sizes on the order of the wavelength.

As to the size, there are four regimes:

- Macro-particle >>> wavelength
- Micro-particle ~ wavelength
- Nano-particle (solid) <<< wavelength

In some cases the scattering forces are strong. If the trapping forces are not strong enough to dominate the scattering pushing forces, then trapping may not be achieved. In this case, one solution is to use two counter propagating laser

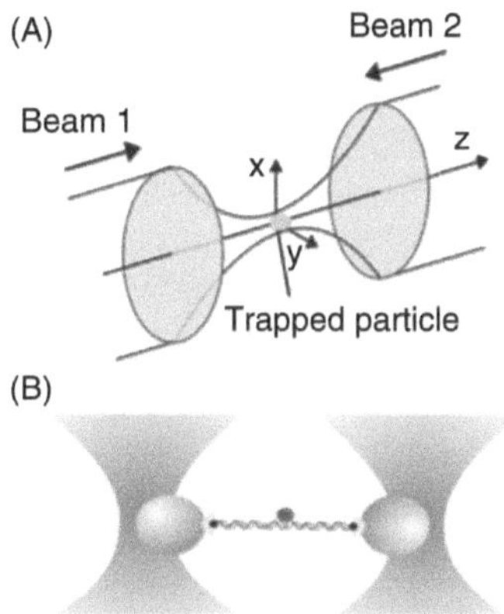

FIGURE 3.18 Advanced trapping configuration. (A) Two counter propagating laser beams trapping configuration that allows scattering forces to cancel out. (B) A system of two (dual) optical tweezers (2OTs).

beams as shown in Fig. 3.18A [15]. This configuration allows scattering forces to cancel out.

Another interesting configuration is the use of two optical tweezers, dual optical tweezers (2OTs). First, one attaches one nanoparticle on each end of the bio component, followed by trapping each particle by an optical tweezer. This allows stretching of the molecule using the two tweezers as shown in the Fig. 3.18B [16]. In the figure a single protein labeled with a fluorophore (red dot) interacting with DNA suspended between the double optical tweezers can be localized and tracked with high accuracy (drawing not to scale). The bio component may include DNA, proteins, and enzymes, and it may be labeled by a florescent marker.

Example 3.10 – Derivation of the Restoring Trapping Force [17]

Understanding the electromagnetic nature of the trapping force requires several steps of EM vector manipulation begins. We begin with calculating the response of the particle to the electric field of the light. The interaction produced an induced dipole moment. The electric field of the laser will then interact with the induced dipole moment to produce a force.

1. The electric field of the laser beam E induces a dipole moment in the particle, which is proportional to the polarizability of the particle. The polarizability, *a*, was given up in terms of the radius and refractive index of the particle.

$$\vec{p}(\vec{r},t)=\alpha\vec{E}(\vec{r},t)$$

2. The electric field of the laser will then interact with the induced dipole moment, inducing electrostatic energy and exerting a force. The energy is –**P.E.** Taking the gradient of the energy gives the force:

$$\vec{F}(\vec{r},t)=(\vec{p}(\vec{r},t)\nabla)\vec{E}(\vec{r},t)$$

3. This expression of the force is now re-written in terms of the energy the electric field produces. The total electromagnetic energy in the nanoparticle due to the presence of the laser light is related to E^2. To do the transformation, we first take the gradient of E^2:

$$\nabla\vec{E}^2=2(\vec{E}\nabla)\vec{E}+2\vec{E}\times(\nabla\times\vec{E})$$

Note that according to Maxwell's equations

$$\nabla\times\vec{E}=0$$

Thus when we substitute in the expression for the force we get

$$\vec{F}(\vec{r},t)=\frac{1}{2}\alpha\nabla\vec{E}^2(\vec{r},\vec{t})$$

We can now average over time because the electric field oscillates with time. It yields

$$\vec{F}_{grad}(\vec{r})=\left(\vec{F}(\vec{r},t)\right)_t=\frac{1}{2}\alpha\nabla\left(\vec{E}^2(\vec{r},t)\right)_t=\frac{1}{4}\alpha\nabla\left|\vec{E}(\vec{r})\right|^2$$

4. We can now write the expression in term of the intensity of the laser beam The light intensity ***I*** associated with the field is [13]

$$I(\vec{r})=\frac{n_m\varepsilon_0 c}{2}\left|\vec{E}(\vec{r})\right|^2$$

where $\boldsymbol{n}_m$ is the refractive index of the exposed material, $\mathbf{e}_0$ is the permittivity of free space, and **c** is the speed of light. This leads to the force relation

$$\vec{F}_{grad}(\vec{r})=\frac{1}{2n_m\varepsilon_0 c}\alpha\nabla I(\vec{r})$$

Example 3.11 Tweezer Gradient Trap Forces.Pdf File [18]

Consider a silica sphere of 50 nm in diameter and refractive index 1.46. It is placed in water (refractive index of 1.33). Therefore the relative refractive index $m = n_{silica}/n_{water}$= 1.46/1.33 = 1.10. A Guassian-shaped laser beam (the wavelength of the laser is 1064 nm) of total power of 200 mW focuses to Guassain beam with a waist of focal spot of 1.5 μm as shown in Fig. 3.19.

Microsphere in air (n_{air}= 1.00). NA = 0.22. The Rayleigh range of the laser is 6.6 μm.

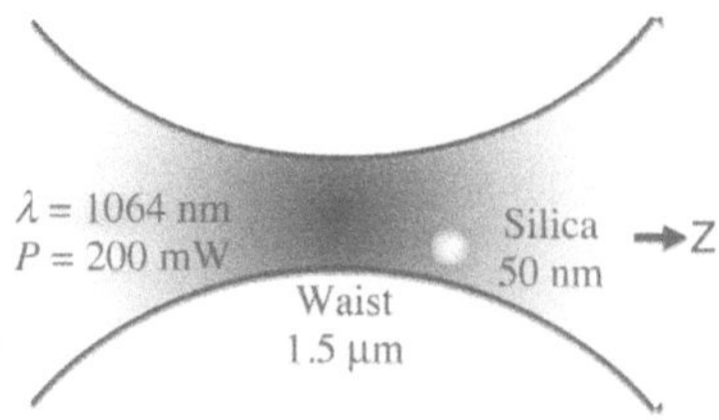

FIGURE 3.19 Layout of the interaction and conditions in a numerical example of trapping nanoparticles.

Fig. 3.20A–B shows the calculated trapping force, as well as the pushing scattering forces along the x-direction (perpendicular to the propagation (z-axis)). The scattering force is zero in the radial direction (Fig. 3.20A) and is positive along the axial direction (Fig. 3.20B). The gradient force is negative at positive coordinates, and positive at negative coordinates. Thus it will always pull back the nano sphere to the center of the trap.

Fig. 3.20C–D shows the corresponding calculated potential. As the refractive index of air is very close to the refractive index of vacuum, the potential and force

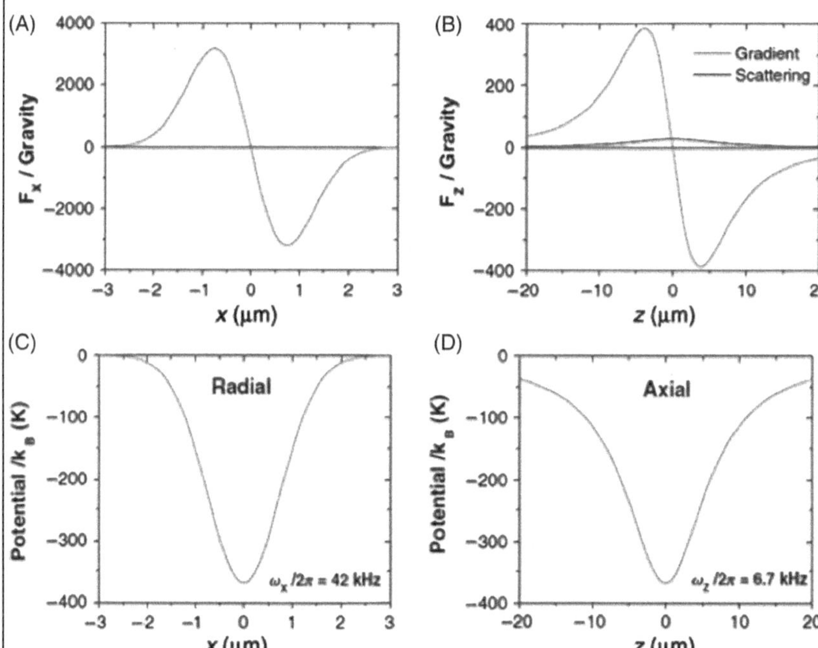

FIGURE 3.20 The trapping forces and scattering (pushing), as well as potentials for the conditions in Fig. 3.19. (A and B) The calculated trapping force, as well as the pushing scattering forces along the x-direction (perpendicular to the propagation (z-axis)). (C and D) Corresponding calculated potential.

on a nanosphere in air is practically the same as that in vacuum. The calculated well depth of the trap is 367 K. The laser will therefore only be able to trap a 50 nm nanosphere at a temperature much lower than room temperature. The potential is approximately harmonic near the bottom of the trap. The oscillation frequency is about 42 kHz in the radial direction and 6.7 kHz in the axial direction for a 50 nm nanosphere trapped near the bottom of the potential.

As exercises, here are some questions Determine the force that would act on the nanoparticle when it enters the laser beam at the waist. Is it repulsive or attractive? What is the force at the center of beam spot, namely at $x = 0$ and $y = 0$? What is the scattering force? What is the trapping potential and how does it compare with kT? The well depth should be at least 10 times larger than the average kinetic energy of the nanosphere.

3.4 TRAPPING OF ULTRASMALL NANOPARTICLES: NANO (TRAP) TWEEZERS

Though the optical trap worked for the smallest particle (atoms) and large object (sails), the question of trapping ultrasmall nanoparticles or quantum dots is intriguing. The difficulty stems from the fact that light forces in this regime scales approximately with particle polarizability hence its volume. This can be clearly seen from the expressions for the trapping force given in Eq. Going from 3 μm particle to 3 nm, the trapping force drops by a factor of 10^9. Thermal fluctuations would easily dominate trapping forces at the nanoscale as they are proportional to the volume of the object. In fact, this difficulty arises for particles in which the linear dimensions are significantly smaller than the trapping wavelength λ ($a << \lambda$). This difficulty arises irrespective of the type of material: metal, semiconductor, and composites structure and of the shape: nanoparticles, quantum dots, nanowires, nanotubes, grapheme, and two-dimensional crystals.

To offset or counter the volume (or polarizability effect), a more intense laser beam must be employed. Moreover, more tightly focusing elements must be used.

3.4.1 Higher NA

Using high-quality microscope objective lens with a high-numerical aperture, such that the optical gradient forces become much more significant. A laser power of about a milliwatt is sufficient to trap a single small transparent particle in three dimensions.

3.4.2 Plasmonic (Metal) Nanolenses

A new type of light focusing utilizes plasmonic nano lenses made of metal arciyutures. Plasmoning lenses will be discussed in Chapter 8. Plasmonic nano lenses utilize nano apertures, as well as solid nanostructures. The structures concentrate and absorb light, resulting in "plasmonic hotspots" much smaller than the diffraction limited focal spots. The "focal length" of these type of lenses is just 10–30 nm depending on the details of the nanostructure. This makes it possible to manipulate nanometer-scale objects suspended in a fluid, enabling what is called plasmonic "nanotweezer."

A bowtie plasmonic nano tweezer is built by a single-mode optical fiber [19]. The fiber is first tapered into a cone, by laser pulling for example, and the coated with 200 nm aluminum to prevent light leakage through the sides of the cone. A flat facet is created by cutting the tip extremity using focused ion beam (FIB) milling. The diameter of the facet is 1 μm. The facet is then metallized by depositing a 100-nm-thick gold film. Finally, FIB is used to mill the facet such that it produces an aperture in the gold film in the shape of a bowtie nanoaperature (BNA) with prescribed dimensions. An SEM image of the aperture is shown in Fig. 3.21A. The aperture is an 85-nm-gap pattern. The fibers

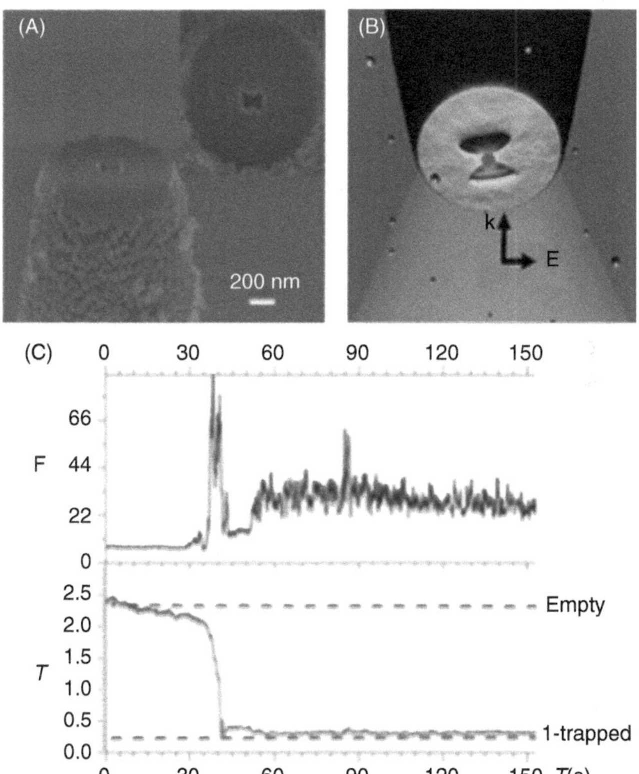

FIGURE 3.21 A bowtie plasmonic nano tweezer using a single-mode optical fiber. (A) SEM image of the aperture. (B) Mode of operation of trapping a bead by the BNA, illuminated from side opposite to the fiber, namely from the water side. (C) Portion of the time traces of the detected transmitted (T) signal (in red color) and the fluorescence (F) from the bead (in blue color).

are mounted on a three-dimensional piezoelectric scanner and introduced into a modified fluidic chamber containing a solution in which 50 nm luminescent polystyrene beads are suspended.

The device is quite versatile with respect to illumination and collection of light. Just like an NSOM tip (see Fig. 4.18B) the bowtie may be illuminated from either side. In the mode of operation, shown in Fig. 3.21B, the BNA is illuminated from side opposite to the fiber, namely from the water side. An Nd–Yag infrared laser beam at 1064 nm is used because water has a minimal absorption at this wavelength. The beam is focused through a $\times$ 40 (0.65 NA) objective lens. As can be seen the polarization of the incident beam is parallel to the BNA gap (perpendicular to the axis of the bowtie). Such a configuration excites the transverse mode of the BNA. In this configuration, the incident 1064 nm beam is detected by transmission through the aperture and the fiber for detection using a silicon photodetector. On the other hand the fluorescence from the nanoparticles is detected in the backward direction through the objective lens. Fig. 3.21C shows some results. It gives a portion of the time traces of the detected transmitted (T) 1064 nm signal (in red color) and the fluorescence (F) from the bead (in blue color) detected through the objective lens. The results show that it took one of the polystyrene bead 35 s to enter the aperture area and it was stably trapped after 10 s. Trapping may stay stable for several minutes.

The signature for trapping is derived from both the transmitted and fluorescence signals. When a bead is captured by the aperture, it blocks/scatters the incident light. Because of its subwavelength size, the nanobead scatters light over a wide range of **k**-vectors, thus decreasing the amount of light coupled into the fiber mode. Thus the trapping event is associated with a decrease in the transmitted signal. At the same time, there would be a simultaneous increase in the fluorescence signal, confirming the presence of the nanobead within the BNA. It is to be noted that the particle is released when the incident laser is switching off or interrupted, which physically verifies the trapping process. This configuration has some issues. For example, one drawback is the fact that the trap position is fixed since the oncoming light is controlled by an external objective. Moreover, as the collection objective is physically independent from the tip, the system is susceptible to vibrations and drifts relative to the tip. The noise associated with such effects is problematic with regard to trapping. Therefore to alleviate these problems it is best that both the trapping and collected light be handled through the tip.

Another interesting configuration is depicted in Fig. 3.22 [19]. In this configuration, an external white light is used to scatter off from the trapping region to allow the monitoring of the trapping event. The tip system is mounted onto an inversed microscope, equipped with a ($\times$40, 0.5) objective, for imaging the light spot at tip apex with a CCD camera. A second (horizontal) imaging channel involving a ($\times$60, 0.7) objective, a filter (to reject the laser) and a CCD camera is added to record the particle optical trapping directly with white light imaging.

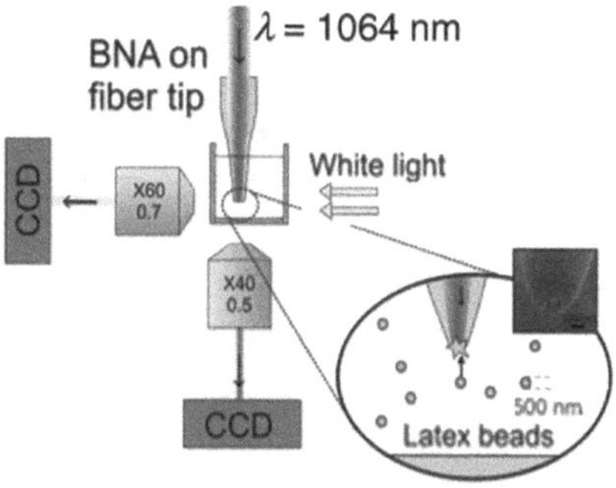

FIGURE 3.22 Nano plasmonic trap with an external white light being used to scatter off from the trapping region to allow the monitoring of the trapping event.

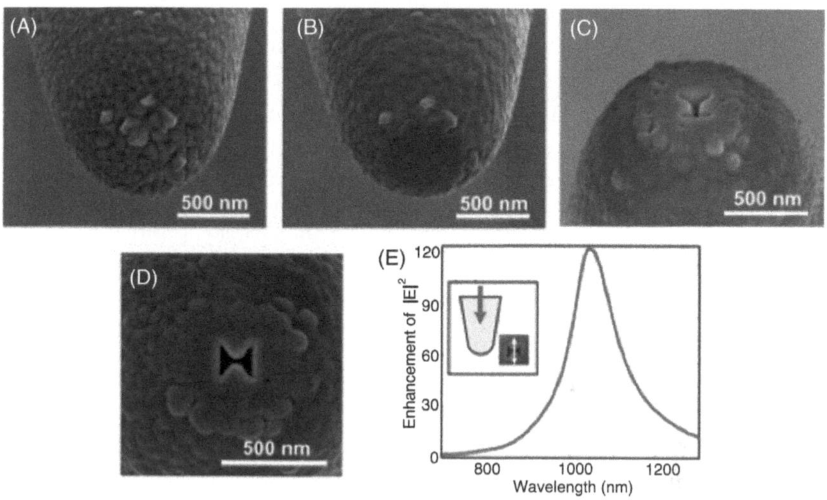

FIGURE 3.23 Construction of a plasmonic tweezer. (A) SEM image of the tip before a bowtie is constructed. (B) Side SEM image of the polished end. (C and D) Side and top SEM images of the BNA at the tip apex, respectively. (E) Theoretical resonance spectrum of 1064 light passing through the BNA when immersed in water. The inset gives the excitation configuration.

The tip in this application is prepared as follows [19]. The process starts with a single-mode optical fiber is cleaved. A polymer tips (30 μm long, radius of curvature of 0.5 μm at tip apex) is grown at the cleaved end facet by photo polymerization. Next, the probes are metal coated with a few nanometer thick titanium adhesion layer followed by a 150 nm thick aluminum layer. The SEM image of the tip given, shown in Fig. 3.23A, shows that the tip has a rough metallic surface with a rounded apex. FIB processing is then used to flatten and polish the tip apex. Fig. 3.23B gives a side SEM image of the polished end. FIB is now used to mill the bowtie aperture. Fig. 3.23C–D are side and top SEM images of the BNA at the tip apex, respectively. It is 165 nm wide with square gap about 35 nm large. Fig. 3.23E present a theoretical resonance spectrum of

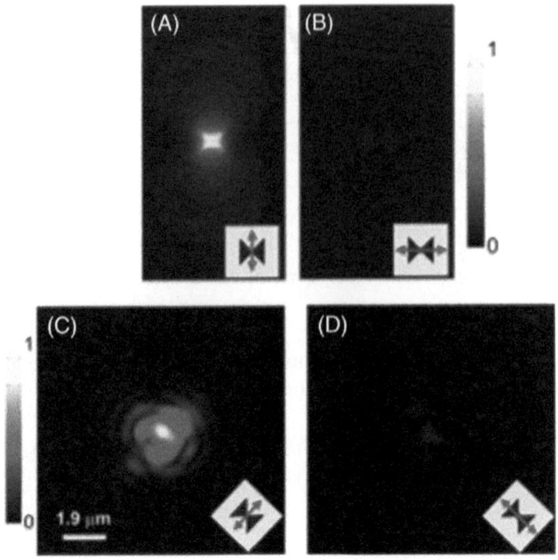

FIGURE 3.24 Display the simulation optical and focusing characteristics of the aperture of Fig. 3.23. (A and B) Near field response on a screen placed at a distance of only 10 nm away from the aperture. (A) A hot spot forms for linear polarization along (in resonance) the polarization axis of the BNA. (C and D) Transmitted light is very weak for (off resonance) polarization rotated by 90°.

1064 light passing through the BNA when immersed in water. The spectrum shows the plasmonic electric intensity enhancement due to the BNA.

It is interesting to display the results of a simulation of the optical and focusing characteristics of the aperture. First, Fig. 3.24A–B shows the near field response where a screen is placed at a distance of only 10 nm away from the aperture. An excitation Gaussian beam is injected into the 2 μm long end portion of the tip considered in the simulation. For linear polarization along (in resonance) the polarization axis of the BNA (Fig. 3.24A)) a hot spot formed. On the other hand, when the polarization is rotated by 90° (off resonance) the transmitted light is very weak. Fig. 3.24C–D shows the corresponding far field results.

Fig. 3.25 demonstrates the three dimensional (3D) optical trapping in the system described earlier. In the demonstration large 0.5 μm latex beads are used. Fig. 3.25A–C shows a particle being trapped at the tip apex (labeled by an arrow). The process takes 15 s. Fig. 3.25D–F shows when the input polarization is turned by 90° (BNA is off-resonance), the trapped particle is released and leaves the tip apex.

A new type of light focusing utilizes plasmonic nano lenses made of metal a nanorod of gold. The device, fabricated at Purdue University's Birck Nanotechnology Center, uses a cylindrical gold "nanoantenna" with a diameter of 320 nm [20]. The structures concentrate and absorbed light, resulting

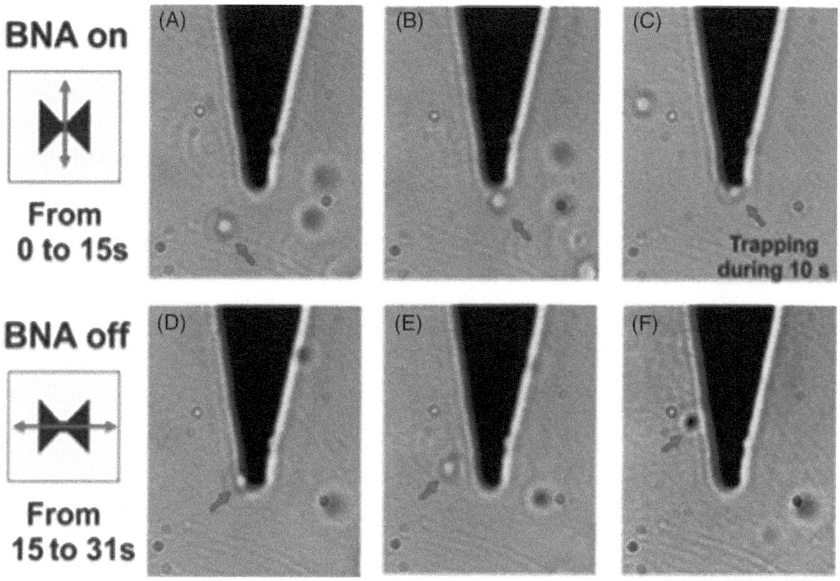

FIGURE 3.25 Demonstration of the three dimensional (3D) optical trapping of 0.5 μm latex beads in the system described in Fig. 3.23. (A–C) Particle being trapped at the tip apex. (D–F) Particle being released by rotating the polarization by 90° (BNA is off-resonance), https://www.osapublishing.org/oe/fulltext.cfm?uri(oe-22-8-10072.

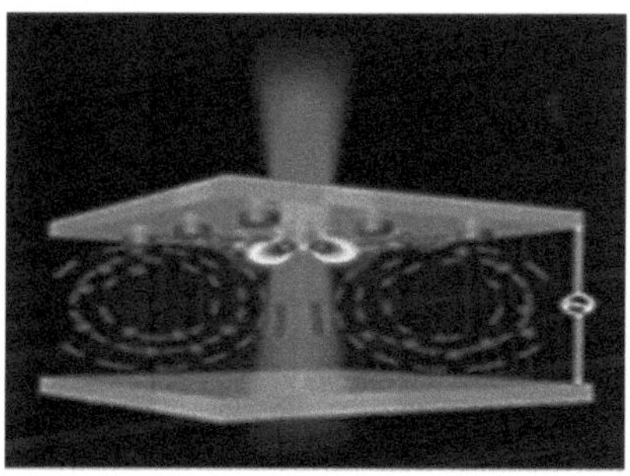

FIGURE 3.26 Nano tweezer made from plasmonic nano lenses (nano rod of gold), creating plasmonic hotspots and making it possible to manipulate nanometer-scale objects.

in "plasmonic hotspots" and making it possible to manipulate nanometer-scale objects suspended in a fluid,"nanotweezer" (shown in Fig. 3.26).

Example 3.12 – Dynamic Light Scattering of a Si_{29} Nano Particle [21]

A highly diluted colloid of 1 nm silicon nanoparticles is prepared. The viscosity $\eta=10^{-3}$ kg/ms. The colloid is filtered to remove any stray larger entities, hence simplify the analysis. The raw before and after filtering, through a 200 nm filter, show that the brightest events are dramatically reduced. Further filtering using a 10 nm filter demonstrated no detectable difference in the time traces. As the luminosity of the larger entities is very high, we believe that they were actually clusters comprised of smaller luminescent particles. A Guassian laser beam at 780 nm is focused in the colloid. The beam has a Guassian profile and there is a two-photon quadratic absorption dependence as measured. Take w as the waist of the beam, and z is the depth of the focal volume.

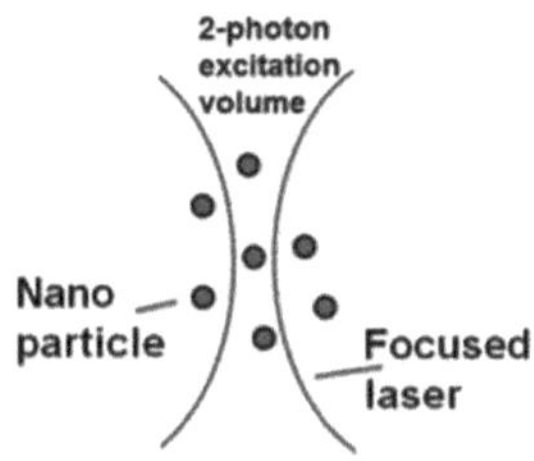

- The scattering light I (t) at the two photon emission in the blue/UV 390nm is measured.
- The correlation function Int{$I(t1t)I(t)*dt$} for the particle sample is calculated.
- The correlation function G (t) is fitted to Gaussian diffusion function
- *The geometry of the interaction spot is measured using a calibration sample.* The response of fluorescein, taken under the same beam focusing conditions, gives $w = 300$ nm and $z = 2.5$ μm, or a volume of 0.68 pcm^3.
- A self-consistent analysis of the autocorrelation spectra gives the number of particles in the focal volume N, and the diffusion coefficient D (cm^2/s).
- Using the Stokes–Einstein relation, $d = kT/3\pi\eta D$, allows the size of the particles d to be estimated, where $T = 300$ K, the viscosity $\eta = 1023$ kg/m s, and k is the Boltzmann constant.

The analysis gives 0.9 for the particle diameter and $D = 9.5 \times 10^{-6}$ cm^2/s

3.5 THERMAL TRAPPING OF PARTICLES

Recent studies have shown that convection forces generated by spatially confined heat sources (using micro-scale electric thermal heaters) can trap micrometer-sized dielectric particles, such as polystyrene spheres and *Escherichia coli* [22]. The heat-induced trapping phenomenon was confirmed using steady-state heat equation along with the Navier–Stokes equation. Temperature gradients are found to lead to a steady-state particle concentration c gradient according to: $dc/dz = -c\, S_T dT/dz$, where S_T is the ratio of the diffusion coefficients; for negative S_T, particles are driven to the hot region while for positive S_T, particles are driven to the cold region. The studies as depicted in Fig. 3.27A showed that the magnitude and sign of this coefficient depends on both temperature regimes as well as on the size of the particle; with smaller particles in general having negative coefficients, which drives them to hot regions.

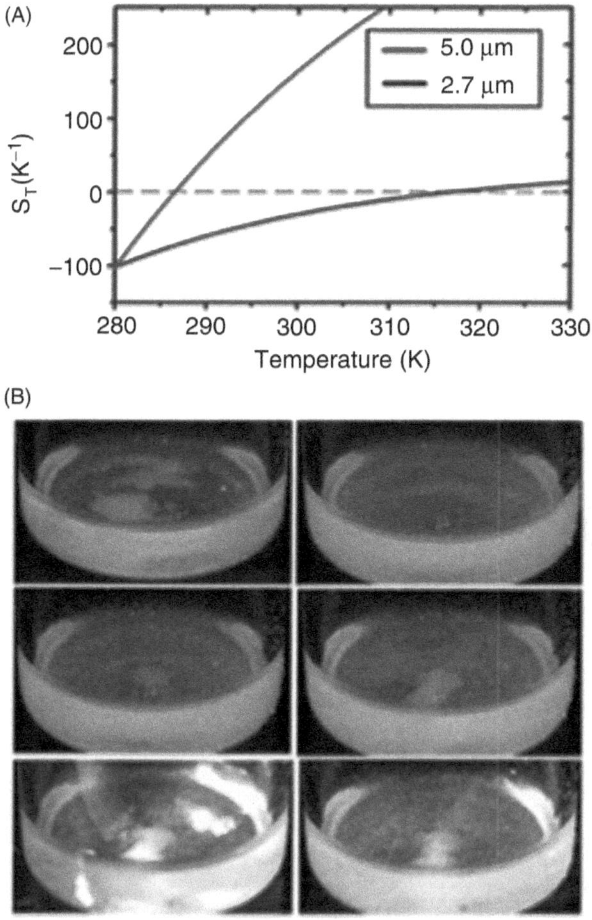

FIGURE 3.27 Thermal trapping of particles. (A) S_T is the ratio of the diffusion coefficients as a function of temperature and for different sizes; (B) clusters of ASTP polymers decorated with silicon nanoparticles in a vial under heating at the bottom, giving several frames showing trapping in the center (hottest region) and at the side (coldest region).

This was demonstrated using composite nanostructures of silicon nanoparticles and aromatic thermoset polymers (ATSP) [23]. The polymer is high- to low-k tunable material, and thermally stable from cryogenic to high temperatures, making them useful for Si microelectronics and spacecraft mission applications. First, red luminescent 3-nm H-terminated silicon nanoparticles were interacted and attached to the precursor oligomers via H-mediated interactions. This results in formation of nanostructures as small as 50–100 nm, decorated with 10–15 nm complexes composed of Si and carbon. Fig. 3.27B shows results for which dot/ds = 8K, and $T = 297$–305 K; convection motion and trapping effects start. The figure gives several frames of the motion under thermal treatment. The figure shows trapping in the center and at the side. Trapping develops at the center of the vial first but later in time another trapping takes place at the side. The trapped material at the center (hottest region) is the smaller clusters; while the trapped material at the side (coldest region) is the larger clusters.

REFERENCES

[1] McInnes CR. Solar Sailing Technology, Dynamics and Mission Applications. Chichester, UK: Praxis Publishing; 1999.

[2] Michaelis MM, Forbes A. Laser propulsion: a review. South Afr. J. Sci. 2006;102(7/8):289–95.

[3] Forward RL. Roundtrip interstellar travel using laser-pushed lightsails. J. Spacecraft Rockets 1989;21:187–95.

[4] http://www.space.com/26011-solar-sail-tech-space-exploration.html#sthash.zbUIcn2S.dpuf.

[5] Dina P. An Introduction to the Theory of Stellar Structure and Evolution. Cambridge, UK: Cambridge University Press; 2000.

[6] Lubin P, Hughes GB, Bible J, Isabella JN. Directed energy for relativistic propulsion and interstellar communications. JBIS 2015;68:172–82.

[7] Cooling and stopping and trapping of atoms: Reviews of Modern Physics, Vol. 71, No. 2, Centenary 1999 0034-6861/99/71(2)/253(10)/$17.00 ©1999 The American Physical Society S253.

[8] Foot CJ. Bose-Einstein condensates with rubidium atoms. Atomic Physics. Oxford, UK: Oxford University Press; 2005. ISBN 978-0-19-850696-6 The Nobel prize in physics 1997". Nobelprize. org. October 15, 1997. Retrieved December 11, 2011.

[9] Ashkin A. Acceleration and trapping of particles by radiation pressure. Phys. Rev. Lett. 1970;24:156–9.

[10] Maragò S O.M., Jones S P.H., Gucciardi S P.G., Volpe S G., Ferrari S A.C. Optical trapping and manipulation of nanostructures. Nat. Nanotechnol. 2013;8:807–19.

[11] Hecht E, Zając A. Optics. Boston, MA, USA: Addison-Wesley, Pub., Co; 1974.

[12] Grier DG. A revolution in optical manipulation. Nature 2003;424:810–6. copyright, 2003.

[13] https://www.physik.hu-berlin.de/de/nano/lehre/Gastvorlesung%20Wien/Seminar/Talk%20 Gerlich.

[14] Donato MG, Monaca MA, Faggio G, Stefano LD, Jones PH, Gucciardi PG, Marag OM. Optical trapping of porous silicon nanoparticles. Nanotechnology 2011;22:8.

[15] Burns JA, Philip PL, Cnrs-Las L, Soter S. Radiation forces on small particles in the solar system. Icarus 1979;40:1–48.

[16] Shawn D, Loïc R, Lukas N. Two beams: cancellation of non-conservative scattering forces in optical traps by counter-propagating beams. Opt. Lett. 2015;40(9):1900–3.

[17] Carina M, Marco C, Gionata B, Francesco V, Francesco SP. Optical methods to study protein-DNA interactions in vitro and in living cells at the single-molecule level. Int. J. Mol. Sci. 2013;14(2):3961–92.

[18] Li T. Fundamental Tests of Physics with Optically Trapped Microspheres. New York: Springer Science+Business Media, New York 2013. doi: 10.1007/978-1-4614-6031-2_2.

[19] Berthelot J, Ac'imovic' SS, Juan ML, Kreuzer MP, Renger J, Quidant R. Three-dimensional manipulation with scanning near-field optical nanotweezers. Nat. Nanotechnol. 2014;9: 295–9.

[20] Ndukaife JC, Kildishev AV, Nnanna AGA, Shalaev VM, Wereley ST, Boltasseva A. Long-range and rapid transport of individual nanoobjects by a hybrid electrothermoplasmonicna notweezer. Nat. Nanotechnol. 2016;11:53–9.
[21] Ackakir O, Therrien J, Belomoin G, Barry N, Muller J, Gratton E, Nayfeh M. Detection of luminescent single ultrasmall silicon nanoparticle using fluctuation spectroscopy. Appl. Phys. Lett. 2000;76:1857–9.
[22] Chen J, Cong H, Loo F-C, Kang Z, Tang M, Zhang H, Wu S-Y, Kong S-K, Ho H-P. Thermal gradient induced tweezers for the manipulation of particles and cells. Sci. Rep. 2016;6(1):35814.
[23] A. Kocyigit, M.B. Osman, S. Cifci, B. Enders, I. Jasiuk, M.H. Nayfeh. Imparting optical functionality to aromatic thermosetting copolyester by luminescent silicon nanoparticles cross-linked via *in situ* thermal polymerization reaction, To be published.

Chapter 4

Seeing Atoms and Clusters on Surfaces

4.1 SEEING NANOPARTICLES AND WHAT'S INSIDE WITH FREE ELECTRONS

Optical microscopes utilize photons in light beams. Their resolution is limited by light diffraction to 200–250 nm. In 2014, optical microscopy was brought into the nanodimension as a result of the work of Eric Betzig, William Moerner, and Stefan Hell. They were awarded the Nobel Prize in chemistry for this development, namely the development of super-resolved fluorescence microscopy [1]. On the other hand, in electron microscopes free electrons are used [2,3]. Thus in principle electron microscopy is similar to conventional light microscopy, with the exception that the wavelength of the electron beam λ_e used for imaging is much smaller than the wavelength of light λ_l. In fact in contrast with the wavelength of visible light λ_l, which is hundreds of nanometers, λ_e can approach an angstrom (~0.1 nm), which makes it close to the spacing of atoms in solid matter. In an electron microscope, short wavelength is achieved by accelerating electrons such that they acquire high kinetic energy, hence high speed and momentum to the degree that the quantum mechanical (de Broglie) wavelength associated with them, h/p, is short enough such that they are smaller than the electronic size of atoms (<1 Å). The short wavelength provided beats by 3–4 orders of magnitude the wavelength of any visible light or near UV used in optical or fluorescence microscopes. In fact it matches the size of the smallest nanoparticles available and even the size of atoms. Fig. 4.1 highlights the differences between election and light microscopes. There are two modes of imaging as explained later.

4.1.1 Transmission Electron Microscope (TEM)

In transmission electron microscope (TEM), a beam of electrons passes through a sample followed by a series of lenses, forming a highly magnified image of the sample on a screen as schematically shown in Fig. 4.2A. The sample is required to be thin enough to allow transmission of electrons. A scanning transmission electron microscope (STEM) is a variation on the standard TEM. In this case the impinging electron beam is focused to a small spot, which is then

Fundamentals and Applications of Nano Silicon in Plasmonics and Fullerines.
http://dx.doi.org/10.1016/B978-0-323-48057-4.00004-9

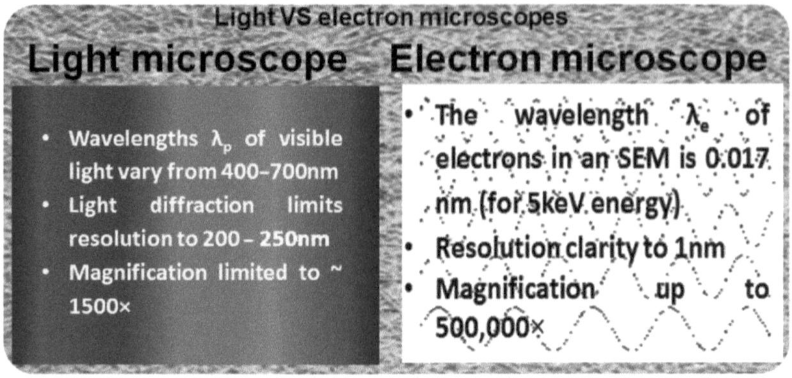

FIGURE 4.1 Differences between election and light microscopes.

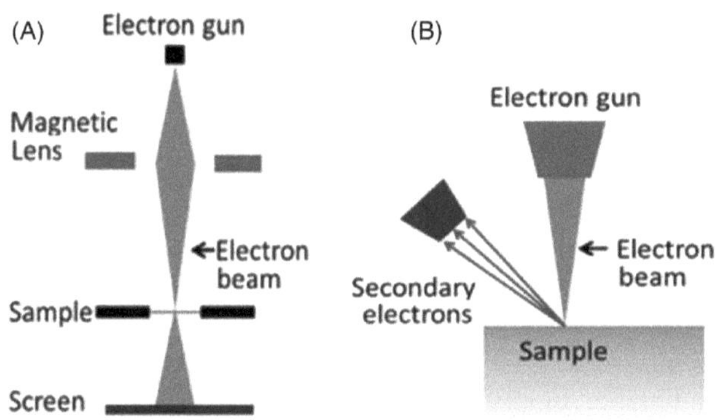

FIGURE 4.2 Electron microscopy. (A) Schematic of the transmission electron microscope (TEM). (B) Schematic of the scanning electron microscope (SEM).

scanned over the sample in a raster manner. High-resolution STEMs require exceptionally stable room environments. Vibration, temperature fluctuations, electromagnetic waves, and acoustic waves must be limited in the room housing the microscope.

A scanning electron microscope (SEM) is another electron microscope. In this instrument a specimen is irradiated by an electron beam, which excites the surface atoms. This excitation produces electrons (secondary electrons) from the surface and collected on the same side of the sample as schematically shown in Fig. 4.2B. Thus, unlike TEM that detected transmitted electrons, secondary electrons emitted by the excited surface atoms are detected in SEM. Thus, the sample for SEM need not be thin. The number of secondary electrons that can be detected depends on specimen topography. By scanning the sample and collecting the secondary electrons that are emitted an image displaying the topography of the surface is created, detailing the morphology,

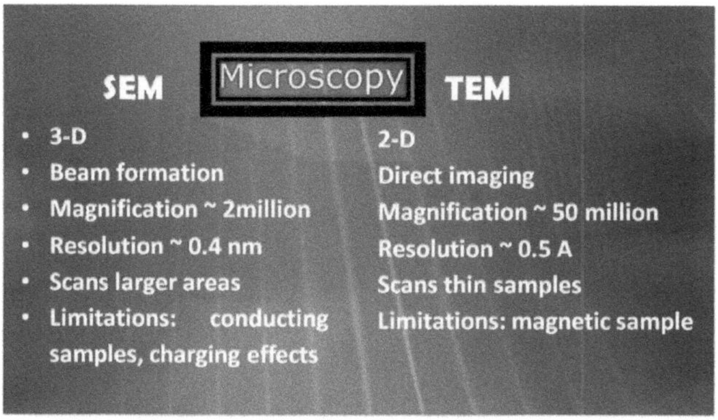

FIGURE 4.3 Contrast of the features of SEM and TEM.

shape, size, or size distribution of materials at the micro and nanoscale. The spatial resolution of high-resolution SEM can identify nanoparticles down to below 10 nm. Although the instrument cannot resolve atoms or molecules; it can achieve resolution better than 1 nanometer (10 Å). The SEM and the TEM are contrasted as follows.

We give here three examples: silicon, gold, and platinum. In 2001, HRTEM was used at the University of Illinois to image silicon nanoparticles of sizes in the range 1–5 nm in diameter [4]. Fig. 4.4 shows it was possible to resolve 3.1 Å atomic planes inside all of the sizes except for the 1 nm particle.

Researchers at Stanford University studied gold nanoparticle of 1.1 nm in diameter and contains 68 gold atoms [5]. They revealed the three-dimensional structure the nanoparticle in which all gold atoms are observed using high-resolution electron microscopy (HTEM). The microscope was operated in very-low current mode in which very few electrons are used to avoid perturbing the structure of the nanoparticle. The success of this approach opens the way to the determination of many more nanoparticle structures and to both fundamental understanding and practical applications. The results are in close agreement with a structure predicted at the University of Jyväskylä, Finland, on the basis of theoretical modeling and infrared spectroscopy (see Fig. 4.5). The result was supported by small-angle X-ray scattering done in Lawrence Berkeley National Laboratory, USA, and by mass spectrometry done at Hokkaido University, Japan.

Atomic resolution was extended to non-stationary particles. Free-floating platinum nanoparticles were studied using electron microscopy and liquid cells. Two 3D structures of individual platinum nanocrystals were reconstructed at near-atomic resolution. The procedure utilizes an algorithm for single-particle 3D reconstruction, which was originally developed for analysis of biological molecules. The method derives the 3D structure from images of individual nanoparticles rotating freely in solution [6].

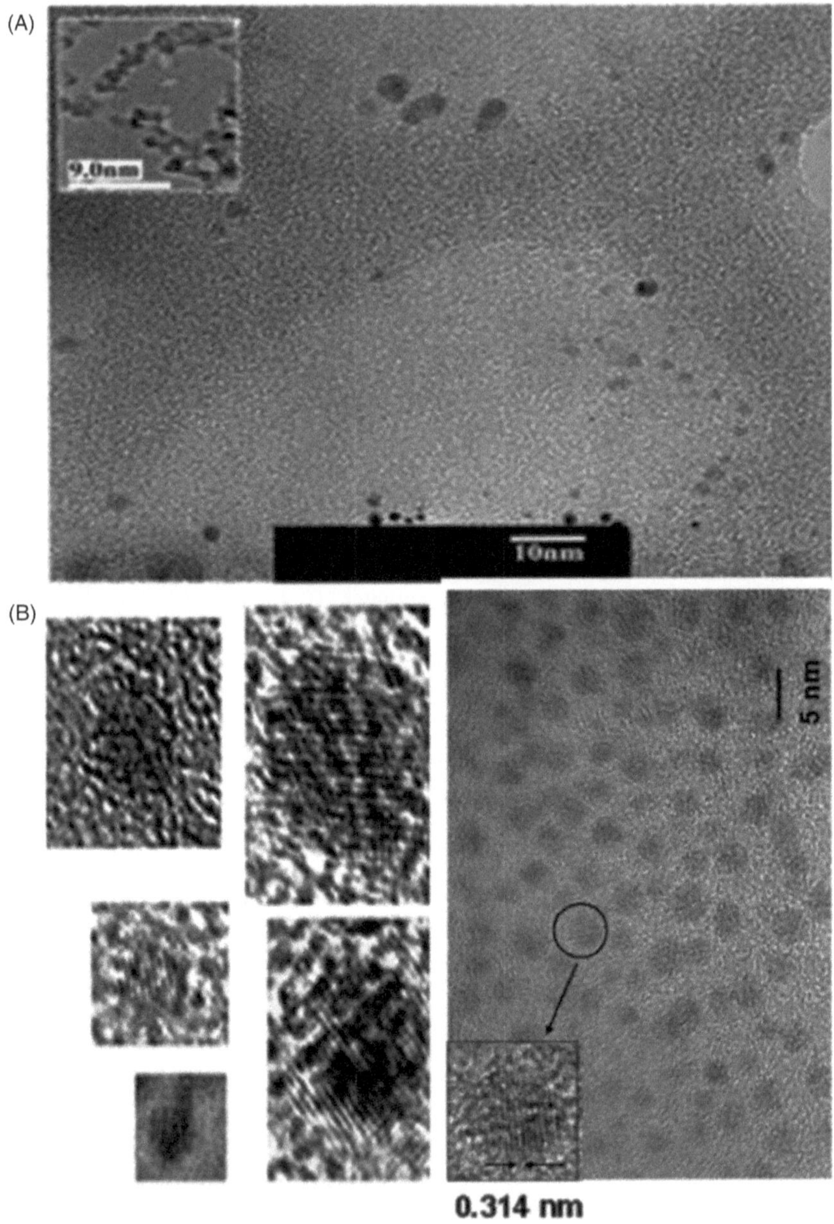

FIGURE 4.4 Example of the use of TEM in imaging ultrasmall silicon nanoparticles placed on a graphite grid. (A) Large area of the grid incubated in a mixture of nanoparticles in the range 1–5 nm in diameter. (Inset) a small area of a grid incubated in a colloid of nanoparticles enriched to have predominantly 1nm diameter particle. (B) (*right*) a small area of a grid incubated in a colloid of nanoparticles enriched to have predominantly 2.9 nm diameter particles (Inset) a zoom on a single 2.9 m particle which resolved atomic planes at 0.31 nm spacing, characteristic of silicon. (*left*) enlargement of TEM images of five single particles of 1, 1.7, 2.2, 2.9 and 3.9 nm particles; it was possible to resolve 0.31 nm atomic planes inside all of the sizes except for the smallest (1-nm) particle.

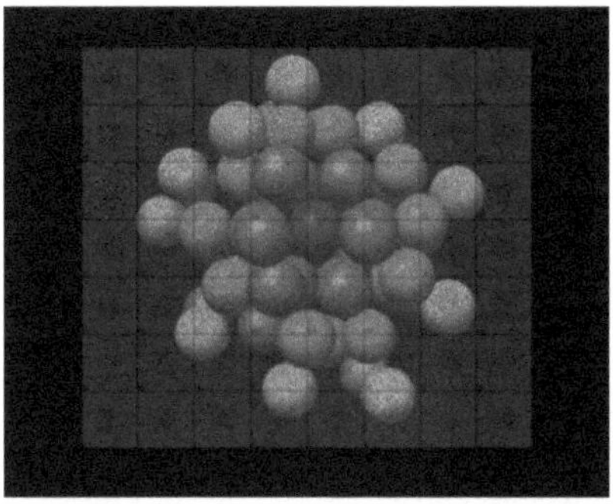

FIGURE 4.5 High-resolution electron microscopy (HTEM) of gold nanoparticle of 1.1 nm in diameter and contains 68 gold atoms. The three-dimensional structure the nanoparticle in which all gold atoms are observed. *(Copyright Maia Azubel, Stanford University)*

Example 4.1 de Broglie wavelength of an electron with kinetic energy of 120 eV

Consider free electrons of kinetic energy, E_k= 120 eV.

Using Planck's constant, h = 6.6 × 10^{-34} Js; Mass of an electron, m = 9.1 × 10^{-31} kg; and Charge on an electron, e = 1.6 × 10^{-19} C

(a) For the electron, we can write the relation for kinetic energy as E_k = 1.2 mv^2, where v = speed of the electron, therefore

$$\therefore v^2 = \sqrt{\frac{2eE_k}{m}}$$

$$= \sqrt{\frac{2 \times 1.6 \times 10^{-19} \times 120}{9.1 \times 10^{-31}}}$$

$$= \sqrt{42.198 \times 10^{12}} = 6.496 \times 10^{6}\ \mathrm{m/s}$$

The momentum of the electron, $p= mv$ = 9.1 × 10^{-31} × 6.496 × 10^{6} = 5.91 × 10^{-24}kg ms^{-1}. Therefore, the momentum of the electron is 5.91 × 10^{-24}kg ms^{-1}; hence

(b) Speed of the electron, v = 6.496 × 106 m/s

(c) De Broglie wavelength of an electron having a momentum p is given as:

$$\lambda = \frac{h}{p}$$

$$= \frac{6.6 \times 10^{-34}}{5.91 \times 10^{-24}} = 1.116 \times 10^{-10}\ \mathrm{m}$$

$$= 0.112\mathrm{nm}$$

Therefore, the de Broglie wavelength of the electron is 0.112 nm.

4.2 SEEING ATOMS, MOLECULES, AND NANOPARTICLES ON SURFACES USING TUNNELING ELECTRONS (STM IMAGING)

We have seen that TEM may be used to observe individual electrons on surfaces of thin enough samples to allow significant transmission of electrons. Typically it uses a hot source of electrons that produces free electrons, which are subsequently accelerated to high speeds at the target, which is located far away from the source. The faster the electrons the tighter they can be focused. Fast electrons can be focused tighter than light; but they could not be focused on an individual atom because at some point they start repelling each other (space charge limited focusing) and cause damage to the sample.

Thus to observe and investigate individual atoms on surfaces new tools which operate on different principles must be developed. A step in this direction was achieved in 1982 with the advent of the scanning tunneling microscope (STM) [7]. Fig. 4.6A depicts a cartoon of an STM microscope [8]. It consists of a sharp metal tip (needle). The tip can be moved in a controlled fashion in three directions (up and down and sideways) by a piezoelectric drive, which allows scanning across the sample. The diameter of the tip is as sharp as the diameter of a single atom (see the enlargement). Beneath the tip there is a conducting sample (substrate) placed at an extremely short distance (nearly a diameter of an atom). The sample is electrically grounded while the tip is biased negatively at a small voltage of less than 1 V. Under these conditions electrons start to move (quantum mechanically tunnel) between the tip and the substrate, effectively producing a tunneling current, which can be recorded. The probability of electrons jumbling is extremely sensitive to the tip-surface spacing. In fact it is an exponential dependence. There setup provides a number of innovative features

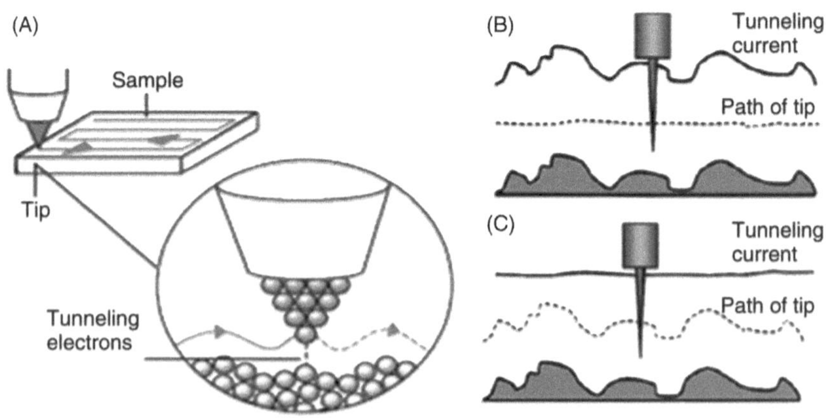

FIGURE 4.6 Use of STM in imaging atoms on surfaces. (A) Depicts a cartoon of STM imaging, having a sharp metal tip (needle). (B) Imaging using constant height mode in which the distance between the tip and the sample is kept fixed, while the tunneling current is measured. (C) Imaging in the constant current mode while the height of the tip is measured.

that are not found in SEM or TEM imaging. We briefly contrast the characteristic of the electrons with those in a TEM microscope:

- Cold electron source (in the form of a sharp metallic needle, as sharp as diameter of an individual atom).
- Very-slow electrons (a factor of ~million slower).
- Non-free electrons motion, dependent on both of the tip and surface (they get sucked in by the surface by quantum tunneling processes).
- Source to target distance is on an atomic scale where electrons will not be emitted till the sample gets to distances of angstroms, the size of atoms.
- Electron current is highly sensitive (exponential dependence) to the tip-surface interdistance. The spatial resolution arises from the exponential sensitivity/dependence.
- Electron current is sensitive to the dimension and shape of individual atoms, which allows the recording of electronic images of atoms.

The STM Ruska, Binning, and Rohrer

Ernst Ruska
Prize share: 1/2

Gerd Binnig
Prize share: 1/4

Heinrich Rohrer
Prize share: 1/4

The Nobel Prize in Physics, 1986 was divided, one half awarded to Ernst Ruska *"for his fundamental work in electron optics, and for the design of the first electron microscope,"* the other half jointly to Gerd Binnig and Heinrich Rohrer *"for their design of the STM"*.

Imaging of the surface topology may then be carried out in one of two modes of operation of the microscope:

1. Constant height mode (in which the tunneling current is monitored as the tip is scanned parallel to the surface)

2. Constant current mode (in which the tunneling current is maintained constant as the tip is scanned across the surface).

In the constant height mode, the distance between the tip and the sample is kept fixed. When the tip is scanned across the surface, the tunnel current flowing between the tip and sample changes. For example, as shown in Fig. 4.6B, it increases as we go over an atom and then goes down between atoms because the top of the atom is closer to the tip. In fact, quantum mechanically the magnitude of the current is determined by the overlap of the quantum electronic wave functions of the tip and sample. An STM image is essentially a map of the electronic local density states of the surface, which mimics the atoms. In this mode, the tip or/and surface are susceptible to damage. To avoid damage this mode is best applied when the surface of sample is very smooth. But, if the sample is rough, or has some large particles on the surface, the tip may contact with the sample and damage the surface. In this case, the constant current mode is applied.

In the constant current mode (Fig. 4.6C), the tip is moved close to an object a small constant voltage is applied to allow electrons to "tunnel" or jump across the remaining gap. The device has a feedback loop such that it can be used to control the current, that is, setting the current to a certain value. The tip is now scanned across the sample. The needle would necessarily move up when it goes over an atom and then goes down when it gets between atoms to maintain the set current. Plotting the change in the distance of the tip from the surface essentially plots the topography of the surface, that is, images of atoms and molecules.

Several examples of scans in two dimensions, shown in Fig. 4.7, display the ability of the device to see individual atoms. Practically, the needle tip can be precisely moved mechanically in three dimensions but controlled electrically by the use of piezoelectric stages to allow observation of individual atoms. The theory of operation of the microscope is discussed in Example 4.2.

Example 4.2 STM can see and resolve individual atoms

Fig. 4.8 shows the gap region of an STM device with a gap width W and operated at a biasing voltage V. The gap is along the z-axis. Consider an electron in the tip with a total energy *E*. Fig. 4.8A defines three regions. Region I is inside the tip. Region II is in the gap between the tip and the sample surface. Region III is inside the substrate. In the classical case, the wavefunction in regimes (II) and (III) would be equal to zero, however when one solves the Schrödinger equation, indeed we will have a travelling wave in regions (I) and (III)) and an exponential decay within the barrier (regime (II)).This means the electron has tunneled from the tip to the substrate.

Further quantum analysis can show that the probability for an electron to tunnel from the tip to the substrate across the vacuum barrier *P* and the current *I* is

$$P \alpha \, e^{-2kW}$$

$$k = \frac{\sqrt{2m(U-E)}}{h}$$

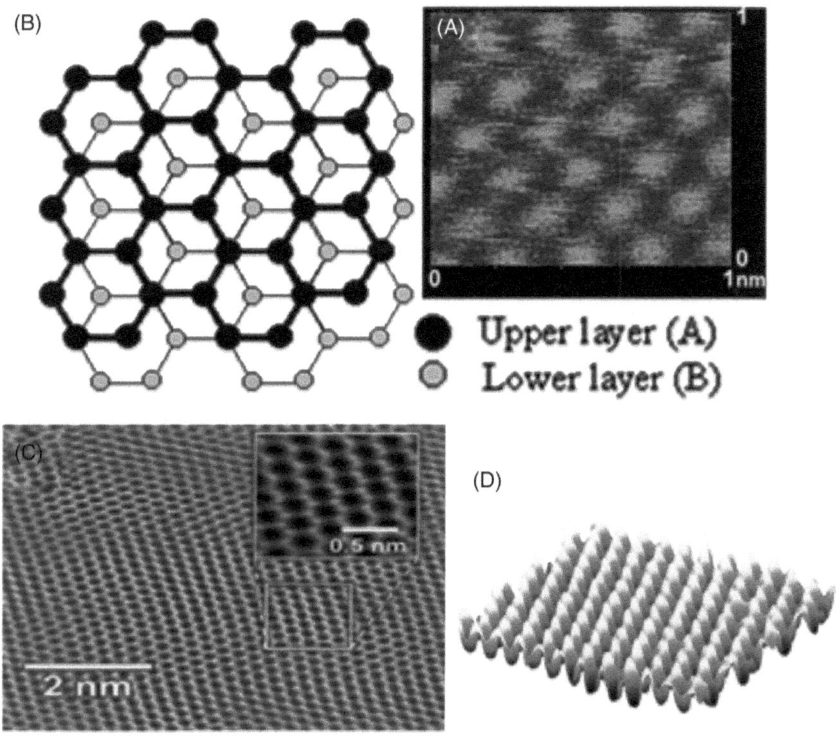

FIGURE 4.7 Several examples of scans in two dimensions display the ability of the device to see individual atoms. (A) Image of a highly ordered pyrolytic graphite. (B) Positional relationship between two identical graphene planes A and B. Graphite structure can be described as an alternate succession of these planes ...ABABAB. (C–D) Atom-resolved imaging of surfaces.

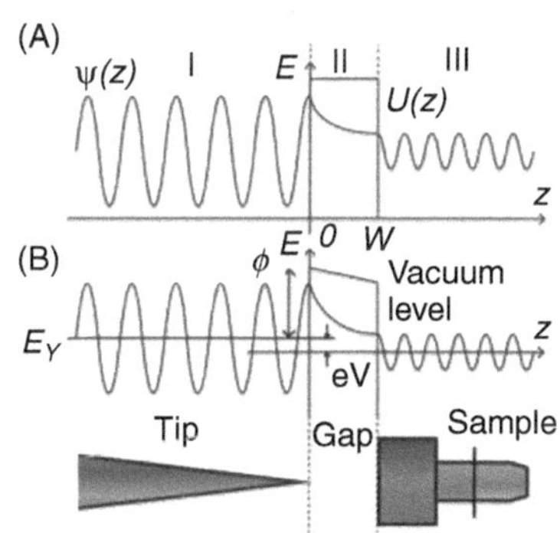

FIGURE 4.8 Theory sketch of scanning tunneling microscope imaging. (A and B) Defines three regions. Region I is inside the tip. Region II is in the gap between the tip and the sample surface. Region III is inside the substrate. It also sketches the wave-functions in the three regions. (C) cartoon of the top, surface, and the gap.

$$I \alpha \sum_{E_f - eV}^{E_f} \rho_s(z,E) e^{-2kW}$$

The sum the probability over an energy range, such as between E_F and E_V, and E_F, gives the number of states available in this energy range per unit volume, that is, the local LDOS, $\rho_s(z,E)$.

This shows that the measured current is an exponential function with the gap width. This is principle behind the unprecedented resolution of STM.

The first example shown in Fig. 4.7A is the image of a highly ordered pyrolytic graphite. Graphite belongs to lamellar materials and consists of identical staked planes. Carbon atoms within a single plane interact much stronger than with those from adjacent planes. That explains the lamellar behavior of graphite. Each atom within a single plane has three nearest neighbors. The network of carbon atoms connected by shortest bonds looks like honeycomb. This two-dimensional and single-atom thick plane is called "grapheme." Fig. 4.7B shows positional relationship between two identical graphene planes A and B. Graphite structure can be described as an alternate succession of these planes ...ABABAB...

4.2.1 Nanoparticles

It is possible to use STM image large structures (three orders larger than an atom), such as nanoparticles. The STM height of a nano crystal is reliable and it serves a rather good indication of its size. However, STM imaging always over estimates the width; it is always larger than the height for a spherical particle. This discrepancy is due to the fact that the tip has an effective radius of curvature which convolutes. By carrying scans of different speed, range and gain one can avoid artifacts especially when imaging in the constant current mode. Some of these effects are illustrated in Fig. 4.9 for imaging spherical gold nanoparticles of 6 nm diameter.

4.2.2 Spectroscopy Using Tunneling Scanning (STS): I–V Spectroscopy

We have seen above that a STM records topographical details of conductive surfaces, namely provides images of surface atoms with real space atomic resolution. The STM is actual more versatile than just topography recording. It can be used to record some information about the electronic behavior of the material. For example, it can be used to determine the number of states available to be occupied. In fact it records the density of those stated per unit energy and a given energy of choice (density of states or DOS). It can also measure the changes of the DOS due to the presence of local perturbations or distortions (local density of states or LDOS). For this function, we use the constant height mode. The microscope tip is parked over a certain point of the surface. The tunneling

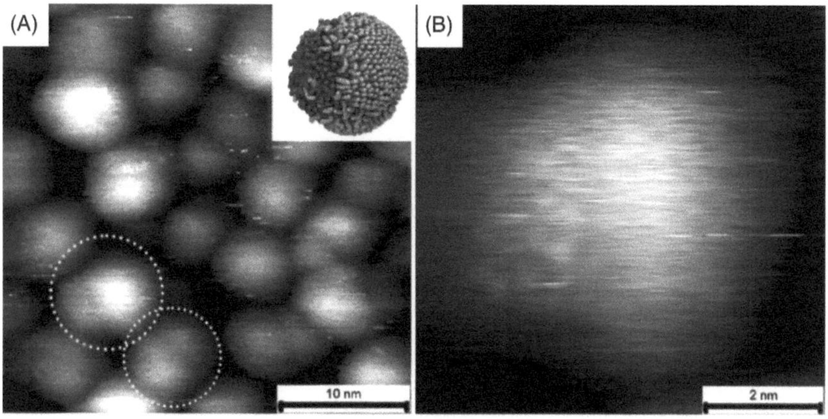

FIGURE 4.9 **STM imaging of spherical 6 nm gold nanoparticles, displaying over estimation of the width.** (A) Large area with numerous nanoparticles; (B) close-up of a single particle. Width always appears larger than the height for a spherical particle.

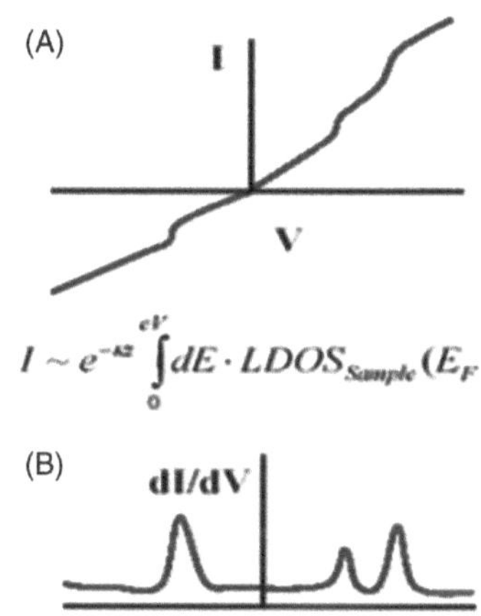

$$I \sim e^{-kz} \int_0^{eV} dE \cdot LDOS_{Sample}(E_F + E)$$

$$\left.\frac{dI}{dV}\right|_V \sim LDOS_{Sample}(E_F + eV)$$

FIGURE 4.10 **Spectroscopy using tunneling scanning (STS).** *I–V* spectroscopy, (A) an example of *I–V* spectrum. It exhibits steps or kinks. (B) Differentiation of the *I–V* spectrum with respect *V*, which turns the kinks into peaks or bands.

current is then set at a certain value and the biasing voltage at a certain value in a constant current mode, giving a tip height of several angstroms. The feedback loop is then disengaged allowing the tip to be held in a constant height mode. Now the biasing voltage is scanned while the tunneling current is recorded as a function of the voltage (*I–V* spectra). This operation is called scanning tunneling spectroscopy (STS) [11]. Fig. 4.10A shows an example of *I–V*. It exhibits steps or kinks. The next step is to differentiate the spectrum with respect to *V* as shown in Fig. 4.10B, which turns the kinks into peaks or bands. Theoretically, this response is described by an energy integral over the DOS as given by equations while the differential is proportional to the DOS. In the language of electrical devices, this is a two-terminal arrangement; the tip acts as the source electrode, while the substrate acts as the drain electrode.

This technique was applied to a film of 1 nm (Si_{29}) silicon nanoparticles placed over a device quality silicon wafer. The measurement allowed the recording of the energy levels of the particle (see Example 4.3) [12].

Example 4.3 Numerical STM resolution

For a typical U–E = 4 eV, the characteristic decay length W = 1 Å. This gives $\kappa \approx$ 1.025Å^{-1}. If the difference between the distance from the sample of the closest tip atom, d_1, and the second closest atom, d_2, is d_2–$d_1 \approx$ 1 Å, then the ratio between the current flowing through atom 1 to that flowing through atom 2 is $I_1/I_2 \approx 10$, that is, a tip atom only 1 Å closer to the surface collects 90% of the current flowing between the surface and tip. It is this exponential dependence of the tunneling current on the gap width which gives the STM its high resolution: in the horizontal plane only those atoms closest to the apex contribute significantly to tunneling; and in the vertical direction, a small 1 Å difference in height changes the current by 10-fold.

http://www.maths.tcd.ie/~bmurphy/thesis/thesisse3.html

4.3 MECHANICAL ATOMIC IMAGING (ATOMIC FORCE MICROSCOPE – AFM)

We can see from the discussions above that STM is only capable of examining material that conduct electricity, leaving out a lot of other material out that are not electrically conductive. The device that can fill this short fall is the Atomic Force Microscope (AFM) which does not rely on the flow of electrons but actually touches the material being examined. This versatility allows us to examine non-conductive material such as plastics and ceramics.

An AFM uses an atomically sharp non-conducting tip [13]. Thus, unlike an STM tip, there is no flow of electrons between the tip and the surface. Fig. 4.11 is a schematic of the AFM microscope [14]. The tip is attached to the end of a cantilever with a low spring constant. The tip is brought very close to the surface. When it is in near-contact with the surface, it results in a mechanical force between them. The force is caused by the van der Waals forces between the tip and sample. The contact force causes the cantilever to bend to accommodate changes in topography. The bending is measured using a laser beam. A laser beam strikes the cantilever. The reflected beam is detected by a position-sensitive photodetector. When the cantilever bends, the laser beam shifts at the detector. The degree of bending is measured by the degree of shift the laser beam suffers [15]. To illustrate the deflection analysis, we use a four quadrant

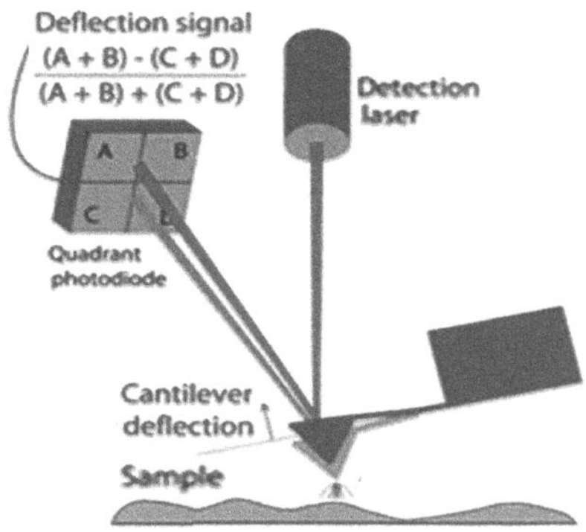

FIGURE 4.11 **Schematic of the AFM microscope.** *(Courtesy of JPK Instruments AG, From www.jpk.com)*

detector panel, called A, B, C, and D. The deflection is defines as ((A + B) − (C + D)).This is then normalized to the sum ((A + B) + (C + D)). The force correlates with topographical variations (roughness or nanostructures for example) across the surface. Those parameters are then correlated with the force between the tip and the sample.

Example 4.4 Contrasting AFM and STM

- STM measures tunneling current; thus the tip and sample for STM must be conducting. AFM measures the deflection of the cantilever caused by the van der Waals forces between the tip and sample. Thus, in general any kind of sample can be used for AFM.
- STM tunneling current has exponential relation with distance. The force in AFM is not exponential (less sensitive)
- STM has a better resolution than AFM. STM actually "see" an individual atom, while in AFM it is almost impossible, and the quality of AFM image is largely depended on the shape and contact force of the tip.
- AFM measured signal may sometime be rather complicated to interpret into morphology or other properties. On the other side, STM can give straight forward electric property of the sample surface.

The process is problematic if the tip makes real contact with the surface. In such a case the two materials may fuse together on atomic scale, resulting in hysteresis when it is pulled back with some damage to both. Moreover, it may not be easy to retrieve and separate to the previous condition. To avoid this problem several designs have been considered to avoid getting the tip in real physical contact with the surface. Fig. 4.12A–C shows and contrasts the three possible modes of operations. We qualitatively discuss the modes, leaving the details of the atomic/molecular basis of the modes for presentation in Example 4.5.

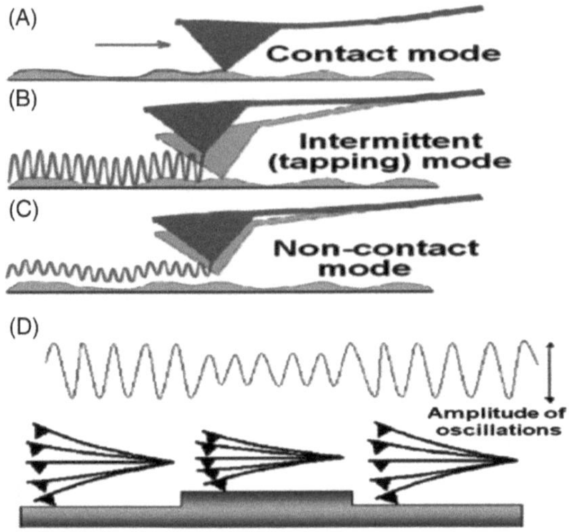

FIGURES 4.12 Modes of operation of an AFM microscope. (A) Contact mode, (B) intermittent contact mode, (C) non-contact (NC) mode. (D) The amplitude of oscillation with the distance of the tip from the surface drops when the tip gets closer to the surface.

Example 4.5 Atomic nature of the force between tip and surface

The nature of the force between a surface and a tip is of atomic/molecular origin. It is described in Fig. 4.13. When a very sharp tip gets very close to a surface, the neutral atom at the apex of tip interacts with that on the surface just beneath the tip apex. Although there are other atoms on either side may also interact, but this interaction dominates. Each of those act as a bound electric dipole, thus the interaction between the two atoms (atom–atom interaction) can be approximated as a dipole–dipole interaction. The dipole–dipole interaction produces an attractive force which depends sensitively on the inter distance. Now if the tip is pushed further towards the surface such that it begins to make contact, then the two atoms begin to compress against each other hence a strong repulsive force develops. This is effectively like pushing hard spheres against each other.

Scientist sought to come up with an approximate force that might represent to a good degree of accuracy both the dipole–dipole attraction and the hard sphere repulsion in one force. Fig. 4.13 shows an example of a combined force that runs smoothly over large separation to contact distances. We should mention that traditionally this process is performed first with potential energies. The Lennard–Jones model consists of two "parts"; a steep repulsive term, and smoother attractive term. It is worth mentioning that the Lennard–Jones model is not the most faithful representation of the potential energy surface, but rather its use is widespread due to its computational expediency. The Lennard–Jones potential is given by the following equation: $V(r) = 4\in[(\sigma r)^{12} - (\sigma r)^{6}]$, where V is the intermolecular potential between the two atoms, $\in$ is the well depth and a measure of how strongly the two particles attract each other and σ is the distance at which the intermolecular potential between the two particles is zero. Sometimes the potential is expressed

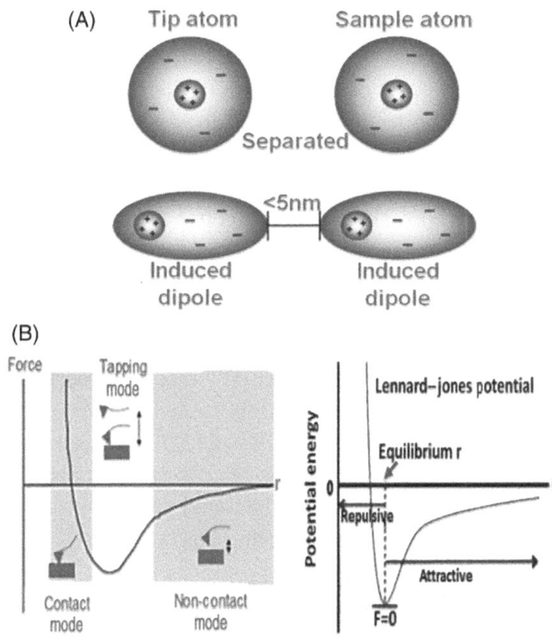

FIGURE 4.13 Atomic nature of the force between tip and surface. (A) Interaction results in induced charge separation (polarization) at close distance. (B) The Lennard–Jones model force is an approximate force. It consists of two "parts"; a steep repulsive term, and smoother attractive term. *(From Bullen, R.A.W.a.H.A., Lecture notes: Introduction to Scanning Probe Microscopy)*

as V(r) = Ar12−Br6 There are three regimes of operation with respect to the probe-surface seperation

1. Contact AFM <0.5 nm probe–surface separation
2. Intermittent contact (tapping mode AFM), 0.5–2 nm
3. Non-contact AFM, large separation

http://chemwiki.ucdavis.edu/Core/Physical_Chemistry/Physical_Properties_of_Matter/Atomic_and_Molecular_Properties/Interm

http://chemwiki.ucdavis.edu/Core/Physical_Chemistry/Physical_Properties_of_Matter/Atomic_and_Molecular_Properties/Interm

http://www.bioinformatics.utep.edu/agriculture/MDSim-module.php

Bullen, R.A.W.A.H.A., *Lecture notes: Introduction to Scanning Probe Microscopy*

- Contact mode: It is depicted in Fig. 4.12A. It is also known as static mode. In this mode, the AFM tip is pushed into the sample to make a soft "physical contact." The force in this mode is the van der Waals short range repulsive force between the tip and the surface.
- Intermittent contact mode: It is depicted in Fig. 4.12B. It is also known as tapping mode. In this mode, a piezoelectric element mounted in the AFM

tip holder is used to make the cantilever to oscillate vertically at a frequency near its resonance frequency. The amplitude of this oscillation is set between 100 and 200 nm. A tapping AFM image is produced by detecting the force of the oscillating contacts of the tip with the sample surface.

- Non-contact (NC) mode: It is depicted in Fig. 4.12C. It is also known as dynamic mode. In this mode, the cantilever is vibrated without contacting the sample surface at a constant frequency slightly above its resonance frequency (100–400 kHz). The amplitude is few nanometers. Because the tip is separated from the surface, there are no repulsive forces involved. Only the longer range of van der Waals force, which is an attractive force are involved (see Example 4.5). When the tip dips closer to the surface, within 10 nm, the attractive van der Waals force ($\sim 10^{-12}$N) results in a decrease of the resonance frequency of the cantilever. The feedback loop maintains the constant frequency condition by adjusting the average tip-to-sample distance. The amplitude of oscillation drops (Fig. 4.12D) Measuring the tip-to-sample distance at each data point allows the microscope to construct a topographic image of the sample surface. It is to be noted that this mode permits imaging "soft" samples as the tip does not contact the sample.

4.3.1 Imaging Nanoparticles

When imaging nanoparticles, the lateral dimensions are strongly distorted by tip convolution effects as sketched in Fig. 4.14. The situation is worse in the contact mode. In this case there is profile broadening due to the tip-sample convolution, as well as height lowering effect due to elastic deformations. In the tapping non-contact mode, the elastic deformation is eliminated so that the height measurement would be representative of the size. Thus, in the lateral direction; a nanoparticle measures much larger than what is expected.

To illustrate the distortion feature we present AFM images of 3-nm silicon nanoparticles nanoparticle [4]. First silicon nanoparticles in isopropyl alcohol are spin coated on thermally grown SiO_2. They were then imaged by an AFM operated in NC tapping mode explained above. Fig. 4.15A–B shows the images of the nanoparticles. Fig. 4.15C–D give a line profile through an individual nanoparticle. A vertical corrugation of 3 nm is measured. The size of the

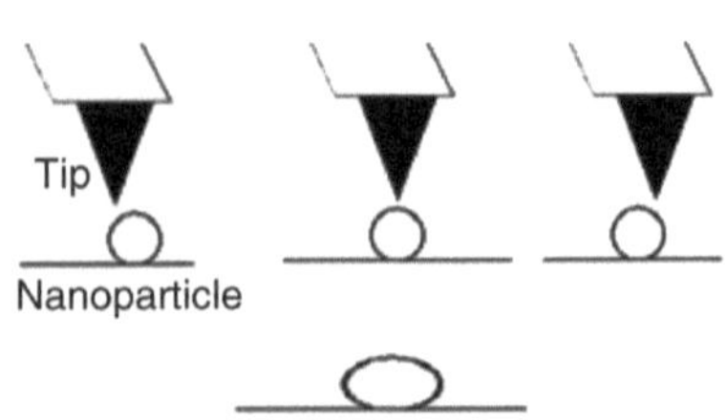

FIGURE 4.14 Sketch of distortion in imaging nanoparticles: lateral dimensions are strongly distorted by tip convolution effects.

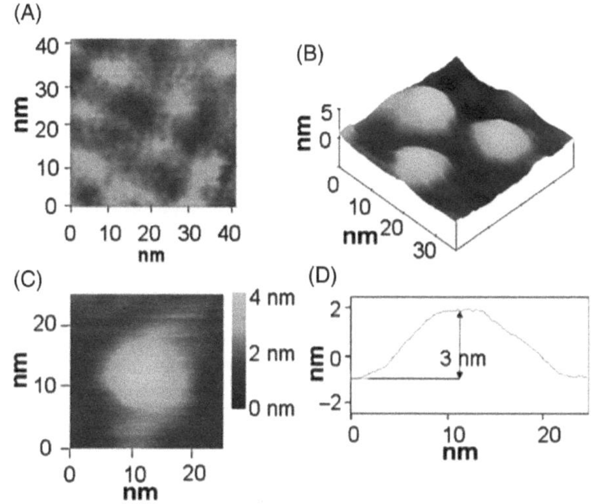

FIGURE 4.15 (A and B) AFM images of 3 nm silicon nanoparticles.in a graphite surface. (C and D) Line profile through an individual nanoparticle. A vertical corrugation of 3 nm is measured. In the lateral direction; a nanoparticle measures 10 nm. The size of the nanoparticles is therefore estimated by vertical depth profiling to be 3 nm.

nanoparticles is therefore estimated by vertical depth profiling via AFM to be 3 nm. However, in the lateral direction; a nanoparticle measures 10 nm, much larger than what is expected. It is known that AFM is only accurate vertically because laterally it is dominated by the tip-size effect, typically 9–10 nm. The AFM measurements were compared to transmission electron microscopy. TEM measurements, which measures true lateral dimension, gave a size of 3-nm diameter.

Example 4.6 Determine the force between a palladium AFM tip and a graphite substrate. The limit of resolution of an AFM

LDA, as early as 1990, was used to determine the Pd-graphite interaction and elastic constantsof graphite. In Fig. 4.16 they gave the force acting between graphite and a Pd atom in the on-top and hollow site.

Force *F* between ai-atom Pd AFM tip and graphite, as a function of the Lip height *z* above the surface of hexagonal graphite. The solid and dashed lines correspond to the sixfold hollow (II) and the on-top (1′) sites, respectively. The inset shows the adsorption geometry in top view.

In order to estimate the limits of atomic resolution in AFM, we use the calculated Pd–graphite interaction to determine the equilibrium tip height *z* at inequivalent surface sites as a function of applied load. We define a load per tip atom *fext*= *Fext/N,* where *fext*is the total load on the tip and *N* is the number of tip atoms in contact with the graphite surface. For loads *fext* $>2.0 \times 10^{-9}$ N, we find *z* == *Zon-top* − *Zhollow* > 0. Atomic resolution can be achieved experimentally when

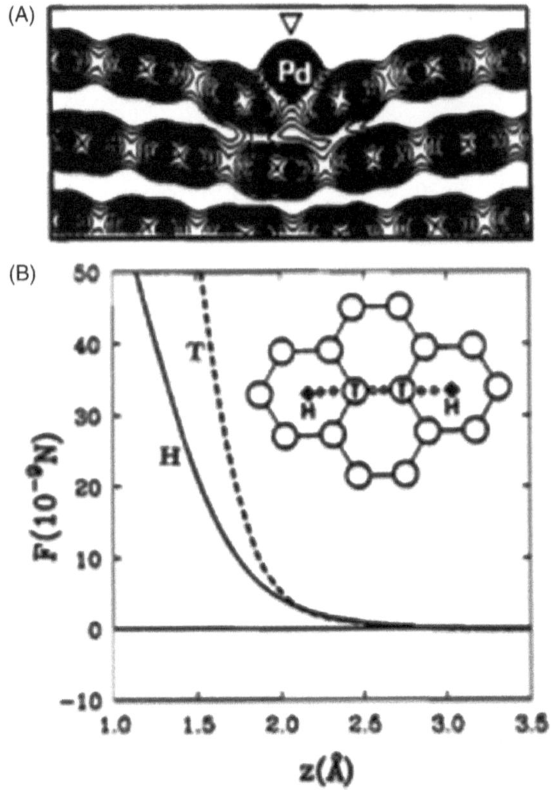

FIGURE 4.16 The force between a palladium AFM tip and a graphite substrate. (A) Schematic of an AFM tip with a palladium atom in its apex close to a graphite surface. (B) Force as a function of interdistance of two sites H and T. (Inset) Adsorption geometry in top view. *(From W. Zhong, G. Overney, D. Tománek, Limits of Resolution in Atomic Force Microscopy Images of Graphite, EPL. Europhys. Letters, 15, (1), https://web.pa.msu.edu/cmp/csc/eprint/DT041.pdf)*

delta z is sufficiently large to be detected, i.e. typically *Delta z* > 0.1 A, which corresponds to *fext* > 10^{-8}N.

On the other hand, our calculation of elastic graphite deformations for *fext* > 10^{-8}N indicates that such large local forces can rupture the graphite surface, in agreement with Experiment 7.

4.4 UNVEILING THE SURFACE TOPOGRAPHY: THE OSCILLATING CANTILEVER

Unveiling the actual fine details of the surface topography using AFM is not straightforward with a limited resolution. The interaction of an oscillating cantilever anchored on one end and with the other end being in the proximity of a surface, where mechanical and material interaction can take place is a rather complicated problem. It involves nonlinear differential equations. Moreover, due to the time dependence of the force between the tip and sample caused by the vibrations, this task is not only difficult, but in fact only possible to solve for certain limiting cases. In many cases the general solution to the cantilever equation of motion cannot be obtained in an analytical form. Therefore, the goal is to determine frequency and phase shift at given oscillation amplitude, which can then be related to a specific tip-sample interaction. In the procedure an appropriate and novel mathematical models to correctly predict the vibrations of an AFM micro cantilever probe must be selected. The model must then be validated with simulations and experimental data. Finally, analysis of the response must show stability, sensitivity, and convergence. These studies are meant to be used to home in on procedures to eliminate instability, decrease nonlinearities in the dynamic response, and simplify the solution and analysis without compromising the solution dynamics. Minimizing nonlinearities is important because nonlinearities tend to introduce parametric resonances and make the resonance frequency of the cantilever depend on the amplitude of the oscillation and exhibit jump phenomena. Several groups, including the group of the late Ali Nayfeh at VPI in Blacksburg, Virginia have analyzed various aspect of the problem using a variety of mathematical/simulation procedures. We give here several references for further reading [19–21].

4.5 VARIETY OF TIP-BASED IMAGING

There is a variety of other tip-based imaging devices. We will here briefly introduce them.

4.5.1 Conductive AFM (CAFM)

In a conductive AFM, both topography and conductivity profile can be simultaneously measured. To convert an AFM to a conductive AFM, two things must be modified. First the non-conductive top is changes to a conducting tip by coating the tip with a metal. Second the AFM is operated in a contact mode. Naturally the sample must be conducting to complete the electric circuit. Third an electric circuit is provided. It is to be noted that the contact mode is necessary because the device does not rely on quantum tunneling to make an electrical connection between sample and tip. Because, unlike the STM, there is no quantum tunneling, the CAFM provides less resolution than the STM. Recently, some manufacturers started providing the option of using the CAFM in a semi-contact mode.

4.5.2 AFM Tip with Single Molecule Apex

The resolution capability of a plain AFT tip was enhanced dramatically by simply attaching a single molecule, such as a carbon-monoxide molecule to its end (apex) as shown in Fig. 4.17A [22]. Pentacene molecules were placed on

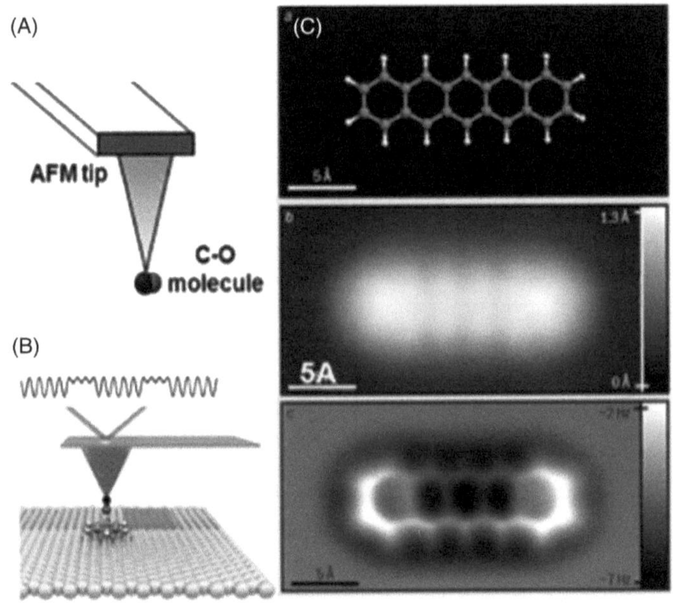

FIGURE 4.17 **Atomic imaging with AFM.** (A) Single CO molecule, attached to end (apex) of AFM tip. (B) Imaging of Pentacene molecules placed on a substrate. (C) Generated images reveal the "architecture" of the pentane molecule, clearly resolving the internal bonds.

a substrate (Fig. 4.17B). But it was found that in order to achieve high-quality intramolecular resolution, it was necessary to place the tip very close to the surface into the repulsive regime of the interaction, namely into Pauli-exclusion component of the interaction potential. Therefore, the AFM was operated in the contact mode discussed in Fig. 4.12A. The use of the tip in the contact mode allowed researchers turned out to be pivotal to achieve high-quality intramolecular resolution, effectively acting as a means to sharpen the tip.

Fig. 4.17C shows the generated image, which reveals the "architecture" of the pentane molecule, clearly resolving the internal bonds. This level of resolution is far better than the resolution of the topography that can be achieved with an STM (also shown in the figure). Meanwhile, researcher RuslanTemirov have also shown that high-resolution imaging in the Pauli-repulsion regime is also possible using STM.

4.5.3 Magnetic Force Microscopy

The magnetic force microscope (MFM) is another variety of atomic force microscope. The tip used in this application is a magnetized tip. The sample to be examined is a magnetic sample. The device is often used in a NC mode. The goal of this device is to detect the tip-sample magnetic interactions and upon scanning, the information is used to construct the magnetic structure of the sample surface. One important magnetic interaction detected by MFM is the magnetic dipole–dipole interaction.

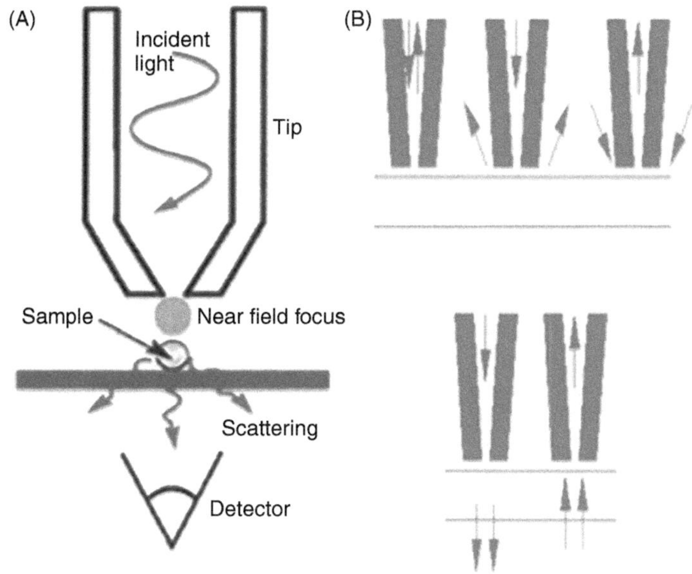

FIGURE 4.18 Near-field scanning optical microscopy (NSOM). (A) Schematic of the optical fiber tip of NSOM. (B) Several possible variation modes of operation (excitation/collection).

4.5.4 Near-Field Scanning Optical Microscopy (NSOM)

A near-field scanning optical microscopy (NSOM) is another device that provides nanoresolution [23]. A schematic is shown in Fig. 4.18A uses an optical fiber tip [24]. The fiber is tapered such that it provides a very small aperture. Visible light is guided through the optical fiber to strike the sample. And the tip is scanned relative to the sample. The scattered light is detected by a photon detector in the far field, from which an image of the sample is constructed.

The aperture diameter is chosen such that it is much smaller than the wavelength of visible light. In other words, the diameter is below the diffraction limit of light. For a light wavelength of 500 nm, a tip diameter of 10 nm is adequate. For a tip diameter of 10 nm, for example, the fiber tip is positioned within 10 nm from the sample surface. As a result of plasmonic effects, which will be discussed in Chapter 8, the amplitude of the exiting light near the tip (near-field light) decays exponentially, effectively focusing below the aperture to a hot spot of less than 10 nm diameter. The tight focusing in the near field beats the diffraction limited focusing in standard dielectric optical microscopy and provides nano resolution. For instance, light diffraction limit of conventional optical microscopy: $\lambda/2$. However, practically it is worse, approaching $\sim\lambda$, which is $\sim$500 nm for visibly light. In contrast, NSOM offers higher resolution $<$30 nm, depending on tip aperture size. However, the aperture size cannot be too small because the transmitted intensity would be too low. Therefore, there exists a tradeoff between resolution and sensitivity (light intensity).The device is quite versatile with respect to illumination and collection of light. Fig. 4.18B shows several self-explanatory possible variation modes of operation. http://www.eng.utah.edu/~lzang/images/Lecture_16_NSOM.p

4.5.5 Scanning Electrochemical Microscopy (SECM) Ultra-Microelectrode (UME) Tip

Nanofabrication of electrodes can be performed by etching a metal wire with sodium cyanide and sodium hydroxide. Etched metal wires can then be coated with wax, varnish, molten paraffin or glass, poly(a-methyl styrene), polyimide, electro polymerized phenol, and electrophoretic paint. Nano tips produced by these methods are conical. Typical preparation of a microscale electrode is performed by heat sealing a microware or carbon fiber in a glass capillary under vacuum.

Fig. 4.19A is the schematics of an ultra-microelectrode. The exposed metal is the active part of the electrode. Fig. 4.19B is an optical micrograph of the tip. The platinum wire is sealed inside a glass sheath. Typical preparation of a microscale electrode is performed by heat sealing a microware (or carbon fiber) in a glass capillary under vacuum. This tip can be connected to a larger copper electrode through the use of silver epoxy then polished to yield a sharpened tip. Nanofabrication of electrodes can be performed by etching a metal wire with sodium cyanide and sodium hydroxide. Etched metal wires can then be coated with wax, varnish, molten paraffin or glass, poly(a-methyl styrene), polyimide, electro polymerized phenol, and electrophoretic paint. Nano tips produced by these methods are conical.

Scanning electrochemical microscopy (SECM) is a technique within the broader class of scanning probe microscopy (SPM) that is used to measure the local electrochemical behavior of liquid/solid, liquid/gas, and liquid/liquid interfaces as shown in Fig. 4.19C. Interpretation of the SECM signal is based on the concept of diffusion-limited current. Two-dimensional raster scan information

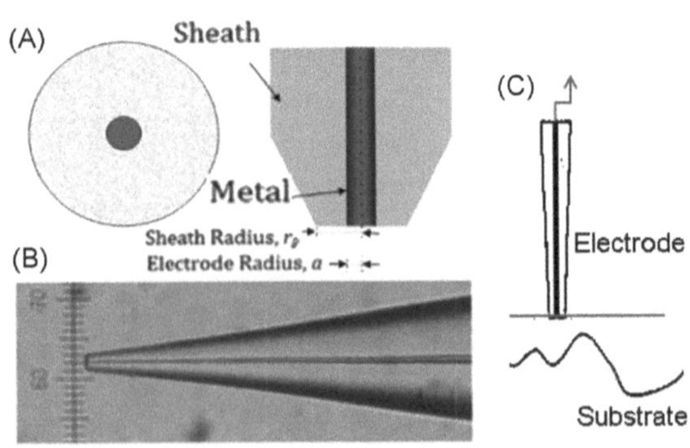

FIGURE 4.19 Scanning electrochemical microscopy (SECM). (A) Schematics of an ultra-microelectrode (UME) tip, (B) optical micrograph of the tip, (C) operation in a constant height mode in which the tip is immersed in a liquid above a surface to measure the local electrochemical behavior of liquid/solid interface.

can be compiled to generate images of surface reactivity and chemical kinetics. Topology and reactivity information can be measured by moving the tip across surfaces and measuring tip current. Both constant-height mode is the most common scanning modem, where the tip height is unchanging and is scanned across the surface in the *x*–*y* plane. Alternatively, a constant-current mode can be used, where the device attempts to maintain a constant current by changing the substrate to tip distance, *d*, and recording the change in *d*.

Fig. 4.20 is a summary diagram of the SPM techniques often used. They include from left to right STM, AFM, scanning electrochemical microscopy (SECM), and NSOM. In addition, scanning electrochemical cell microscopy (SECCM) has been used. All of those are presently used in nanoparticle research providing important contributions for their characterization.

Fig. 4.21 also shows the three tip devices and summarizes the basic principles of the operation. The STM uses tunneling current I between the outermost atoms of a conducting probe within an atomic distance from a substrate

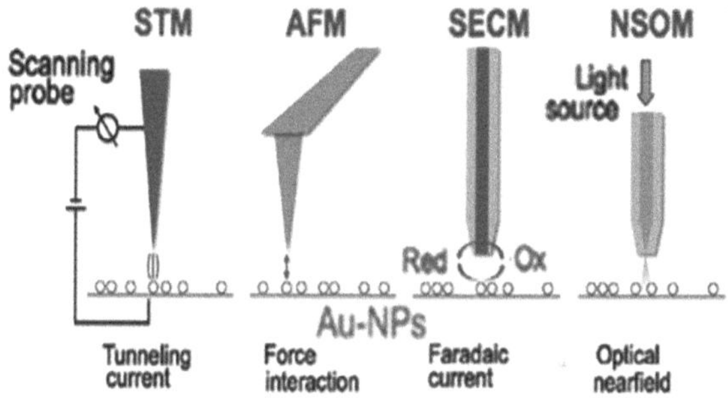

FIGURE 4.20 **Summary diagram of the SPM techniques often used.** *(From ChristineKranz, BorisMizaikoff*
Chapter 6 - Microscopic Techniques for the Characterization of Gold Nanoparticles, Comprehensive Analyt.Chem.,66, 2014, 257-299)

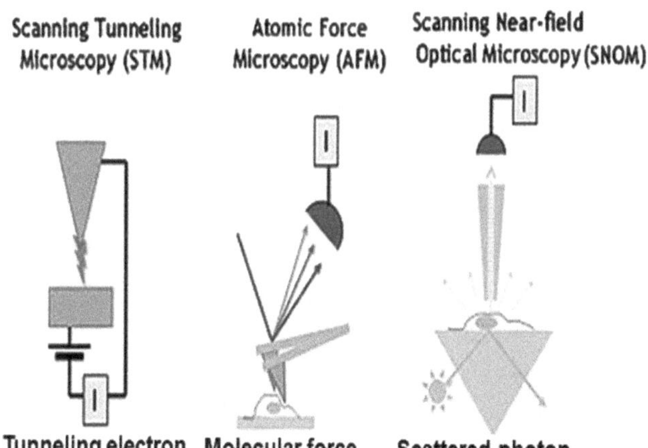

FIGURE 4.21 **Three tip devices summarizes the basic principles of the operation.**

to map out the sample topography and electrical properties. AFM uses the van der Waals molecular forces or contact forces between a tip and the sample to measure the sample topography or mechanical properties. Scanning near-field optical microscope (SNOM) uses the scattered photons (light) through a sub-wavelength aperture to form an image.

4.6 TIP PREPARATION AND EFFECT OF SHAPE

In this section we discuss procedures for making several types of tips, including tungsten, gold/indium/platinum, multiple wicker, carbon cone, metal-coated silicon, and carbon nanotube (CNT).

4.6.1 Metal Tips for STM [26]

The tip in a STM is pivotal in obtaining good images. The ultimate tip, that is, the sharpest tip would have at its apex a single atom.STM tips are usually made from tungsten metal or gold or a platinum–iridium alloy. Tungsten tips are usually made by electrochemical etching. On the other hand, platinum–iridium or gold tips can be made by mechanical shearing, as well as by electrochemical etching.

Fig. 4.22 shows the etching cell for preparation of tungsten tips and three example tips. A piece of low-cost polycrystalline tungsten wire is partially dipped in a bath of KOH in water. A counter platinum wire is dipped in the solution. The hydroxide is 3–5 M.The tungsten wire is biased positive and the platinum is biased negative with 5–10 V voltage difference. A current of ~1–3 mA is set up as shown in the figure. After some time, the end of the wire breaks off. The

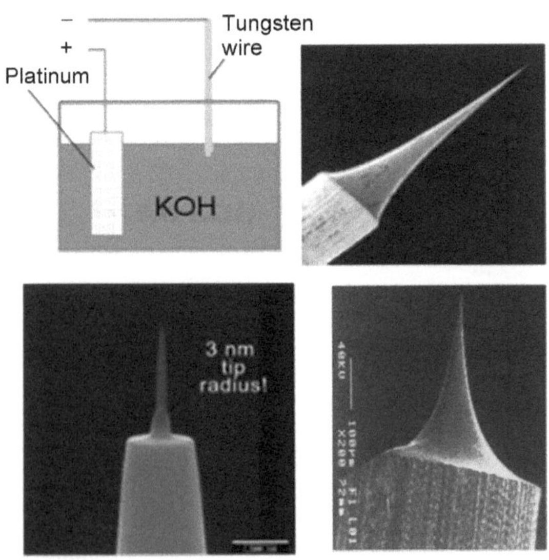

FIGURE 4.22 **Etching cell for preparation of tungsten tips and three example tips.**

tungsten wire is then removed and deionized DI water is used to clean residual KOH on the tip followed by drying in hot air.

Additional *in-situ* treatments would be needed to sharpen it further to reach atomically resolved imaging. This may be accomplished by a number of treatments under ultra-high vacuum conditions, including high temperature annealing, electric field restructuring, co-axial ion milling, controllable crashing using voltage pulses, or a contact between the tip and the sample. It is to be noted that if tungsten tips are left in air for an extended period of time (~5–10 h) they oxidize, resulting in a thin tungsten oxide coating which renders the tips non-conducting. Flushing with KOH solution, or a quick repeat of the electrochemical etch, removes the oxide coating.

4.6.2 Platinum–Iridium or Gold Tips

Gold tips can be obtained from wires by electrochemical etching in 0.8 M–97% KCN solution. The wires are biased by applying a voltage in the range 8–20 V. The process produces tip radius varying between 200 and 500 nm. Sharp edge tips may also be prepared from these materials because these materials are rather soft. A razor blade is placed on a 0.25 mm diameter wire at approximately a 45 degree angle. The wire is then cut. This results in a sharp tip. One can also cut the wire with a wire cutter. The tip quality is sufficient to obtain quality images.

4.6.3 Multiple Whisker Tips

If the tip is damaged due to contact with the substrate or picks up a lot of debris during imaging, the tip may be sharpened in situ under vacuum. This can be accomplished by pulling the tip quite away from the substrate and applying a large voltage pulse of ~100 V across. This tends to anneal and sharpen the tip. Such action creates a sharp tip but sometimes produces a multiple whiskers on the tip. Imaging using those produces doublets or triplets depending on the number of whiskers as shown in Fig. 4.23 [27].

4.6.4 Solid Carbon Cone Tapping Mode

AFM tips of 3-nm radius consisting of solid carbon cone for use in tapping mode probes are commercially available. They are available with length of 125 μm, spring constant of 40 N/m, and resonant frequency of 300 kHz.

4.6.5 Metal-Coated Silicon AFM Tips [28]

AFM tips are made either of pure silicon or of silicon nitride. Conventional photolithography is used to synthesize the tips. One starts with a 1cm × 1cm piece of a silicon wafer for example. After standard cleaning a 1 μm thick SiO_2 layer is grown by wet oxidation. The pattern for the tip is defined using standard photolithography with a Mask aligner. This is followed by removal of the exposed silicon oxide (using BHF solution). The unexposed photoresist is stripped by immersing the samples in acetone. KOH solution is now used to perform bulk

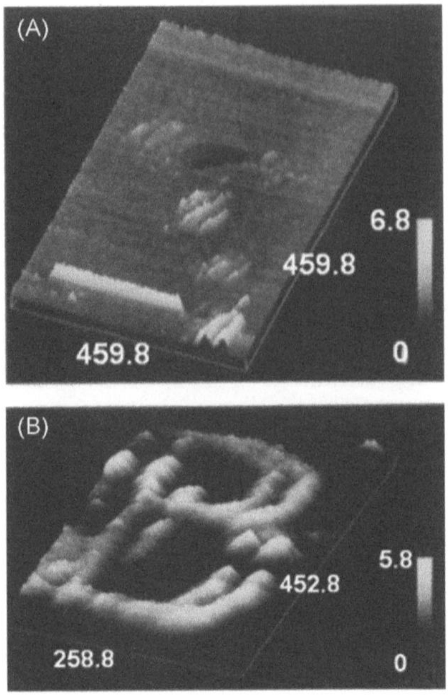

FIGURE 4.23 Multiple whisker tips. (A) Fabrication using tips produces doublets or triplets depending on the number of whiskers. (B) Fabrication letter B using tip with double whiskers.

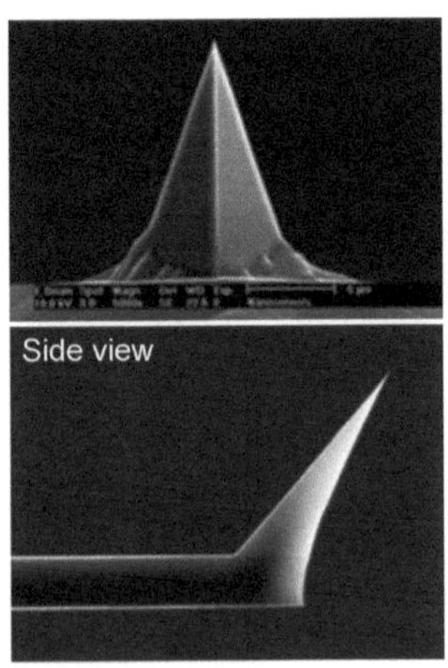

FIGURE 4.24 AFM silicon tip structure made by photolithography procedure.

micromachining of silicon. This step results in making the AFM tip structure. The tip is imaged by SEM to record the tip diameter and sharpness. Fig. 4.24 is an example of tips made.

For use in a harsh environment the tips can be coated with either platinum, iridium or gold. AFM probes can be made with different force constants and

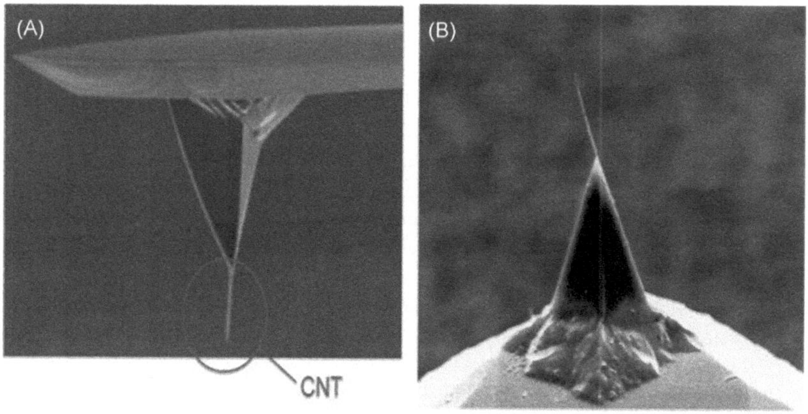

FIGURE 4.25 (A) and (B) are examples of Carbon nanotube tips, deposited on tungsten tips. A tungsten tip is subjected to microwave plasma-enhanced chemical vapor deposition, which fabricates a carbon nanotube (CNT) directly attached to its apex.

resonant frequencies for the three modes of imaging: contact mode, NC/tapping mode, and force modulation mode measurements. Harsh environment include applications in a variety of wet as well as radiation intensive environments. The gold-coated tip is especially interesting for applications in the fields of biology and life sciences.

4.6.6 Carbon Nanotube Tips [29]

A tungsten tip is subjected to microwave plasma-enhanced chemical vapor deposition. The process fabricates a carbon nanotube (CNT) directly on it as shown in Fig. 4.25. The reduction of the density of plasma ions directed to the apex of the tungsten tip suppresses the sputtering effect, and makes a successful fabrication of CNT tip. The length of CNTs is proportional to the time of treatment. Lengths below 1 μm can be achieved. The tip was used to resolve the graphite structure.

Single-walled carbon nanotubes (SWNTs) are useful as tips because they have nano metric diameter, high-aspect-ratio with high-mobility electrical conduction, and chemically and mechanical robustness.

REFERENCES

[1] Chang K. Americans and a German Are Awarded Nobel Prize in Chemistry. New York, NY: New York Times; 2014. Retrieved 8 October 8 October 2014.

[2] McMullan D. Scanning electron microscopy 1928–1965. Scanning 2006;17(3):175–85.

[3] Erni R, Rossell MD, Kisielowski C, Dahmen U. Atomic-resolution imaging with a sub-50-pm electron probe. Phys. Rev. Lett. 2009;102(9):096101. Bibcode:2009PhRvL. 102i6101E. PMID 19392535.

[4] Nayfeh Osama M, Antoniadis Dimitri, Mantey Kevin, Nayfeh MH. Uniform delivery of silicon nanoparticles on device quality substrates using spin coating from isopropyl alcohol

colloids. Appl. Phys. Lett. 2009;94:043112. Belomoin G, Therrien J, Smith A, Rao S, Chaieb S, Nayfeh MH. Observation of a magic discrete family of ultrabright Si nanoparticles. Appl. Phys. Lett. 2002;80:841.

[5] Azubel M, Koivisto J, Malola S, Bushnell D, Hura GL, Koh AL, Tsunoyama H, Tsukuda T, Pettersson M, Häkkinen H, Kornberg RD. Electron microscopy of gold nanoparticles at atomic resolution. Science 2014;345:909. www.sciencemag.org/content/345/6199/909.abstract (stanford).

[6] Park J, Elmlund H, Ercius P, Min Yuk J, Limmer TD, Chen Q, Kim K, Hoon Han S, David A, Weitz A, Zettl A, Alivisatos P. 3D structure of individual nanocrystals in solution by electron microscopy 2015;349(6245):290.

[7] Binnig G, Rohrer H. Scanning tunneling microscopy. IBM J. Res. Dev. 1986;30(4):355–69.

[8] Stroscio JA, Kaiser WJ. Scanning Tunneling Microscopy. Cambridge, MA: Academic Press; 2013.

[9] Theory tunneling: http://www.maths.tcd.ie/~bmurphy/thesis/thesisse3.html

[10] Ong QK, Reguera J, Silva PJ, Moglianetti M, Harkness K, obardi ML, Mali KS, Renner C, De Feyter S, Stellacci F. High-resolution scanning tunneling microscopy characterization of mixed monolayer protected gold nanoparticles 2013;7:8529–39. https://www.researchgate.net/publication/256499730_High-Resolution_Scanning_Tunneling_Microscopy_Characterization_of_Mixed_Monolayer_Protected_Gold_Nanoparticles.

[11] Wolf EL. Principles of Electron Tunneling Spectroscopy. Oxford, UK: OUP; 2012.

[12] Therrian J. Size dependence of the electrical characteristics of silicon nanoparticles. Champaign, IL: University of Illinois at Urbana-Champaign; 2002.

[13] Binnig GK, Quate CF, Gerber C. Atomic force microscope. Phys. Rev. Lett. 1986;56(9):930–3. PMID 10033323.

[14] Haugstad G. Atomic Force Microscopy: Understanding Basic Modes and Advanced Applications. Hoboken, NJ: John Wiley & Sons; 2012.

[15] REF:Afm: http://usa.jpk.com/what-is-atomic-force-microscopy.432.us.html REF types of forces: Bullen, R.A.W.a.H.A., Lecture notes: Introduction to Scanning Probe Microscopy. http://www.bioinformatics.utep.edu/agriculture/MDSim-module.

[16] Israelachvili JN. Intermolecular and Surface Forces. Cambridge, MA: Academic Press; 2015. Tseng AA, Li Z. Nanofabrication using atomic force microscopy. J. Nanosci. Nanotechnol. 2015; www.cc.ntut.edu.tw/~smchen78/news/AFMLIntro10.pdf.

[17] W. Zhong, G. Overney, D. Tomanek Of the Department of Physics and Astronomy and Center for Fundamental Materials Research, Michigan State University, East Lansing, Michigan 48824-1116 Force om palladium.

[18] de Oteyza DG, et al. Direct imaging of covalent bond structure in single-molecule chemical reactions. science 2013;340:1434–7. (New York).Read more at: https://phys.org/news/2013-05-first-ever-high-resolution-images-moleculereforms.html#jCp.

[19] Raman A, Melcher J, Tung R. Cantilever dynamics in atomic force microscopy. Nanotoday 2008;3:1–2. FEB-APR http://csmres.co.uk/cs.public.upd/article-downloads/raman.pdf.

[20] Delnavaz SA, Mahmoodi N, Jalili MN, Ahadian M, Zohoor H. Nonlinear vibrations of microcantilevers subjected to tip-sample interactions: Theory and experiment. J. Appl. Phys. 2009;106:113510.

[21] I. Manoubi, F. Najar, S. Choura, A.H. Nayfeh, Nonlinear dynamical analysis of an AFM tapping mode microcantilever beam, in: The International Conference on Structural Nonlinear Dynamics and Diagnosis, MATEC web of conferences, 1 (2012) 040002 (DOI:10.1051/matecconf/20120104002; Contact force identification using the subharmonic resonance of a contact-mode atomic force microscopy, Eihab M Abdel-Rahman and Ali H Nayfeh, Published

7 January 2005, IOP Publishing Ltd, Nanotechnology, Volume 16, Number 2. REF:Afm: http://usa.jpk.com/what-is-atomic-force-microscopy.432.us.html

[22] L. Gross, et al., The chemical structure of a molecule resolved by atomic force microscopy, Science 325 (2009), p. 1110; F.J. Giessibl, Advances in atomic force microscopy, Rev. Mod. Phys. 75 (2003), p. 949; F.J. Giessibl, et al., Sub-atomic features on the Si(111)-(7 × 7) surface observed by atomic force microscopy, Science 289 (2000), p. 422.

[23] Liu J. Development and Application of a Near-Field Scanning Optical Microscope in the Studies of Compound Semiconductor Materials. Madison: University of Wisconsin; 1996.

[24] NSOM: http://www.photonics.com/EDU/Handbook.aspx?AID=25127

[25] Scanning electrochemical microscopy, https://en.wikipedia.org/wiki/Scanning_electrochemical_microscopy

[26] Williams RD. The Fabrication of Specialized Probes for Surface Metrology. Austin, TX: University of Texas; 2007.

[27] Zheng X, Hetrick J, Yau S-T, Nayfeh MH. Parallel fabrication on chemically etched silicon using scanning tunneling microscopy. Ultramicroscopy 1992;1303:42–4. M. Nayfeh, Fabrication of Nanometer Scale Structures, Technology of Proximal Probe Lithography C. Marrian SPIE Institutes IS 10 (1993) 200.

[28] NANOSENSORS™ introduces AdvancedTEC AFM Probes with Gold and Platinum Coating: http://www.openpr.com/news/44798/NANOSENSORS-introduces-AdvancedTEC-AFM-Probes-with-Gold-and-Platinum-Coating.html

[29] CMT tip Properties and Applications of Composites and Nanocomposites, Published by Alina Delman: http://slideplayer.com/slide/1554643/; https://www.nccrnano.org/nccr/nanoscience/pictures/gallery_01/gallery_01_03/index_html_print

Chapter 5

Manipulation and Patterning of Surfaces (Nanolithography)

5.1 STANDARD VISIBLE AND NEAR UV PHOTOLITHOGRAPHY

In standard photolithography (optical or UV lithography) [1,2], a photomask is first created. It is basically an opaque screen in which holes or transparencies that allow light to pass through are created. A substrate is coated with a thin film of an organic material (photoresist). The coating is chosen such that its solubility changes upon exposure to light. The photomask is placed above the coated substrate as shown in Fig. 5.1. Visible or UV light is now shined through the photomask. The transmitted light defines the hole pattern on the photoresists, which modulates or varies its solubility. The substrate is then chemically developed to selectively dissolve and hence remove either the exposed areas or the unexposed area. Subsequent series of chemical treatments may be used to engrave the exposure pattern into the material underneath the photo resist. On the other hand, chemical treatments can be designed to enable deposition of a new material in the desired pattern upon the substrate. One application of this technology is creating complex integrated circuits. For instance, a CMOS wafer goes through the photolithographic cycle up to 50 times, which creates extremely small patterns from micrometer down to the nanoscale of a few tens of nanometers.

5.2 CURRENT PHOTOLITHOGRAPHY (SHORTER WAVELENGTH)

An *excimer laser* is an UV laser [3]. The wavelength of an excimer laser depends on the molecules used, and is usually in the ultraviolet. It is commonly used in the production of microelectronic devices, semiconductor based integrated circuits or "chips," eye surgery, and micromachining. Fig. 5.2A shows a table of the possible lasers. Current (state-of-the-art) photolithography tools use excimer laser light sources with wavelength shorter than what was used in standard photolithography as depicted in Fig. 5.2C. The spatial resolution is diffraction limited by the wavelength of the light that is used. Moreover, using a reduction lens system that captures more of the diffraction orders from the

Fundamentals and Applications of Nano Silicon in Plasmonics and Fullerines.
http://dx.doi.org/10.1016/B978-0-323-48057-4.00005-0

FIGURE 5.1 **Photolithography process using a mask, photoresist and a light source.**

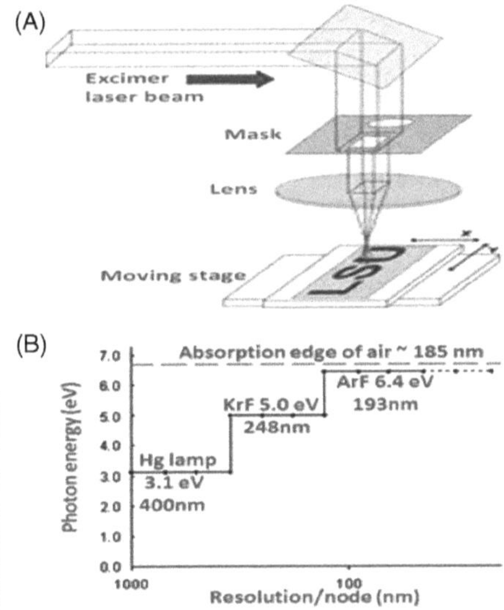

(C)

Excimer	Wavelength	Excimer	Wavelength
Ar_2^*	126 nm	KrF	248 nm
Kr_2^*	146 nm	XeBr	282 nm
F_2^*	157 nm	XeCl	308 nm
Xe_2^*	172 & 175 nm	XeF	351 nm
ArF	193 nm	KrCl	222 nm

FIGURE 5.2 **Excimer lasers used in photolithography.** (A) Layout of the setup for excimer-based photolithograpy, (B) decrease of the wavelength used in lithography improves the resolution. However if the wavelength is taken below 185 nm, the absorption edge of air, operation under ambient conditions, becomes problematic as air absorption becomes significant below 185 nm. (C) A table of the possible excimer lasers.

illuminated mask enhances the resolution. For instance deep ultraviolet (DUV) light from excimer lasers with wavelengths of 248 and 193 nm are becoming the dominant lithography technology today, and been also called excimer laser lithography, which allow minimum feature sizes down to 50 nm.

5.2.1 Wavelength-Independent Resolution

Further decrease of the wavelength used in lithography improves the resolution. However if the wavelength is taken below 185 nm (see Fig. 5.2B), the absorption edge of air, operation under ambient conditions becomes problematic because air absorption becomes significant below 185 nm. In this regime operation must be conducted under vacuum conditions. There are several procedures for improving lithography resolution without shortening the lithography wavelength. Those are presently fully integrated in the standard current CMOS lithography technology. They include improving the projection lenses using immersion in a liquid, and multiple patterning procedures.

5.2.2 Immersion Lithography

One method for improving resolution is to use an immersion liquid medium which has a high refractive index [4]. The higher refractive index affords the use of optics with numerical apertures exceeding 1.0. In this scheme, the projection (objective) lens is immersed in the liquid as shown in Fig. 5.3A. Fig. 5.3B shows the basic principle of the effect by displaying the ray tracing for both vacuum and water. The minimum feature size CD that a projection system can print is proportional to the wavelength λ and inversely proportional to the

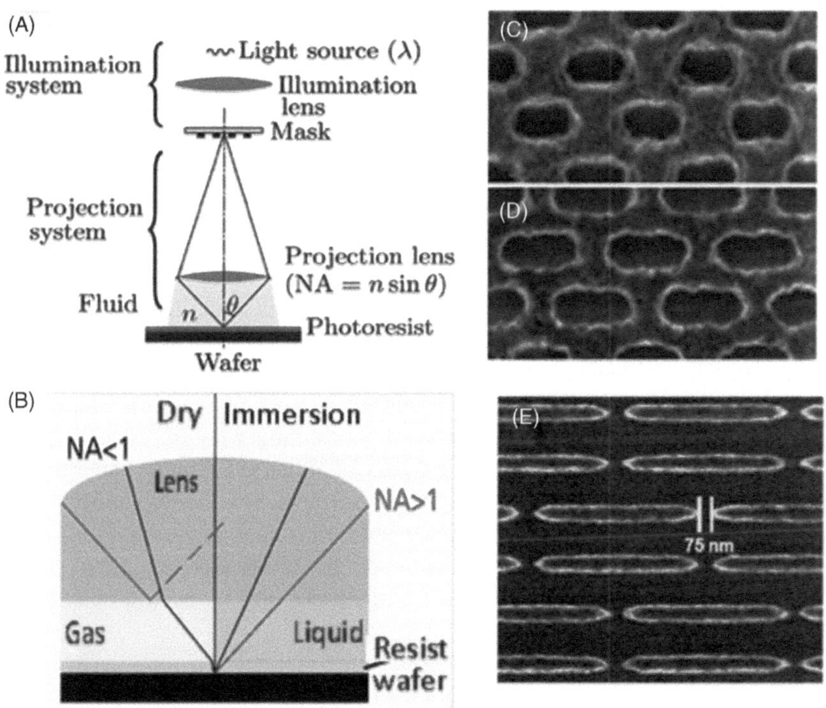

FIGURE 5.3 Immersion lithography. (A) The projection (objective) lens is immersed in a liquid with high refractive index affords the use of optics with numerical apertures exceeding 1.0. (B) The basic principle of the immersion effect by displaying the ray tracing for both vacuum and water. (C–E) Demonstration of the advantages of the immersion procedure.

numerical number NA of the lens (given approximately by: CD = λ/NA). That is by decreasing the wavelength, and increasing the numerical aperture we can achieve a tighter focused beam and a smaller spot size. However, these conditions reduce the depth of focus DF according to DF = λ/NA^2, which places a competing constraint, limiting how far one can push the parameters.

As to the liquid, water is used as the immersion medium for the projection lens. The liquid used is typically ultra-pure, deionized water, which provides for a refractive index above that of the usual air gap between the lens and the wafer surface. The water is continually circulated to eliminate thermally induced distortions. Water will only allow *NA* of up to ~1.4, but materials with higher refractive indices will allow the effective *NA* to be increased.

A demonstration of the advantages of the immersion procedure is shown in Fig. 5.3C–E [5]. Contact bars were printed using (C) 0.93NA lithography and (E) 1.2NA immersion. The figure shows that there is 1.4X improvement in the aspect ratio upon the use of 1.2NA process compared to the use of 0.93NA lithography, as well as there is substantial improvement in resolution. In fact IBM was the first to accomplish industrial implementation of the technology in 2006, by utilizing it in the 45-nm node technology. When 1.2 NA at a wavelength of 193 nm was used SEM images showed gate to gate tip distance of 75 nm compared to 88 nm achieved by 0.93NA dry lithography. In addition, immersion technology competes well with double-exposure techniques by reducing design complicities cost, as well as overcoming potentially higher defect density, especially migration from 65 nm to 45 nm GRs. Due to all of these advantages and more immersion lithography has become the most advanced form of photolithography as early as 2009.

5.2.2.1 Double Patterning: Increasing Feature Density

Double patterning or multiple patterning in general is a photolithography procedure developed to enhance the feature density [6]. The simplest case of multiple patterning is *double patterning*, where a conventional lithography process is enhanced to produce double the expected number of features. The procedure requires little change in the standard photolithography infrastructure. For example, conventional lithography process is enhanced double patterning in to produce double the expected number of features. It is being applied to the 45 nm and 32 nm nodes. Moreover, several companies, including IBM, Intel, and TSMC, have prepared for the continued use of current lithography, using double patterning, for the 22 nm and 16 nm nodes, and extending double patterning beyond 10 nm.

There are several types of double patterning. Fig. 5.4A–B shows schematics showing the two processes: Litho-Etch-Litho-Etch(LELE) and Litho-Freeze-Litho-Etch (LFLE), respectively. We briefly describe the LELE process. It consists of a total of five steps after the deposit of two masks (hard mask 1 and mask 2). In the first step (In Litho 1) a pattern is exposed onto hard mask 1. This is followed by step 2 in which an etching is performed (Etch 1) that imprints

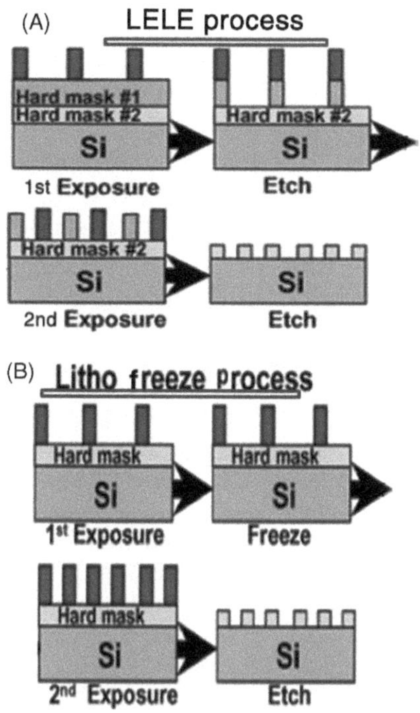

FIGURE 5.4 Schematic of double patterning. (A) The Litho-Etch-Litho-Etch (LELE) *process* and (B) Litho-Freeze-Litho-Etch (LFLE) process. *(From Paul Zimmerman, Double patterning lithography: double the trouble or double the fun? http://spie.org/newsroom/1691-double-patterning-lithography-double-the-trouble-or-double-the-fun?SSO=1)*

first pattern into the hard mask1. In step 3, a second pattern is exposed onto hard mask 2. The final double density pattern is engraved into hard mask 2. The double density pattern is engraved in silicon and the remaining mask is washed away. It is to be noted that double patterning is being applied to the 45 nm and 32 nm nodes. Moreover, several companies, including IBM, Intel, and TSMC, have prepared for the continued use of current lithography, using double patterning, for the 22 nm and 16 nm nodes, and extending double patterning beyond 10 nm.

5.3 NEW GENERATION OF LITHOGRAPHY (NGL)

The use of X-rays, extreme UV, and charge particles in lithography defines the concept of new generation of lithography (NGL) [7]. It is expected that current laser photolithography even when it is optimized by multiple patterning and index matching to run out of advantages once we approach the 7-nm node. In fact, it is becoming clear that smaller and smaller node technologies would no longer be cost effective as predicted by Moore's law. For example, the only known technique for single exposure 5 nm node (22 nm pitch) patterning is a complex and expensive procedure, namely what is called self-aligned multiple patterning enabled by sidewall profile modulation. Although NGL is slated to replace the current photolithography technology. However NGL is not without problems:

- Infrastructure requires significant changes to the infrastructure
- Research and development may require significant degree of research and development.
- Cost: EUV is more expensive than three 193i exposures.

5.3.1 Extreme Ultraviolet Lithography (EUV-Lithography)

As was discussed earlier (see Example 5.1), excimer lasers-pumped tin or xenon produces radiation in the extreme ultraviolet regime. *EUV* is one of the next-generation lithography technologies. One popular emission is at 13.5 nm. This EUV source is expected to be used in the future in combination with immersion lithography at the 7 nm node (or as called 32 nm pitch resolution). It may also be used at 5 nm node, but this would require a higher numerical aperture than currently available and would also require multiple patterning to a greater degree than immersion lithography at 20 nm node.

Example 5.1 Light Sources For Lithography

We summarize the present and future light sources relevant to the photolithography process against the transistor size.

Gas discharge: Mercury lamp produce light across a broad spectrum with several strong peaks in the ultraviolet range. In the 1960s through the mid-1980s, Hg lamps 436 nm, 405 nm, and 365 nm lines were in use.

Excimer lasers: Excimer lasers are gas-based lasers that use inert and halide gases (Kr, Ar, Xe, F, and Cl) that are charged by an electric field as gain media. Since 1982, the krypton fluoride (KrF) laser at 248 nm and the argon fluoride (ArF) laser at 193 nm.

The use of 157 nm wavelength from the F2 excimer laser has been discontinued as a replacement of the 193 nm laser due to persistent technical problems with the laser and economic considerations. A much shorter UV wavelength at 126 nm (for Ar_2*) has been demonstrated.

Plasma sources: Laser-pumped metals or gases can generate non-coherent extreme UV (EUV) light at 13.5 nm for extreme ultraviolet lithography. For example tin or xenon can be excited by an eximer laser, producing plasmas which radiate the EUV. Fabrication of feature sizes of 10 nm has been demonstrated in production environments, but not yet at rates needed for commercialization.

Free electron laser: The *free-electron laser* (laser for an X-ray device) can be used when wavelengths needed continue to decrease to extreme UV or X-ray. Those, which utilize accelerated electrons as source of the radiation can produce high quality beams at arbitrary wavelengths.

Note: Air absorbs below ~185 nm; thus utilizing sources with wavelength <185nm is problematic and requires use of expensive vacuum conditions.

It is to be noted that the development of EUV lithography is behind schedule for a number of reasons. First, the power available at 13.5 nm is limited. Second, immersion lithography is still more than four times faster than EUV. Due to these limitations, multiple patterning with immersion lithography has already been used in projects which were previously planned for EUV use.

5.3.2 X-Ray Lithography

X-rays and deep X-ray with wavelengths as short as 1 nm and 0.1 nm are shorter by more than an order or two orders of magnitude respectively from that of EUV wavelength of 13.5 nm. If X-ray-sensitive resistance is used, then the current optical/UV lithography process may be effective with such short wavelengths. Moreover, a big advantage of X-rays is that X-rays diffraction limit is so small that it easily allow smaller feature sizes. A major disadvantage, however, is that elementary collimating mirrors or diffractive lenses must be used in the place of the currently used refractive visible or UV optics. Thus X-ray lithography is a good candidate for next-generation lithography for the semiconductor industry. For 0.1 nm wavelengths, modified procedures, such as the LIGA process, are used to fabricate deep and even three-dimensional structures.

5.3.3 Electron-Beam Lithography

Electron-beam lithography (EBL) scans a focused beam of electrons in a patterned fashion across a surface in order to create very small structures, such as integrated circuits or other nanostructures. In other words, electron-beam (e-beam) lithography utilizes an e-beam instead of a visible or UV light beam in the lithography process. Just as UV, energetic electrons can change the chemical characteristics or the solubility of the resist, enabling selective removal of either the exposed or non-exposed regions of the resist. Electrons have advantages and disadvantages. The primary advantages and disadvantages of electrons are:

- Mode of fabrication mask less direct-write format
- Focusing much more tightly than light, which makes much finer patterns with sub-10 nm resolution.
- Resolution: Higher resolution than light, limited by lateral scattering of (secondary) electrons from the beam-defined location (Fig. 5.5)
- Proximity effect: Spilling over exposure of an adjacent region into the exposure of the currently written feature, effectively enlarging image, and reducing its contrast.
- Fast VS slow electrons: Fast (higher energy) electrons provide higher resolution. Slow (low energy) electrons especially in the range $<0–5$ eV compromise resolution as they have significant range (as far as 100 nm), suffering very little loss on colliding resist molecules. Low energy electrons are weakly interacting because their kinetic energy falls below the ionization potential.

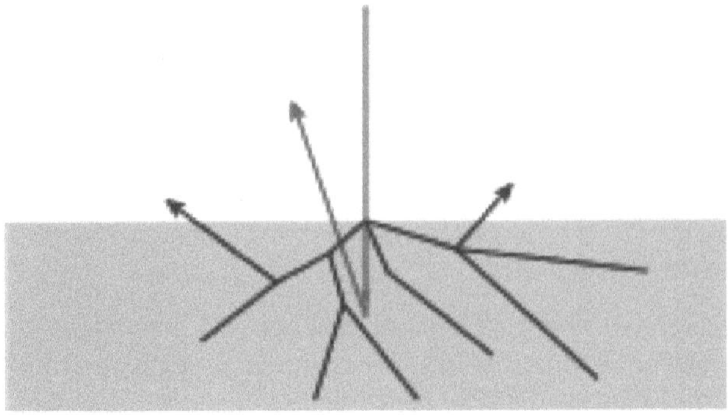

FIGURE 5.5 **Lateral scattering of (secondary) electrons from the beam-defined location.**

- Throughput: Low
- Processing: High level of complexity, expense, and maintenance intensive.
- Speed: 10 million times slower than current optical lithography tools, especially when writing dense patterns over a large area.

EBL systems and sources may be classified according to the application they are being used in: commercial VS research. For commercial applications, EBL systems are dedicated e-beam writing systems that cost more than $1M. EBL systems may be further classified according to beam characteristic: shape, deflection strategy, and energy. Older systems used Gaussian-shaped beams and scanned them in a raster fashion. Newer systems use shaped beams, which may be deflected to various positions in the writing field (known also as vector scan). As to beam energy, systems may use low-energy, offering sub-100 nm resolution (dependent on dose); extremely high electron energies (at least 100 keV); however, this mode is very slow, inefficient and could drill and cause damage.

Example 5.2: Electron-Beam Lithography (The State Of The Art) [8]

Generally for EBL in most resists, it is difficult to go below 25 nm lines and spaces, with a limit of 20 nm lines and spaces has been found [9]. Recently an order of magnitude improvement has been achieved. In the demonstration hydrogen silsesquioxane resist was used. Moreover, aberration corrected STEM scanning transmission electron microscope. It is a state-of-the-art system located at the Cavendish Laboratory that uses an e-beam of diameter 4 nm and energy up to 100 kV. Fig. 5.6A shows example of the capability of this system. It shows 2 nm isolated feature size and 5-nm half pitch. It shows also the point spread function of the system. Moreover, it is capable of patterning substrates of up to 200 mm diameter with a resolution as high as 10 nm. Fig. 5.6B shows a micrograph of a grid formed in ZnO demonstrating a sub-10 nm resolution. The facility is a Leica VB6 UHR EBL machine located at the Cavendish Laboratory.

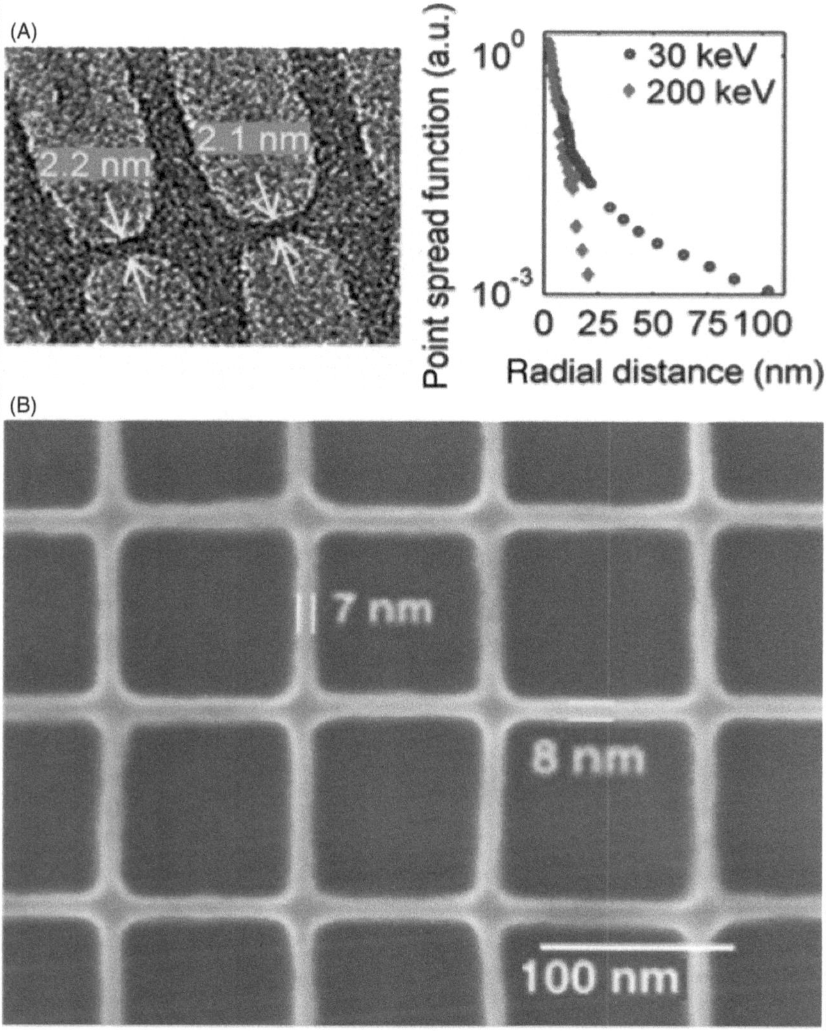

FIGURE 5.6 The state of the art of electron beam lithography with aberration-corrected STEM scanning transmission electron microscopy. (A) Example of the capability of this system (2 nm isolated feature size and 5 nm half-pitch). (B) Micrograph of a grid formed in ZnO demonstrating a sub-10 nm resolution. The facility is a Leica VB6 UHR electron beam lithography machine located at the Cavendish Laboratory.

For research applications, lower cost retrofitted systems are usually used. For example, it is very common to convert an electron microscope into an EBL system using low cost accessories (<US$100K). Linewidths of ~20 nm has been achieved using such converted systems, while current dedicated systems have produced line widths on the order of 10 nm or smaller.

5.3.4 Focused Ion Beam (FIB) Lithography

In ion beam lithography, a focused beam of ions is used instead of a focused beam of electrons. It has several advantages over EBM and has high potential to play an important role in nanometer technology. Because ions have much heavier mass than electrons by 3–5 orders of magnitude, they have much less back scattering, which results in less proximity effects. Second, because of the heavier particles and more momentum they suffer less forward scattering; hence it affords higher resolution patterning than UV, X-ray, or EBL. The large mass and momentum gives the ion beam a smaller wavelength than even an e-beam and therefore almost no diffraction. An example displaying the resolution of ion beam is shown in Fig. 5.7A–B [10].

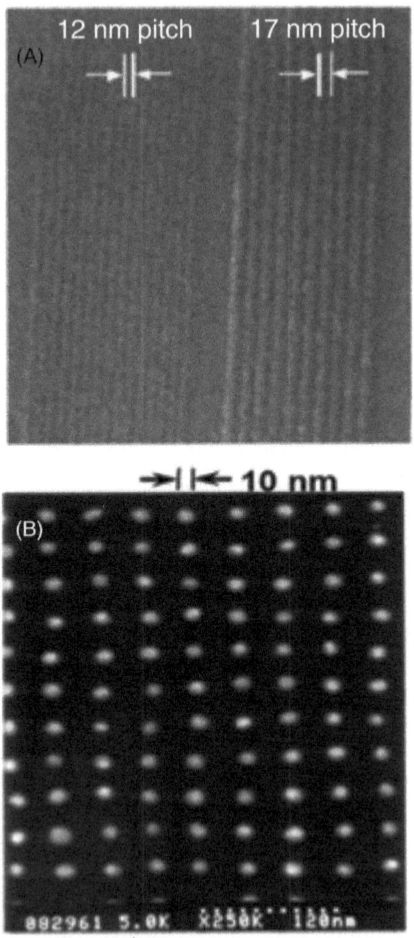

FIGURE 5.7 Focused ion beam (FIB) lithography. (A and B) Examples displaying the high-resolution lithography using ion beam. Large mass and momentum of ions gives the ion beam a smaller wavelength with diffraction nearly eliminated. *(A, From http://ieeexplore.ieee.org/ieee_pilot/articles/96jproc02/96jproc02-pease/article.html; B, From Stephen Y. Chou, Peter R. Krauss, Wei Zhang, Lingjie Guo, Lei Zhuang, Sub-10 nm imprint lithography and applications, J. Vac. Sci. Technol. B 15, 2897 (1997); https://doi.org/10.1116/1.589752)*

Focused IBL can be used in direct writing format, or it can be used with a mask. In masked IBL the beam passes through ion-transparent membrane, patterned with absorber material, positioned close to the resist coated substrate. The most common resists used in this application is polymethylmethacrylate (PMMA).

The advantages and disadvantages of ion beam lithography processing can be summarized as follows:

- High exposure sensitivity
- Two or more orders of magnitude higher resolution than that of EBL
- Negligible ion scattering in the resist and low back scattering from the substrate
- Penetration of the particle beam is small, compared to e-beam, reducing blurring resulting from beam scattering (reduce proximity effect; increase localization, precision)
- Can be used as physical sputtering etch, milling, and chemical assisted etch (see Chapter 4 for use in preparation of plasmonic nano tweezer tips)
- Can also be used as direct deposition or chemical assisted deposition, or doping
- Lower throughput, extensive substrate damage

There are gas, as well as metal ion beam sources. Hydrogen and helium gas sources are used. The most common metal source is Ga. Moreover, Au/Si and Au/Si/Be alloys have also been used for lithography because these sources can supply lighter mass ions.

5.4 VARIETY OF NON-CONVENTIONAL LITHOGRAPHY

There is a number of emerging non-conventional lithography process, including nano mold-based printing, plasmonic-assisted lithography, laser interference, object-shadow lithography.

5.4.1 Nanoimprint (Mold) Lithography

Nanoimprint lithography does not use light or charge particles [11]. Therefore there are no limitations due to light diffraction or charge beam scattering. In fact, it uses a pre-prepared mold of a hard material in combination with a mechanical press. A cartoon of the process is shown in Fig. 5.8. Topographical nano features are first created on a mold. The mold is then pressed against the resist-coated wafer (monomer or polymer) with an appropriate force. This action negatively imprints the topographical features onto softer resists. After the removal of mold, the wafer is processed with a standard reactive ion beam etching (RIE) to remove (clean) the compressed resist.

As to resolution, the procedure has demonstrated 25-nm feature size, 70 nm pitch, vertical and smooth sidewalls, and nearly 90 degree corners. The ultimate

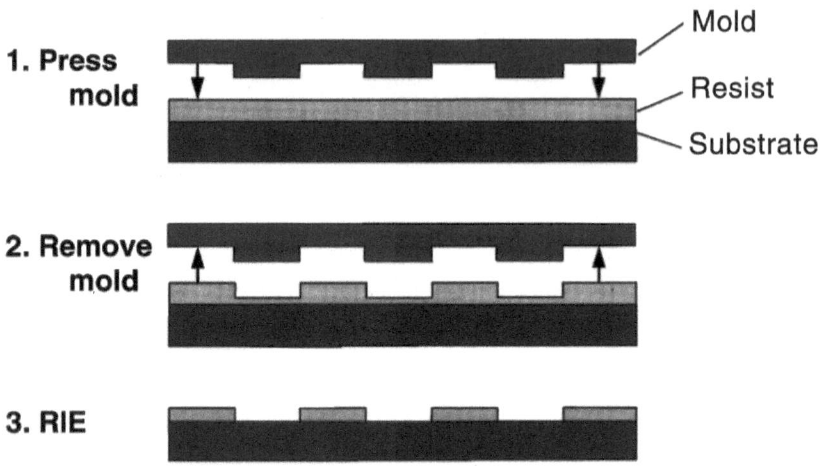

FIGURE 5.8 A cartoon of the nanoimprint lithography process using a mold. (*From Stephen Y. Chou, Peter R. Krauss, Preston J. Renstrom, Nanoimprint lithography, J. Vacuum Sci. Technol. B 14, 4129 (1996); https://doi.org/10.1116/1.588605*)

resolution of nanoimprint lithography could be sub-10 nm. As to applications, nanoscale photodetectors, silicon quantum-dot, quantum-wire, and ring transistor have been fabricated using this technology. The process has been added into the International Technology Roadmap for Semiconductors (ITRS) for 32 and 22-nm nodes. Currently, state-of-the-art nanoimprint lithography can be used for patterns down to 20 nm and below. From a commercial and mass production point of view, the process is expected to be low cost, and high throughput. The imprint resist can also be simply cured by heat or UV light during the imprinting.

5.4.2 Plasmonic-Assisted Lithography

Thin films decorated with patterns of nanoparticles, holes, grooves, cuts, dimples, small gaps, or slits focus visible light to sub 50 nm spots due to plasmonic effects (see Chapters 4 and 8 for details). Plasmonic lithography utilizes such focusing in photolithography. One disadvantage of this technology is the fact that the light gets focused very close to the surface, 10–15 nm focal length with very small focusing depth. This poses difficulties in precision alignment and control of the substrate as will be discussed later. Secondly, metal apertures are not simply circular (pin) or rectangular hole. Rather they are made of two off-center lopes in the shape of a bow-tie. Moreover, because of absorption losses in metal generating heat, the most used conductor materials are silver and gold for wavelengths in the visible and near infrared spectral range, which afford the least losses. The theory of plasmonics and plasmonic lenses, as well as general applications will be presented in Chapter 8.

Fig. 5.9A [12] shows an SEM image of a bowtie aperture. The aperture is in the shape of a single bowtie. It is drilled in an aluminum film that is 120 nm thick. The focal plane of this aperture in the near field falls at a distance of

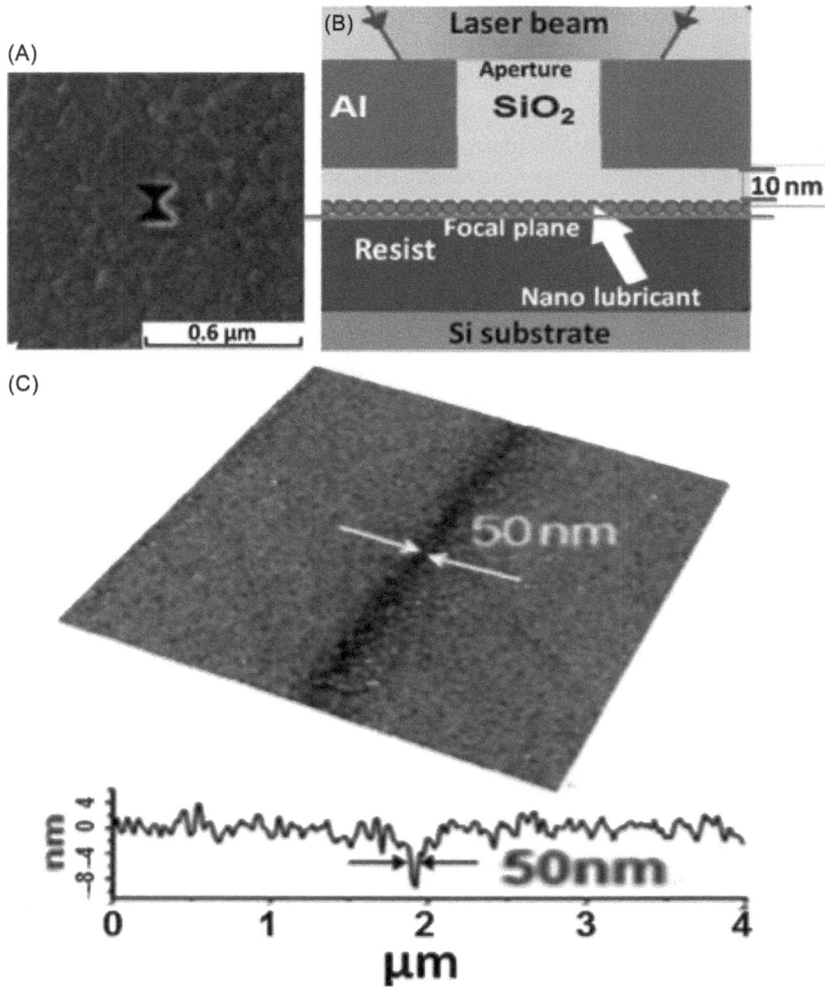

FIGURE 5.9 Plasmonic-assisted lithography. (A) SEM image of a bowtie aperture drilled in an aluminum film. (B) Schematic of the entire interaction region. (C) A line pattern 50 nm wide produced when the wafer is scanned linearly.

10 nm from the exit of the aperture. Fig. 5.9B shows a schematic of the entire interaction region. The aperture is depicted as a circular hole with the laser beam is focused on it. It is filled with SiO_2 (glass), and below it there is a glass spacer 10 nm thick. On the bottom surface of the glass spacer there is a lubricant consisting of a layer of nanoparticles, deposited using self-assembly procedures. A silicon wafer, coated with a photoresist, is placed against the glass spacer, such that the photoresist is in contact with lubricant layer. The system is designed such that the near field plasmonic focal plane of the aperture falls on the photoresist; that is, the photoresist is placed at the hot spot where the laser

light focuses. Now the silicon wafer is scanned under the focal spot according to a programed pattern, while keeping the laser beam and the aperture stationary. Fig. 5.9C shows a line pattern 50 nm wide produced when the wafer is scanned linearly. Using contact probe, 50 nm width line patterns at 10 mm/s can be recorded. This rate is 500 times faster than conventional near-field scanning optical microscope (NSOM) lithography described in section.

The quality of focusing, as well as the intensity at the focal spot in plasmonic lithography are enhanced by two processes. The SEM image given in Fig. 5.10A [13] shows how the bow-tie aperture is improved. Three thin ring apertures surrounding the bow-tie are added. Second, the transmitted intensity falling on the photoresist is enhanced by adding more apertures. This is accomplished by replacing the single aperture device with a 2-D regular array of hexagonally packed apertures as shown in Fig. 5.10B. This allows more light to go through, making the focusing spot smaller and sharper. Fig. 5.10C shows an image of the focal spot when a 355 nm UV light is incident on the array. The intensity has been normalized to the incident intensity. It is to be noted that adding the rings or using an array will not change the distance of focal plane from the aperture plane, that is, it is still located 10 nm away from the array surface. The half-circular shaped side lopes in the intensity are the direct transmissions through the three rings and their intensities are far below the exposure threshold of the resist. Fig. 5.11A–C shows an example of how the enhanced intensity

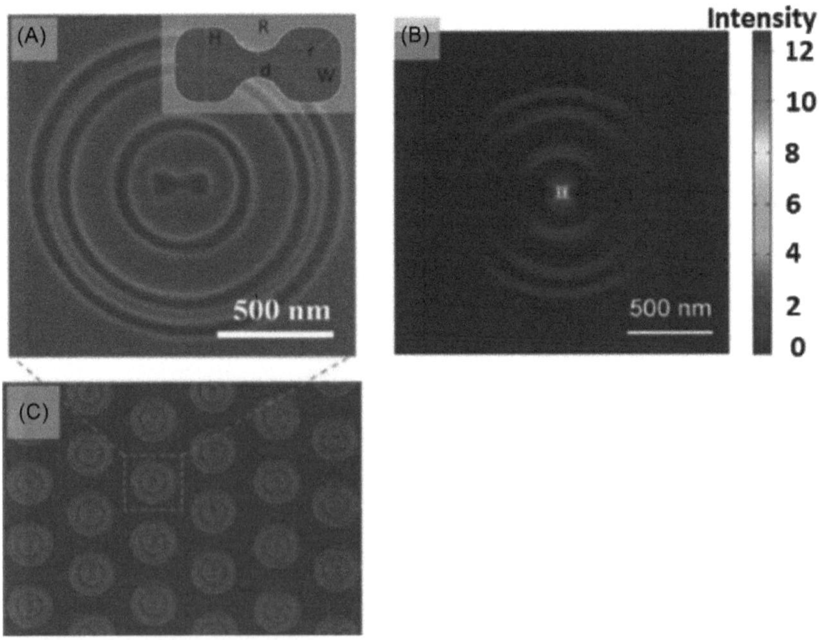

FIGURE 5.10 Improved bow-tie aperture. (A and B) SEM images of a 2-D regular array of hexagonally-packed apertures. This allows more light to go through (C) an image of the focal spot when a 355 nm UV light is incident on the array. The intensity has been normalized to the incident intensity.

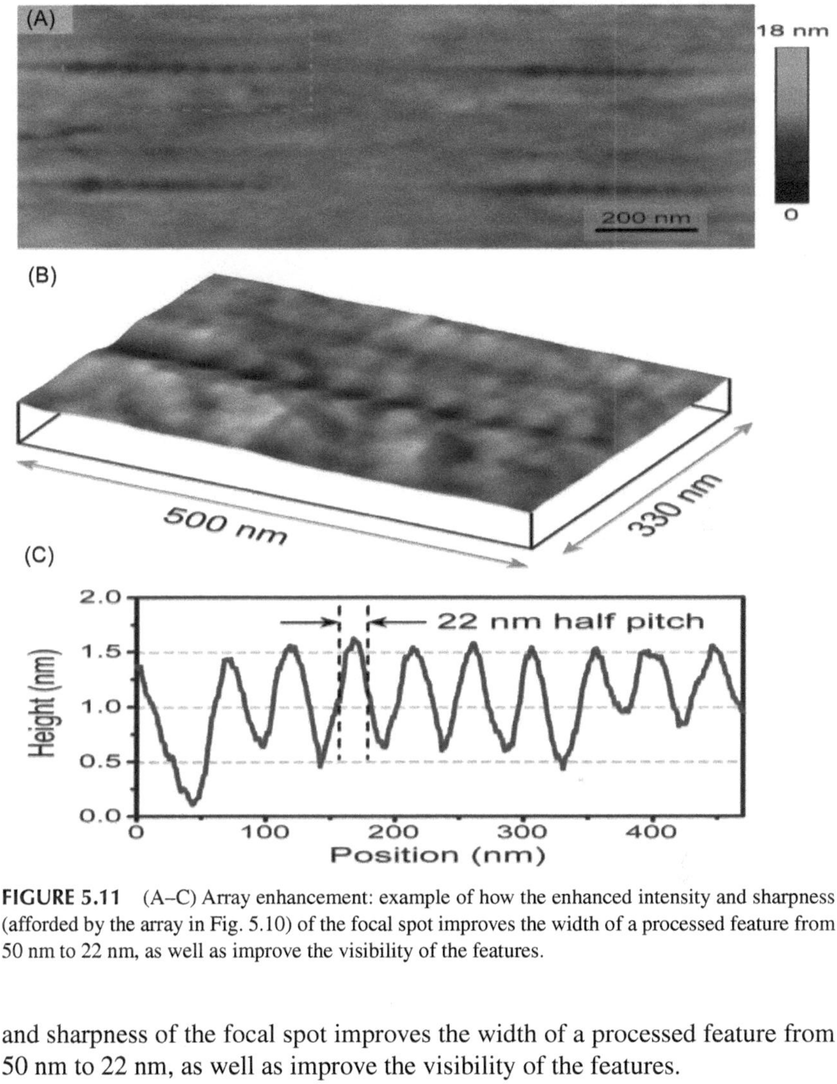

FIGURE 5.11 (A–C) Array enhancement: example of how the enhanced intensity and sharpness (afforded by the array in Fig. 5.10) of the focal spot improves the width of a processed feature from 50 nm to 22 nm, as well as improve the visibility of the features.

and sharpness of the focal spot improves the width of a processed feature from 50 nm to 22 nm, as well as improve the visibility of the features.

5.4.2.1 Plasmonic Hyperlens–Based Lithography

There are other configurations of plasmonic metal–based focusing (lensing) systems that can be used for nanolithography. One configuration consists of an alternating multilayer of a conductor and insulator. The conductor is silver and the insulator is aluminum oxide (Al_2O_3). Each layer is 10 nm thick, but with a varying radius of curvature. The outermost layer had a radius of curvature of 1400 nm, and the innermost layer had a radius of curvature of 600 nm. Fig. 5.12 [14] shows a simulation of the concept of a hyper lens using UV illumination wavelength of 365 nm. The color in the figure is a false color.

FIGURE 5.12 **Plasmonic hyperlens–based lithography: simulation of the concept of a hyper lens (plasmonic metal–based focusing (lensing)) using UV illumination wavelength of 365 nm.**

5.4.3 Laser Interference Lithography [15]

In laser interference lithography, a coherent UV laser beam goes through a spatial filter for cleanup and expansion as shown. A mirror and a tilted polymer resist-coated substrate are assembled perpendicular to each other. The expanded laser beam strikes a tilted mirror and the tilted polymer resist–coated substrate simultaneously as shown in Fig. 5.13. The reflection from the mirror also strikes the substrate. The incoming light and the reflected light interfere on the substrate creating a standing wave 2-D regular pattern of bright and dark spots. This constitutes a basic setup of the well-known Lloyd's mirror interferometer.

5.4.4 Nanosphere Shadow Lithography [16]

One interesting lithography procedure utilizes pre-made monosize spheres of a certain material to prepare masks. The procedure is described in Fig. 5.14A. For example, a monolayer of polystyrene nanospheres are assembled on a surface. The spheres form a dense packing configuration. The uniformity of the spheres and the dense packing provide a uniform pattern of voids through which desired material can be deposited. Fig. 5.14B–C shows results for depositing gold nano pattern on the underlying substrate using polystyrene spheres. In other words, this monolayer effectively constitutes a colloidal crystal mask. After deposition, the sphere mask is removed, keeping the ordered deposited patterning. It noted that due to the nature of close packing of spheres, the pattern produced is usually limited to triangular structures.

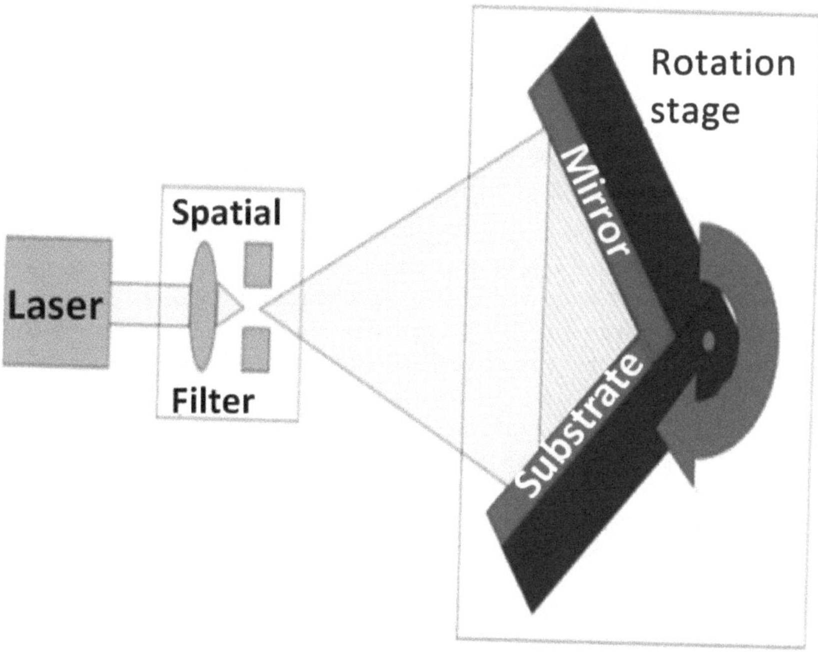

FIGURE 5.13 Schematic of laser interference lithography.

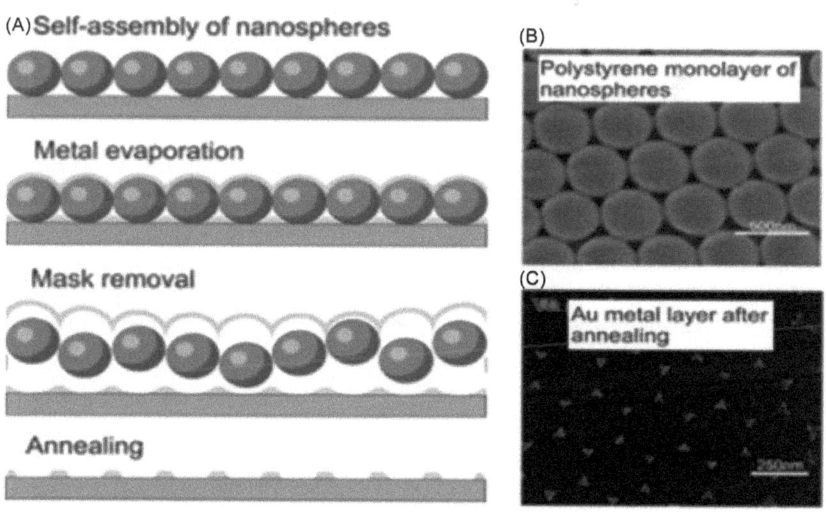

FIGURE 5.14 Nanosphere shadow lithography. (A) Procedure of the process. (B and C) Depositing gold nano pattern on the underlying substrate using polystyrene spheres. *(From Fig. 1: 'ZnO nanowire arrays – Pattern generation, growth and applications' by Margit Zacharias, Kittitat Subannajui, Andreas Menzel, Yang Yang, https://doi.org/10.1002/pssb.201046399)*

5.4.5 Resists [17]

We have seen earlier that a resist is a thin organic layer used to transfer a circuit pattern to the semiconductor substrate which it is deposited upon. It can be patterned via lithography to form a (sub) micrometer-scale, temporary mask that protects selected areas of the underlying substrate during subsequent processing steps. We summarize some of their properties for completeness:

- Resists are organic and inorganic polymer nanofilms
- They undergo chemical bond dissociation or formation upon interaction with secondary electrons or with photons. The chemical alteration changes the solubility, hence affords different paths for developing the exposed regions
- Positive resist becomes more soluble in the photoresist developer upon UV exposure
- Negative resists become extremely difficult to dissolve upon UV exposure
- Resolution afforded by positive photoresists is higher (developer does not permeate the exposed areas) than that of negative resists (developer permeate the exposed areas)
- Resist materials, solvents, and developer are the same for electron and ion beam lithography
- Examples of resists: The first EBL was PMMA (a positive resist with highest resolution). Other positive resists used are Toray EBR-9; ZEP series NEB-31 is a negative e-beam resist. Another negative E beam resists is Microposit SAL601.

5.5 SCANNING PROBE NANOFABRICATION (TIP-BASED FABRICATION) [18,19]

We have seen how several tip-based devices have been used to image or/and gather spectroscopic information about individual atom/molecules or corrugations or surfaces. Those include: atomic force microscopy (AFM) (see Fig. 4.11); scanning tunneling microscopy (STM) (see Fig. 4.6); magnetic force microscopy (MFM), etc. Once the capability of these devices to observe individual atoms or nanostructures has been established, some attention was shifted to the feasibility of using them as a means to manipulate and fabricate nanoscale structures, especially for directly inducing selective modifications on a surface with a precise localization of the defined pattern. Several processes have been demonstrated, including electro/magnetic/mechanical manipulation of individual atoms/molecules/nanoparticles, resist modification, chemical modification, material transfer, deposition, material removal, and etching. Many scientists around the globe use it to move single atoms and molecules to build new structures. Concurrently, a number of other devices have been additionally developed including Kelvin probe force microscopy (KFM); NSOM; scanning electrochemical microscopy (SECM); scanning ion conductance microscopy (SICM). Our discussion start out with the two most widely used, namely STM and AFM.

5.5.1 Tip-Based Devices

The needles or tips of the STM and the AFM made it possible to machine the surface of many materials. In addition, the tips make it possible to place assemblies of hybrid organic–inorganic nanoscale structures from gases or liquid on those surfaces. The tips effectively act in this function as nano electrochemical "pens."

5.5.2 STM Fabrication

To convert the STM from an imaging mode (quantum tunneling) to a lithography/fabrication mode, the STM is placed in a tunneling mode by biasing the tip negatively at a low voltage with respect to the substrate and increase the magnitude of the tip bias to moderate voltage values in the range of 12–35 V. This translates the tip further away from the substrate. Under these conditions electrons are emitted from the tip and accelerated towards the substrate without quantum tunneling. This mode is called a field emission mode where electrons are emitted from the tip freely and act as an electron source for lithography purposes. The tip is now moved latterly according to a programed pattern to modify a resist coated substrate or even a bare substrate.

5.5.2.1 Comparison of STM With e-Beam Lithography

The extraordinary spatial and magnitude sensitivities of the tunneling current represent the basic advantages of using STM for nanofabrication or nanolithography (STML). In fact, the STM current can be considered as a miniature e beam to achieve most direct writing functions, normally performed by EBL, even at diminishing scales. Direct writing on resists or substrates, with or without chemical reactions, results in material modification, deposition, removal, and manipulation, which consist of the major functions of STM for nanofabrication. The following are some highlights when comparing the STM nanofabrication with e-beam fabrication:

- Just as EBL, STM uses electrons instead of light and as such it is a direct-write, mask-less approach that can reach resolutions below 10 nm.
- It is a miniaturized electron source – substrate configuration with no need for focusing system.
- The typical current, voltage, and scanning rate of the e-beam induced by STM are ~1 nA, 10 V, and 1 mm/s, respectively, compared to 1 nA and 50 kV for EBL.
- The *current density* of STM is about 3 orders of magnitude higher because the spot size in STM is much more compact than an e-beam spot size. This makes it easier for STM to expose the high resolution resists, because high resolution resists normally have less sensitivity.
- The applied voltages of STM are typically 4 orders of magnitude lower than that of EBL. The low voltage of STM produces less back scattering and

minimizes the proximity effect, which normally dictates the resolution of EBL

- Low voltage operation in STM may limit STM to be used to expose very thin resist films.
- STML puts the tip so close to the target surface hence *electrons* do not have a chance to diverge, while in direct writing EBL, various stages of electromagnetic lenses are used to deflect and accelerate electrons, which forms a relatively large e-beam spot.
- STML can be operated in ambient air, while EBL is always conducted under high vacuum conditions. STML in an ambient condition can be used for modifications of metallic and ceramic materials.
- STML avoids or eliminates the problems associated with backscattering and secondary electrons.
- STML can reverse polarity (positively biased tip); hence electrons are pulled from the substrate, which eliminates electron induced desorption or ionization of atoms and molecules if the substrate.

5.5.2.2 Fabrication Modes

There are several modes of operation for a sharp conducting needle in proximity to a conducting plane. These cases, outlined in Figs. 5.15–5.18, involve the electric field between the tip and sample, the inelastic tunneling electrons, or mechanical contact between tip and atom/molecule.

(1) The tip is placed over a surface with no biasing.

(2) The tip is placed at a large distance from the substrate (Fig. 5.15). A potential difference is set between the substrate and the needle such that there is no electron flow. In this case there is only an electric field between the two electrodes, with a direction either from the tip to the substrate or vice versa depending on the sign of the biasing voltage difference. In this configuration

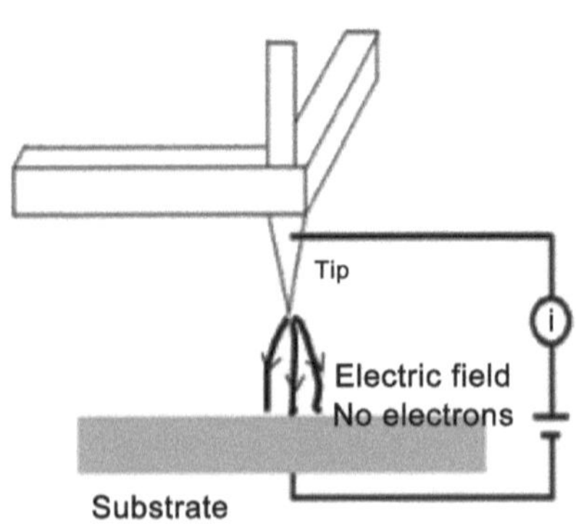

FIGURE 5.15 **Biasing mode with an electric field (no current) of operation for an STM.**

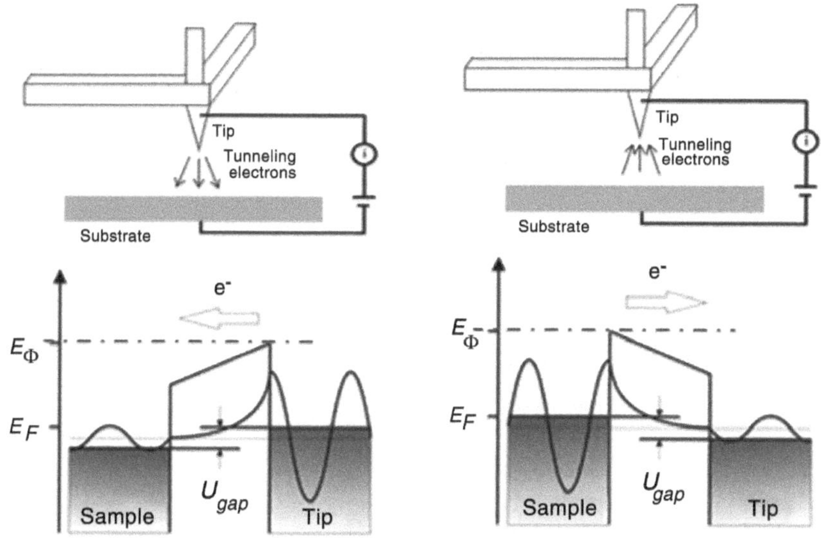

FIGURE 5.16 Tunneling mode of operation for an STM. (A) schematic of electrons from tip to surface along with model of quantum operation. (B) Schematic of electrons extracted front the surface into the tip along with electron wavefunctions and model of quantum tunneling.

the biasing voltage difference needs to be several volts while the spacing on the order of several nanometers.

(3) Another biasing configuration involves tunneling electrons from the tip to the substrate or from the substrate to the tip depending on the sign of the biasing voltage difference (Fig. 5.16). In this case, current lines replace the electric field lines. In this configuration the biasing voltage difference needs to be small <1 V while the spacing on the order of 1 nm. Fig. 5.16A (bottom) shows the energy scales and the electron quantum wave function. When the tip is biased negative relative to the substrate, the electron wavefunction is oscillatory and large, tunnels through the vacuum gap (exponential) and into the substrate as oscillatory but with a small amplitude. The opposite flow takes place when the bias is reverses as shown in Fig. 5.16B. The conduction band states of the substrate and the vacuum level at E_ϕ dip down upon going from the tip to the substrate, making electrons tunnel from the tip to the substrate.

(4) The fourth configuration involves free electron flow (Fig. 5.17) with a direction either from the tip to the substrate or vice versa depending on the sign of the biasing voltage difference. In this configuration the biasing voltage difference needs to be large above the field emission voltage (>10 V) while the tip-substrate is large >10–25 nm

(5) The fifth configuration is when the tip-substrate spacing collapses to zero, namely the tip touches the substrate (Fig. 5.18).

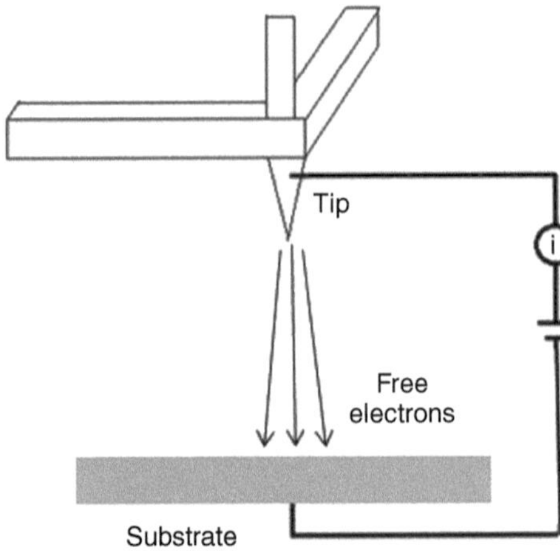

FIGURE 5.17 **Free-electron flow mode of operation for an STM.**

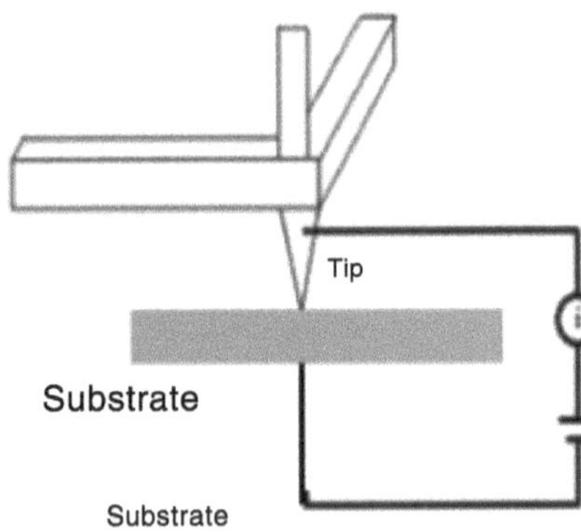

FIGURE 5.18 **The contact modes of operation for an STM (the tip-substrate spacing collapses to zero, namely the tip touches the substrate).** *(From P. Zeppenfeld, C.P. Lutz, D.M. Eigler, Manipulating atoms and molecules with a scanning tunneling microscope, 42–44, Part 1, July 1992, 128-133)*

5.5.3 Fabrication Mechanisms for STM

5.5.3.1 Dragging Atoms

The first demonstration of the concept of dragging atoms was carried out in 1990 by Eigler and Schweizer of IBM. A nickel substrate in an STM was cooled with liquid helium to a temperature less than 4 °K above absolute zero. Xenon atoms were sprayed on the nickel substrate, with 35 of them freezing on it. Low temperatures are needed because xenon atoms cannot be held in place due to their natural thermal motion. When the needle is brought close to a particular atom, the needle attracts the atom by Van der Waals attractive forces. The needle is then moved, which drags the atom along behind it. The atom is released by raising the tip. This process was used to arrange the 35 atoms from a random configuration into the three IBM

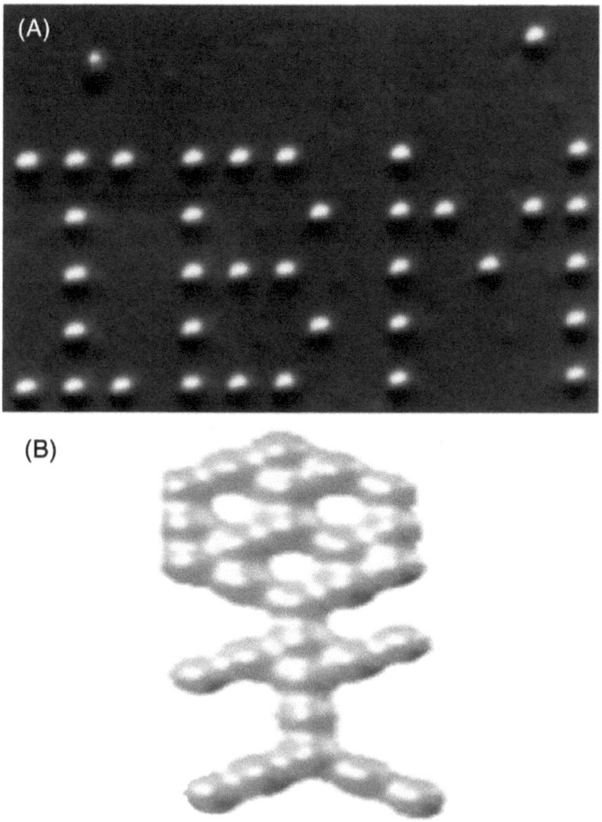

FIGURE 5.19 Dragging atoms with an STM. (A) Arrangement of 35 xenon atoms from a random configuration into the three IBM letters on a cold nickel surface. (B) Assembling carbon monoxide CO molecules in the shape of a molecular man.

letters as shown in Fig. 5.19A [9,20]. Besides manipulating these atoms to spell out I.B.M., Fig 5.19B shows that they assembled carbon monoxide CO molecules in the shape of a molecular man. The molecular man is built from 28 CO molecules. The structure measures 45 Å from head to foot. The sliding of the CO molecules on the Pt substrate was achieved by lowering the tip of an STM, thereby increasing the interaction strength between the tip and the adsorbed molecules [9]. The process requires cold temperatures and a vacuum environment so that gas molecules do not disturb it. This makes the whole apparatus very bulky and practical for room temperature device applications.

Extraction of strongly bound atoms on a surface was also accomplished using the tip of a scanning electron microscope. Fig. 5.20 gives an example [20]. In Fig. 5.20B a groove was etched on a silicon surface at the University of Illinois. The groove is 5 millionth of a millimeter wide [20].

Example 5.3 How Does an STM Tip Pick And Release Individual Atoms? [21]

Let us consider an STM tip with nickel substrate cooled to liquid He as was discussed in Chapter 4. A small number of Xe atoms are frozen on the substrate. The Xe atoms are weakly bound to the surface. The STM tip is placed over one of the xenon atoms at a large imaging distance say ~6 Å as shown in Fig. 5.20A. An electric field is now applied by biasing the tip (Fig. 5.15). Then the atom gets pulled up by the tip and gets attaches to it. The tip may now be moved laterally under these condition, carrying the atom around at will. When we want to release the atom back to the surface at any location, we simply reverse the direction of the electric field by reversing the biasing voltage.

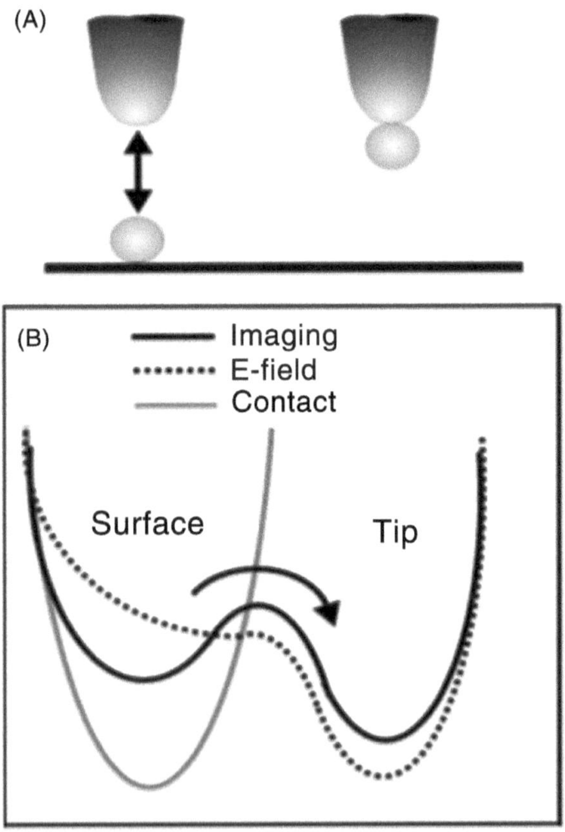

FIGURE 5.20 Mechanism of STM tip picking and releasing atoms/molecules. Vertical manipulation of atoms/molecules using an STM tip. (A) Schematic drawing shows atom-tip combination in the process. (B) The double-potential well model of the process. (*Black-solid*) the shape of potentials at an image-height, (*dash*) the shape of potentials under an electric field and (*gray*) the shape of potentials at the tip-molecule contact.

In another process, the tip is moved down till it crashes into the atom sitting on the surface. The height is essentially reduced to zero, allowing the tip to make contact with the atom (Fig. 5.18). Once this happens the atom/molecule easily sticks to the tip and can be easily carried around on the tip.

These processes may be understood using the potential energy of the tip-atom-surface triplet system. Fig. 5.20B gives a sketch of the potential energy of the system. It shows that it has two minima. In another words there are two potential wells that separated by a barrier (heavy solid curve). The potential well on the left represents the atom being stable on the surface. The well on the right represents the atom being sitting stable on the tip. Both of these are stable positions because of a high barrier between them, which is much higher than the thermal agitation energy. Now the application of the external field can lift the well at the surface to an energy above the barrier (dotted curve) and the atom can then easily slide into the well at the tip (meaning the atom moved to the tip). If one reverses the electric field, one can lift the right well above the barrier allowing the atom to slide back. In the case in which the tip is crashed into the atom, the two potential wells overlap and appear as one well (light solid curve). Thus the atom/molecule is then easily transferred to the tip.

5.5.3.2 Induced Deposition of Metal – Nano CVD [22,23]

A technique that works at room temperature and is able to select the atom type was developed by Nayfeh and collaborators at the University of Illinois. The STM chamber is filled with trimethyl aluminum gas, which puts molecules near the atomically sharp metal tip (tungsten or gold tip) of the STM, as shown in Fig. 5.21A. Below the tip is a conducting substrate at a distance of 1 nm. When the tip is biased by 1–3 V, an electric field is setup in the gap which can reach more than 100 MV/cm. An intense pulsed laser light bathes the metal tip and the gap and the substrate. The electric field in the vicinity of the gap is very strong due to the dc field of the tip as well as to plasmonic field of the metal tip. The combined effect of the light and electric field is confined to a nano space and atoms or molecules falling in this space near the tip will be subjected to very intense electric field. The processes involved may be outlined as follows.

- The laser field extracts aluminum atoms by dissociation of the trimethyl aluminum molecules in the gap between the tip and the silicon substrate.
- The intense pulsed laser light is tuned to resonate with a certain state in an atom or molecule, selectively promoting them to high-lying excited states while the combined electric field can assist in stripping the excited electron to produce a single ionization event of a free electron and a free ion. As light can resonate with specific atoms or molecules then the process is highly selective and can pick a certain atom from a mixture of atoms or molecules

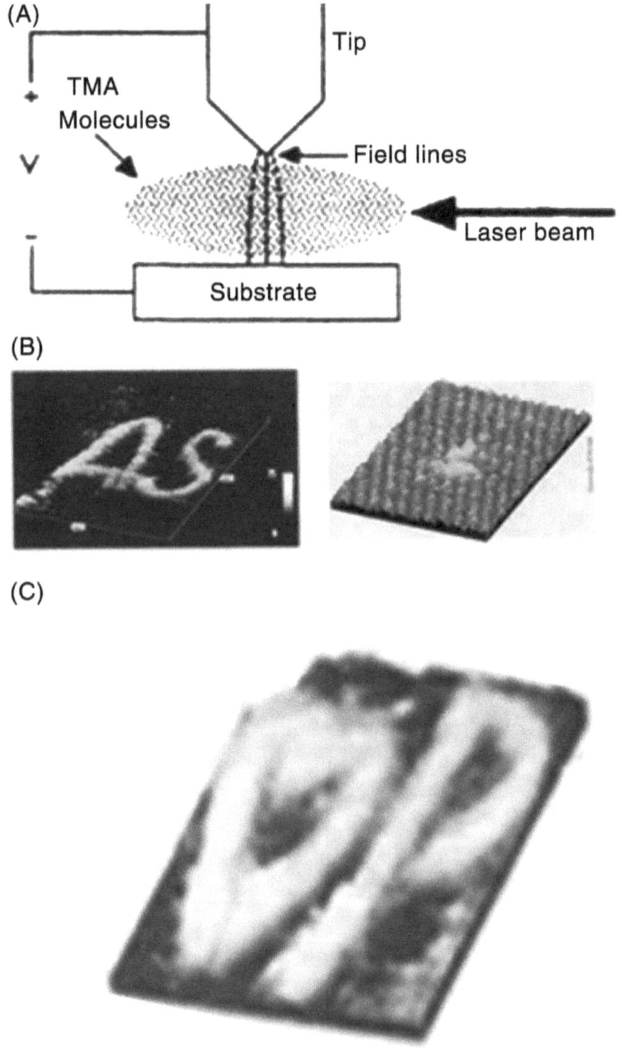

FIGURE 5.21 **Induced deposition of metal nano CVD.** (A) Layout of room temperature technique that selects the atom type using laser-induced resonance ionization of trimethyl aluminum gas, with the produced ion being pinned down on the surface by the tip. (B and C) Examples of patterns using the procedure. A single molecule of TMA has been picked and pinned to a graphite surface; the initials of Arthur Schawlow "*AS*"; dedication of heart to P.

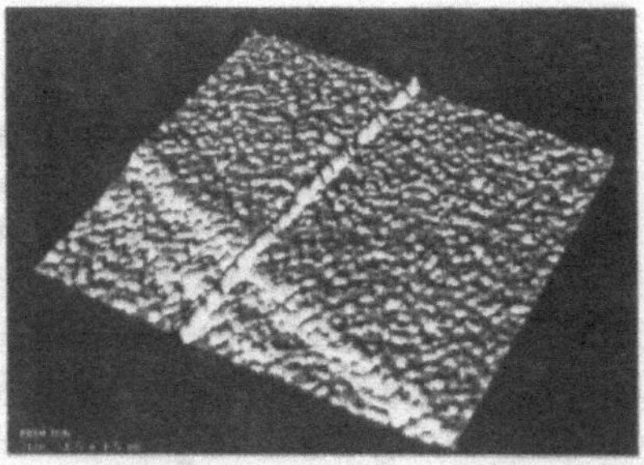

FIGURE 5.22 Scanning the tip over a surface with a thin coating of nickel carbonyl can make lines of 50 nm across and 95% metallic nickel.

- As the electric field is focused or confined to nano scale then the atomic ion is accelerated, pushed and guided to the surface with nano scale resolution.
- Programming the position of the tip allows making nano patterns on the surface with atomic resolution.

Fig. 5.21 demonstrates the process. A single molecule of TMA has been picked and pinned to a graphite surface as shown in Fig. 5.21A (*right bottom*). In 1992 on 70th birthday of Arthur Schawlow, the inventor of the laser, the author presented him with the nanoscale of his initials *"AS"* shown in Fig. 5.21A (*left bottom*). The initials were written with trimethyl aluminum molecules on a silicon wafer. In Fig. 5.21B, Nayfeh dedicated his heart to P. Thus one sees that the STM tip has been effectively used as a pen to write patterns on surfaces, while the laser-liberated metal ion constituted the ink.

Another approach, developed by Alex de Lozanne from the University of Texas at Austin [24], involves forming a thin coating of a material on a surface containing metal atoms (nickel carbonyl for example). A sufficiently high voltage is applied to the STM tip, which emitted a beam of electrons toward the surface. The flow of electrons broke up the gas on the surface leaving a deposit of nickel. Scanning the tip can make lines of 50 nm across and 95 per cent metallic nickel as shown in Fig. 5.22.

Analogous procedures were used to deposit Cd structures down to 20 nm in size from dimethyl cadmium gases; Si- and Ge-based nanostructures with lateral sizes of about 10 nm were successfully deposited on Si 111 substrates starting from SiH_4, SiH_2Cl_2, and GeH_4 precursors. Magnetic parallel Fe-based nanowires with a lateral size of about 10 nm are deposited on an Ag substrate from ferrocene precursor gas. Aluminum and iron nanostructures from highly reactive dimethylaluminum hydride and ferrocene gaseous precursors have been deposited by Mitsui et al. [25] and Adams [26] et al., respectively. Other metals, including Cd, Ni, and Pd, have also been successfully deposited by various STML techniques.

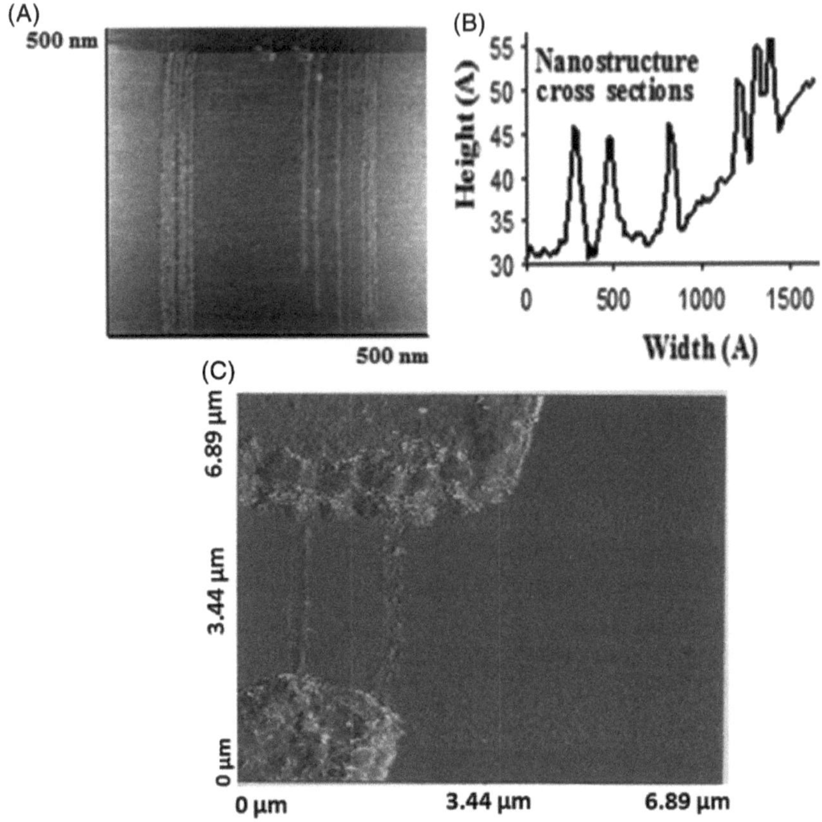

FIGURE 5.23 Deposition of aluminum atoms from trimethyl aluminum is used to deposit aluminum quantum nanowires. (A) An AFM image of several aluminum quantum wires fabricated on silicon. (B) A line profile across the multiple wires. (C) AFM image of quantum wires fabricated between the pads of a MOSFET.

The procedure used above to deposit aluminum atoms is used to deposit aluminum quantum nanowires [27]. Fig. 5.23A is an AFM image of several aluminum quantum wires fabricated on silicon. Decomposition of an organometallic (e.g., TMA) gas by applying a voltage across the tip-sample gap is used to deposit aluminum. The organometallic gas molecules become dissociated and their fragments become deposited on the substrate. Fig. 5.23B gives a line profile across the multiple wires. Fig. 5.23C is an AFM image of quantum wires fabricated between the pads of a MOSFET. Periodicity of 10 nm has been achieved. As to shape, that is, size continuity and sharpness, the images show the nanofabricates to be continuous. There is evidence of metallic content in the fabricated wires. The conductivity may be improved on by doing multiple passes with the STM tip. A large-range scan STM allows one

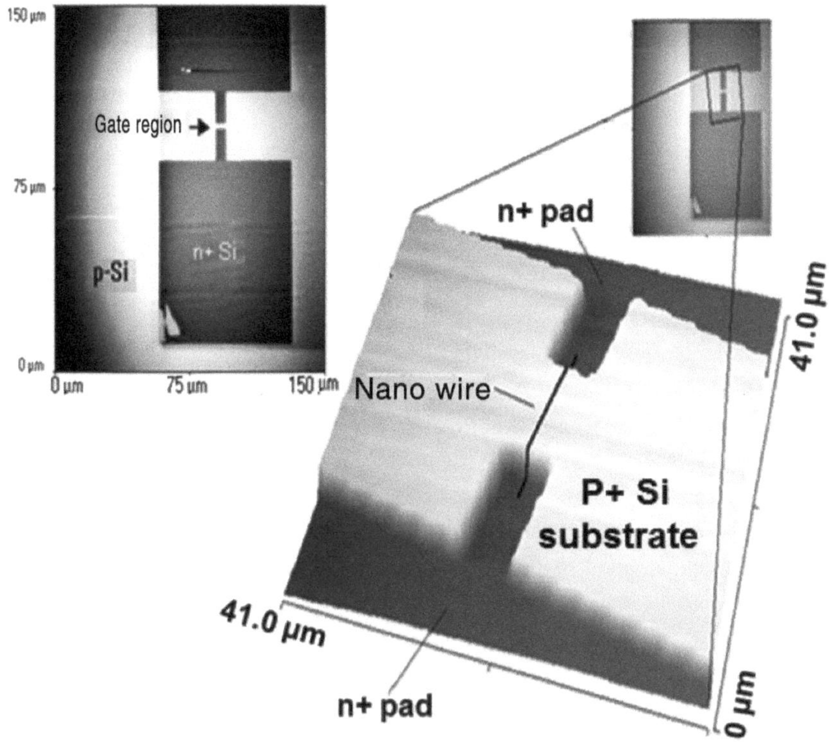

FIGURE 5.24 A sketch of Proto-MOSFET structure, with two closely spaced *n*+ pads on a *p*-Si substrate studied by researchers at the University of Illinois at Urbana Champaign. The Proto-FET structure is used as a nano laboratory to probe the nano-wire.

to integrate the nanostructures into micrometer-scale devices and ultimately the outside world.

The capability of embedding the wires in macroscopic devices allows for interfacing nano devices with macroscopic biasing and characterization systems, as well as exploits quantum effects in advanced devices. Fig. 5.24 gives a sketch of Proto-MOSFET structure, with two closely spaced n^+ pads on a p-Si substrate studied by researchers at the University of Illinois at Urbana Champaign. The Proto-FET structure is used as a nano laboratory to probe the nano-wire.

5.5.3.3 Cluster Emission From Soft Tips

Another procedure uses the tip of an STM as a source of material. In this application, the material of STM tip is chosen to be a soft metal, such as gold, silver, gold-palladium, or platinum. Also it is adequate to use a trip that is coated with a soft metal. A nano-second long voltage pulse is applied to the tip. This causes

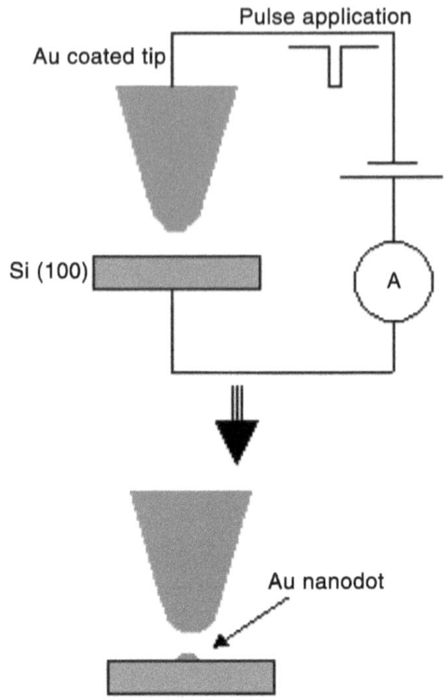

FIGURE 5.25 Schematic of cluster emission from soft tips. A nano-second long voltage pulse is applied to the tip. This causes a small blob of metal to break off and deposit on the substrate.

a small blob of metal to break off and deposit on the substrate as schematically shown in Fig. 5.25 [28]. Ag-coated W tip was also used to deposit sequential Ag dots. The deposition works also even if the tip in on the move, that is, as it scans over the surface. This versatility allows fast formation of patterns. It is to be noted that the deposition can be reversed where a voltage pulse in the opposite direction can cause the metal atoms to fly back onto the tip; hence erasing the deposition action. This is analogous to the case of moving atoms between the tip and the substrate (see Example 5.3). In fact, pushing the tip of an STM very close or into the surface can result in a hole, but the tip can be damaged in the process.

In addition to gold and silver, a variety of Pt patterns, including dot arrays and lines on *n*-type Si substrates were made using Pt tips as shown in Fig. 5.26A–B [20,28], respectively. Dots were deposited by using sequential pulses with the tip maintained at a fixed position on the target while the lines were constructed by sequential dots, in which each dot is connected with the neighboring dots by being slightly overlapped. Fig. 5.26A shows the resulting 639 Pt dot array, where the average values of dot height and width are 4 and 22 nm respectively. Because the associated standard deviations of the dot sizes are relatively high, at about 10% of the mean, this may imply that the deposition process is not

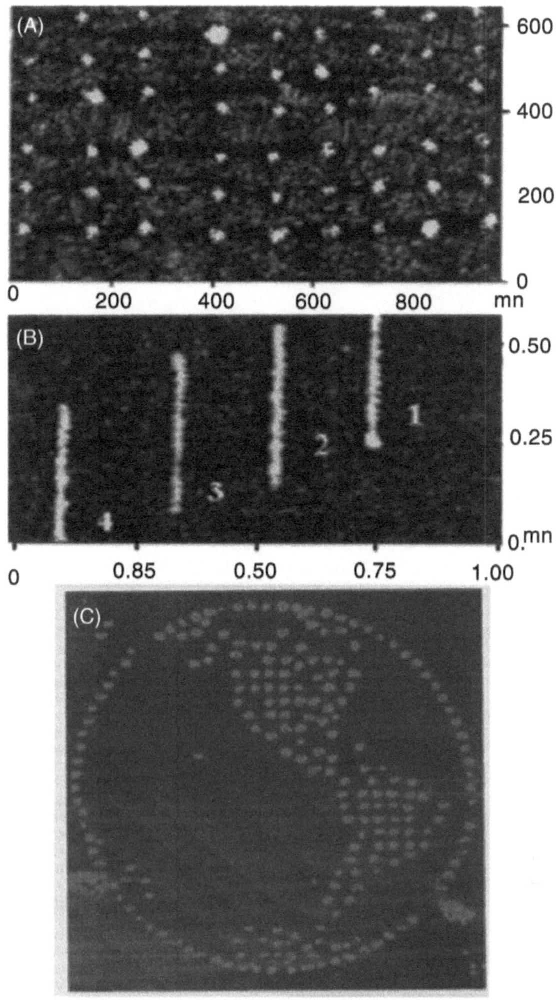

FIGURE 5.26 A variety of Pt patterns *n*-type Si substrates were made using Pt tips. (A) Dot arrays and (B) lines arrays. (C) IBM group used the phenomenon to draw a cartoon of the map of the world using gold clusters.

consistent enough for making a reliable nanoscale device. The IBM group used the phenomenon to draw a cartoon of the map of the world using gold clusters (see Fig. 5.26C).

The mechanism of ejection and suction of material clusters is interesting. First, both positive and negative polarity of the tip relative to the surface work; however the tip negative polarity is usually more stable with higher transfer probability and can yield smaller feature sizes. Second, there is a few-volt threshold for the process to proceed, which depends on the material of the tip and the target, the distance of the tip/target gap, the tip/target configuration, and the polarity. The physical mechanism of tip–target material transfer is not fully understood, but it has been conjectured that the process works because:

- Ionization induced ejection: Voltage pulses cause metal atoms of the tip to become ionized and the electric field between the tip and surface makes them fly off onto the substrate.
- Field-enhanced diffusion (of silver atoms) to the tip apex rather than by enhancement of tip evaporation caused by the high-electric field between the tip/substrate gap.

5.5.3.4 Termination/Passivation Removal Electron-Induced Desorption of Hydrogen and Oxygen Passivation [29]

Two technologically important surface condition in the microelectronic industry is oxygen (SiO_2) and hydrogen (SiH). Thin films of silicon oxide are used in transistor amplifier, as well as in memory cells. Si surface with an H termination or passivation is the starting point for a number of processes that transform the chemical and physical nature of the semi-conductor interface. In this case, silicon atoms covalently bonded to hydrogen, which reduces organic pickup from air and can last for hours. It is relatively inert and can be handled in air without any special care for several minutes.

- UV or ozone (O3) in air degrades the H-termination.
- High-energy electrons in *e*-beam with energy in the range 30–50 keV can desorb oxides and hydrogen when focused on target.
- Low-energy electrons from an STM tip can also be used to desorb oxides and as such can be used define an oxide pattern for subsequent CVD.
- STM electrons can also be used to remove or de-passivate hydrogen atoms from a surface that has been hydrogen terminated H passivated.

STM is parked at a large distance from a silicon substrate of ~100 nm under UHV conditions as shown in Fig. 5.27A. [30] The silicon wafer has an oxide layer coating. The tip is biased at a high voltage such that the tip operates in the field emission mode, i.e., electrons are emitted. Now when the wafer is heated to a temperature in the range 300–700 °C the incident electrons begin to trigger decomposition of SiO_2, hence remove oxygen from the silicon surface as demonstrated by Li. Fig. 5.27B shows example of the process: concentric ring pattern with aluminum linewidth of 25 nm fabricated by STM tip scanning using computer controller courtesy of Hiroshi Iwasaki of Osaka University, Japan. In another development, Lyding et al. at the University of Illinois used the STM to remove the H passivation under vacuum on a silicon wafer according to a preprogrammed pattern. This was followed by exposure to an oxygen environment, which produced an oxygen termination in places where hydrogen was removed. The hydrogen layer has effectively act as a resist system for nanolithography. This also confirmed the fact that H-terminated surfaces is of great interest

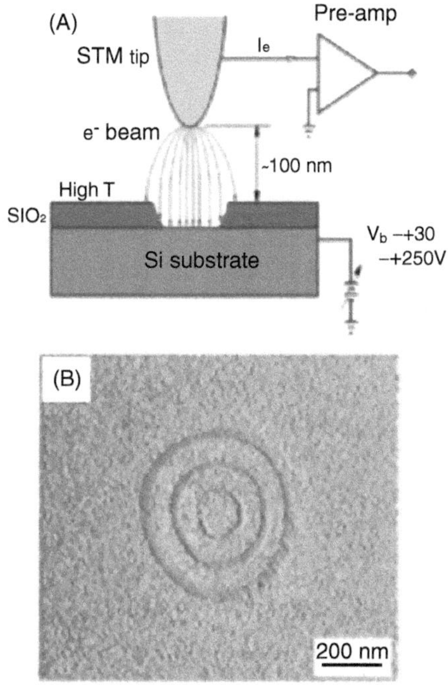

FIGURE 5.27 Fabrication by electron-induced desorption of H/O passivation. (A) Schematic of setup for removal by electrons of oxygen layer overlaid on a heated silicon wafer. (B) Example of the process: concentric ring pattern with a minimum linewidth of 25 nm fabricated by STM tip scanning using computer controller courtesy of Hiroshi Iwasaki of Osaka University, Japan.

because of its stability. As all surface Si atoms are fully coordinated, hydrogen termination leads to enhanced stability in ambient environments unlike unpassivated surface atoms, or dangling bonds.

5.5.4 Modification of Resist

We give here a number of examples for using an STML in traditional e-based lithography. For example, a 50-nm-thick SAL-601 resist was spin coated on a silicon wafer. This resist is sensitive to electrons. The wafer is uploaded in the STM system and the tip of the STM is negatively biased while the silicon wafer substrate is grounded. A tip-sample voltage of −25 V is used to emit electrons. As electrons from the tip strike the SAL-601 they induce a change in the molecular structure of the resist. The tip is moved on the surface according to a preprogrammed pattern. The substrate is then removed and processed is the same way as in the EBL case. The pattern is transferred and etched 100 nm into GaAs substrate by reactive ion etching. Fig. 5.28A shows an example pattern [31].

It is to be noted that the proximity effects, which are usually observed in EBL are found to be absent. This is consistent with the fact that the electrons emitted by the tip are of low energy. But the voltage needed to expose the resist

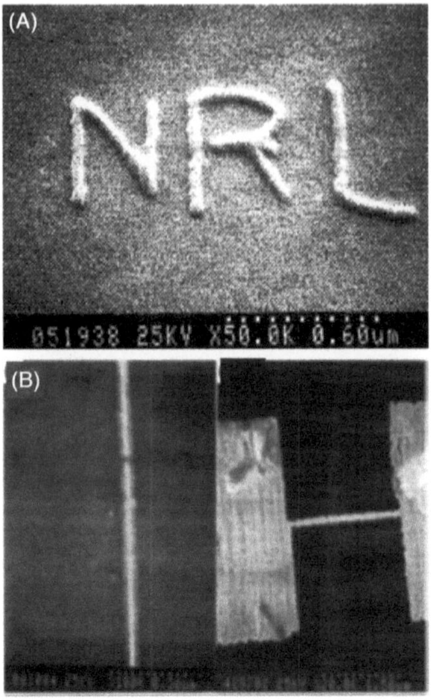

FIGURE 5.28 Modification of resist using an STML in traditional e-based lithography. (A) "NRL" pattern written. (B) Metal line on the silicon substrate.

is found to be larger in magnitude than 12 V (<−12 V), while the minimum tip-sample current needed is 3 pA. The minimum line dose sufficient to expose the resist is (30 nC/em).

In another example, PMMA film was used as a resist by McCord and Pease [32]. This material is a positive resist. A 20-nm thick line was drawn by rastering the STM tip over the resist film. The STM delivered a 100 pA beam current with the tip negatively biased at −20 V at a writing speed of 1 mm/s. An Au–Pd film was then deposited on the resist film. After depositing the metal film, the sample is placed in a beaker filled with acetone and allowed to sit in the acetone for approximately a day. Acetone dissolves the photo-resist. After this soaking, the sample is physically agitated by an ultrasonic treatment or by simply stirring with a glass stirring rod to ensure the complete removal of the resist. This liftoff procedure leaves behind the metal line on the silicon substrate as shown in Fig. 5.28B. The same liftoff procedures is used for Au, Pd, Cr, Cu, and Al. It is to be noted that reversing the polarity at the STM tip to a positive value also works. In this case, electrons are field emitted from the sample up through the resist. One advantage of this configuration is that electrons move across the resist without scattering in elastically and can quickly gain enough energy to cause exposure.

Example 12 Current and Field Distribution

A sharp conducting tip is brought to a distance *d* from a a ground conducting plane. Estimate the range of the field lines R. To estimate we assume that the apex of the tip is as sharp as a single atom of a point charge.

One can represent the problm as a pont charge at a distance d from a grounded substrate. The method of images may be used to silve for the electric field.

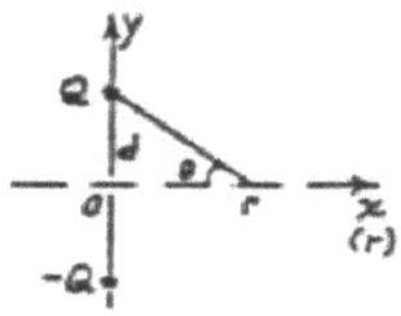

$$\bar{E}\Big|_{y=0} = -\bar{a}_y \frac{Q}{4\pi\epsilon R^2}(2\sin\theta) = -\bar{a}_y \frac{Qd}{2\pi\epsilon (d^2+r^2)^{3/2}}$$

$$a)\ \rho_s = \bar{a}_y \cdot \epsilon \bar{E}\Big|_{y=0} = -\frac{Qd}{2\pi (d^2+r^2)^{3/2}}.$$

$$b)\ \int_0^\infty \rho_s\, 2\pi r\, dr = -Q.$$

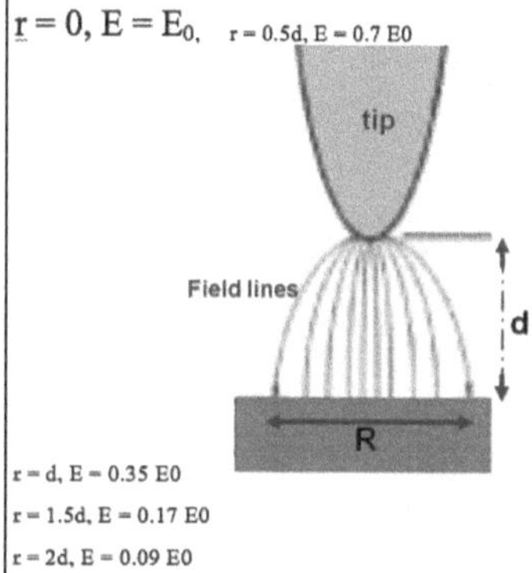

A silicon-based resist was also used. In this example both STM, AFM, and carbon nanotube were utilized to produce a final product. Thin polysilane (polymethylphenylsilane (PMPS)) films with thicknesses of 6–8 nm were prepared by spin coating on n-type Si wafers. Then it was subjected to a heat treatment at 150°C to obtain a smooth surface. Masks were prepared by an AFM tip with a carbon nanotube tip attached to it (see next section below for AFM nanofabrication). The contact mode and tapping mode lithographic processes were used.

The processed Si substrates were chemically etched using a solution of $NH_4/H_2O_2/H_2O$. This is followed by removal of the PMPS mask using toluene. Now the etched samples are observed by AFM operated in a tapping-mode. A fresh nanotube tip was used in this imaging step to obtain a higher resolution as the tube used in the fabrication may have picked some debris. Fig. 5.29 shows images of (A) grooves formed using tapping-mode and (B) grooves on a Si substrate chemically etched using the polysilane mask with grooves [33]. Line profiles of the grooves in the PMPS film and on the Si substrate are shown in (C).

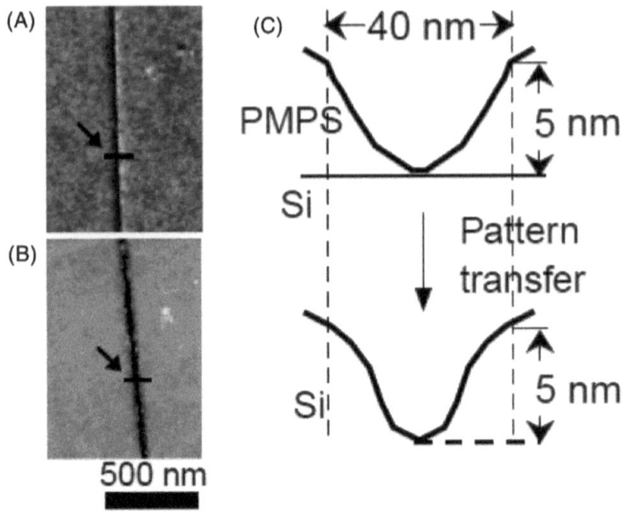

FIGURE 5.29 **AFM fabrication.** (A) Grooves formed using tapping-mode. (B) Grooves on a Si substrate chemically etched using the polysilane mask with grooves. (C) Line profiles of the grooves in the PMPS film and on the Si substrate.

5.5.5 Nanofabrication With Atomic Force Tip (Conducting and Non-Conducting) [34]

In addition to STM, AFM is also widely used in nano-science. AFM is used in imaging but it provides moderate resolution. For instance, it can image nanoparticles but is almost impossible to image individual atoms, and the quality of AFM image is largely depended on the shape and contact force of the tip. In some cases, the measured signal would be rather complicated to interpret into morphology or other properties of sample. Early stages of development involved the use of non-conducting tips, but later developments adopted the use of conducting tips along with electric circuitry to provide biasing or current to the tip. This widened the application of AFM.

In addition to imaging, an AFM tip can be used to fabricate structures on surfaces at the nanoscale (nano fabrication) through several functionalities that include: material removal, such as mechanical scratch; material modification, such as anodic oxidation; nano-manipulation of nanoparticles, and nanocluster; material deposition and growth, such as electric field-induced mass-transfer, and electrochemical deposition. These processes or operations are not unlike what is normally used in daily life, such as push, pull, move around, bend, deform, cut, indent, scratch, plow, and induce chemistry, except all of these are done in the nanoscale. Moreover, the procedures are amenable to computer control and automation such that the tip can serve as a computer-controlled robotic arm with capability to precisely position and assemble nano components into pre-planned systems.

5.5.5.1 Conductive Atomic Force Microscopy (C-AFM)

In conductive atomic force microscopy (C-AFM), the tip of the AFM is coated with a conducting material. Thus, when the tip-surface is placed at voltage difference, current flows through the metal-coated tip of the microscope and the conducting sample. This variation, therefore allows the device to use electrical current to construct (STM-mode) the surface profile of the sample. In addition, the AFM can still be operated in its standard mode by vibrating the tip simultaneously with the conducting mode to acquire a topography image. Such dual functionality enables correlation of the spatial features with its conductivity. It is to be noted that the tips used are conventional silicon tips coated with a metal or metallic alloy, such as Pt–Ir alloy (see Section 4.5). Moreover, the device is operated not only in the imaging mode, but also in spectroscopic mode (I-V spectroscopy).

There is a variety of AFM tips for use in the conductive mode: (i) metal coated probes; (ii) probes with conductive diamond coating; and (iii) highly conductive platinum silicide which unites high conductivity (higher than conductive diamond coating and as good as metal-coated tips) with a high wear resistance (much higher than metal-coated probes and almost as good as diamond coated probes). Additionally PtSi probes are now available with a slightly decreased tip radius compared to standard metal coated AFM probes. They can be used for several types of electric or electrostatic AFM measurement. Fig. 5.30 shows a variety of platinum PtSi silicide conductive tips.

5.5.5.2 Moving Nanoparticles Around [35]

Fig. 5.31A gives four cartoons showing the steps of moving nanoparticles with an AFM tip. NanoMove is a commercial user interface software, which allows control and operation for both imaging and manipulation. Fig. 5.31B shows a 30-nm gold particle before (top) and after (bottom)) being pushed over a 10 nm high step along the direction indicated by the arrow. Image sizes are both 1 μm × 0.5 μm. Fig. 5.31C shows an example of organizing a random number of nanoparticles into the letters USC.

[36] shows how one can use the tip to slide, as well as to lift/drop atoms or nanoparticles. Shows that the tip of an AFM in the tapping mode is placed oscillating relatively far from the surface over the particle. It is then lowered closer and closer to the particle. The amplitude of oscillation continues to decrease. When it runs into the particle or becomes sufficiently close, the vibration amplitude goes to zero. It is at this point the particle slides, accompanied with a cantilever deflection. It is sufficient to use the change in vibration amplitude and cantilever DC deflection as an effective monitor of the operation without further imaging. Shows the tip lifts and drops nanoparticles. LMR researchers developed these methods for positioning colloidal nanoparticles (typically gold colloids with diameters 5–30nm) accurately and reliably on mica and silicon substrates, in ambient air or liquid environments.

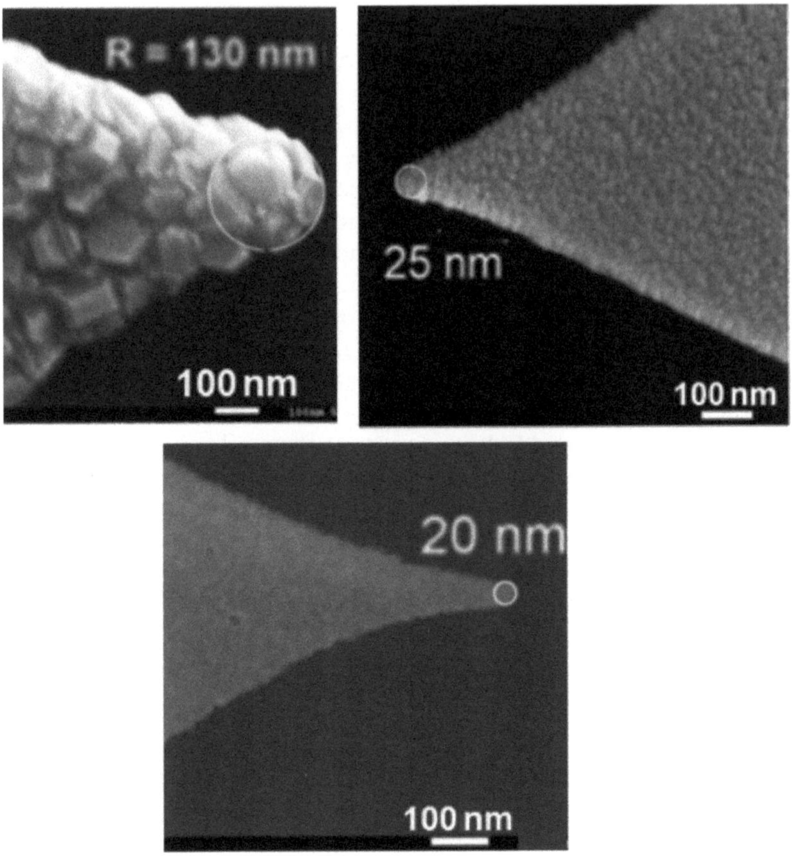

FIGURE 5.30 Conductive AFM tips: a variety of platinum PtSi silicide conductive tips as sharp as 20 nm.

5.5.5.3 *Mechanical Scratching/Plowing/Bending/Deforming/Cutting [37]*

The AFM tip may be used in a scratching mode. In this technique, strong loading forces are applied to the top such that the tip make contact with the substrate. Now scanning the tip removes material from the substrate or from the resist surface as shown in Fig. 5.32A. Moreover, the AFM may be operated in a semi-contact mode in which them tip vibrates. The surface can be modified by indenting it with this vibrating tip. This procedure is called dynamic plowing lithography (DPL). Thomas Jung of the University of Basel used the process to write "Heureka" between the tracks of a compact disc (Fig. 5.32B).

5.5.5.4 *Mechanical Press for Soft Material [38]*

An AFM tip can also be used to cut or bend soft materials, such as polymers, DNA, and nanotubes. In one demonstration the tip was used to depress, deform, poke or make a hole in the wall of flexible nanotube as shown in Fig. 5.33. The deformation was used to measure the stiffness of a nano tube made from a network of silicon and oxygen.

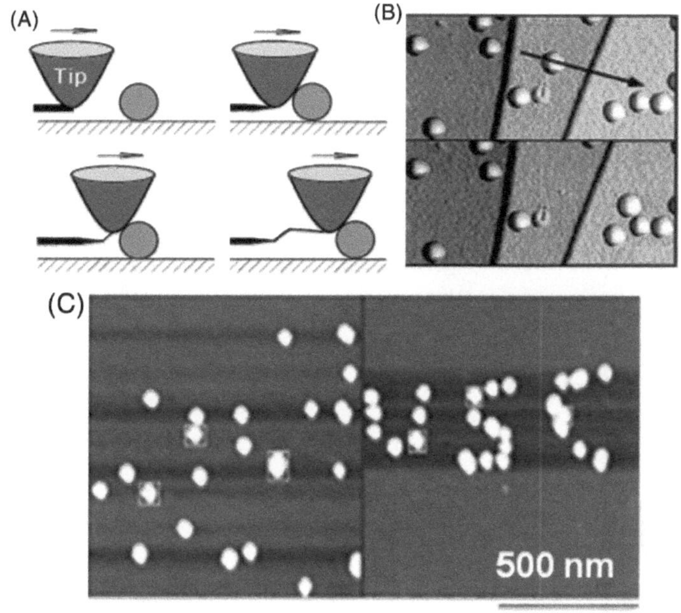

FIGURE 5.31 Moving nanoparticles around with AFM. (A) Four cartoons showing the steps of moving nanoparticles with an AFM tip. (B) A 30-nm gold particle before (top) and after (bottom)) being pushed over a high step. (C) An example of organizing a random number of nanoparticles into the letters USC. *(From C. Baur, B.C. Gazen, B. Koel, T. R. Ramachandran, A.A.G. Requicha, L. Zini , Robotic Nanomanipulation with an SPM in a Networked Computing Environment, Laboratory for Molecular Robotics, University of Southern California, Los Angeles, http://www-bcf.usc.edu/~requicha/LMR/publications/NanoIV/)*

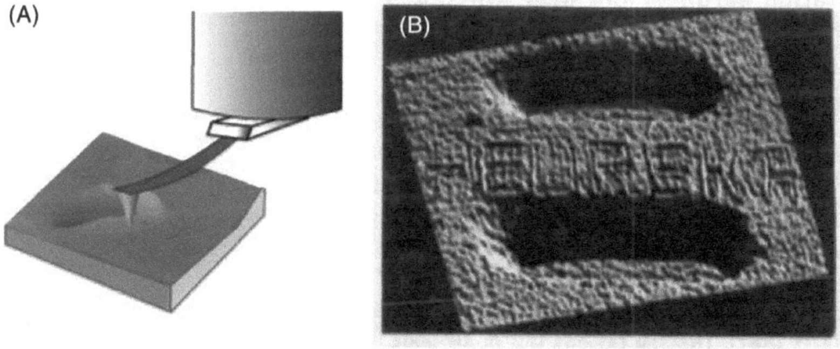

FIGURE 5.32 Mechanical scratching/plowing/bending/deforming/cutting. (A) AFM removes material from the substrate or from the resist surface. (B) To write "Heureka" between the tracks of a compact disc.

The flexible tubes are prepared as follows. A silicon substrate anode electrode is horizontally immersed in a colloid of 1-nm silicon nanoparticles in water/alcohol (1:100 ratio), and a counter 5 mm diameter stainless-steel rod electrode as the cathode. A current of 0.3 mA is established between the electrodes. Under the electric field, particles are driven toward the anode substrate

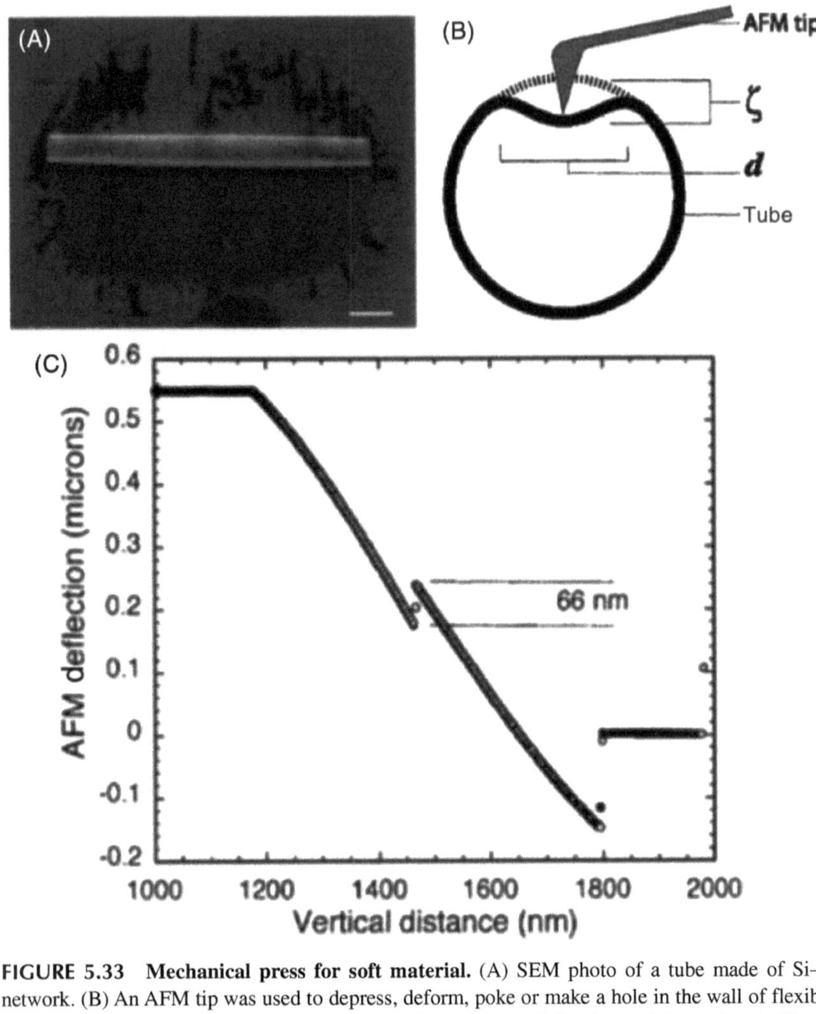

FIGURE 5.33 **Mechanical press for soft material.** (A) SEM photo of a tube made of Si–O network. (B) An AFM tip was used to depress, deform, poke or make a hole in the wall of flexible nanotube. (C) The measurement can be used to determine the deflection and determine the Young modulus of the tube.

at a velocity of 1 m/s where they get deposited. The thin films obtained are near-circular patches of about 100–200 microns in diameter. The sample is then removed from the bath. Upon drying, the film delaminates and rolls up into uniform tubes of a diameter ranging from 0.2 to 5 μm. Figure 5.33A presents an example image of a tube taken with an SEM, which show that the tube is hollow and not solid. The tubes are transparent enough to see through them. The transparency is due to the fact that the band gap of the Si nanoparticles is 3.5–3.7 eV (see Chapter 9), constituting a wide band-gap material.

The arm of the AFM tip, which is of radius ~10 nm is pressed against the tube to provoke a small local deflection of linear size d and depth ζ (<100 nm).

In Fig. 5.33B, we plot the vertical distance of the deflection of the arm (deflection of the film) as a function of the vertical motion of the AFM arm. The measurement shows a vertical deflection of approximately 66 nm. Since this deflection is $\sim 2t$ (twice the thickness of the tube), we can use a linear elasticity analysis to compute Young's modulus of the film. The analysis gives $E \sim 0.06$ GPa. This value of Young's modulus is about 5000 times smaller than that of bulk silicon and is just 30 times larger than that of rubber. The flexibility of these tubes is another confirmation that they are hollow.

5.5.5.5 Writing With Charge [39]

A special substrate must be used to realize writing with charge. For example, a silicon substrate is covered with a thin layer of silicon dioxide and topped off with a layer of silicon nitride. This is one possible substrate. As shown in the conductive tip of an AFM is now placed on the silicon nitride layer. The tip is biased positive with respect to the substrate. A voltage pulse is applied to the tip, which creates an electric field through substrate. The field beneath the tip draws electrons from the silicon substrate, up through the SiO_2 insulating layer and into the silicon nitride layer. The drawn electron gets trapped in the nitride layer. Silicon nitride is known to retain charge. The small zone of trapped charge then acts as a bit of data which can be read. The tip motion and pulsing can be programmed to write or draw a certain image. Shows an example in which the process was used by Robert Barrett, formerly at Stanford University, to inscribe the Stanford logo in trapped charge.

5.6 CHEMISTRY-BASED NANOFABRICATION

5.6.1 Local/Anodic Oxidation/Nanolithography (LON) [40]

AFM or STM devices can be used to deposit or grow nanosized patterns of silicon dioxide on a wafer surface at room temperature under air ambient conditions. The process is called local oxidation nanolithography (LON). An STM is operated in air (in the presence of oxygen and OH groups). The substrate is a hydrogen-passivated Si. The field between the tip and the surface enhances oxidation of the surface. The tip is raster scanned over the surface, such that it generates the patterns shown in Fig. 5.34. It is to be noted that the STM tip cannot identify the grown oxide; it views a generated oxide as a depression on the surface (reduction in the tunneling current). Therefore, in order for the STM to maintain its current constant, the feedback loop pushes the tip towards the surface, risking contact or crash with the oxide, which may damage the tip. This leads to poor reliability for nanooxidation. It is therefore best to use a subsequent AFM imaging step to visualize the pattern which has been generated. Moreover, AFM has much finer control of the tip-surface distance and is therefore seen as an improved tool for nanooxidation.

In addition to silicon and silicon oxide, the process is amenable to a variety of materials. Fig. 5.35A–B [41,42] presents an example of nanometer-scale

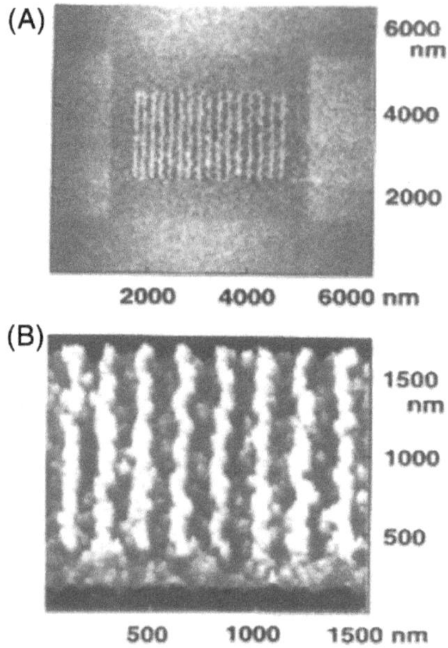

FIGURE 5.34 Local oxidation nanolithography (LON). (A and B) oxidation patterns are generated by raster scanning over a H-passivated Si wafer surface at room temperature under air ambient conditions (presence of O and OH). *(From J.A. Dagata, J. Schneir, Scanning tunneling microscope-based fabrication and characterization on passivated semiconductor surfaces, in Technology of Proximal Probe Lithography, C. Marrian, ed., SPIE Institutes, IS 10, 100 (1993))*

surface oxidation modification using an AFM tip-induced oxidation of the surface of hyperfine titanium film on the silicon substrate. A conductive cantilever (coated with a conducting layer) of AFM tip is used Fig. 5.35A. The surface is positively charged and the tip is negatively charged. In this arrangement, the tip and the surface will electrochemically act as anode and cathode. The substrate to be modified is a silicon wafer with multiple layer coating of silicon oxide, Ti, and TiO_2. It is biased negatively with respect to the silicon surface. In air or other humid atmosphere the probe and the surface of the sample are covered by thin film of absorbed water. The tip is made to approach sufficiently close to the surface. The absorbed layers come in contact with the tip, which forms a water bridge due to capillary effects. The application of a corresponding electric field initiates through the bridge electrochemical reactions in water-surface border, in water and in the probe. Oxide will grow on the point right under the tip. An example pattern is shown in Fig. 5.35B.

5.7 LOCAL ELECTRO ETCHING

5.7.1 HF Silicon Etching [43]

Fig. 5.36A shows an STM tip biased negative with respect to a silicon substrate in the presence of HF liquid. The solution is 0.05% hydrofluoric HF

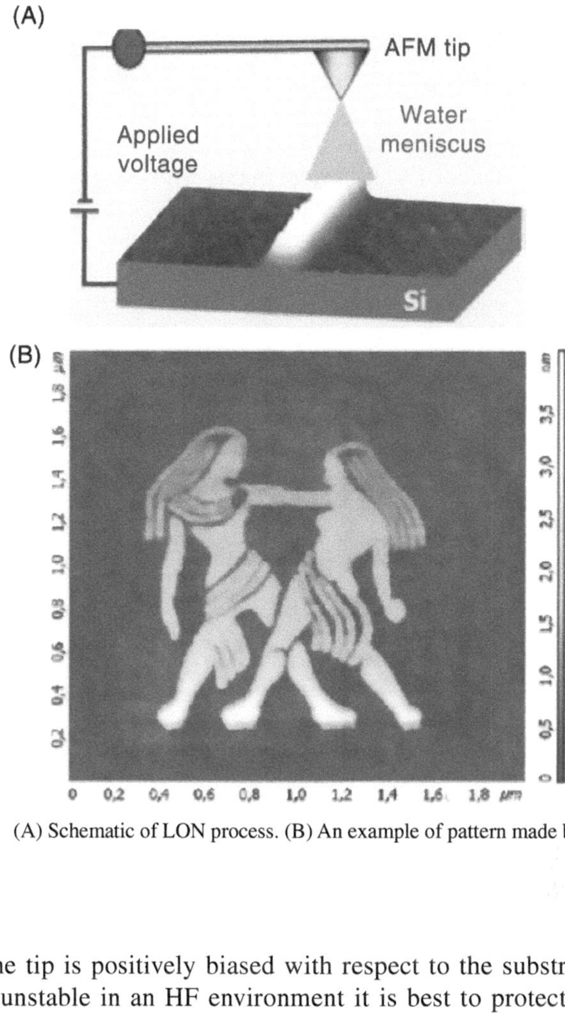

FIGURE 5.35 (A) Schematic of LON process. (B) An example of pattern made by oxide growth.

solution. The tip is positively biased with respect to the substrate. Because tungsten is unstable in an HF environment it is best to protect the tip with a layer of wax as shown in the figure. Alternatively a PtIr tip may be used. Moreover the process proceeds when the tip is biased positive relative to the substrate. Controllable etching can produce holes or etched lines into Si<100> if tip is scanned, which are few tenths of nm wide and a few nm deep. As to the mechanism, oxidation takes place due to the presence of the intense electric field in the tip/target gap, which can reach as high as 10^7 V/mm. The produced oxide gets etched in the presence of HF. Silicon etching using HF is self-limiting since HF passivates silicon with H, making it inert against HF. Thus the presence of the STM tip current effectively enhances or catalyzes the process.

It is to be noted that hole charging of a silicon surface has been known to catalyze etching of silicon in HF to form porous silicon. In the STM system, a tip bias of +1.4 V relative to the substrate and a tunneling current of

1 nA, the tip pull electrons from the substrate, which effectively charges the surface with holes. For the opposite polarity in which negative potential at applied the tip, HF attacks only the defects sites of the H-terminated silicon surface.

Fig. 5.36B [44] shows an STM tip biased negative with respect to a silicon substrate in the presence of water vapor. In this arrangement, nano holes can be produced at a certain threshold of biasing voltage. Some reports showed that nano holes can be also produced in gold substrates. Holes of 3-nm diameter and 0.24-nm depth, corresponding to a removal of about 100 gold atoms have been observed.

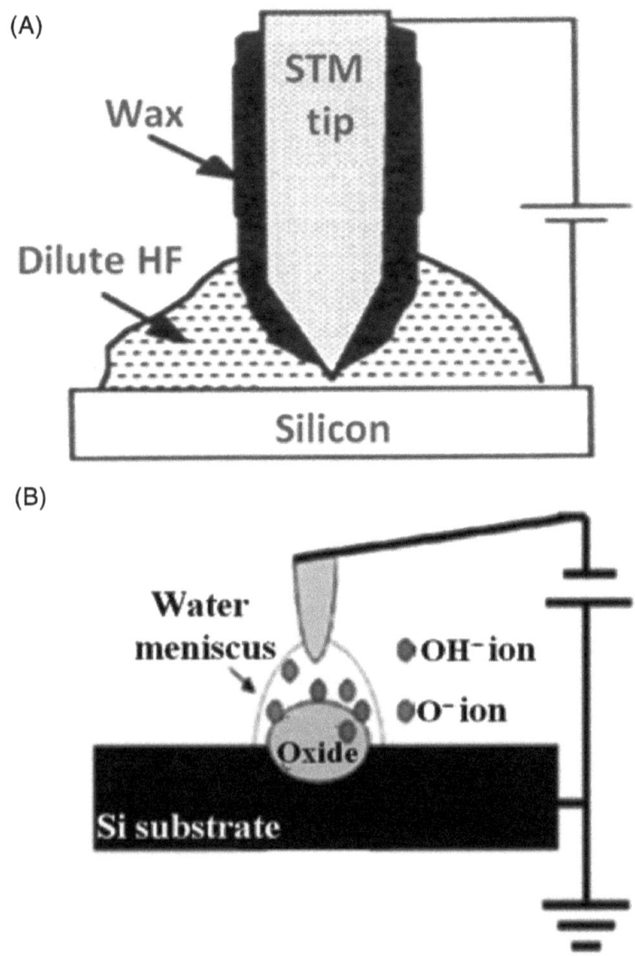

FIGURE 5.36 Local electro HF etching. (A) Schematic of an STM etching in the presence of HF liquid. (B) Schematic of an STM etching in presence of water vapor.

5.8 DIP PEN LITHOGRAPH WITH AFM [45]

Alkane thiolates are put on the tip of an AFM tip. It is placed over the substrate to be written on. In the dip pen lithography process, the AFM tip is placed over the substrate to be written on. A water meniscus (capillary forces) forms under ambient conditions between the tip and substrate (Fig. 5.37A). Transfer

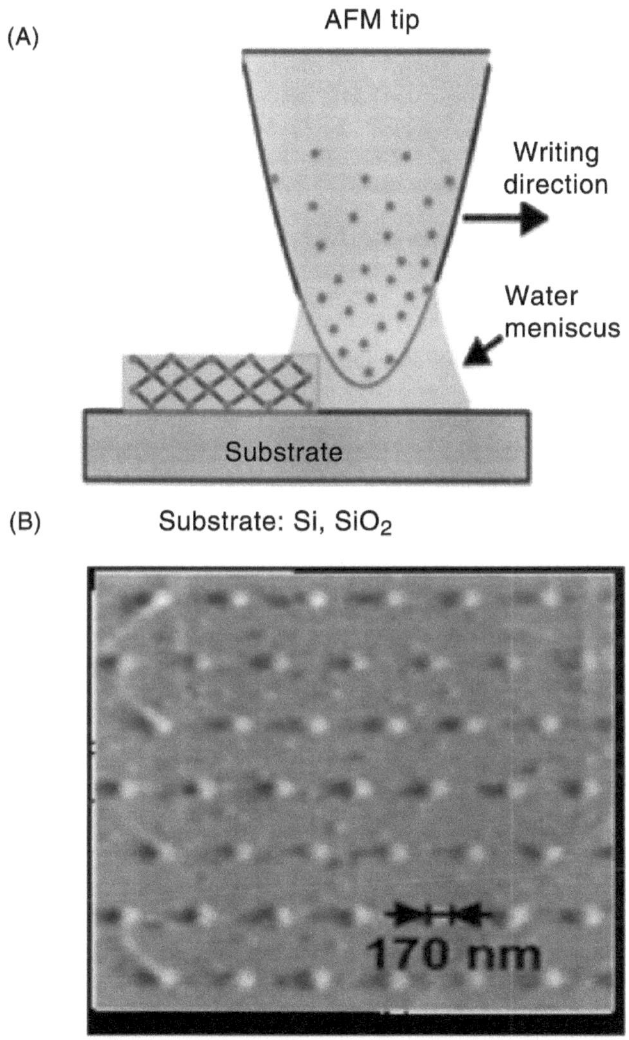

FIGURE 5.37 Dip pen lithograph with AFM. (A) Schematic of the dip pen lithography process in which the conducting tip is covered with alkane thiolates, and under ambient conditions forms water meniscus (capillary forces). (B) Alkane thiolates are imprinted onto a gold surface, which formed well-ordered self-assembled monolayers (SAMs).

of the thiolate occurs by diffusion through the water meniscus. In fact, alkane thiolates were imprinted onto a solid state gold surface, which formed well-ordered self-assembled monolayers (SAMs) (Fig. 5.37B). The process works for all sort of substrates including metals, semiconductors, and insulators. The ink includes polymers, colloidal nanoparticles (e.g., magnetic nanocrystals, carbon nanotubes), sol–gel precursors, small organic molecules, biomolecules (proteins and oligonucleotides) and even single virus particles and bacteria. The procedure is capable depositing molecules and materials on surfaces with sub-50 nm resolution.

5.9 WEAR AND TEAR AND CONTAMINATION OF AFM TIPS [45]

5.9.1 Imaging

Imaging with AFM requires real contact or near contact between the tip and the surface. With time, using the same tip, images show progressive ambiguity and loss of resolution. The problem is inherited because the tip is a very sharp, fragile needle and therefore sucesstible to damage due to contact with the object to be examined. The resulting damage include plastic deformation, pick up of material from the surface or emission of material to the substrate. These damage effects are not unlike the characteristics of near destructive testing. The problem it is not easy to identify since in most cases it is a gradual deterioration with no sharp threshold or signature.

5.9.2 Fabrication

Tips are even more susceptible to damage when they are used in a nanofabrication mode. The effects become much more severe when the tip is necessary brought into contact with the surface or with components on the surface as in scratching, plowing, drilling, pushing/pulling, etc. In such modes tips made of harder materials are used such as diamond-based tips to minimize deformation or lower adhesive material to avoid pickup.

5.9.3 Calibration

To calibrate an AFM, scanning of mm-size features must be made. Moreover, scanning is carried out in the contact-mode. In addition, tip geometry has significant influence on the results when imaging features of several hundreds of nanometers across. As a result, "de-convolution" must be applied which necessary requires details of the tip geometry as well as the changes of the tip geometry during scanning. There are several mechanisms of wear in the AFM-tip due to scanning, which include adhesive wear, abrasive wear, low cycle fatigue, and tribo-chemical wear.

REFERENCES

[1] Chris M. Fundamental Principles of Optical Lithography: The Science of Microfabrication. Hoboken, NJ: John Wiley & Sons; 2011.

[2] Vivek B. EUV Lithography. Bellingham, WA: SPIE Press; 2009.

[3] Andreas G, Claus S. Excimer Lasers The powerful light source in the UV and VUV. Weinheim: WILEY-VCH Verlag GmbH & Co. KGaA; 2005. 46-50, LTJ June 2005 No. 2. http://onlinelibrary.wiley.com/doi/101002/latj.200790037/pdf.

[4] Yayi W, Robert L. Advanced Processes for 193-nm Immersion Lithography. Bellingham, WA: SPIE Press; 2009.

[5] S. Narasimha S., "High Performance 45-nm SOI Technology with Enhanced Strain Porous Low-k BEOL, and Immersion Lithography," IEEE International Electron Devices Meeting (IEDM) Technical Digest, Dec. 11-13 2006. DOI: 10.1109/IEDM.2006.346879 Conference: Electron Devices Meeting, 2006. IEDM '06. International, pp 1-4.

[6] Changes and challenges abound in multi-patterning lithography, in: J. Dorsch (Ed.), Semiconductor Manufacturing and Design. http://www.semi.org/en/node/54491; Double, Triple and Quadruple Patterning and Future Lithography, brian wang June 23, 2009. http://nextbigfuture.com/2009/06/double-triple-and-quadruple-patterning.html.

[7] Materials and Processes for Next Generation Lithography, Elsevier, Nov 8, 2016;

[8] Vitor R, Lihua Z, Dong S, Huigao D, Richard GH, Eric AS, Karl BK. Resolution limits of electron-beam lithography toward the atomic scale. Nano Lett. 2013;13(4):1555–8. http://www.phy.cam.ac.uk/research/research-groups/sp/facilities/ebl.

[9] P. Zeppenfeld, C.P. Lutz, D.M. Eigler, Scanning-Probe-Based Science and Technology. http://www.pnas.org/content/94/24/12749/F5.expansion.html#ref-14.

[10] H.D. Wanzenboeck, S. Waid, Ion beam Focused Ion Beam Lithography, http://cdn.intechweb.org/pdfs/24490.pdf.

[11] Kooy N, Mohamad K, Pin LT, Guan OS. A review of roll-to-roll nanoimprint lithography. Nanoscale Res. Lett. 2014;9:320. https://doi.org/10.1186/1556-276X-9-320; (WeiminZhou, Nanoimprint Springer-Verlag Berlin Heidelberg 2013. DOI 10.1007/978-3-642-34428-2).

[12] Yongwoo Kim, Seok K, Howon J, Jae WH. Plasmonic bowtie High throughput plasmonic lithography for sub 50nm patterning with a contact probe Alternative Lithographic Technologies II. In: Daniel J C Herr, editor. In: Proc. of SPIE; 2010. doi: 10.1117/12.848329. Vol. 763776371F-1 © 2010 SPIE CCC code: 0277-786X/10/$18.

[13] L. Pan, Y. Park, Y. Xiong, E. Ulin-Avila, Y. Wang, L. Zeng, S. Xiong, J. Rho, C. Sun, D.B. Bogy, X. Zhang, Maskless plasmonic lithography at 22 nm resolution, Scientific Reports, 1, p. 175. DOI: 10.1038/srep00175.

[14] Super Resolution Imaging. http://circuit.ucsd.edu/~zhaowei/Research.html.

[15] Laser: https://www.imtek.de/professuren/nano/forschung/nanolito?set_language=en.

[16] Nanosphere: https://www.imtek.de/professuren/nano/forschung/nanolito?set_language=en.

[17] Photoresist Composition for Extreme Ultraviolet Lithography, Sandia National Laboratories, 1999.

[18] Christie MRK. Technology of Proximal Probe Lithography. Bellingham, WA: SPIE Optical Engineering Press; 1993.

[19] Nikodem T, KuanEng JG. Scanning Probe Microscopy. Singapore: World Scientific; 2011.

[20] Clery D. Nanotechnology rules, OK! Feature, New Scientist.

[21] Saw-Wai Hla. STM Single Atom/Molecule Manipulation and Its Application to Nanoscience and Technology, Critical Review article. J. Vac. Sci. Tech. 2005. https://arxiv.org/ftp/cond-mat/papers/0506/0506038.pdf.

[22] Yau ST, Saltz D, Nayfeh MH. Laser-Assisted Deposition of Nanometer Structures Using Scanning Tunneling Microscopy. Appl. Phys. Lett. 1990;57:2913. Yau ST, Saltz D, Nayfeh MH. STM-laser fabrication of nanometer structure. J. Vac. Sci. Techol. 1991;B9:1371.

[23] Nayfeh M. Fabrication of nanometer scale structures, in technology of proximal probe lithography. In: Marrian C, editor. SPIE Institutes IS; 1993. p. 200. 10.

[24] Alex. SR, Ehrichs E, de Lozzane A. Direct writing of metallic features with a scanning tunneling microscope. Appl. Phys. Lett. 1987;51:247.

[25] Mitsui T, Hill E, Ganz E. Nanolithography by selective chemical vapor deposition with an atomic hydrogen resist. J. Appl. Phys. 1999;85:522.

[26] Adams DP, Mayer TM, Swartzentruber BS. Selective area growth of metal nanostructures. Appl. Phys. Lett. 1996;68:2210.

[27] Hetrick JM. Degradation studies of ultrathin silicon oxide under electrical stress using a scanning tunneling microscope. Urbana-Champaign: University of Illinois; 1999.

[28] Houel A, Tonneau D, Bonnail N, Dallaporta H, Safarov VI. Direct patterning of nanostructures by field-induced deposition from ascanning tunneling microscope tip. J. Vac. Sci. Technol. B 2002;20:2337. http://www.terpconnect.umd.edu/~phaneuf/STM_nanodot.html.

[29] Lyding JW, Abeln GC, Shen T-C, Wang C, Tucker JR. Nanometer scale patterning and oxidation of silicon surfaces with an ultrahigh vacuum scanning tunneling microscope. J. Vac. Sci. Technol. 1994;B12:3735.

[30] Iwasaki H, Yoshinobu T, Sudoh K. Nanolithography on SiO_2/Si with a scanning tunnelling microscope. Nanotechnology 2003;14:R55.

[31] Maiman C, Dobisz E, Dagata j. Low voltage e-beam lithography with STM technology of proximal probe lithography. In: Marrian C, editor. SPIE Institutes IS; 1993. p. 58. 10.

[32] McCord MA, Pease RFW. Lift-off metallization using poly (methyl methacrylate) exposed with a scanning tunneling microscope. J. Vac. Sci. Technol. 1988;B6:293.

[33] Kazaki Ai O, Seiji KA, Yoshikazu AN. Scanning probe microscope lithography of silicon using a combination of a carbon nanotube tip and a polysilane film as a mask. Jpn. Appl. Phys. 2002;41:4973–5.

[34] NANOSENSORS™ Wear Resistant Conductive AFM Probe. Platinum silicide probes: highly conductive AFM probes with small radii and high wear resistance (nanosensors).

[35] A.A. Tseng. Nanofabrication Using Atomic Force Microscopy. www.cc.ntut.edu.tw/~smchen78/news/AFMLIntro10.pdf.

[36] Tseng AA, Li Z. Manipulations of atoms and molecules by scanning probe microscopy. Nanotechnology 2007;2582–95.

[37] American Scientist DANIEL CLERY, The Smallest Grafitti in the World, New Scientist, March 7, 1992. Irmer B, Blick RH, Simmel F, Gödel W, Lorenz H, Kotthaus JP. Josephson junctions defined by a nanoplough. Appl. Phys. Lett. 1998;73:2051. Rosa JC, Wendel M, Lorenz H, Kotthaus JP. Direct patterning of surface quantum wells with an atomic force microscope. Appl. Phys. Lett. 1998;73:2684. Carpick RW, Salmeron M. Scratching the surface: fundamental investigations of tribology with atomic force microscopy. Chem. Rev. 1997;97:1163. Smirnov V.A., Taganrog Technological Institute of Southern Federal University. AZO Nano. http:http://www.azonano.com/article.aspx?ArticleID=2605.

[38] Chaieb S, Nayfeh M, Smith A. Assembly of silicon nanoparticles roll up into flexible nanotubes. Appl. Phys. Lett 2005;87:062104-1.

[39] Clery D. Nanotechnology rules OK! Feature, New Scientist.

[40] Dagata LONJ, Schneir J, Harary H, Evans C, Postek M, Bennett J. Modification of hydrogen-passivated silicon by a scanning tunneling microscope operating in air. Appl. Phys. Lett. 1990;56(20):2001–3.

[41] Smirnov VA. Taganrog Technological Institute of Southern Federal University. AZO Nano Nanotechnol. 2001;12:273. http://www.azonano.com/article.aspx?ArticleID=2605; http://www.ntmdt.com/spm-principles/view/afm-oxidation-lithography.

[42] Dagata JA, Schneir J. Scanning tunneling microscope-based fabrication and characterization on passivated semiconductor surfaces. In: Marrian C, editor. Technology of Proximal Probe Lithography, vol. 10. SPIE Institutes IS; 1993. p. 100.

[43] Nagahara HFLA, Thunadt T, Lindsay SM. Reviews if scientific instruments. In: Marrian C, editor. Roles of Scanning Prope Microscopes the Development of Nanoelectronic Devices, Technology of Proximal Probe Lithography, vol. 60. SPIE Institutes, IS; 1989. p. 3128. 10, 33 (1993).

[44] Bouchiat V, Faucher M. Josephson junctions and superconducting quantum interference devices made by local oxidation of niobium ultrathin films. Appl. Phys. Lett 2001;79:23.

[45] Ming S, Xiaogang L, Shu-You L, Vinayak DP, Chad MA. Moving beyond molecules: patterning solid-state features via dip-pen nanolithography with sol-based inks. J. Am. Chem. Soc. 2002;124(8):1560–1. http://www.eng.utah.edu/~lzang/images/Lecture_5_DPN.pdf.

Chapter 6

Particle-by-Particle Nanotechnology

6.1 SCHEMES OF BUILDING NANOCOMPONENTS

STM devices provide extremely high writing resolution (atomic resolution) but they are very slow because they require the movement of a tip with a finite mass. The speed of an STM is no match to the speed with which one can deflect light, electrons, or ions. Thus it is said that you can improve the patterning resolution with STM-based devices such that you can store the whole library of congress on one disk; however it would take you forever to do that.

In fact, the slow speed of the STM tip makes it unsuitable for creating large arrays of devices on a single substrate or for mass-producing chips. In an effort to speed up the process, it was suggested that instead of using mechanical manipulation of atom by atom as Feynman suggested, we would start out with pre-assembled nanoparticles, clusters, or quantum dots consisting of 30–1000 atoms as building blocks. Methods, however, must be first developed for pre-assembling those components. Physical- or chemical-based techniques are in use but chemical-based techniques are the most dominant. Two main approaches are used to synthesize, produce, or assemble [1]:

1. "bottom-up" approach where materials and devices are built from atoms or molecules (as Feynman conjectured) which assemble themselves chemically using principles of molecular recognition (Fig. 6.1) (see Chapter 1)
2. "top-down" approach where nano-objects are constructed from larger entities via some chemical etching without atomic-level control (Fig. 6.2)

In top-down, we have the following steps (see Fig. 6.3)

- Divide a large piece of a material to smaller and smaller
- Stop before you get to an atom, say stop when the pieces consist of 30–1000 atoms (in size to 1–3 nm or billionth of a meter)
- The tiny parts begin to have properties that are new/novel and are not found in the larger pieces (novel chemical, optical, magnetic, electronic, or mechanical)
- Rap them if necessary with protective coating to preserve or stabilize them

Fundamentals and Applications of Nano Silicon in Plasmonics and Fullerines.
http://dx.doi.org/10.1016/B978-0-323-48057-4.00006-2

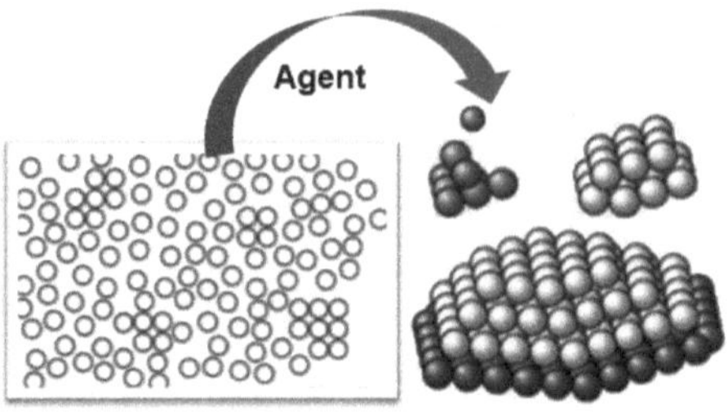

FIGURE 6.1 Feynman scheme "bottom-up" or atom-by-atom (ABA) approach in which materials and devices are built from atoms or molecules.

- Use "identical" tiny parts as building blocks. Build formations as large as the original parent piece

In bottom-up, we have the following steps:

- Start with atoms of a given material
- Assemble a number of them (~30–1000) together
- Rap them with protective coating if necessary
- Use those as building blocks

As to the techniques normally used to produce miniaturized nano material one can choose from among a host of physical, as well as chemical techniques, depending on the material, size, quality requirements, and volume of nanomaterial needed. Those include ball milling, laser ablation, ion sputtering, combustion flame, microwave-induced plasma, and chemical and electrochemical reactions.

One can miniaturize and produce nano particles from basically any solid material, including ceramics, polymers, semiconductors, metal (conducting, magnetic, and superconducting), and its oxides, etc. Here we will describe briefly some popular cases for illustration, such as silicon-, metal-, and carbon-based structures.

As to STM reading, it is also limited by the speed of the tip. STM reads data 100 times slower than a magnetic computer disc. One idea to speed up the

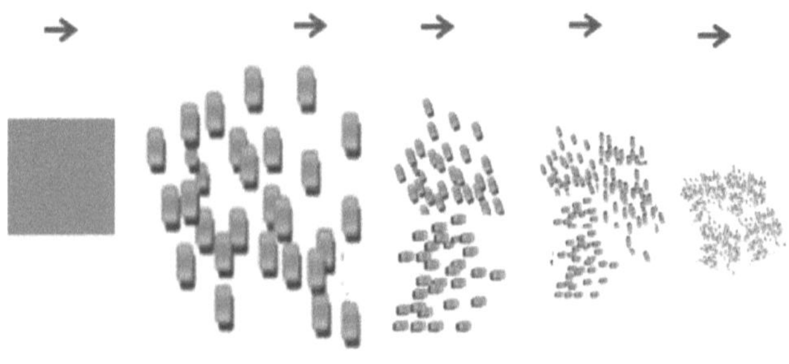

FIGURE 6.2 "Top-down" approach where nano-objects are constructed by miniaturization of larger entities.

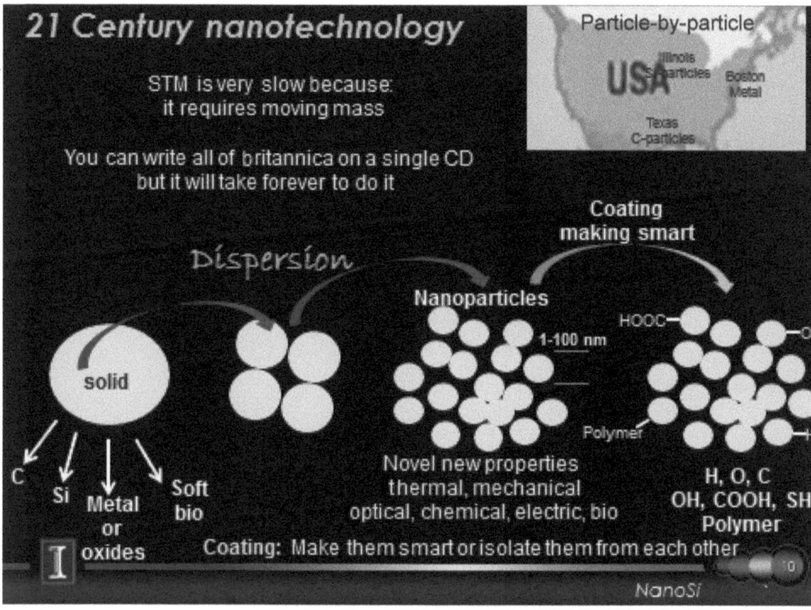

FIGURE 6.3 Detailed steps of the top-down fabrication.

reading was proposed by Rugar and Mamin at IBM [2]. The scheme relies on moving both the sample and the tip simultaneously as shown in Fig. 6.4. The STM head is provided with an extra mechanism to lift the sample up and down very quickly to help the up and down motion of the tip. The tip can then scan at a speed of 1 mm/s, a thousand times faster than a conventional STM tip. With this hybrid head they were able to make films (at two frames a second) of events on the surface.

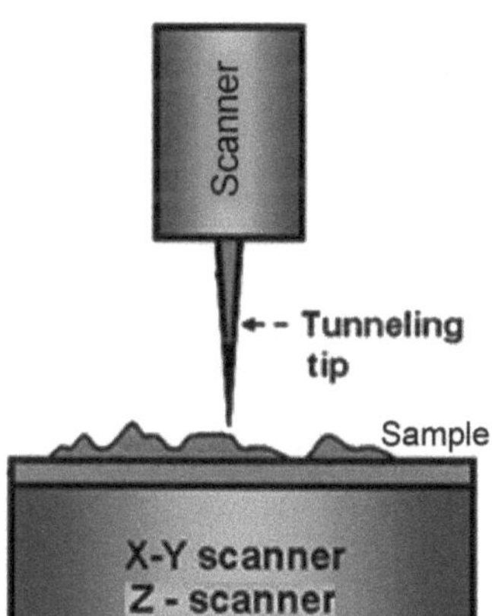

FIGURE 6.4 Scheme relies on providing the STM head with an extra mechanism to lift the sample up and down very quickly to help the up and down motion of the tip.

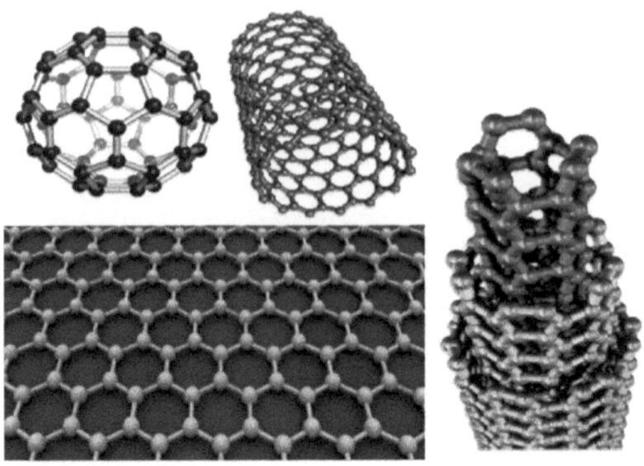

FIGURE 6.5 **Carbon nano components.**

6.2 POPULAR NANOCOMPONENTS

Nanoparticles in the range below 100 nm in diameter are designated as nanocomponents. There are natural and man-made blocks. The most popular nano building blocks or components are made of semiconductor-based materials (silicon, carbon, metals, oxides, etc.).

6.2.1 Carbon Nanocomponents

There are various forms that carbon exists in. Among those nanoparticles [3], nanotubes [4], grapheme [5], diamond and graphite (see Fig. 6.5). Carbon fullerene balls, such as C60 are empty closed shells of ~0.7 nm diameter. Carbon nanotubes (CNTs) are empty cylinders. Their electrical properties span from metallic, semi-metallic to semiconducting depending on their atomic structures. Mechanically, nanotubes have high Young's modulus and tensile strength (~100 times higher than steel). An arc-evaporation procedure is used to make nano tubes. In the procedure, a high current is passed between two graphite electrodes in an atmosphere of helium. This causes graphite to vaporize with some of it condensing on the cathode to form the CNTs.

CNTs offer a variety of diverse application, constituting as such a platform technology. For instance it can be used in construction of transistors and electronics, biosensor, displays, and in hardening and improving of composites. However, it has several shortcoming, such as low-production yield, difficulty in isolating them from each other as they tend to entangle, and multi-walling where often they come out with multiwall configuration. Most importantly they may pose health hazard because they may not flush after entering the human body because they are needle-like in shape. Another structure which was more recently discovered is graphene. It is a one atom thick open structure layer of hexagonal carbon rings. Graphene sheets can be seen by the naked eyed as they absorb 2.3% of white light.

Example 6.1 The first attempt at miniaturization

Flint stone is a hard, sedimentary cryptocrystalline silicate form of the mineral quartz (SiO_2). It is usually dark-grey, blue, black, or deep brown in color, and often has a glassy appearance. The color is due to the presence of impurity. The material is also characterized as composite flint.

Flint fractures into very sharp points. The first attempt to use this property to produce small objects took place in Palestine 20,000 years ago. By knapping flint, they produced objects usually about a centimeter long or less. These were called microliths. Because of this property, it was used by early man for making sharp stone tools, probably used as barbs on arrows, spears, and other composite tools.

Microliths were either produced from small blades (microblades) or by snapping normal big blades in a controlled manner to produce geometric microliths. They can be formed as various kinds of triangles, lunate shaped, trapezes, etc. More precisely, they are typically 1-cm long and half a centimeter wide when finished. These inventions may be termed primitive but in actuality they are extraordinarily beautiful tools, many with edges that are sharper than our finest surgical steel blades. This process constitutes the first attempt by man at miniaturization of matter in a controlled manner. The term micro which we now use is derived from microlith. The term lithography in nano lithography is also derived from this word.

6.2.2 Titanium Oxide

The most popular oxide nanoparticles is titanium oxide (TiO_2) nanoparticles because of ease if synthesis and diversity of applications. TiO_2 nano powder absorbs most of the harmful UV radiation; hence promises to be an important commercial product. The commercial importance stems from the fact it gives greater UV protection in a new generation of sunscreens, cosmetics, plastics, and coatings. The naturally occurring oxide of titanium (TiO_2) is known as Titania. Heating a water solution of titanium-based salt (e.g., $TiOSO_4$) in the temperature range of 100–250°C produces titanium oxide particles. Another example of oxides is silicon dioxide or silica (glass) nanoparticles.

6.2.3 Metal Nanoparticle

Metal nanoparticles, including those of gold [6], silver, and magnetic metal, such as iron are becoming popular. Synthesis and characteristics of those will be covered in Chapter 9. Heavy metals include tungsten (W) and molybdenum (Mo). It is more difficult to synthesize tungsten or molybdenum nanoparticles because they have high-melting-temperature. Thermal decomposition, electrochemical processes, as well as powerful short pulses of infrared laser beams are used to produce W nanoparticles. High-intensity laser beams melt or/and flash evaporate, causing microscopic explosion ejecting chunks of highly heated micron- or nano-scale particles.

6.2.4 Semiconductor Nanoparticles (Direct Material)

One kind of nanomaterial actually consists of two materials: cadmium and sulfur (CdS), or cadmium and selenium (CdSe). Chemists Moungi Bawendi of the Massachusetts Institute of Technology and Paul Alivisatos of the University of California at Los Angeles developed the synthesis and chemical functionalization of these nanocrystals [7]. The resulting material is a direct bandgap material. The nanoparticles or as they are traditionally called quantum dots, produce luminescence under light excitation giving sharp lines spectra whose color depends on the size of the particle. It affords application in displays, and it is popular as a luminescent tag for biomedical applications. However it suffers from bio-incompatibility due to threat of heavy metal poisoning. Its reactivity necessitates a silicon oxide protective coating, which increase the minimum size to ~8 nm. CdS nanoparticles are prepared through a colloidal rout, which results in a size distribution.

6.2.5 Silicon (Indirect Material)

Silicon material is one of the most important materials for nanotechnology applications because it is the backbone of the microelectronics industry as well as the most biocompatible material. It is the least toxic material, promising to be a green technology. It was Canham who first showed in 1990 that visible red photoluminescence light (PL) can come out of silicon wafer crystals at room temperature when electrochemically etched [8]. A magic family of silicon nanoparticles are produced in Nayfeh's lab at the University of Illinois from bulk silicon by a special patented electrochemically dispersion process. The family consists of highly luminescent stable particles of 1, 1.7, 2.2, 2.9 nm in diameter [9]. Unlike bulk silicon, a spectacularly dull material, the silicon nanoparticle family is spectacularly efficient at emitting light in RGB colors, covering the blue, green, yellow, and red part of the spectrum respectively when exposed to dark ultraviolet radiation (see Fig. 6.6). Silicon nanoparticles are akin to the carbon fullerene balls. But unlike the empty shell fullerenes, the silicon nanoparticles are filled and have hydrogen on their surface. The particles may constitute a new phase or "super molecule" that exhibits solid-like behavior as well as molecule-like behavior.

Silicon nanoparticles can also be produced by condensing silicon atoms freed from silicon compounds either in the gas phase or liquid phase. But the

FIGURE 6.6 (A) Suspension of blue and red luminescent silicon nanoparticles of 1 and 3 nm in diameter under excitation with incoherent 365 nm wavelength radiation. (B) Computer prototype of a 1 nm Si ($Si_{29}H_{24}$) nanoparticle.

processes do not lead to the path of the above magic family. Moreover, those need to be post processed with hydrogen or oxygen to stabilize them against environmental attack. In the continued effort to lower the cost of production of nanoparticles, the University of Illinois group, in collaboration with Laila Abuhassan from the University of Jordan and Jon Host of DOW chemicals, were able to synthesize red luminescent silicon nanoparticles from lower grade silicon grain with enhanced yield over that produced from silicon wafers.

Silicon nanoparticle technology has significant technological implications because of its possible integration in existing silicon electronic chip technology. It constitutes a wide platform technology with wide-ranging applications including electronics (low power nanomemory and transistors); photonics (photodetectors (in the UV)); photovoltaic and lighting technologies (Si solar cells and light emitting diodes); substance sensors (glucose and dopamine); catalyst and fuel cells; and biomedical fluorescent tags. In addition to being ultra bright, reconstituted films of the particle family exhibit the phenomenon of stimulated emission, which is a necessary condition for the production of laser beams. Light-emitting Si devices could eventually result in a laser on a chip, new generation of Si chips, and extend the functionality of Si technology from microelectronics into optoelectronics and biophotonics.

Other silicon structure of importance is silicon nanowires (SiNWs). They have unusual quasi one-dimensional electronic structure and are considered as one of the most important one-dimensional materials. They could have a function as building blocks for nanoscale electronics applications including photovoltaics, nanowire batteries, thermoelectrics and non-volatile memory. In particular, important applications such as catalysis, Li ion batteries, solar cells, biological and chemical sensors are discussed. However, unlike carbon which belongs to sp2 coordination, silicon belongs to sp3 coordination, silicon structures are necessarily solid nanostructures with no magic sizes or structures resulting inconsistent

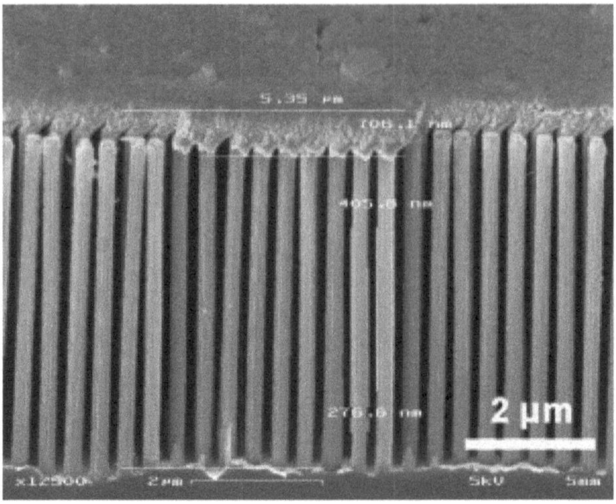

FIGURE 6.7 **Vertical silicon nanowire arrays in channels using metal catalyzed etching.**

nanowire diameters and defects in the shape and size uniformity. Fig. 6.7 shows vertical SiNW arrays in channels using metal catalyzed etching [10]. But it is not possible to produce those dispersion in liquid for later delivery. The figure also shows an Au top contact using tilted e-beam evaporation.

Unlike silicon nanoparticles, which are made of pure silicon, silica nanoparticles are basically spheres made of pure glass (SiO_2). They are produced by simultaneous hydrolysis and condensation reactions of metal compounds, which carry silicon and oxygen (alkoxide). They are of importance to various industrial applications, including catalysts, pigments or pharmaceuticals. They are also used to make electronic substrates, thin film substrates, electrical insulators, thermal insulators, and humidity sensors. Silica particles, especially monodispersed particles are suitable candidates for application in chemo-mechanical polishing (CMP). It is therefore important to develop procedures which produce high-purity silica particles of a narrow size distribution. Present commercial silica has a broad size distribution and varying levels of metal contaminants.

6.3 DISCOVERY OF NOVEL CHARACTERISTICS OF NANOCOMPONENTS AND WHY [11]

Solid nanoparticles of semiconductor, metal, oxides, ceramic, polymer, and alloys do not naturally exist. The nanoparticles are innovative because of several features.

6.3.1 Surface Effects

First, in those the surface to volume ratio is extremely large, that makes their surface a good catalyst for chemical reactions. Moreover, being small allows molecules, surface which get absorbed in their surface get very close to each other, which allows reactions to proceed at lower temperatures. Also surfaces of nanostructures are exotic, being less elastic (less rigid) than those of bulk, which makes them more amenable to modification. Nano surfaces can be chemically modified to design new surface configurations (super molecules).

6.3.2 Fundamental Characteristics

Miniaturization of material lead to novel characteristics because confinements of electrons or positive charge change some of their fundamental parameters, including the mass of electron, dielectric constant, conductivity, purity level, mechanical vibration, etc.

6.3.3 Material Properties

When we go from the large to the nanoscale in material:

- Opaque substances become transparent (copper)
- Dull solids become bright (silicon)
- Inert materials become catalysts (platinum)
- Stable materials turn combustible (aluminum)
- Solids turn into liquids at room temperature (gold)
- Insulators become conductors (silicon)
- Nanoparticles will get cooler faster than larger objects

One famous example is what happens to gold when it is miniaturized. In bulk (at normal scales), gold is chemically inert; but it can serve as a potent chemical catalyst at nanoscales. Another example is that sub-layering reveals mechanical strength (graphite is soft, but single layers are very strong-tubes, sheets).

6.3.4 Derivative Functionalities

Bio/chemical properties may be designed by crafting linkers or bi-linkers on the nano-surfaces, such as functional groups (–SH, –OH, –COOH, –NH) (see Fig. 6.3) to make them smart with regard to bio processes; or attaching polymer tails to enable ultrathin lubrication films, etc.

6.4 CLINTON/BUSH AND DREXLER AND FATE OF NANOTECHNOLOGY [12]

The fate of nanotechnology in terms of atom-by-atom as originally proposed by Feynman and advocated by Drexler versus the emerging particle-by-particle (PBP) was essentially decided during the period 1999–2003 as a result of government intervention. Drexler envisioned a sea of minuscule robots that could move atoms/molecules so quickly and position them so precisely that they could produce almost any substance out of ordinary ingredients in a matter of hours. On the other hand, the PBP essentially uses variations on conventional chemistry to first develop novel nanomaterials in the form of nanoparticles, clusters or quantum dots with new combinations of characteristics, such as strength, toughness, density, conductivity, flame resistance, brightness, and membrane separation characteristics; and using those to build novel nanodevices.

6.4.1 Nanotechnology US Initiatives

The first governmental intervention came along in 1999 with the establishment of a national nontechnology initiative (NNI) as introduced in Fig. 6.8. In a January 2000 speech, 40 years after Feynman proposed ABA fabrication, President Bill Clinton advocated nanotechnology development from the same lecture Hall at the California Institute of Technology. The NNI goals are general goals in nature to advance nanotechnology research and development (R&D) and the transfer into products for commercial and public benefit; and develop relevant educational, training and infrastructure and tool resources. The second government intervention is that of President Bush who signed in 2003 the 21st Century Nanotechnology Research and Development Act, allocating $3.7 billion for molecular-scale R&D. The funds were not allocated to support Drexler's molecular manufacturing, rather to use variations on conventional

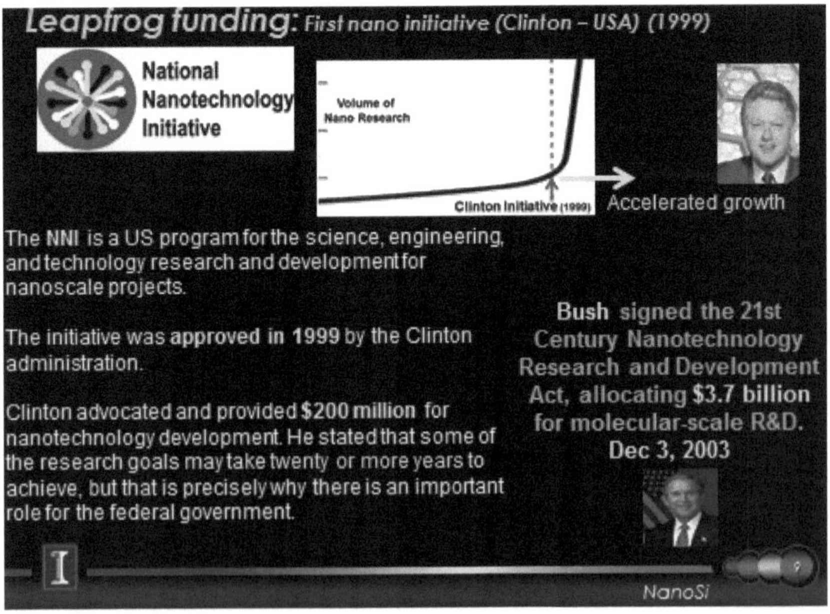

FIGURE 6.8 The Clinton and Bush initiatives and the establishment of the US national nontechnology initiative (NNI).

chemistry to develop novel materials, such as nanoparticles, clusters or quantum dots.

In the late 1980's and early 1990's, nanotechnology research had been conducted only at a small number of laboratories (see Chapter 5). But with the Clinton and Bush's initiatives it exponentially grew (see Fig. 6.8). Although Feynman's ABA concept and Drexler's molecular manufacturing did not win, the Feynman's stature as a Nobel laureate and as an iconic figure in 20th century science helped advocates of present-day nanotechnology gain support, providing valuable intellectual sources. Moreover, Drexler's role has been pivotal in keeping interest in nanotechnology alive during the drought years.

James (Jim) Murday

A Champion of nanotechnology

James Murday is presently associate director for physical sciences of the USC Washington, DC Office of Research Advancement. His specialties over the last 50 years included

- nanoscience and nano-enabled technology,
- surface science
- Federal research funding.

It is not an exaggeration if one says he was one of the most influential scientists/funding agents who helped guide the emerging field of nanotechnology. As early as 1985 he recognized the direct funding of the use of STM in nanofabrication. I for one was one of his recipients ($2 million dollar) for my project: "writing with atoms at room temperature."

Prior to USC, Murday's career at the Naval Research Laboratory (NRL) included leading the Surface Chemistry programs and the Chemistry Division. Additional responsibilities included tenures as Director of Research for the Department of Defense, Research and Engineering; Chief Scientist, Office of Naval Research; Director, National Nanotechnology Coordination Office; and Executive Secretary to the U.S. National Science and Technology Council's Subcommittee on Nanometer Science Engineering and Technology (NSET).

When in May, 2003, the U.S. House approved spending $2.4 billion over the following three years on research to develop nanotechnology (see also Fig. 6.8), it was said "so you can bet James Murday has an attentive audience wherever he goes."

6.5 PRESENT-DAY NANOTECHNOLOGY

Now nanotechnology has been coined as the technology of this century and the meeting ground of all fields, attracting billions of funding dollars, and thousands of scientists and engineers. It is hailed as exciting, novel and ventures into areas of S&T beyond nature where we never been. It captures the imagination of all, and affords us to dream to solve the problems facing the human race, such as acute disease, energy, lighting, water, and food crunch.

The technology works at the limit of size, on a scale no bigger than 100 nm. Due to miniaturization, the surface becomes the dominant factor, with materials often acquiring much improved or even novel properties not found in precursor bulk. Nano materials have improved mechanical, electrical, optical, and chemical properties. The impetus of nanotechnology is the exploitation of nanomaterials as building blocks for manufacturing better products that are lighter, stronger, and cheaper (at lower cost) and with new functions enabling applications in health and agriculture, energy, information, defense and security, providing jobs, economic growth, and improvement of the quality of life. Applications can be mundane, such as using nano engineered particles to make vinyl floors tough enough to last a lifetime; or less mundane but not quite ready for prime time are biomedical applications to speed drugs through the human blood system and improve diagnostics and treatment. Other applications are futuristic, such as nanosize robots programmed to take care of our material needs.

The importance of nanotechnology is highlighted by the billions of dollars that were spent worldwide and by market size projection of $1Trillion in 10 years. Western countries established national initiatives outlining R&D and commercialization policy. Asian countries followed suit and some countries in the Arab and Muslim World entered the race.

REFERENCES

[1] Introduction to Nanoscience and Nanotechnology, Chris Binns, John Wiley & Sons, Oct 19, 2010; https://www.nano.gov/nanotech-101/what/manufacturing

[2] Allegrini M, Garcia N, Marti O. Nanometer Scale Science and Technology. Rugar and Mamin at IBM. IOS Press; 2001. p. 302.
[3] Kroto HW, Heath JR, O'Brien SC, Curl RF, Smalley RE. C60: Buckminsterfullerene. Nature 1985;318(6042):162–3.
[4] Sumio I. Helical microtubules of graphitic carbon (PDF). Nature 1991;354(6348):56–8. (7 November, Bibcode: 1991, Nature 354–356).
[5] Geim AK, Novoselov KS. The rise of graphene. Nat. Mater. 2007;6:183–91.
[6] Obare SO, Jana NR, Murphy CJ. Preparation of polystyrene-and silica-coated gold nanorods and their use as templates for the synthesis of hollow nanotubes. Nano Lett. 2001;1:601. Huang, X., Neretina, S. El-Sayed, Gold nanorods: from synthesis and properties to biological and biomedical applications, M.A., Adv. Mater, 21 (2009) 4880.
[7] Marcus MA, Flood W, Stiegerwald M, Brus LB. Structure of capped cadmium selenide clusters by EXAFS. J. Phys. Chem. 1991;95:1572–6.
[8] Canham LT. Silicon quantum wire array fabrication by electrochemical and chemical dissolution of wafers. Appl. Phys. Lett. 1990;57:1046.
[9] Belomoin G, Therrien J, Smith A, Rao S, Chaieb S, Nayfeh MH. Observation of a magic discrete family of ultra bright, Si nanoparticles. Appl. Phys. Lett. 2002;80:841.
[10] H. Zhao, A. Al-Obeidi, C.V. Thompson, Silicon Nanowires for Chemical Sensing Systems, MTL ANNUAL RESEARCH REPORT, 2013, Massachusetts Institute of Technology, Cambridge, MA. http://www-mtl.mit.edu/wpmu/ar2013/silicon-nanowires-for-chemical-sensing-systems/.
[11] C. Binns, Introduction to Nanoscience and Nanotechnology, John Wiley & Sons, Oct 19, 2010; Science learning Hub: https://www.sciencelearn.org.nz/resources/1676-novel-properties-emerge-at-the-nanoscale.
[12] J.F. Sargent, Jr., National Nanotechnology Initiative (NNI): Overview, Reauthorization, and Appropriations Issues, DIANE Publishing, 2011; The National Nanotechnology Initiative: Present at the Creation, Neal Lane, Thomas Kalil, http://issues.org/21-4/lane/; https://www.nano.gov/.

Chapter 7

Characterization and Simulation Technologies of Nanomaterial

7.1 SIZE DISTRIBUTION: TEM, SEM, STM, AFM, DLS, RAMAN

Size of nanostructures is one of the most important physical properties because much of the novel quantum confinement characteristics of particles depend on the size. In fact particle size or size distributions pivotally affect device operation, quality of products and associated processes, and as such can be critical to the success of manufacturing businesses. Moreover, theoretical and simulation procedures of material properties and device operation, as well as testing models, against experimental results and field operation would require knowledge of the size of the nanoparticle being used. Therefore such measurement must be routinely carried out across a wide range of research and industries. Numerous types of instrumentation are needed to cover a wide range of particle size from sub-nanometer to millimeters.

7.1.1 Scanning Tunneling Microscope (STM)

Scanning tunneling microscope (STM) observes atoms immobilized on conducting surfaces. The microscope was invented by Gerd Binnig and Heinrich Rohrer, who worked at IBM, Europe. It was revolutionary in the basic science and application to the degree that it earned the inventors the 1986 Nobel Prize in physics. It employs an extremely small, sharp, metallic needle positioned a few angstroms above atoms on a conducting surface and measures currents between them, which is converted to height (see Chapters 4 and 5). The magnification with an STM is high enough to allow observation (seeing/visualization) single atoms, as well as it allows manipulation and writing (moved and placed in a desired pattern) of single atoms.

7.1.2 Atomic Force Microscope

The atomic force microscope (AFM) generally uses a non-metallic needle placed over nanostructures on a non-conducting surfaces. Its basic principle is based on measuring mechanical forces between the needle and the surface, from

Fundamentals and Applications of Nano Silicon in Plasmonics and Fullerines.
http://dx.doi.org/10.1016/B978-0-323-48057-4.00007-4

which size is deduced. What is measured is deflection of the tip, which can be recorded and convert to a height (see Chapters 4 and 5). The deflections caused by this force allow the AFM to record topographic features of the surface. The device measures vertical as well as lateral sizes of nanostructures. The vertical dimension of the feature approach the real height; however the lateral dimensions are actually convolution of the dimension of the feature with the diameter of the tip. It should be noted that unlike STM, AFM profiling is not sensitive enough to image individual atoms. Because the tip may damage the surface, two modes of operation are used:

- Contact mode (repulsive mode): the probe is held a few nanometers above the sample. Due to the hard contact, the sample and tip interact in this mode via a repulsive force
- Non-contact mode (attractive mode): the probe is held a few tens of nanometers above the sample and set to vibrate above its natural frequency. In this mode, the tip and sample interact via an attractive molecular force. This mode is well-suited for studying biological and some polymeric specimens. However this mode provides poorer lateral resolution than the contact mode.

7.1.3 Transmission Electron Microscope

Electron microscopy uses electrons instead of photons (light); otherwise in principle they operate in a similar way to conventional light microscope (see chapter 4). The electron source is generally an electron gun. The electron beam is focused, by multiple stages of magnetic lenses, onto the sample. Data is collected from transmitted electrons (passed through the sample). The sample must be very thin to allow transmission. The resulting images are recorded on film or video. The microscope is used to image a wide range of thin solids, including metals and ceramics, as well as biological and organic samples. The device is capable of a resolution of 5 Å but it can be pushed to a resolution of 2 Å.

7.1.4 Dynamic Light Scattering Technology

Dynamic light scattering (DLS) techniques use light scattering to measure sizes much smaller than the wavelength of the light used [1]. To allow subwavelength resolution the particles must be monodispersed in a viscous liquid. The procedure circumvents the diffraction limit by using the diffusion characteristics of the particle in the liquid. The Stokes–Einstein theory relates the particle size to the particle motion in the liquid according to the following expression

$$D_h = \frac{k_B T}{3\pi\eta D_t}$$

where D_h is the particle size or the hydrodynamic diameter; D_t is the translational diffusion coefficient (we find this by dynamic light scattering), k_B is

Boltzmann's constant, T is thermodynamic temperature, and η is dynamic viscosity. For given η in the procedure, both the particle size and D_t can be determined by measuring the random changes in the intensity of light scattered from a suspension or solution. The technique is also called photon correlation spectroscopy (PCS).

DLS is most commonly used to analyze the size of nanoparticles. Examples of nanoparticle material include gold, protein, latex, and colloids. The technique is applicable to a wide range of sizes from submicron to less than a nanometer. In this size regime, and for the purpose of size measurement, the distinction between molecules and solid particle and even a second liquid phase (as in an emulsion) becomes blurred.

7.2 MATERIAL COMPOSITION: XPS, AES, EDX, NMR, RAMAN, FTIR

There are two major methods that can be used to obtain material composition of nanostructures. The first uses electron spectroscopy known as Auger electron spectroscopy (AES) after the French physicist Pierre Auger who studied the phenomenon. The second uses X-ray photoelectron spectroscopy (XPS). AES provides higher spatial resolution over XPS. However, the XPS can provide chemical information.

In the AES process, an electron is ejected from an inner shell of an atom, say in the K shell, the resultant vacancy is soon filled by an electron from one of the outer shells, say the L1 shell. The energy released from this transition could cause another electron of the outer shell, say L2 shell, to be ejected from the atom. This last electron is called a KL1L2 Auger electron. The emitted Auger electron will have the kinetic energy Ekin = EK−EL1−EL2, where the EK; EL1, and EL2 are the energies of the inner level, the first, and second outer shells, respectively. All of these energies are measured from the vacuum level. This kinetic energy, Ekin is characteristic of the emitting atom, thus, allows us to identify the element using some known tables.

In the AES process there is a large electron background due to direct scattering of incident electrons. To circumvent or eliminate this background, the signal is differentiated. This procedure works well because the background signal is expected to be a smooth function of energy.

It therefore is common to analyze the yield in differential form – $dN(E)/dE$ – which basically subtracts the background. It is to be noted that Auger electrons have very short mean free path in solids due to their low energies. Therefore the bulk of detected electrons come from the first few atomic layers of the surface (the top 5-nm layer) but from relatively large area ($\sim$1 mm^2). It is basically a surface sensitive spectroscopic technique yielding compositional information. By simultaneously exposing the surface to an ion flux which leads to sputtering (i.e., removal) of the surface atoms, one can obtain information about deeper layers and in general can plot the variation of composition with depth below the surface of a sample. This technique is known by the name of *Auger Depth Profiling.*

The second method uses XPS. It also used ion etching to arrive at information on layer thickness. For instance, it combines a sequence of ion etch cycles with XPS analyses to provide the information.

In energy dispersive X-ray (EDX) spectroscopy for material analyses, the normal electron beam of an SEM is used as the excitation source. The SEM system is furnished with an auxiliary a sub energy measuring X-ray detector. A thin, low mass element window is added to filter out signals from carbon, thus enables EDS for all elements from carbon on up in atomic mass. In the4 procedure an incident high-energy electron ejects an electron from deep level of an element in the sample. This creates a deep hole state. Soon after, characteristic X-rays are emitted by the element because an electron from a higher binding energy electron level falls into the core hole. The energy of the emitted X-ray is the difference of the electron level binding energies. In practice there are many ways to fill the hole, resulting in several X-ray peak energies when the sample has many atoms. If the identification of one peak is ambiguous, other peaks will often allow a reasonable elemental identification of the peak.

7.2.1 Nuclear Magnetic Resonance (NMR) [2]

The structure of a nucleus under the influence of an external magnetic field of strength B_0 is depicted in Fig. 7.1. Many nuclei act as a small magnets as depicted in Fig. 7.1A. The origin of the magnet is the spin of the nucleus. In fact protons and neutrons, each has spin 1/2, and the spin of the nucleus is the

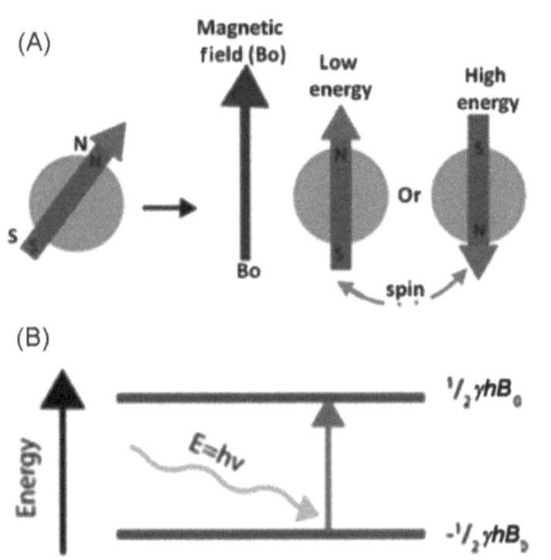

(C)

Nuclei	Gyromagetic Ratio (γ)	Spin
^{1}H	42.576	1/2
^{13}C	10.705	1/2
^{31}P	17.235	1/2
^{27}Al	11.103	5/2
^{23}Na	11.262	3/2
^{7}Li	16.546	3/2
^{29}Si	-8.465	1/2
^{17}O	5.772	5/2
^{15}N	-4.361	1/2

FIGURE 7.1 Nuclear magnetic resonance (NMR). (A) Many nuclei act as a small magnets. (B) Schematic of the two spin energy levels for the case of a spin ½ as in the proton of the hydrogen atom of energy ± ½ γhB_0. (C) A table of γ, the gyromagnetic ratio which characteristic of the nucleus (finger print) for some nuclei.

quantum sum of the spin of all protons and neutron in the nucleus. The presence of the magnetic field split the energy levels into two configurations. Fig. 7.1B shows a schematic of the two spin energy levels for the case of a spin ½ as in the proton of the hydrogen atom. The spin energy levels of nuclei are different from each other and act as finger prints. The energy of the two levels are ± ½ $\gamma h B_0$, where B_0 is the strength of the applied magnetic field, h is Planck's constant, and γ is the gyromagnetic ratio which characteristic of the nucleus (finger print). A table of γ for some nuclei are given in Fig. 7.1C.

When an external oscillatory magnetic field is now applied at a specific frequency as shown in Fig. 7.1B, it can change the state of the spin by an energy transfer and when the spin returns to its base level, energy is emitted at the same frequency. As spin of different nuclei absorb at different frequencies the method may be used to identify the nuclei hence the atom and the material. The most commonly used NMR nucleus include spin-½ nuclei, which include proton (^1H or hydrogen-1), as well as many other nuclei, such as ^{13}C, ^{15}N and ^{31}P. Many nuclei, such as deuterium (^2H or hydrogen-2) have a higher spin and are therefore quadrupole, and although they yield NMR spectra their energy diagram and some of their properties are different.

It is to be noted that the parameter γ, hence the magnetic frequencies, for a certain nucleus changes when it is placed in a chemical environment such as a solid, that is, depends on neighboring chemical environment. This procedure can be used to give information about the molecular environment in which the spin ½ nucleus resides. The external magnetic field is affected by electrons in nearby atoms hence on the chemical environment. The precise resonant frequency of the energy transition is dependent on the effective magnetic field at the nucleus. As a result, information about the nucleus' chemical environment can be derived from its resonant frequency. The chemical shift, using the equation $\delta = (\nu - \nu_0)/\nu_0$, is not dependent on the magnitude magnetic field.

7.3 STRUCTURAL ANALYSIS (DIFFRACTION)

Most of crystal structures are determined by conventional techniques, such as X-ray or neutron diffraction. However there are cases when un-conventional techniques must be used. For instance, electron diffraction enables crystal structures to be determined from sub-micron crystallites.

7.3.1 X-Ray Diffraction (XRD)

X-ray diffraction (XRD) has long been used to determine the atomic-scale structure of ordinary crystal materials. This technique is based on the fact that the wavelength of X-rays is comparable to the distances between atoms in condensed matter. When a material exhibiting a long-range (i.e., at least micrometers), periodic atomic order, such as a crystal, is irradiated with X-rays it acts as an extended, well-defined grating and produces a diffraction pattern showing numerous sharp spots, called Bragg diffraction peaks. By measuring and analyzing the positions and intensities of these peaks it is possible to determine the spatial characteristics of the grating – that is to determine the three-dimensional (3-D) arrangement of atoms in the crystalline material being studied.

7.3.2 Nanomaterial

For ordinary crystals, this information is obtained by traditional (Bragg) X-ray diffraction. It is difficult to use this approach on materials structured at the nanoscale because show few, if any, Bragg peaks, and have a pronounced diffuse component. A non-traditional approach based on high-energy X-ray diffraction and atomic pair distribution function data analysis may be used instead. In this case one needs to go beyond traditional X-ray diffraction. For instance, diffraction data at higher wave vectors ($\sim$20 Å^{-1} and higher) can be obtained using X-rays of shorter wavelengths, that is of higher energy. Such X-rays may be obtained from synchrotron or laboratory sources (e.g., sealed X-ray tubes with a Mo (energy $\sim$17 keV) or Ag (energy $\sim$22 keV) anode). For reference, the energy of Cu K_{α} radiation is only about 8 keV, and hence this is not suitable for high-energy XRD.

X-ray diffraction may be used to obtain the radial electron density distribution. The method provides information on the interatomic distances and coordination numbers in the structure and allows one to detect and estimate the size of nanostructures (phases) less than 3 nm in size, which cannot be detected by conventional powder X-ray diffraction analysis. The method is sensitive to the changes in the local structure measuring 1–3 unit cells; therefore, it is successfully used to identify various defects in nanoobjects of any chemical composition.

7.3.3 Selected Area Electron Diffraction (SAED)

Conventional electron diffraction has been rarely used as a standard tool for crystal identification mainly because the electron interactions with matter are about 10,000 times stronger than that of X-rays. As a result, the diffracted intensities are so much altered that they cannot be trusted, unless the crystal thickness is very thin. Selected area electron diffraction (SAED) patterns are a projection of the reciprocal lattice, with lattice reflections showing as sharp diffraction spots [3]. By tilting a crystalline sample to low-index zone axes, SAED patterns can be used to identify crystal structures and measure lattice parameters. SAED is essential for setting up dark-field (DF) imaging conditions. Other uses of SAED include analysis of: lattice matching; interfaces; twinning and certain crystalline defects. SAED of nanocrystals gives ring patterns analogous to those from X-ray powder diffraction, and can be used to identify texture and discriminate nanocrystalline from amorphous phases.

Precession electron diffraction (PED) is a modification to electron diffraction, which minimizes the effects of dynamical diffraction (which is basically

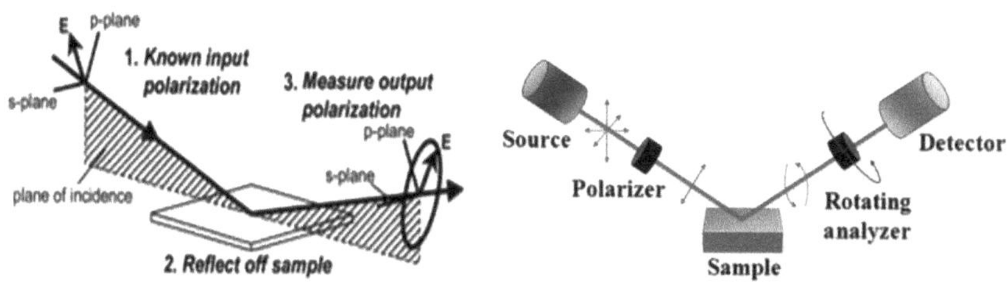

FIGURE 7.2 Typical ellipsometry configuration.

independent of the crystal structure) that can plague structural solution. Precession electron diffraction is applied to a range of different oxides for structure solution and refinement. As a result of this precession movement there is improvement over conventional SAED patterns, reflection diffracted intensity is much closer to the integrated intensity values, and it is much closer to kinematical (like X-Ray case). Precession Electron Diffraction is installed on JEOL 2010F and JEOL 2100LaB6.

7.4 NANOSOLID THICKNESS

There are several methods for measuring the thickness, as well as assessing the uniformity of nanosolid films made from nanoparticles. One procedure uses the optical-based method of spectroscopic ellipsometry. The other methods are mechanically based, including AFM and dektak.

7.4.1 Ellipsometry

Fig. 7.2 is a typical ellipsometry configuration [4]. A linearly polarized light beam is reflected from the sample surface and the polarization change is measured to determine the sample parameters. The first step starts with a light beam. It is linearly polarized by a polarizer with specific p- and s-component polarization in the plane and perpendicular, respectively. It is then reflected from the sample. The reflection process changes the amplitude and phase of both p- and s-polarized light. The new polarization components are then analyzed and measured by a polarizer. The changes are then obtained and related through theoretical analysis to the thickness of the sample. Finally, the thickness is obtained. In the procedure it is required to know the material from which the film is made of. In other words, the optical constants of the material (the refractive index and the absorption coefficient) must be entered.

The change in polarization is the ellipsometry measurement, commonly written as:

$$\rho = \frac{r_p}{r_s} = \left|\frac{r_p}{r_s}\right| \exp(i\Delta) = \tan\psi \exp(i\Delta)$$

where r_p and r_s is the reflection coefficient of the p- and s-polarizations. And Δ and ψ are the useable parameters. Common ellipsometer configurations include rotating analyzer (RAE), rotating polarizer (RPE), rotating compensator (RCE), and phase modulation (PME).

In addition to thicknesses, ellipsometry is used to examine the uniformity of the film, as well as to calibrate the density of the starting colloid. To estimate the film thickness, the optical constants were needed. Such application was employed for studying films made of ultrasmall silicon nanoparticles of average size of 3 nm (see Fig. 12.4). In the analysis, the nanoparticle film was treated as an extension of the SiO_2 layer. This is a good approximation, as the refractive indices (real part) were previously shown to be similar. Moreover, inaccuracy in the absolute thickness due to uncertainties in the optical parameters may not significantly bear on the uniformity of the film derived from ellipsometry measurements. Fig. 12.4 shows the thickness of the oxide film that was grown on a device quality silicon substrate and the measured thickness of the film after the nanoparticles were spin coated (SiO_2 + Si nanoparticles). The measurements give an oxide thickness of 4.2 ± .2 nm and a thickness of (2.5 ± .4 nm) for the nanoparticle film, indicating submonolayer coverage. The measurement indicates good global uniformity or coverage, an important criterion for integration of silicon nanoparticles across entire substrates for advanced device applications.

7.4.2 Profilometry

There is more than one type of profilometers [5]. In stylus profilometers, the device uses a probe to detect the surface. The probe is physically moved along the surface to acquire the surface height. This is done mechanically with a feedback loop that monitors the force from the sample pushing up against the probe as it scans along the surface. A feedback system is used to keep the arm with a specific amount of torque on it. Surface variations cause the stylus to be translated vertically. The changes in the Z position of the arm holder can then be used to reconstruct the surface. One commercial brand in this category is the Dektak stylus profilometer. The AFM is an advanced device that belongs to this class.

The device uses a diamond- or silicon-tipped stylus as shown in Fig. 7.3. The process is electromechanical in nature. The stylus is moved over the sample according to a programmed scan length, speed, and stylus force. The tip is controlled with high precision. The stylus is mechanically coupled to the core of a linear variable differential transformer. The scan drive moves the stage and sample. In some systems, the sample stage moves to allow for measurement, while in others the detector moves and in some both move.

Stylus profilometry requires force feedback and physically touching the surface, so while it is extremely sensitive and provides high Z resolution, it is sensitive to soft surfaces and the probe can become contaminated by the surface. This technique can also be destructive to some surfaces.

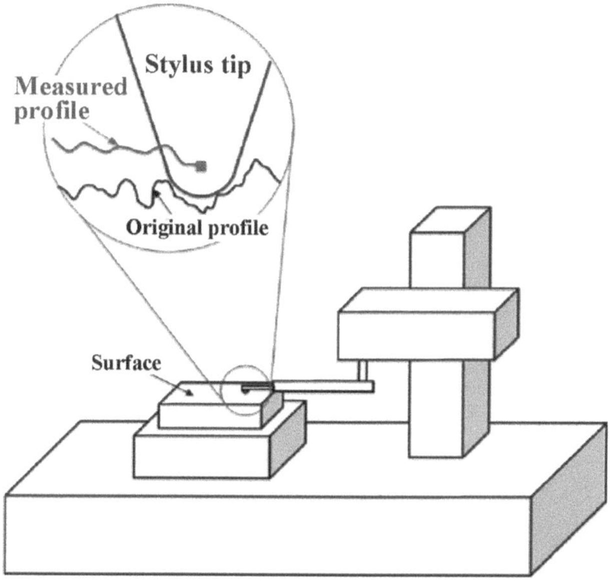

FIGURE 7.3 **Schematic of stylus profilometer used to measure the thickness of films mechanically.**

Because a stylus profilometer involves physical movements in *X*, *Y*, and *Z* while maintaining contact with the surface, it is slower than non-contact techniques. The stylus tip size and shape can influence the measurements and limit the lateral resolution.

7.5 OPTICAL, THERMAL, AND ELECTRICAL

Fluorescence or luminescent spectroscopy uses a beam of ultraviolet light to excite electrons in molecules. This causes them to emit light of lower energy (red-shifted). The emitted light can be within the visible and non-visible range. Detailed measurements of samples is performed by excitation using different light sources, including lasers, photodiodes, lamps, xenon arcs, mercury vapor, etc. A detector is attached at a viewing angle (usually around 90 degrees) as shown in Fig. 7.4A–B, which prevents incident light from polluting the detected fluorescent light.

In fluorescence microscopy, the sample or slide can either be fluorescing in its natural form like chlorophyll and some minerals, or treated with fluorescing chemicals. The excitation light is focused at the specimen and then the much weaker emitted light is detected and used to make up the image. First, the microscope has a filter on the excitation arm (or excitation monochrometer) that only lets through radiation with the desired wavelength. The radiation excites to a higher energy level atomic electrons in the sample. When the excited electrons relax to a lower level, they emit red-shifted light. The emitted light is separated

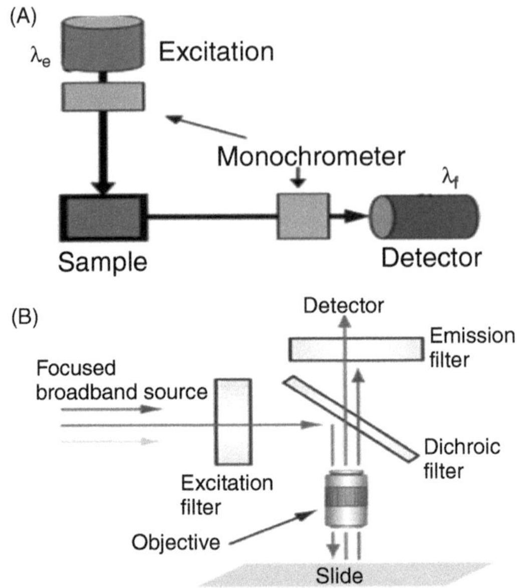

FIGURE 7.4 (A) Configuration of fluorescence or luminescent spectroscopy. It uses a beam of ultraviolet light to excite electrons in molecules. Emission is usually measured in a direction normal to the excitation beam. (B) Configuration of fluorescence microscopy: a filter on the excitation arm (or excitation monochrometer) only lets through radiation with the desired wavelength. Emitted light is separated from the much brighter excitation light using a second filter (emission monochrometer) in the emission arm.

from the much brighter excitation light using a second filter (emission monochrometer) in the emission arm.

Absorption spectroscopy is another optical process, which determines how much light is absorbed into the substance versus the amount of light that is reflected. These processes provide information about the excited states and hence the molecular structure as well as non-radiative relaxation processes of excited states.

7.5.1 Capacitance–Voltage (C–V, CV) Characteristics

In this technique a voltage is applied across a specimen, and the capacitance is recorded as a function of the applied voltage. The applied voltage may be DC voltage, or a sum of DC and a small-signal AC signal (the *conductance method)*, or a large-signal transient voltage. This measurement is used to determine an oxide thickness, oxide charges, contamination from mobile ions, and interface trap density, as well as shift in device operation threshold voltage (see Chapter 14 for example from nanomemory). The procedure applies to a wide range of electronic and photonic devices, such as transistors, memory cells, photovoltaic cells, etc.

7.5.2 Current–Voltage (*I–V*) Characteristic

In this application, a voltage or potential difference is applied across a specimen. The corresponding electric current through the sample is measured and the relationship between the current and voltage (*I–V* or *IV*) is recorded by varying the voltage applied. Such measurements are used to characterize electronic circuits, devices, or material. For example, a resistor exhibits a linear relationship.

7.5.3 Electro- and Cathodoluminescence

In electroluminescence (EL), low-energy electrons are used to bombard a sample to excite luminescence. In (cathodoluminescence CL), high-energy electrons are used instead of low energy electrons. Unlike photoexcitation, which can be tuned into resonance with the transitions of the sample, electron excitation is non-selective/non-resonant. Moreover, electron injection is weak, resulting in low luminescence yield and resolution. It is to be noted that the amenability of nanostructures or devices to electron excitation, enables integration on a silicon chip. Electroluminescence is the result of radiative recombination of electrons and holes in a material, usually a semiconductor. The excited electrons release their energy as photons (light). The phenomenon is therefore is a simultaneous optical phenomenon and electrical phenomenon.

7.5.4 Thermoluminescence

In certain crystalline materials, such as minerals, absorbed high energy radiation creates electronic excited states, which may get *trapped* or *arrested.* Such trapping remain for extended periods of time [6]. Energy gets stored by localized defects, or imperfections. But when the material is heated the trapped states are triggered to interact with lattice vibrations (phonons). The interaction causes them to rapidly decay, which results in the emission of photons.

7.6 THEORETICAL SIMULATION

Nanosystems do not comfortably fit traditional paradigms, such as periodicity and band structure used for describing the basic properties of solids or molecular orbital picture of traditional molecular physics. The mentioned variety of phenomena makes the theoretical investigations difficult and challenging. In particular, very often details in atomic and electronic structures can appreciably influence optical, magnetic, dielectric and conductivity properties, chemical reactivity, and stability of nanosystems.

The atomic and electronic structures are often investigated by a combination of methods [7–9]. For very large clusters, molecular dynamics and/or Monte Carlo methods with interatomic pair potentials are typically employed. For intermediate sizes say, clusters with the number of atoms between several tens and hundreds TB methods are used, sometimes combined with sophisticated optimizations, such as genetic algorithms in order to efficiently find the most promising candidates for structures of ground states. The density functional methods based on density functional theory (DFT) [10] are commonly used for sizes up to a few dozens of atoms with the time-dependent density functional theory (TDDFT) applied to study excited states and optical spectra. More advanced methods for excited states are based on the GW approach augmented with Bethe–Salpeter equation which captures excitonic (two-particle) effects [11,2]. Some of the most accurate calculations that have been carried out for these systems are based on the quantum Monte Carlo (QMC) methods [13–17]. Let as briefly describe the DFT and QMC approaches from the perspective of their applicability to nanosystems.

7.6.1 DFT Methods

In the two seminal papers of the DFT [10] Walter Kohn and his collaborators have shown that the ground state total energy of the system of electrons in the external field can be expressed as a functional of the single-particle electron density. Such an exact functional is claimed to exist and to be universal but it is unknown. Although at present the exact functional is out of reach, a number of approximate functionals have been suggested and proved very useful in practice. Historically the first one is the LDA based on the homogeneous electron gas model. In LDA the electron density is locally represented by the homogeneous electron gas of the same density. The properties of the homogeneous electron gas and its energy, in particular, are known, so one can write the total energy as an integral over the energy density of the homogeneous electron gas. The approach was then improved by incorporating the gradient of the electron density into the total energy in the form of the so-called generalized gradient approximation (GGA) [18].

The GGA functionals take into account, in an approximate way, the spatial variation of electron density and in most cases provide significantly improved estimations for energy differences and other quantities. Unfortunately, these improvements are not systematic and the quality of predictions can vary from system to system. The DFT methods proved to be remarkably productive and established themselves as a "sweet spot," which very successfully balanced the conflicting demands on computational speed and accuracy of predictions. The DFT methods have been tested on a full scale of electronic structure systems, such as molecules, solids, surfaces, etc. over the last three decades. It is also clear that the success of the approximate DFT functionals stems from the cancellation of errors between exchange, correlation, and the component of kinetic energy related to the electron–electron interaction. It is not difficult to verify that these three terms tend to compensate each other as the sum of exchange and correlation is smoother than each of them separately and the kinetic energy contribution from electron–electron interaction decreases the amplitude of the exchange–correlation having the opposite sign. The compensation works very well when comparing systems which have similar bonding patterns, charge densities, etc., such as a solid or a molecule close to equilibrium geometry. The DFT methods therefore often provide excellent description for equilibrium geometries. The cancellation is less effective when the variation in electron densities in the system is significant or when strongly localized states (such as *d*-electrons in transition elements) energetically compete with states which are more delocalized. In such cases, with typical examples being transition metal oxide solids, the DFT methods need to be modified either by *ad hoc* approaches such as LDA + U or by high-level perturbation methods, such as self-consistent GW approach. Very often, even change of bonding pattern, bond-breaking or non-equilibrium geometries can produce significant errors in the DFT predictions [19–23].

The DFT theory was originally presented as a ground-state approach, nevertheless, its use was extended to estimations of excited state properties as well. For example, one particle eigenvalues of the Kohn–Sham formalism can be interpreted as approximate quasi particle energies and therefore enable to extract estimations of band gaps. Although it was early recognized that the DFT gaps are lower than the true ones, typically by 30–50%, they provided reasonable starting guesses for many sp insulators. The extension of DFT to the TDDFT is based on similar ideas as the random phase approximation and configuration interaction with single (CIS) excitations in the post-Hartree–Fock methods [24]. The key difference is in the effective potential which in CIS method is the Fock potential with the corresponding orbital occupation while in the TDDFT the Fock potential is replaced by the corresponding GGA/LDA effective potential. In the simplest version the TDDFT theory exhibits similar underestimations of gaps as the quasi particle eigenvalues. To correct for the DFT bias for excited states it is necessary to include more sophisticated approaches, such as the GW method and beyond [11,12].

7.6.2 Quantum Monte Carlo

The QMC method is based on solving the quantum many-body problem, or, more precisely, on using the stochastic techniques for sampling the wave functions and for solving the corresponding quantum many-body problem, that is, the stationary Schrodinger equation [11,12]. This approach enables one to achieve high accuracy by describing the electron correlation effects explicitly and directly. In particular, QMC treats with an unprecedented efficiency the dynamical correlation, such as electron–electron cusp, which is very difficult to describe by more traditional methods based on expansions in basis sets. In a certain sense the QMC methods are complementary to the traditional correlated wave function approaches developed in quantum chemistry community, since what is easy in QMC is difficult in traditional approaches and vice versa [11,12]. A brief description of the two basic QMC approaches is given below.

7.6.3 Variation Monte Carlo

In the variation Monte Carlo (VMC) method, the expectation values are estimated from stochastic samples of electron configurations in real space. For example, the variation energy of an arbitrary trial wave function is given by an average of the local energy. The stochastic method of integration is the key ingredient as it enables exploring explicitly the correlated trial wave functions. The correlated trial wave function depends on the interelectronic distances and includes electron–electron cusps which are so difficult to capture by traditional basis set expansion methods. For very accurate calculations the VMC correlation part includes also the most important triple correlations of the electron–electron–ion type [12]. The commonly used VMC trial wave function is a product of linear combination of one (or a few) Slater determinant(s) and a correlation factor. Calculations done by the Mitas group of North Carolina State University and others demonstrate that this procedure works for a whole range of systems, such as atoms, molecules, solids, surfaces, etc. In many cases already this level of correlation is capable of providing very unique information such as accurate cohesive energies of diamond and silicon solids [13]. However, the choice of the trial function and its optimization using stochastic samples of electron configurations could be a source of variation bias. This is particularly important for comparison of two different systems (e.g., molecule and isolated atoms for estimation of atomization energies) because the bias can and often does vary from system to system.

7.6.4 Diffusion Monte Carlo

In order to eliminate a major part of variation bias the diffusion Monte Carlo (DMC) method is employed. It is based on the projection operator, which projects out the ground state of a given symmetry from any trial function with nonzero overlap. It is to be noted that DMC works also for excited states which are ground states of different symmetries. It can be applied also to genuine excited states, that is, which are not the ground states of the given symmetry, although the computational demands are significantly higher.

REFERENCES

[1] Schärtl W. Light Scattering from Polymer Solutions and Nanoparticle Dispersions. Berlin, Germany: Springer Science & Business Media; 2007.
[2] Trela KK. Nuclear Magnetic Resonance. London, UK: Royal Society of Chemistry; 2015.
[3] Selected Area Diffraction, Bartholomei Timotheos Crispinus Vertpress, 2011.
[4] Fujiwara H. Spectroscopic Ellipsometry: Principles and Applications. Hoboken, NJ: John Wiley & Sons; 2007.
[5] Balasubramanian Muralikrishnan Control and Implementation of Rapid Stylus Profilometerm, University of North Carolina at Charlotte.
[6] Pagonis V, Kitis G, Furetta C. Numerical and Practical Exercises in Thermoluminescence. Berlin, Germany: Springer Science & Business Media; 2007.

[7] Kumar V. Nanosilicon. Amsterdam, Netherlands: Elsevier; 2011.
[8] Tapan KS, Andrey LR. Complex-Shaped Metal Nanoparticles: Bottom-Up Syntheses and Applications. Hoboken, NJ: John Wiley & Sons; 2012.
[9] Hashem R-T. Computational Physics of Carbon Nanotubes. Cambridge, UK: Cambridge University Press; 2008.
[10] P. Hohenberg, W. Kohn, Inhomogeneous electron gas, Phys. Rev. 136, B864 (1964); W. Kohn, L.J. Sham, Self-consistent equations including exchange and correlation effects, Phys. Rev. 140, A1133 (1965).
[11] M.S. Hybertsen, S.G. Louie, Electron correlation in semiconductors and insulators: band gaps and quasiparticle energies, Phys. Rev. Lett. 55, 1418 (1985); Electron correlation in semiconductors and insulators: Band gaps and quasiparticle energies, Phys. Rev. B 34, 5390 (1986).
[12] M. Rohlfing, S.G. Louie, Excitonic effects and the optical absorption spectrum of hydrogenated Si clusters, Phys. Rev. Lett. 80, 3320 (1998); J.C. Gorssman, M. Rohlfing, L. Mitas, S.G. Louie, M.L. Cohen, High accuracy many-body calculational approaches for excitations in molecules, Phys. Rev. Lett. 86, 472 (2001).
[13] D.M. Ceperley, M.H. Kalos, Quantum many-body problems, in: K. Binder (Ed.), Monte Carlo Methods in Statistical Physics, Springer, Berlin, 1979, pp. 145–194; K.E. Schmidt, D.M. Ceperley, Monte Carlo techniques for quantum fluids, solids and droplets, in: K. Binder (Ed.), Monte Carlo Methods in Statistical Physics II, Springer, Berlin, 1984, pp. 279–355.
[14] Hammond BL, Lester Jr. WA, Reynolds PJ. Monte Carlo Methods in Ab Initio Quantum Chemistry. Singapore: World Scientific; 1994.
[15] Anderson JB. Exact quantum chemistry by Monte Carlo methods. In: Langhoff SR, editor. Understanding Chemical Reactivity. Dordrecht: Kluwer; 1995. p. 1–45.
[16] Ceperley DM, Mitas L. Quantum Monte Carlo methods in chemistry. In: Prigogine I, Rice SA, editors. Advances Chemical Physics. New York: Wiley; 1996. p. 1–38. XCIII.
[17] Foulkes MWC, Mitas L, Needs RJ, Rajagopal G. Quantum Monte Carlo simulations of solids. Rev. Mod. Phys. 2001;73:33.
[18] Ernzerhof M, Perdew JP, Burke K. Density functionals: Where do they come from, why do they work? In: Nalewajski RF, editor. Topics in Current Chemistry. Berlin: Springer; 1996. p. 180.
[19] Grossman JC, Mitas L. High accuracy molecular heats of formation and reaction barriers: essential role of electron correlation. Phys. Rev. Lett. 1997;79:4353.
[20] Torelli T, Mitas L. Electron correlation in C_{4N+2} carbon rings: aromatic versus dimerized structures. Phys. Rev. Lett. 2000;85:1702.
[21] Mitas L, Martin RM. Quantum Monte Carlo of nitrogen: atom, dimer, atomic, and molecular solids. Phys. Rev. Lett. 1994;72:2438.
[22] J.C. Grossman, L. Mitas, Quantum Monte Carlo of nitrogen: atom, dimer, atomic, and molecular solids, Phys. Rev. Lett. 74, 1995, 1323; L. Mitas, J.C. Grossman, I. Stich, J. Tobik, Silicon clusters of intermediate size: energetics, dynamics, and thermal effects, Phys. Rev. Lett. 84, 2000, 1479.
[23] Grossman JC, Mitas L, Raghavachari K. Structure and stability of molecular carbon: importance of electron correlation. Phys. Rev. Lett. 1995;75. 3870.
[24] S. Hirata, M. Head-Gordon, R.J. Bartlett, Configuration interaction singles, time-dependent Hartree–Fock, and time-dependent density functional theory for the electronic excited states of extended systems, J. Chem. Phys. 111, 1999, 10774; I. Vasiliev, S. Ogut, J.R. Chelikowsky, *Ab Initio* absorption spectra and optical gaps in nanocrystalline silicon, Phys. Rev. Lett. 86, 2001, 1813.

Chapter 8

Nanometal

8.1 NATURE OF INTERACTION OF LIGHT WITH METAL

In a dielectric, all electrons are bound in atoms; and each atom interacts with light individually through the interaction of a single bound electron. The total effect in a sample is simply the sum of the individual atomic responses. In a metal solid, some electrons are bound to atoms and others are not bound in specific atoms or ions. Upon interaction with light, two processes can take place in metal. In inter band transition, bound electrons, that is, electrons in the valence band can be promoted due to light absorption to an empty level in the conduction band where they become not bound to a specific ion. Interband absorption form a significant loss mechanism in metal at optical frequencies. On the other hand in a metal solid, electrons that are already in the conduction band form a see or cloud of electrons not bound to specific atoms or ions. The cloud can interact with and move under an external electric force collectively at the same time.

8.1.1 Plasma Model

Fig. 8.1A shows a schematic of the lattice of bulk metal where electrons are subjected to a force created by the electric field of light. Free electrons opposite to the direction of the electric field while positive ions are stationary, thus causing a shift between the center of distribution of the negative charge and the positive charge. In the absence of the force the centers of the two distributions coincide. Fig. 8.1B shows the same situation for a very small sphere of metal, namely a nanoparticle.

Electrons in a small sphere are described by a simplified model of a gas of free electrons that moves against a fixed background of positive ion cores [1–3]. This has been called a *plasma model*. In the model, details of the lattice potential and electron–electron interactions are not included explicitly; rather they are simply incorporated into the effective optical mass m of each electron. Under irradiation, the negative electrons sea or cloud gets pushed back and forth relative to the fixed background of positive ion at the frequency of oscillation of the electric field of the electromagnetic light wave. In addition, the cloud

Fundamentals and Applications of Nano Silicon in Plasmonics and Fullerines.
http://dx.doi.org/10.1016/B978-0-323-48057-4.00008-6

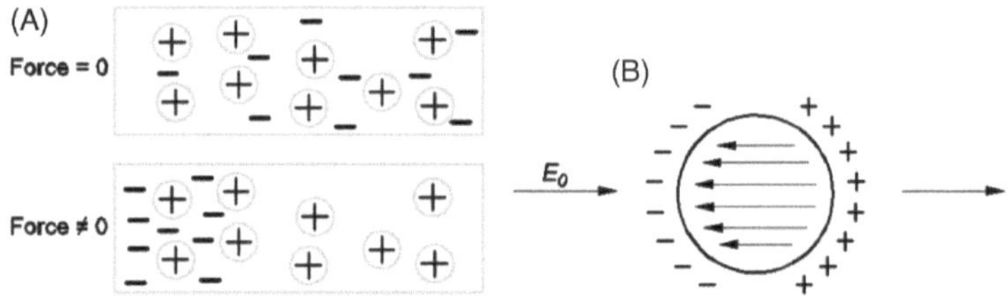

FIGURE 8.1 Sketch of metal lattice subjected to a force created by an electric field. (A) Bulk and (B) nanoparticle

experiences an effective drag force due to collisions with the positive ion core which tends to slow it down. According to the plasma model, electrons oscillate and their motion is damped with a characteristic collision frequency γ as high as 10^{14} per second (100 THz) (corresponding to the frequency of infrared light). The time elapsed between two consecutive collisions, $\tau = 1/\gamma = 10^{-14}$ s, is known as the relaxation time of the free electron gas. The balance of these two forces gives the response or what is called the real part of the dielectric function of the cloud, $\in(\omega)$, to the light wave:

$$\in(\omega)=1-\frac{ne^2}{\in_0 m\omega^2}$$

where n is the density of electrons in the cloud, e and m are the charge and the effective mass of a single of electron, $\in_0$ is the dielectric function of vacuum, and ω is the frequency of the incoming light wave. It is customary to define a useful quantity called plasma frequency or plasmon frequency ω_p in terms of the following group of constants:

$$\omega_p=\sqrt{\frac{ne^2}{\in_0 m}}$$

In terms of the plasma frequency, the dielectric function takes the simple form $1-\omega^2/\omega_p^2$. For a metal with a free electron density of $10^{23}/cm^3$, ω_p corresponds to the frequency of ultraviolet light (plasmon energy $\hbar\omega_p \sim 10eV$) which is 100 or even 1000 times larger than the relaxation frequency γ.

Fig. 8.2 plots $\in(\omega)$ as a function of frequency. At resonance with the plasma frequency ($\omega = \omega_p$), $\in(\omega)$ drops to zero. This is a plasma oscillation resonance. For light with a frequency below the plasma frequency, the dielectric function is negative and the light cannot penetrate the sample, rather it is totally reflected. Above the plasma frequency the light waves penetrates the sample as sketched in the figure [4].

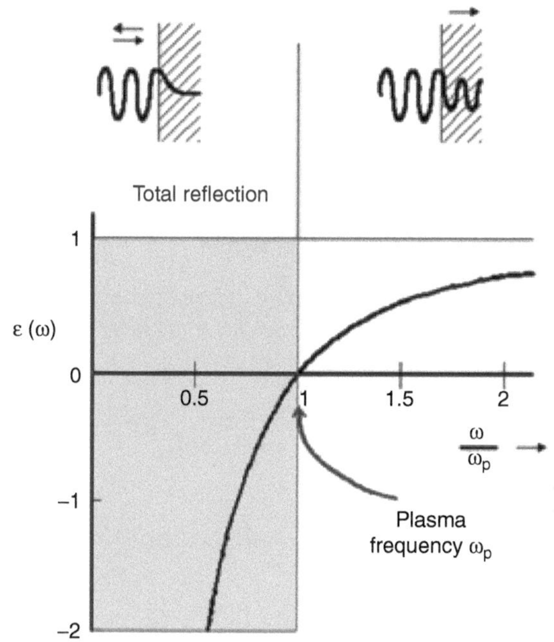

FIGURE 8.2 Plot of the dielectric function $\in(\omega)$ as a function of frequency for a metal. $\in(\omega)$ drops to zero at resonance with the plasma frequency, ($\omega = \omega_p$). (Image from E. Margapoti, Plasmonics: Fundamentals and Applications, 2012, Technical University München, Munich, Germany, Summer semester; https://www.coursehero.com/file/10546609/Plasmonics/).

Example 8.1 Plasmon frequency (Drude model) [4–5]

Consider an electric field polarized along the x-axis with a harmonic time dependence

$$E = E_0 \hat{x} e^{-i\omega t}$$

For an ensemble of electrons of mass m and charge e and density n, the electric field exerts a force of $F = -eE\hat{x}$ on each electron. In addition, let us assume there is a drag force acting on each electron. $F_d = -m\gamma\hat{x}dx/dt$.

The Newton's equation of motion of an electron under the influence of these two forces is

$$md^2x/dt^2 + m\gamma\, dx/dt = -eE$$

$$md^2x/dt^2 + m\gamma\, dx/dt = -e\, E_0\, e^{-i\omega t}$$

The particular forced solution response of the electron must have the following form: $x = x_0\, e^{-i\omega t}$. Substituting this expression in the equation of motion and factoring out the exponential factor $e^{-i\omega t}$ gives an expression for x_0/E_0

$$\frac{e}{m\left(\omega^2+i\gamma\omega\right)}$$

hence

$$\vec{x}(t)=\frac{e}{m\left(\omega^2\gamma\omega\right)}\vec{E}(t).$$

The electric field therefore results in an induced electric dipole – ex. We can now sum up the effect for all electrons within a unit of volume, namely *n* write down expression for the polarization

$$\vec{P}=-ne\vec{x}(t)=-\frac{ne^2}{m\left(\omega^2+i\gamma\omega\right)}\vec{E}(t).$$

$eE = D = e0\ E + e0P$.

From the polarization expression P we can get the dielectric permittivity

$$\varepsilon(\omega)=\frac{D(\omega)}{\epsilon_0 E(\omega)}=1-\frac{P(\omega)}{\epsilon_0 E(\omega)}$$

Substituting for *P* we get

$$\varepsilon(\omega)=1-\frac{\omega_p^2}{\omega^2+i\gamma\omega},$$

where

$$\omega_p^2=\frac{ne^2}{\epsilon_0 m},$$

According to the plasma model, electrons oscillate and their motion is damped with a characteristic collision frequency γ as high as 10^{14} per second (100 THz), (corresponding to the frequency of infrared light). The time elapsed between two consecutive collisions, $\tau = 1/\gamma = 10^{-14}$ s, is known as the relaxation time of the free electron gas. This is the time scale in the problem. There are three time or frequency regimes:

1. for low frequency, such that $\gamma\omega << \omega^2$ or $\omega\tau >> 1$ one can neglect the imaginary term $i\gamma\omega$ and then we have

$$\varepsilon(\omega)=1-\frac{\omega_p^2}{\omega^2}.$$

In Fig. 8.2, a sketch of the dielectric function versus the frequency is shown. Electromagnetic waves propagate without damping when $\in$ is positive and real, while is totally reflected when ε is negative.

2. In general, the dielectric function can be written as: $\in(\omega) = \in_1(\omega) + i\in_2(\omega)$, where $\in_1(\omega)$ is the real part and $\in_2(\omega)$ the imaginary part. At optical frequencies, $\in(\omega)$ can be experimentally determined via reflectivity studies.
We can also define the complex refractive index as

$$\tilde{n}(\omega)=n(\omega)+ik(\omega)$$

also defined as $\tilde{n}(\omega)=\sqrt{\varepsilon(\omega)}=\frac{\varepsilon_2}{2n}$,

known as *extinction coefficient* and determine the optical absorption of electromagnetic waves propagating trough the medium: $\alpha(\omega)=\frac{2k(\omega)\omega}{c}$, which is known as Beer–Lambert law.

Example 8.2 Effect of interband transitions in metals (silver and gold)
(Drude model and Lorentz-oscillator model)

In the case of Gold, this validity is already broken at the boundary between visible and near-infrared. The optical properties of gold and silver at visible frequencies are described by the function:

$$m\frac{d^2\vec{x}}{dt^2}+m\gamma\frac{d\vec{x}}{dt}+m\omega_0^2\vec{x}=-e\vec{E}.$$

Interband transitions are described using the classical picture of a bound electron with resonance.

Combination of Drude model and Lorentz-oscillator model generates a good agreement between theory and experiments.

8.2 MINIATURIZED METAL: SUB-WAVELENGTH CONCENTRATION OF LIGHT

We now discuss how the interaction of metal with light manifest itself as we reduce the size of the metal sample from large bulk samples to small nanoparticles. We will compare three kinds of structures sketched in Fig. 8.3: bulk, which is labeled as three dimensional (3D); a sheet or quantum well, which is labeled

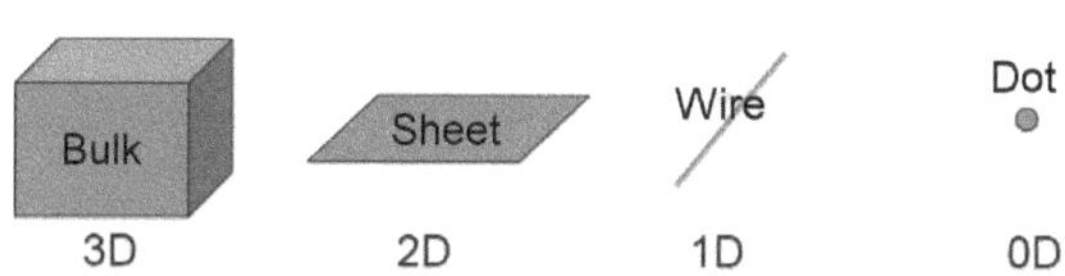

FIGURE 8.3 Schematic of four kinds of structures: bulk, a sheet or quantum well, quantum wire, and nanoparticle or quantum dot.

as two dimensional (2D); quantum wire, which is labeled as one-dimensional (1D); and nanoparticle or quantum dot, which is labeled as zero-dimensional (0D).

8.2.1 Bulk Material (3D)

Because free conduction electrons can move over very large distances in bulk metal objects, electrons on average do not oscillate against a certain localized ions with a restoring force. Instead the motion is actually similar to a mass being dragged in a viscous fluid. If the light has a frequency above the plasma frequency (in the ultraviolet (UV) range for metals), electrons will not even oscillate and the light will simply be transmitted or absorbed in interband transitions of the metal. If the light has a frequency below UV, electrons will oscillate out of phase with the incident light, causing a strong reflection. At the plasma frequency, the dielectric function is 0.

8.2.2 Thin Film or Sheet (2D)

When bulk metal is shrunk to a thin film of a few nm thicknesses, electrons cannot move very far in the direction normal to the film. In this case there will be electron oscillations upon light exposure; but they will only exist at the surface of the film. This is accompanied with propagation of charge waves along the film. This is known as surface plasmon polaritons (SPPs) as sketched in Fig. 8.4A. In fact, detailed Maxwell's theory shows that the surface waves can propagate along the surface with a broad spectrum of frequencies from 0 up to $\omega_p/\sqrt{2}$, where ω_p is the plasma frequency defined above. At $\omega_p/\sqrt{2}$ the dielectric function is -1 with a charge-wave wavelength shorter than the wavelength of the incident light. The combined effect is a mixed or hybrid light-electron-wave-state. This results in intense light–matter

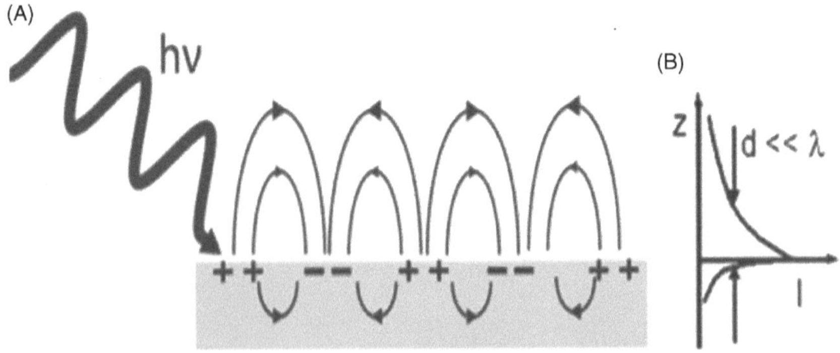

FIGURE 8.4 Schematic of light interaction with a few nm-thin metal films (A) shows electron oscillations at the surface of the film, accompanied with propagation of charge waves along the film. (B) Sketch of the field intensity with distance from the surface *(Adapted from https://en.wikipedia.org/wiki/Surface_plasmon)*.

interactions with unprecedented optical response. As shown in Fig. 8.4B, the intense optical field resulting from the hybrid field is a local one, extending outside the thin film into the dielectric only few nanometers, a distance much less than the wavelength of the incident light ($d \ll \lambda$). Thus visible light, which has a wavelength of approximately half a micrometer can be concentrated by a factor of nearly 100 to travel through metal films just a few nanometers (nm) thick.

8.2.3 Nanowire (1D)

A variety of metal nanowires has been fabricated and studied. They may come as solid nanowire of metal with a surrounding dielectric jacket, that is, glass

Example 8.3 Confinement and plasmon frequency

Detailed analysis of EM scattering using the drude model can yield the plasmon frequency for several geometries: bulk, surface and sphere according to the following table [5].

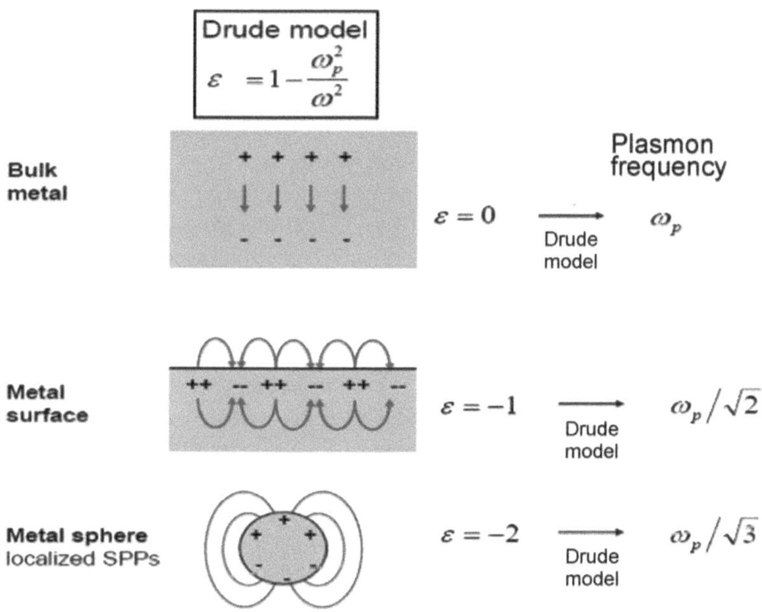

For dielectric media, the permittivity $\in$ (hence the dielectric constant) is a complex quantity; that is we can write it in terms of real and imaginary quantities: $\in = \in_1 + i\in_2$. $\in_1$ governs the refraction while $\in_2$ governs the absorption of light in the material. A bulk plasmon occurs at ω_p when $\in_1 = 0$. A surface plasmon occurs at $\omega_s^2 = \omega_p^2/2$ when $\in_1 = -1$.
http://uw.physics.wisc.edu/~himpsel/551/Lectures/Plasmonics.pdf [5].

or air. For example, chemically prepared silver nanowires ~100 nm diameters were fabricated [6]. The wires were found to support surface plasmon modes propagating along the wires. The wavelength of the propagating charge wave is shortened to about half the wavelength of exciting light. The propagation length of SPP is about 10 μm. The reflectivity at both ends of the wire is about 25%. Those characteristics are sufficient for the wires to be used as an optical instrument, namely surface plasmon Fabry–Perot resonators. Wires can also be in the form of a cylindrical shell of metal. The shell may have an inner dielectric jacket, such as glass (silicon oxide). The bore of the system may be filled with a semiconductor, such as Si or CdSe. More complicated variations of these architectures have been fabricated to accommodate ports for light entry or extraction. Semiconductor wires with hemispheres of metal on both ends have also been fabricated. The schemes involving hybrid semiconductor-plasmonic (metal) architectures will be discussed in Section 4.1 on integrating optics with electronics.

8.2.4 Nanoparticles/Dot (0D)

Bulk is now shrunk in three dimensions to form particles, such that the dimension of the particle is less than the wavelength of incident light. Because the size of the particle is small compared to the wavelength the incident electric field will be constant across the nanoparticle, inducing a uniform displacement of the electron density, making the electron motion in phase (collective cloud motion). Because electrons are confined within the small particle, they will oscillate while being accompanied by a strong restoring force from specific positive ionic core background (Fig. 8.5A). The restoring force leads to non-propagating (called also evanescent) collective oscillation of the surface cloud with a characteristic oscillation frequency similar to a simple harmonic oscillator. This is unlike bulk which does not have a specific oscillation frequency. Fig. 8.5B shows representative field lines, effectively resembling those of an oscillating electric dipole. Another striking effect is that the collective oscillations lead to a large absorption and scattering cross-section, as well as an amplified local optical electromagnetic field. The electric field has its maximum strength just outside the nanoparticle and drops rapidly with distance extending only to ~30 nm. For small particles less than ~15 nm, light absorption dominates; whereas for nanoparticles greater than ~15 nm scattering dominates. In fact, detailed Maxwell's theory shows that surface waves can be resonantly sustained with frequency at $\omega_p/\sqrt{3}$, which corresponds to dielectric function of −2. We should stress because the evanescent light is non-propagating it is localized within the approximate radius of the particle, even if the size of that particle is much smaller than the incident light wavelength (Fig. 8.5C). The region where the non-propagating evanescent light is localized is related only to the object's size and not to the wavelength, which can be considered as light with no diffraction limit. The nanoparticle practically confines light into super intense "hot-spot."

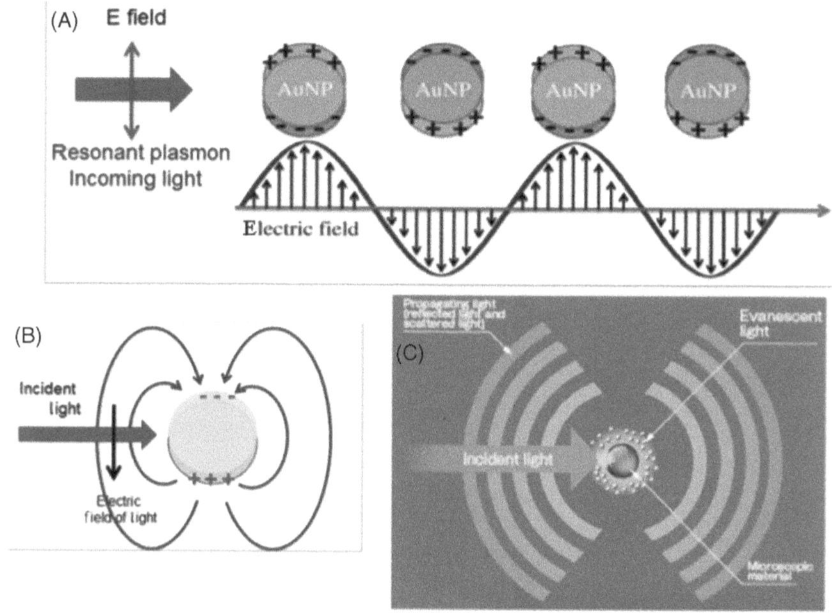

FIGURE 8.5 Schematic of the interaction of a gold nanoparticle with linearly polarized light beam. (A) Electrons oscillate while being accompanied by a strong restoring force from specific positive ionic core background. (B) Representative field lines, effectively resembling those of an oscillating electric dipole. (C) Region where the non-propagating (evanescent) light is localized, related only to the size of the nanoparticle, not to the wavelength. *(Adapted from E. Yasun, H. Kang, H. Erdal, S. Cansiz, I. Ocsoy, Y.-F. Huang, W. Tan, Cancer cell sensing and therapy using affinity tag-conjugated gold nanorods, Interface Focus 3 (2013) 0006: http://dx.doi.org/10.1098/rsfs.2013.0006; Image adapted from V.G. Veselago, The electrodynamics of substances with simultaneously negative values of ε and μ, Sov. Phys. Usp., 10 (1968), 509–514).*

Example 8.5 Plasmonic frequency of a gold nanoparticle [7]

This is an electromagnetic exercise, which the reader may find in many EM books. Light scattering by spherical particles in the quasistatic limit may be understood in terms of a conducting sphere in an external electric field. It is addressed by solving the Laplace equation for the scalar electric potential

$$\nabla^2\phi = 0 \text{ and } E = -\nabla\phi \tag{1}$$

with continuous boundary conditions

$$\phi_1|_a = \phi_2|_a \text{ and } \varepsilon_p \frac{\partial\phi_1}{\partial r}\bigg|_a = \varepsilon_m \frac{\partial\phi_2}{\partial r}\bigg|_a \tag{2}$$

where $E = E_0\vec{z}$, a, ϵ_p and ϵ_m are the electric field along z direction, radius of particle, dielectric function of the particle, and that of the medium, respectively. By comparing the scattering potential from the Laplace equation with that of a dipole, the effective dipole moment can be expressed as

$$P = 4\pi\varepsilon_m a^3 \frac{\varepsilon_p - \varepsilon_m}{\varepsilon_p + 2\varepsilon_m} E_0$$

The cross-sections for scattering (C_{sc}) and adsorption (C_{abs}) are obtained from the scattering field radiated by this dipole, which is induced by the incident plane wave (i.e., the dipole moment P). The resulting expressions are

$$\begin{aligned} C_{sc} &= \sigma_{geom} Q_{sc}, Q_{sc} = \frac{8}{3} q^4 \left| \frac{\varepsilon_d - 1}{\varepsilon_d + 2} \right|^2 \text{ and} \\ C_{abs} &= \sigma_{geom} Q_{abs}, Q_{abs} = 4q\mathrm{Im}\left[\frac{\varepsilon_d - 1}{\varepsilon_d + 2} \right] \end{aligned} \quad (3)$$

where $\sigma_{geom} = \pi a^2$ is the geometrical cross-section. We introduce the dimensionless cross-sections of scattering (Q_{sc}) and adsorption (Q_{abs}), the dimensionless size $q = ka$ and the relative dielectric function $\in_d = \in_p/\in_m$ for convenience in the discussion that follows. In the formula, k is the wave vector in medium.

The above formula implies that as the particle size is decreased, the efficiency of adsorption will dominate over the scattering efficiency. Therefore, particles with very small size will be difficult to detect by light scattering. In addition, one may note that there is a resonant enhancement for scattering and adsorption when the condition $Re(\in_d) = -2$ (the Fröhlich condition) is satisfied. This resonance is due to resonant excitation of the dipole surface plasmon. With the Drude model of dielectric function, the frequency of the dipole surface plasmon can be expressed as $\omega_{sp} \approx \omega_p/\sqrt{3}$, if the frequency of electron collisions γ is small. The spectrum for the scattering cross-section then has a Lorentz line shape,

$$Q_{sc}^{(Ra)} = \frac{8}{3} \frac{\omega_{sp}^4}{\left(\omega^2 - \omega_{sp}^2\right)^2 + \omega^2\gamma^2} q^4 \quad (4)$$

Light scattering and surface plasmons on small spherical particles Open
Xiaofeng Fan, Weitao Zheng and David J Singh
http://www.nature.com/lsa/journal/v3/n6/full/lsa201460a.html

8.3 MINIATURIZATION-INDUCED COLORATION OF METALS

Fig. 8.6B shows a large block and a nanoparticle of gold under illumination by white light. When light strikes the large block, the red and green components are reflected; while the blue component is absorbed and scattered. Interband transitions of bulk gold give it a yellow color. But a metallic luster is also added to the

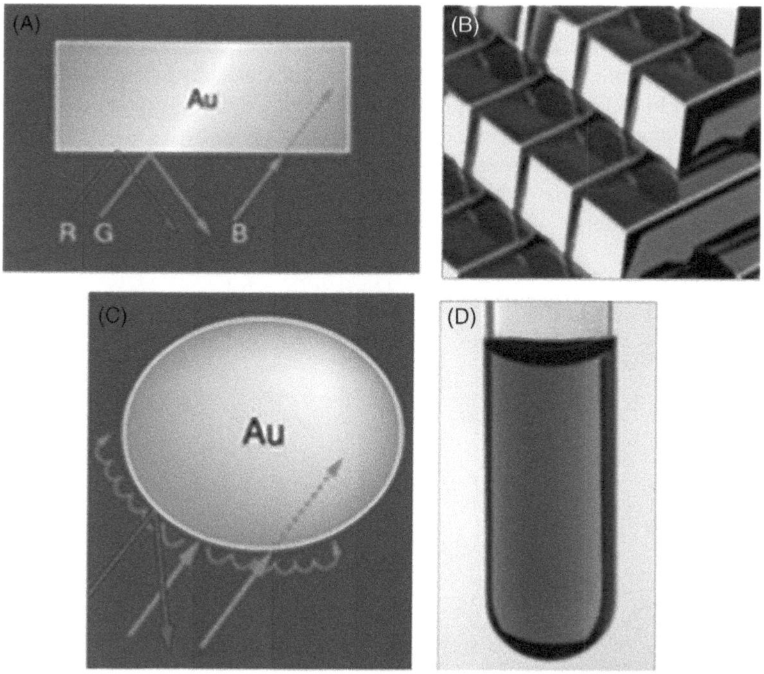

FIGURE 8.6 Schematic of the interaction (reflection vs absorption) of gold with white light. (A) A large block of gold under illumination by the red, green and blue components of white light. (B) True color of large blocks of gold in white light. (C) A gold nanoparticle under illumination by the red, green and blue components of white light. (D) True color of a colloid of gold nanoparticles in white light. *(Adapted from: V.G. Veselago, The electrodynamics of substances with simultaneously negative values of* ε *and* μ*, Sov. Phys. Usp., 10 (1968), 509–514).*

yellow color, which appears as a golden color (Fig. 8.6B). Charge oscillations in bulk do not have a specific frequency to cause specific coloration.

When white light strikes, on the other hand, the nanoparticle shown in Fig. 8.6C, the color will be different. This is due to a specific plasmon resonance due to a confined collective oscillation of electrons. A plasmon resonance in gold nanoparticles occurs over a thin slice of light at a frequency in the green. Thus not all of the components of incoming white light resonates and is absorbed. Among the RGB components, only the G (green) component resonates with the electrons and is absorbed in the gold nanoparticle. The B (blue) light is absorbed and scattered, and only the remaining R (red) component gets reflected or passes through. This is the reason that stained glass mixed with gold nanoparticles appears red to the naked eye. Fig. 8.6D demonstrates that, when light strikes a gold nanoparticle colloid, only the red color is reflected or passes through [8].

Example 8.6 Mie theory for metal nanoparticles/color [9]

Consider a spherical particle of radius r_0, complex electric permittivity $\in = n^2$ embedded in a dielectric medium of permittivity $\varepsilon_m = n_m^2$. This particle is illuminated by a plane wave of angular frequency $\omega = 2\pi c/\lambda_0 = kc/n_m$. Let us define from now on a set of useful dimensionless parameters: $m = n/n_m$, $v = kr_0$ and $\omega = mv$. In these conditions, the extinction, scattering, and absorption cross sections are given by the formulae:

$$\sigma_{ext} = \frac{2\pi}{k^2}\sum_{j=1}^{\infty}(2j+1)\,\mathrm{Re}\left(a_j + b_j\right)$$

$$\sigma_{sca} = \frac{2\pi}{k^2}\sum_{j=1}^{\infty}(2j+1)\left(\left|a_j\right|^2 + \left|b_j\right|^2\right)$$

$$\sigma_{abs} = \sigma_{ext} - \sigma_{sca}$$

$$a_j = \frac{m\psi_j(w)\psi'_j(v) - \psi_j(v)\psi'_j(w)}{m\psi_j(w)\xi'_j(v) - \xi_j(v)\psi'_j(w)}$$

$$b_j = \frac{\psi_j(w)\psi'_j(v) - m\psi_j(v)\psi'_j(w)}{\psi_j(w)\xi'_j(v) - m\xi_j(v)\psi'_j(w)}$$

In these expressions, ψ_j and ζ_j are Ricatti–Bessel functions defined as:

$$\psi_j(x) = \sqrt{\frac{\pi x}{2}}\, J_{j+\frac{1}{2}}(x)$$

$$\xi_j(x) = \sqrt{\frac{\pi x}{2}}\left[J_{j+\frac{1}{2}}(x) + iY_{j+\frac{1}{2}}(x)\right]$$

Jv and Yv are the Bessel functions of first and second order, respectively. They are standard Matlab functions, named respectively besselj and bessely. This theory and a Matlab code can be found in

http://www.guillaume.baffou.com/pdf/Mie_Theory.pdf
Mie theory for metal nanoparticles, Guillaume Baou
J. Phys. Chem. C119, 28586–28596(2015).

Example 8.7 The Mie scattering theory for dielectric particles

Consider a dielectric sphere of radius *a*, and refraction index m_p embedded in a uniform infinite medium of refractive index m_h, with a relative refractive index $m = m_p/m_h$. The magnetic permittivity of the sphere and the surrounding medium is assumed to be 1. The sphere is non-absorbing, hence the dielectric permittivity is real and is taken $\varepsilon_p = m_p^2$. The permittivity of the medium is $\varepsilon_h = m_h^2$.

The sphere is illuminated with a plane light wave of frequency ω, which is linearly polarized along the *x*-axis and propagating along the *z*-axis as shown in the Fig. 8.7.

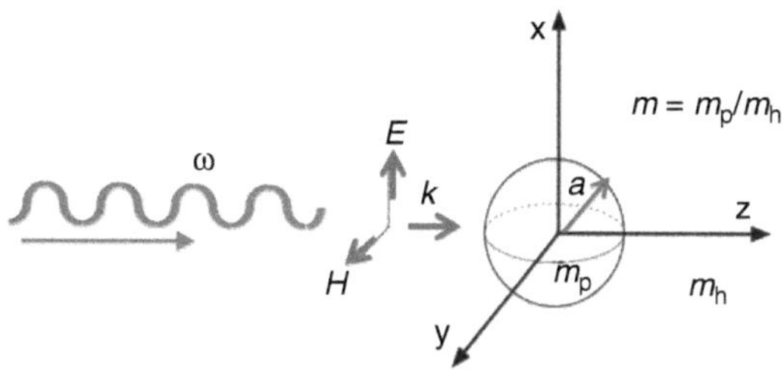

FIGURE 8.7 Presentation of Mie scattering theory for dielectric nanoparticles.

The incident electric and magnetic fields *E* and *B* can be written as:

$$E = |E_0 u_Z e^{ikX} e^{-i\omega t}, B = B_0 u_Y e^{-ikX} e^{-i\omega t}$$

where $k = m_h\omega$, $k = 2\pi/\lambda$ and λ is the wavelength in vacuum and $B_0 = \mu_0 H_0 = (m_h/c)E_0$.

According to the Mie treatment, the field scattered by the sphere can be decomposed into a multipole series:

$$a_n = \frac{1}{2}\left(1 - e^{-2i\alpha_n}\right) = i \sin\alpha_n e^{-i\alpha_n}$$

$$b_n = \frac{1}{2}\left(1 - e^{-2i\beta_n}\right) = i \sin\beta_n e^{-i\beta_n}$$

Where a_n and b_n are the coefficient of the expansion, which depend on phase angles α_n and β_n, and *n* is integer that spans the range 1, 2, ... It is important to note that the coefficients a_n and b_n are associated with the electric and magnetic fields, respectively. Note that in absence of absorption, that is, for real m, the phase angles α_n and β_n are real.

The angles are given by the following complicated expressions in terms of the spherical Bessel and Neumann functions $j_n(x)$ and $y_n(x)$ and the primes indicate differentiation with respect to the argument:

$$\tan\alpha_n = \frac{m^2 j_n(y)\left[x j_n(x)\right]' - j_n(x)\left[y j_n(y)\right]'}{m^2 j_n(y)\left[x y_n(x)\right]' - y_n(x)\left[y j_n(y)\right]'},$$

$$\tan\beta_n = \frac{j_n(y)\left[x j_n(x)\right]' - j_n(x)\left[y j_n(y)\right]'}{j_n(y)\left[x y_n(x)\right]' - y_n(x)\left[y j_n(y)\right]'},$$

Where $x = ka$, the size parameter and $y = mx$. The extinction and scattering cross sections, σ_{ext} and σ_S, respectively, are infinite sums in terms of these coefficients and angles

$$\sigma_S = \sigma_{ext} = \frac{2\pi}{k^2}\sum_{n=1}^{\infty}(2n+1)\left\{\sin^2\alpha_n + \sin^2\beta_n\right\} = \sum_{n=1}^{\infty}\left\{\sigma_{E,n} + \sigma_{M,n}\right\}$$

This is called Mie's expansion; it is an exact result. It is characterized by the electric and magnetic Mie coefficients a_n and b_n, respectively; (a_1 and b_1 being proportional to the electric and magnetic dipoles, a_2 and b_2 to the quadrupoles, and so on). The above expression can now be evaluated numerically for any set of input parameters: refractive indexes (m = mp/mh), frequency, wavelength or wave number of the light ($k = 2\pi/\lambda$), and the size of the particle (a). Small Matt lab programs can be used to plot. Mathematica may be used also.

Note: For objects whose dimensions are much larger than the wavelength of visible light (>1 μm) geometric optics is applicable. However, when the particle size is on the order of the wavelength of light or smaller. The formal solution to this problem was first given by Gustav Mie in 1908 and is now known as Mie theory. The Mie scattering theory has been very instrumental in understanding the response of small sub-micron to nano-scale particles whether metallic or dielectric to light radiation. The theory explained several phenomena including the more than 1000 year old phenomenon of stain glass and pottery phenomena.

Example 8.6 Gold nanoparticles: size dependence plasmon [10]

We examine scattering and absorption of gold nanoparticles as a function of size from 50 nm to 450 nm by evaluating the Mie results numerically (Fig. 8.8).

Note that as the size of the particle increase, the peak moves to longer wavelengths and becomes broader. This peak is the dipole plasmon resonance. *Dipole* means that the wave set up on the surface of the particles has *one node*. On further increase in particle size, a second peak begins to emerge where the dipole resonance occurred for small particles. This is the quadrupole plasmon resonance. *Quadrupole* means that the wave set up on the surface of the particles has *two nodes*. One can use the on-line "Mie Theory Calculator" It calculates the extinction per particle wavelength. See the following link:

https://nanocomposix.com/pages/tools

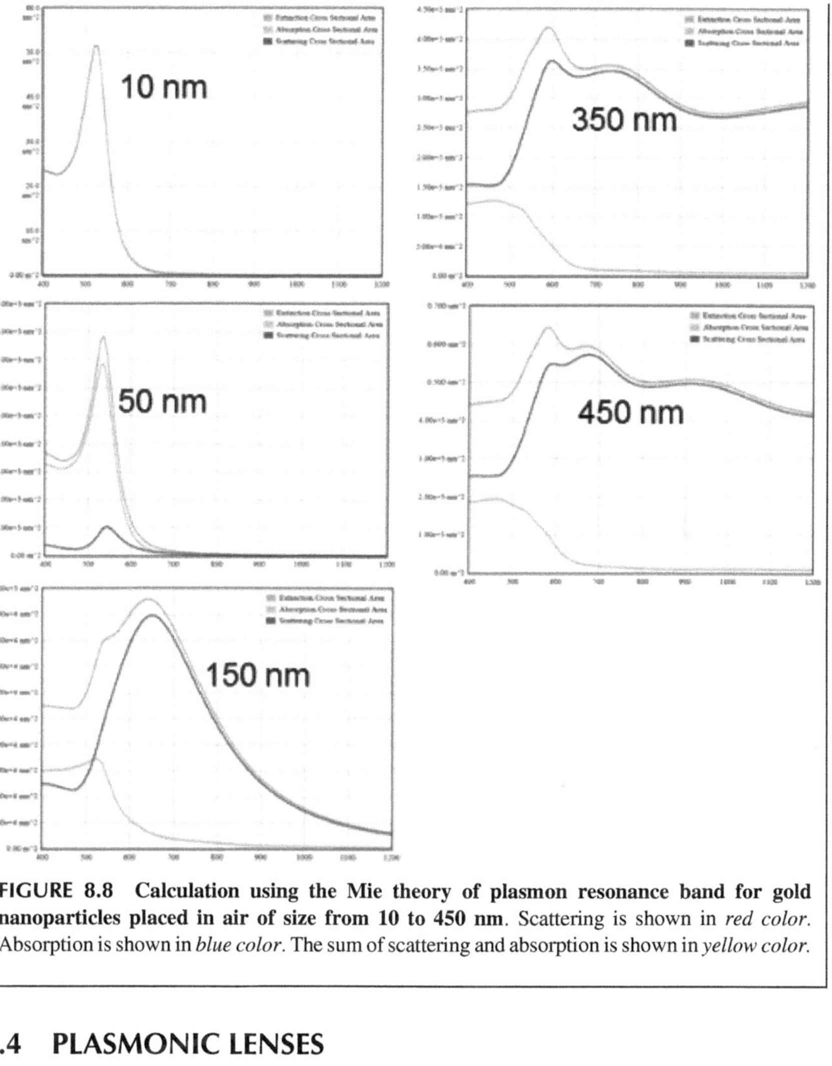

FIGURE 8.8 Calculation using the Mie theory of plasmon resonance band for gold nanoparticles placed in air of size from 10 to 450 nm. Scattering is shown in *red color*. Absorption is shown in *blue color*. The sum of scattering and absorption is shown in *yellow color*.

8.4 PLASMONIC LENSES

Having a negative refractive index is the basic principle behind plasmonic lenses. This is a very common property for noble metal at specific frequency. There are two types of such lenses. One type is based on confinement enhancement. Another is based on transmission enhancement of evanescent waves. In this section we present examples of metal-based lenses.

8.4.1 Confinement-Based Lensing

We discuss two configurations for confinement-based lensing: a continuous conic waveguide concentrator and discontinuous chain of nano indentation (particles). (i) Consider the hollow cone of metal shown in Fig. 8.9A. The radius of the cone

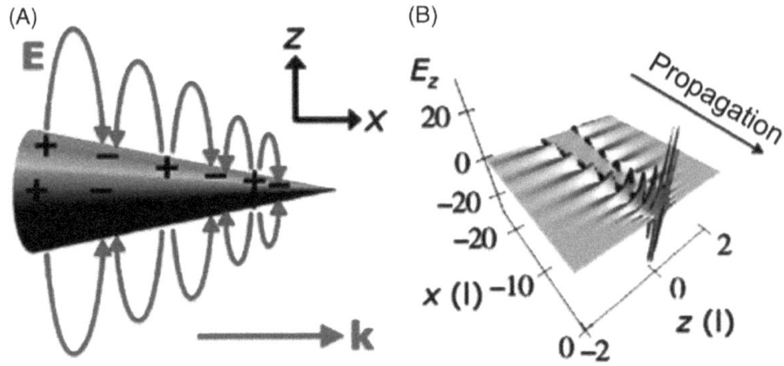

FIGURE 8.9 (A) Geometry of a conic nanoplasmonic waveguide showing propagation of a charge oscillation wave. (B) Snapshot of instantaneous Ez fields in the longitudinal cross section (xz) plane, normalized to far-zone (excitation) field. *(Adapted from "M.I. Stockman, Nanofocusing of optical energy in tapered plasmonic waveguides, Phys. Rev. Lett. 93 (2004) 137404).*

gradually decreases from 50 nm to 2 nm for example. When light strikes the cone and a plasmonic resonance is excited at the opening of the cone, a surface plasmonic polaritron (SPP) propagates toward the tip. This causes accumulation of energy resulting in a giant local field at the tip as shown in Fig. 8.9A. Fig. 8.9B displays the amplitudes of the local optical field in the cross section of the cone for the normal and longitudinal (with respect to the axis) components of the optical electric field. The magnitude of the field grows significantly as the oscillations approach the tip. The transverse x component grows by an order of magnitude, while the longitudinal z component, which is very small far from the tip, grows relatively much stronger at the tip. This causes the local field to increase by nearly 3 orders of magnitude in intensity and four orders in energy density. It should be noted that the propagation is slowed down and asymptotically stopped when it approaches the tip. It never actually reaches the tip (the travel time to the tip is logarithmically divergent) [11]. The cone as such may represent a tapered plasmonic waveguide. (ii) Consider a chain of nanoparticles shown in Fig. 8.10A. Fabrication starts with a thin gold film. Nano indentations (nanoparticles) are made on the film in a parabolic chain configuration. The radius of curvature of the chain is 10 μm. The particle diameter and inter-particle distance are ~350 nm and 850 nm, respectively, while the particle height is 300 nm. When light, in the wavelength range of 700–860 nm, for example, is normally incident on the thin film on the right of the chain, charge oscillations is induced on the surface of the film which travel or propagate to the left toward the chain (SPP). When the waves hit the gold nanoparticles they get focused to a submicron spot as shown in Fig. 8.10B [12].

8.4.2 Transmission-Based Lensing

Another concept of lensing involves transmission through holes or slits. An array of holes is drilled in opaque metallic film in an arc formation as shown in Fig. 8.11A. The diameter of the holes is of nano-scale. For a plane wave

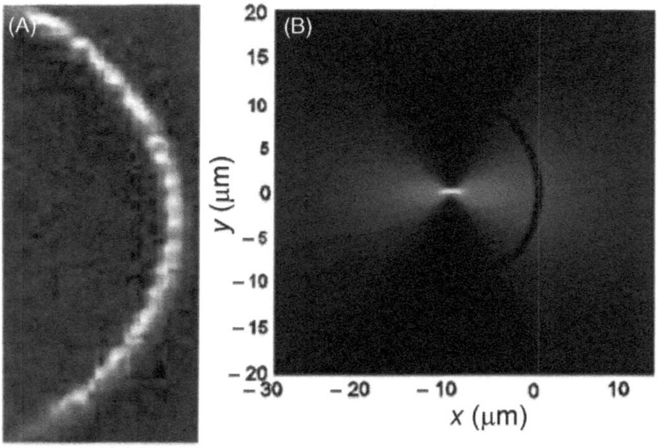

FIGURE 8.10 Focusing of light by a curved chain of spheroid gold nanoparticles indented in a thin gold film. (A) Top view image of the chain using a scanning electron microscope (SEM). (B) Magnitude of scattered electric field calculated above the gold surface. The illuminating beam has a wavelength of 800 nm and is incident perpendicular to the gold surface and polarized along x-direction. *(Adapted from" A.B. Evlyukhin, et al., 2007, Focusing and directing of surface plasmon polaritons by curved chains of nanoparticles, Opt. Exp., 15 (2007) 16667–16680).*

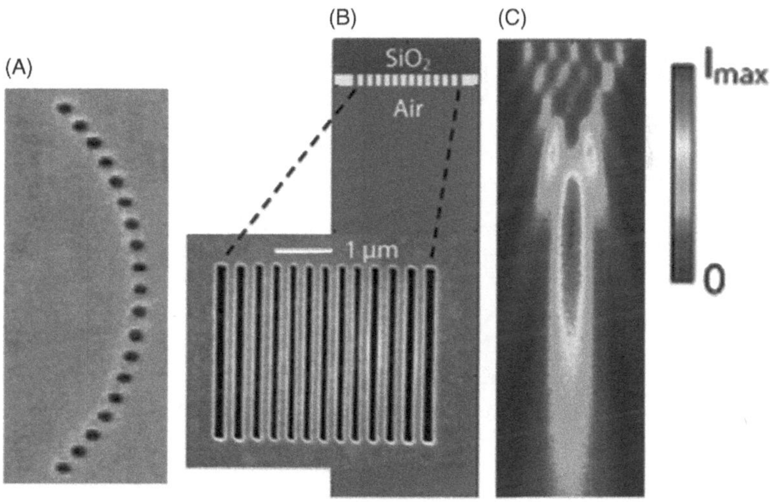

FIGURE 8.11 Focusing of light by transmission through holes or slits. (A) Chain of holes in a metal plate. (B) Nanoscale slit array drilled in gold film on a fused silica substrate (dark blue). The film is 400nm thick (yellow). The air slits are different in widths (80–150 nm) (light blue). The inset shows a scanning electron micrograph of the structure as viewed from the air-side. (C) Focusing pattern measured by confocal scanning optical microscopy (CSOM). *(Adapted from L. Verslegers et al., Planar lenses based on nanoscale slit arrays in a metallic film, Nano Lett. 9 (2009) 235–238).*

incident upon such a structure, the phase shift experienced by light as it passes through each individual hole is sensitive to either the length, width, or even the materials inside the hole. With the adjustment of the properties of individual holes, it becomes possible to achieve a focusing action [13].

A slit-based lens is built by first depositing a 400 nm thick flat gold film on silicon oxide substrate. Then an ion beam is used to mill 13 rectangular slits in the thin film as shown in the left panel of Fig. 8.11B. The slits increase in widths from 80 nm at the center slit to150 nm on the left or right end. Fig. 8.11C shows the measured field intensity in a cross section through the center of the slits (along the *x*-direction). The measurement demonstrates focusing of the wave. Thus this configuration acts as a far-field cylindrical lens for light at optical frequencies [14].

8.4.3 Shape Effects in Plasmonic: Rods Versus Spheres[15]

Fig. 8.12A shows the surface plasmon oscillations in spherical gold nanoparticles (AuNPs) [15]. On the other hand Figure shows plasmaon oscillations in gold nanorods (AuNRs). For spherical AuNPs, there is only one SPR frequency, around 520 nm in the visible region, which is responsible for the intense red color of the spherical AuNPs. For gold nanorods (AuNRs), there are two SPR frequencies, known as the longitudinal and transverse bands as shown in Fig. 8.12B. The transverse band is for electron oscillations that take place along the transverse direction and is a weak absorption band in the visible region similar to the SPR frequency of the spherical AuNPs. The longitudinal band is related to the electron oscillations along the longitudinal direction and is a strong absorption band in the vis-NIR region.

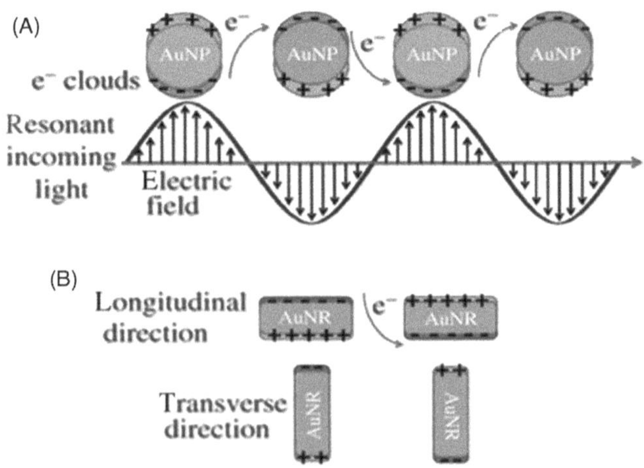

FIGURE 8.12 Shape effects in plasmonic metal. (A) Gold nanoparticles (B) gold nanorods.

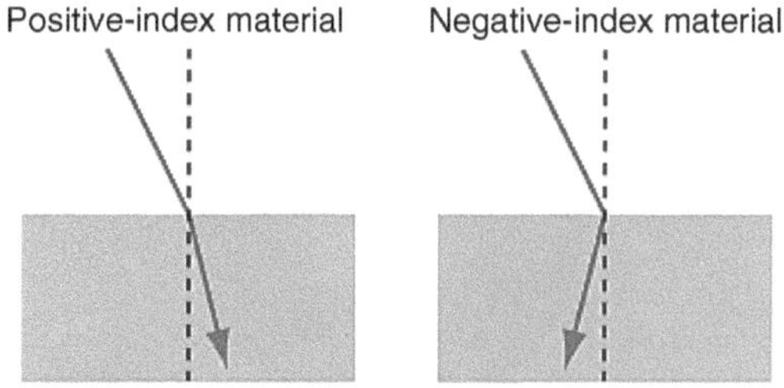

FIGURE 8.13 Schematic of light refraction at a plane interface of dielectric – dielectric interface and at a dielectric – metamaterial (negative index) interface.

8.5 METAMATERIALS: NEGATIVE REFRACTIVE INDEX

The above discussion shows that metal dielectric interface, with features smaller than the wavelength of light separated by distances smaller than the wavelength of light are very special. At plasmon resonances they have a negative index. In fact those may be considered as a class of more general material called metamaterials that exhibit properties beyond those found in nature. In 1967 [16], Victor Veselago theorized that material with negative refractive index would exhibit optical properties opposite to those of dielectrics, such as glass or air. Contrary to dielectrics, when light propagates in metamaterial (i) light refracts on the other side of the normal to the interface that is energy is transported in a direction opposite to the dielectric case as shown in Fig. 8.13; (ii) light produces negative pressure, which pulls metamaterial toward it instead of positive pressure which pushes away as in conventional material. The basic principle behind all of the opposite effects is closely related to the above plasmonic effect. They are due to the collective interaction of the light with the electron clouds at the surface of the conductor. This photon–plasmon interaction generates intense, localized optical fields. The waves are confined to the interface between metal and insulator. This narrow channel serves as a transformative guide that, in effect, traps and compresses the wavelength of incoming light to a fraction of its original value.

8.6 HEAT LOSS: ARE PLASMONIC-BASED DEVICES PRACTICAL?

One pivotal problem inherit in plasmonic technology is heat loss. Because plasmonic devices involve light interaction with metal, it necessarily involves energy loss due to heat dissipation. Metals are plagued by large losses due to strong electronic interband absorption, especially in the visible and UV spectral ranges. Among all metals, noble metals of gold and silver show the least energy loss. That is why gold and silver are used in most common plasmonics. Even in gold and silver, losses in the optical range, including near- and mid-infrared (IR) regions are effectively still too high to make practical plasmon-based devices.

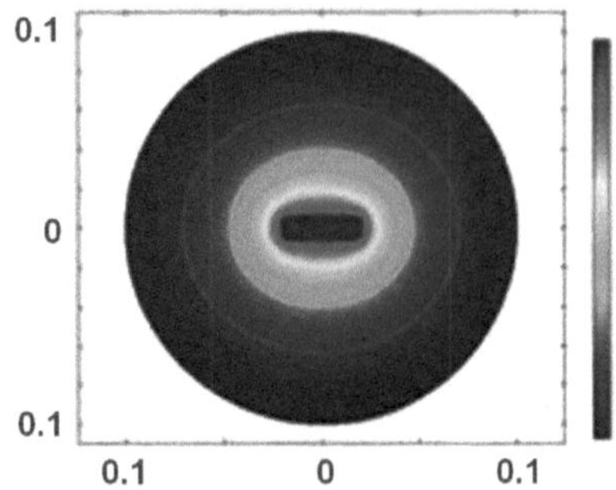

FIGURE 8.14 2D model of a laser activated gold nanorod. The temperature map after 2 s irradiation with a 0.5 mW/cm^2 diode laser.

Therefore, the search is continuing to find different approaches, such as creation of alloys and composites to make plasmonic materials that exhibit lower losses [17].

The 2D model of a laser activated gold nanorod is shown in Fig. 8.14 [18]. The temperature was mapped after 2 s irradiation with a 0.5 mW/cm^2 diode laser.

Example 8.8 Trade-off between localization and plasmonic heat loss [19]

The most simple geometry sustaining SPPs is that of a single, flat interface (Fig. 8.15) between a dielectric, non-absorbing half space ($z > 0$) with positive real dielectric constant $\in 2$ and an adjacent conducting half space ($z < 0$) described via a dielectric function $\in_1(\omega)$. The requirement of metallic character implies that $Re[\in_1] < 0$. The geometry for SPP propagation at a single interface between a metal and a dielectric is shown in Fig. 8.4.

The ideal plasmonic effects assume that there is no absorption in the conductor. Having no absorption means that the dielectric permittivity $\in_1$ is real and has no imaginary part ($\text{Im}\ [\in_1] = 0$). In actual situation there is loss of energy due to the fact that conduction electrons of real metals however suffer both from free-electron and inter band damping. Therefore, $\in_1(\omega)$ is complex. Detailed electromagnetic analysis gives the following result for the electric field associated with SPP perpendicular to the interface for $z < 0$, that is, inside the conductor:

$$E_Z(z) = -A_1 \frac{\beta}{\omega \varepsilon_0 \varepsilon_1} e^{i\beta x} e^{k_1 z}, \beta = k_0 \sqrt{\frac{\varepsilon_1 \varepsilon_2}{\varepsilon_1 + \varepsilon_2}}. k_1^2 = \beta^2 - k_0^2 \varepsilon_1,$$

The traveling SPPs are damped with an energy attenuation length (also called propagation length) $L = (2\text{Im}[\beta])^{-1}$, typically between 10 and 100 μm in the visible regime, depending upon the metal/dielectric configuration in question. On the other hand $\hat{z} = 1/|kz|$, defines the evanescent decay length of the fields perpendicular to the interface, which quantifies the confinement of the wave.

Let us consider SPPs at a silver/air interface at $\lambda_0 = 450$ nm. For this case we find have $L \approx 16$ μm and $\hat{z} \approx 180$ nm. At $\lambda_0 \approx 1.5$ μm, however $L \approx 1080$ μm and $\hat{z}$

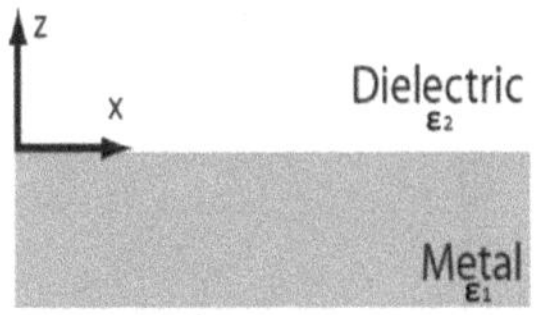

FIGURE 8.15 Heat loss effects in plasmonic material and trade-off between localization and plasmonic heat loss.

≈2.6 μm. The better the confinement, the lower is the propagation length. This characteristic trade-off between localization and loss is typical for plasmonics. We note that field confinement below the diffraction limit of half the wavelength in the dielectric can be achieved close to ω_{sp}. In the metal itself, the fields fall off over distances on the order of 20 nm over a wide frequency range spanning from the visible to the infrared.

Chapter 2 SURFACE PLASMON POLARITONS AT METAL ... - Springer www.springer.com/cda/content/document/cda.../9780387331508-c2.pdf?...0-0–45.

Example 8.9 Ultra-directional scattering by plasmonic particles: Forward versus back [20]

Scattering properties of a silicon (Si) nanoparticle with radius of 75 nm in free space calculated by Mie theory. Forward (F curve) and the backward (B curve) scattering cross-sections, and the forward-to-backward ratio (F/B curve) of the Si nanoparticle. The grey dashed line represents the values of forward-to-backward ratio equal to unity. Inset into (left) is the calculation scheme: the nanoparticle is excited by a plane wave from top while the scattered light is integrated over the upper and lower hemispheres for backward and forward scattering. (Right) Far-field scattering radiation patterns in the four spectral points: 660 nm – maximum forward-to-backward ratio; 564 nm – minimum forward-to-backward ratio; 500 nm and 603 nm – equal forward and backward scattering at electric (500 nm) and magnetic (603 nm) dipole resonances (Fig. 8.16).

Applications [21]: More specifically, there have been fascinating developments in regard to the light scattering by nanosized particles, including metal particles and surfaces, where localized surface plasmons can be excited leading to optical resonance phenomena. Small particles with surface plasmons can be used to detect the fluorescence of single molecules, enhance Raman scattering, resonantly transfer energy of excitons and create nanosized quantum amplifiers of optical energy. Potential practical uses include small-scale sensing techniques, numerous biomedical applications, manipulation of light for solar energy technologies and others.

Light scattering and surface plasmons on small spherical particles, Xiaofeng Fan, Weitao Zheng[1] and David J Singh *Light: Science & Applications* (2014) **3**, e179; doi:10.1038/lsa.2014.60 Published online 6 June, 2014. Nanotweezers jets

http://www.nature.com/lsa/journal/v3/n6/full/lsa201460a.html?foxtrotcallback (true#bib14

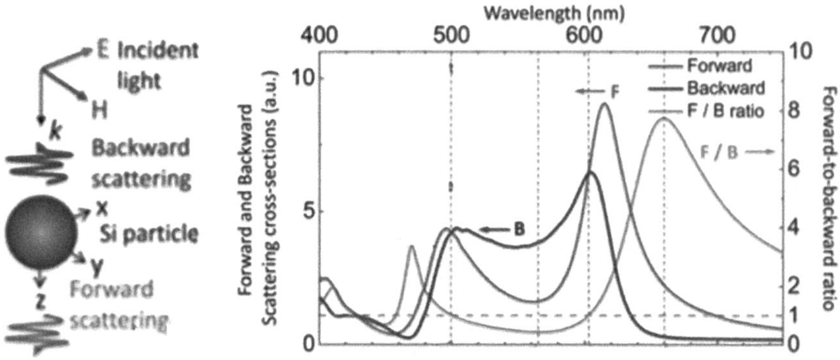

FIGURE 8.16 Calculation of forward vs back scattering as a function of the wavelength of the incident light of a silicon nanoparticle with radius of 75 nm in free space using Mie theory. (*Left*) Sketch of the interaction. (*Right*) Curve F is forward scattering. Curve B is backward scattering. Curve F/B is the ratio of forward scattering to backward scattering. *(From Directional visible light scattering by silicon nanoparticles Yuan Hsing Fu, Arseniy I. Kuznetsov, Andrey E. Miroshnichenko, Ye Feng Yu, Boris Luk'yanchuk, http://www.lukiyanchuk.ru/publ/80.pdf)*

8.7 SYNTHESIS OF METAL-BASED NANOSTRUCTURES

In this section, we present brief discussions of the procedures for synthesis of metal nano structures [22], including particles and rods. Metal of interest include (i) noble metal, such as gold, silver, and platinum, as well as (ii) magnetic metals. There are produced using a variety of procedures.

8.7.1 Physical and Semi-Physical Procedures

In physical methods metal atoms are first generated from bulk metals using, for example, high temperatures or laser ablations. The atoms are then allowed to coalesce to form nanoparticles. Appropriate precursor complexes which bear the metal of interest have also been used. In this case, thermal decomposition is used to liberate the metal atoms. Also soft ligand removal from organometallics has been used.

8.7.2 Chemical Procedures

Wet chemistry is used in chemical methods. The cartoon in Fig. 8.17 outlines the steps used for the case of gold. The procedures involve two elements: a source of metal ions (salt for example) and a reducing agent to produce the metal atoms by neutralizing the ions. An optional chemical agent may be used as an isolating (capping) agent to control the size or shape and prevent agglomeration.

8.7.2.1 Gold Nanoparticles

Noble metals are of great interest, since their salts or complexes can easily be reduced to the oxidation state 0. Moreover, the lack of sensitivity towards

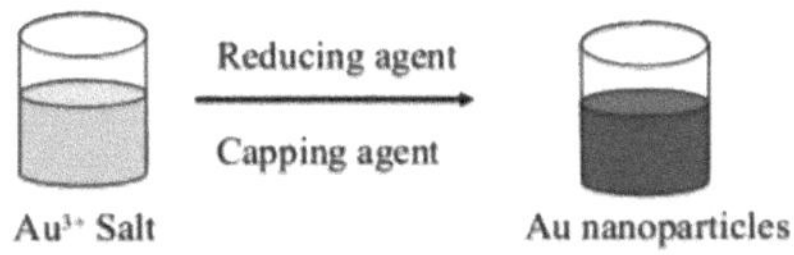

FIGURE 8.17 General protocol for synthesis of gold nanoparticles using *chemical procedures*.

oxygen, especially gold makes them ideal for a wet route. Nanoparticles of less noble metals can also be prepared with some care. We focus on gold to introduce the procedures. Variations may be needed for the other metals. Auric acid or salt is the most popular source of gold ions. But several types of reducing agent have been used.

In the procedure, chloroauric acid ($H[AuCl_4]$) is first dissolved. The reducing agent is then added while the solution is rapidly stirred. The neutralization process proceeds as follows:

- Au^{3+}ions is reduced to Au^{+}ions
- Three Au^{+} ions give rise to Au^{3+} and 2 Au^{0} atoms (a disproportionation reaction)
- Au^{0} atoms act as center of nucleation around which further Au^{+} ions gets reduced
- A stabilizing agent (citrate for example) that sticks to or caps the nucleate to stop growth may be used.

Fig. 8.18 gives examples of SEM images of gold nanoparticles made by this method. It shows an assortment of shapes including triangular, trapezoid, and pyramidal.

8.7.2.2 Reducing Agents

There are already several reducing agents available for gold. But there is still concerted ongoing research aimed at finding other convenient, environmentally friendly and efficient reducers.

- In metal-based route, there is $HBCl_2$, $NaBH_4$ (sodium borohydride), SMB (sodium metabisulfite), copperas (iron II sulfate), sulfur dioxide gas (SO_2), and hydrazine sulfate ((N_2H_5) HSO_4.
- In organic-based route, citric acid, CS2, hydroquinone (C_6H_4 $(OH)_2$), formaldehyde (CH_2O), and more recently oxalic acid ($H_2C_2O_4$ or HOOCCOOH) have been used.
- In the biochemical route, bio-oxidation and bio-sorption processes have been used; diverse bio-sorbents, such as bacteria, yeasts, fungi, actinomycetes, algae, bio-polymers and some bio spent materials have been reported.

8.7.2.3 Capping Agents: (Stability of Dispersion)

Wet procedures do not provide control over the size of the nanoparticles. The reaction and growth continue to increase the size. In addition, nanoparticles

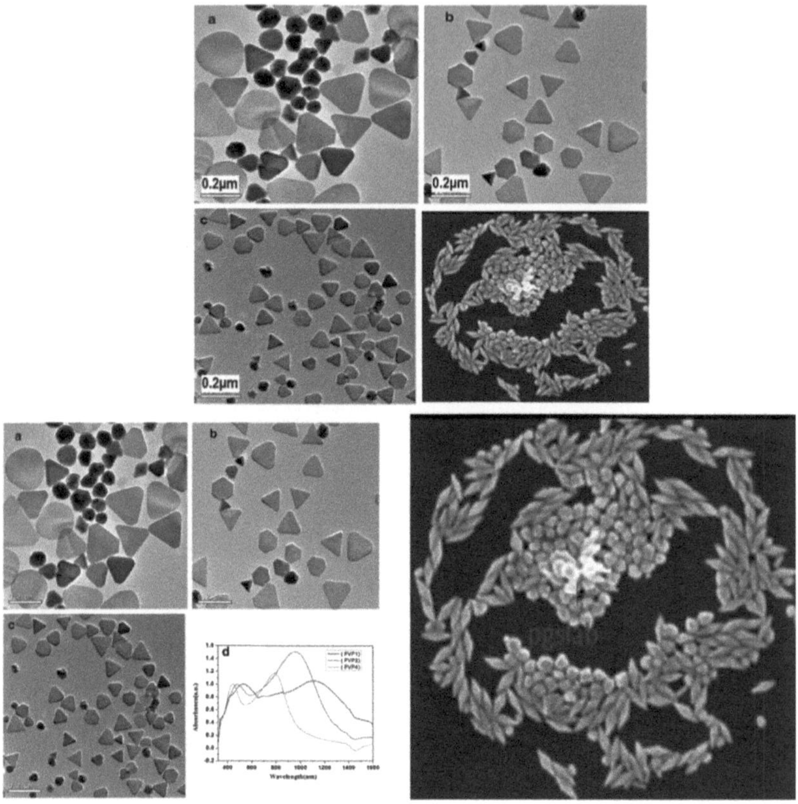

FIGURE 8.18 **Examples of gold nanoparticles made by this method using SEM imaging.**

tend to agglomerate in wet environment. For successful synthesis of monodispersed metal nanoparticles of a narrow size distribution, protecting materials must be added to the solution. Those tend to coat the nanoparticles preventing more gold growth, as well as protecting and isolating the nanoparticles from each other, hence prevent their agglomeration. There are several mechanisms for the effect: electrostatic repulsion-based of charged species or polymer-, surfactant-, or covalently bonded ligand molecules based. Examples may include the directional (one-dimensional) organic growth agents (cetyltrimethylammonium bromide, CTAB) and the stabilizing agent citric acid that sticks to or caps the nucleate to stop growth.

8.7.2.4 Metal Source and Mechanism

The most popular source of gold for use in wet chemistry procedures is chloroauric acid ($H[AuCl_4]$). In the procedure, the acid is first dissolved. The reducing agent is then added while the solution is rapidly stirred. This step causes Au^{3+} ions to extract electrons from the reducing agent and hence be reduced to neutral

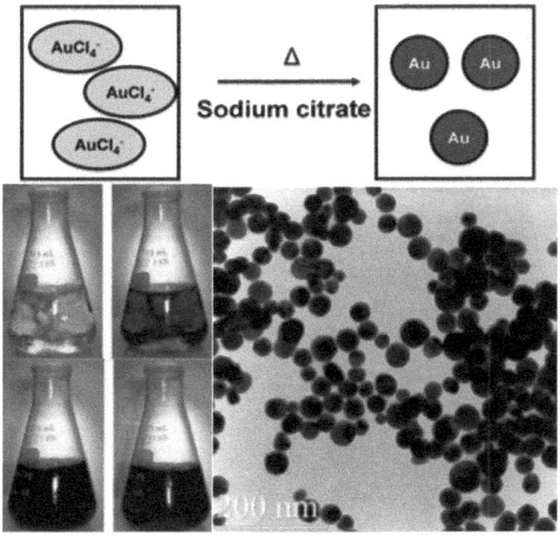

FIGURE 8.19 Presentation of Turkevich method for synthesis of gold nanoparticles using gold Salt and citrate.

gold atoms. As more and more gold atoms form, the solution becomes supersaturated, and gold starts to nucleate, forming nanoparticles or other structures.

8.7.2.5 Salt/Citrate (Turkevich Method) [23]

A popular method, developed by Turkevich uses a gold salt and sodium citrate is depicted in Fig. 8.19. $HAuCl_4$ is first dissolved in deionized water. The solution shows a faintly yellowish color. The solution is then heated until it boils. While boiling and stirring vigorously, a small volume of sodium citrate solution is added. The color of the solution gradually changes from faint yellowish to clear to grey to purple to deep purple, until it finally settles on wine-red in a total period of 30 min.

Sodium citrate is convenient as it serves both as a reducing agent and a non-aggregation agent. In the solution, the citrate provides negatively charged citrate ions, which get absorbed onto the gold nanoparticles. The negative surface change introduced on the surface repels the particles from each other, hence preventing them from aggregating and forming a stable dispersion.

There is another function of the citrate. It acts as a stop agent of the reaction, thus reducing the amount of sodium citrate increases the average size of the nanoparticlcs. In other words, reducing the amount of sodium citrate reduces the citrate ions available for stabilizing the particles, which causes small particles to aggregate into bigger ones.

8.7.2.6 Salt/Inorganic(Martin Method) [24]

In this method (depicted in Fig. 8.20A), $HAuCl_4$ is first dissolved in deionized water. This is followed by adding metal-based reducing compounds, such as

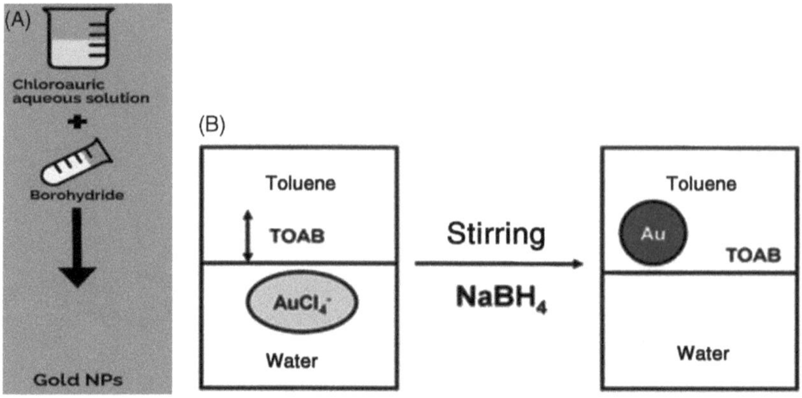

FIGURE 8.20 (A) Presentation of the Martin method for synthesis of gold nanoparticles using gold Salt and an inorganic reducer. (B) Presentation of the Brust method for synthesis of gold nanoparticles using gold salt an inorganic reducer and an organic solvent.

$HBCl_2$ or $NaBH_4$. The procedure produces nanoparticles, such as nano cubes, prisms, and spheres form.

8.7.2.7 Salt/Inorganic/Organic Solvent (Brust Method) [25]

In the Brust method (shown in Fig. 8.20B), $HAuCl_4$ is dissolved in water and tetraoctylammonium bromide (TOAB) is dissolved in toluene. TOAB acts as a phase transfer catalyst and the stabilizing agent. The $HAuCl_4$ solution is then added to TOAB while stirring vigorously for about 10 min. Toluene is an organic liquid which is not miscible with water. It is to be noted that the color of the aqueous phase becomes clear, and the color of the organic phase (toluene) should become orange. The reducing agent $NaBH_4$ is then added while stirring for several hours causing a change in color from orange to white to purple to eventually reddish. The organic phase can now be separated and washed with dilute sulfuric acid (H_2SO_4) to neutralize it, followed by washing with distilled water. The resulting gold nanoparticles are about 5–6 nm in diameter.

8.7.2.8 Silver Nano Spheres Synthesis

A method analogous to the Turkevich method [23] for making gold nanoparticles can be used to make silver nanoparticles as shown in Fig. 8.21. A silver salt is used as a source of silver ions and a sodium citrate is used as a reducing agent to produce silver atoms. Silver nitrate is first dissolved in deionized water. This is followed by adding a small volume of sodium citrate solution. This results in citrate-coated silver nanoparticles.

Silver nanoparticles may also be prepared by another method. Silver acetate ($AgC_2H_3O_2$) is used as a source of silver ions, 1–2 dichlorobenzene (C6H4C12) as a solvent, and oleylamine ($C_{18}H_{37}N$) as a reducer (neutralizer). Oleylamine simultaneously acts as a reducing agent and capping agent. An amine group

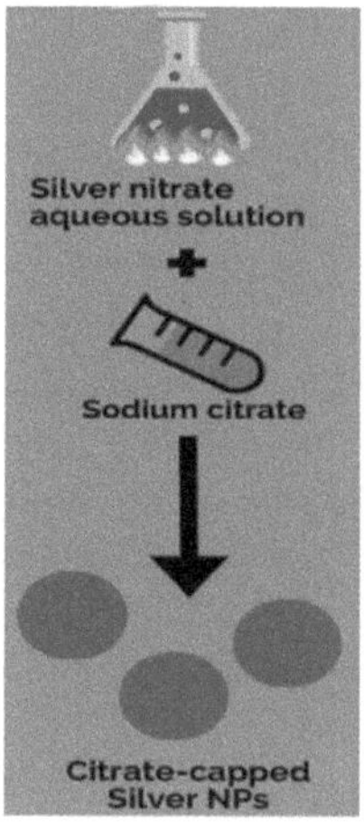

FIGURE 8.21 **Procedure for synthesis of silver nanospheres.**

in oleylamine molecule acts as the reducing group, which can undergo metal ion-induced oxidation. This technique produces ~20 nm silver nanospheres with UV-VIS absorbance peak at ~421 nm. This technique can be modified for production of different metallic nanospheres. It can also be used to make gold nanoparticles.

8.7.2.9 Magnetic Particles [26]

Magnetic nanoparticles are composed of (i) pure magnetic metals: iron (Fe), cobalt (Co), nickel (Ni), as well as magnetic bimetallic alloy-like material. In addition, magnetic nanoparticles may be composed of (ii) magnetic oxides, such as iron oxides. Magnetic nanoparticles are a class of nanoparticle that can be manipulated using magnetic fields. The two types are contrasted as follows:

- Metallic nanoparticles have higher magnetic moment than oxide nanoparticles of the same size
- For the same magnetic moment, metallic nanoparticles can be smaller than their oxide counterparts
- Unlike oxide nanoparticles, metallic nanoparticles are pyrophoric and can be reactive to oxidizing agents and difficult to synthesize in wet aqueous environment
- Metallic nanoparticles are used typically in dry technical applications, whereas oxides (maghemite, magnetite) are used in biomedical applications.

8.7.2.10 Architecture, Size, and Functionalization

Generally characteristics of nanoparticles depend strongly on size, structure, and shape. In fact, the magnetic characteristics of nanoparticles depend much more strongly on size, structure, and shape, which make uniformity of magnetic particles critical. The ranges available for sizes are as follows. (i) *Microbeads* are 0.5–500 μm in diameter. (ii) Magnetic nanoparticles smaller than 1 μm in diameter (typically 5–500 nm). (iii) Magnetic nanobeads composed of a number of individual magnetic nanoparticles with a diameter of 50–200 nm. (iv) *Magnetic nanochain assemblies* of magnetic nanoparticles.

Because of the sensitive to the environment, metallic nanoparticles are produced with a protecting shell. Passivated by gentle oxidation, surfactants, polymers, and precious metals has been employed. In an oxygen environment, Co nanoparticles for example develop a thin coating of CoO. Cobalt oxide is an anti-ferromagnetic material. Co–CoO coreshell nanoparticles are protected with a gold outer shell. Finally, nonreactive shell made of graphene has been synthesized on iron or cobalt nanoparticles.

As to *ferrite* nanoparticles or *iron oxide nanoparticles* (*iron oxides* in crystal structure of *maghemite* or *magnetite*), clusters may be synthesized with narrow size distribution consisting of superparamagnetic oxide nanoparticles (~80 maghemite superparamagnetic nanoparticles per bead). Finally ferrites nanoparticles with a functional shell may be designed. Being relatively inert they are first coated with a layer of *silica*. Silica is amenable to straightforward modification with various surface functional groups using covalent bonding between organo-silane molecules and silica. This enables attachment to *fluorescent dye* molecules using covalent bonding to the functionalized *silica* shell. Such particles exhibit double functionalities, magnetic and optical functionalities, providing enhanced detection and discrimination.

Synthesis of magnetic metal nanoparticles mainly follow recipes known from noble metals. Co-precipitation is used to synthesize iron oxides (either Fe_3O_4 or γ-Fe_2O_3) from aqueous Fe^{2+}/Fe^{3+} salt solutions by the addition of a base under inert atmosphere at room temperature or at elevated temperature. Magnetic nanocrystals with smaller size can essentially be synthesized through the thermal decomposition of alkaline organometallic compounds in high-boiling organic solvents containing stabilizing surfactants. Using the microemulsion technique, metallic cobalt, cobalt/platinum alloys, and gold-coated cobalt/platinum nanoparticles have been synthesized in reverse micelles of cetyltrimethlyammonium bromide, using 1-butanol as the co-surfactant and octane as the oil phase.

8.7.2.11 Applications

Magnetic nanoparticle may further assemble into magnetic nanochains. Magnetic nanoparticles exhibit attractive properties, which afford potential use in extremely wide range of application including catalysis, biomedicine and tissue specific targeting, photonic crystals, microfluidics, magnetic resonance imaging (MRI), magnetic imaging, data storage, environmental remediation, nanofluids and optical filters, defect sensor, and cation sensors.

8.7.2.12 Copper Nanoparticles [27]

There are several methods that have been used to synthesize copper nanoparticles. Most of the methods use wet chemical procedures. For example, copper chloride is used as precursor for copper ions. Sodium citrate is used as reducing agent. Another reducing agent is myristic acid in an aqueous solution containing sodium formaldehyde sulfoxylate (SFS). This produces fairly consistent pure copper nanoparticle powder. Non-spherical shapes, such as rod and cube-shaped nanoparticles may be produced from copper (II) acetylacetonate when reduced in oleyl amine and oleic acid. Another method involves the reaction of salt with radical hydrogen. In the procedure, copper (II) hydrazine carboxylate salt is used. Radical hydrogen is produced by ultrasound heating. The mixture undergoes a radical reaction that forms copper nanoparticles. Copper oxide nanoparticles may be produced by using reflux. Green chemistry *may also be used. In this procedure*, copper chloride is reduced using L-ascorbic acid, which produces stable copper nanoparticles.

8.7.2.13 Gold Nanorod

Physical procedures have been used to make well defined gold nanorods. Lithography or deposition techniques were employed. Those techniques however are expensive and time consuming which made them not feasible in large scale. Another physical procedure utilizes a template. Electrodeposition into the pores of an anodic alumina membrane (AAO), for example results on the formation of nanorods. This method can be used to fabricate Au nanorods of different lengths but constant diameters (so one can get different aspect ratios).

A wet chemistry technique (shown in Fig. 8.22) was subsequently developed which is amenable to lager throughput. This technique first introduced at the University of Illinois by Catherine Murphy [28] and later developed at Georgia Institute of Technology by Nikoobakht and Al-Seyyed [29]. This method is called seed-mediated growth synthesis method. The technique consists of two steps: a seed step and a growth step. In the seed step a solution is prepared from HAuCl4 (source of gold ions), sodium borohydride (reducer), and CTAB (surfactant). The seed step results in small gold nanoparticles. In the growth step,

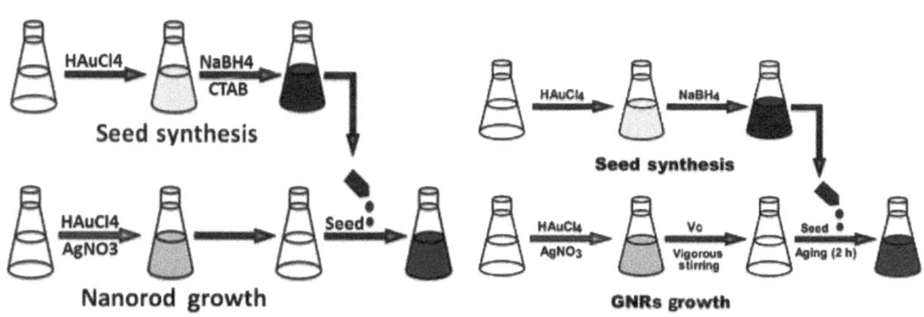

FIGURE 8.22 Procedure for synthesis of gold nanorod using wet chemistry techniques.

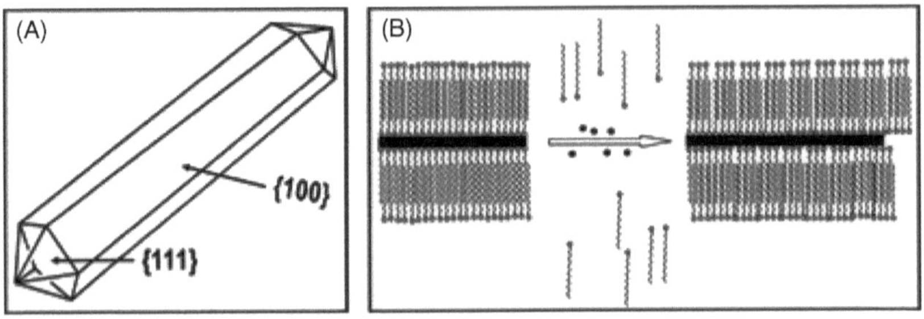

FIGURE 8.23 Presentation of hypothesis for explanation of the selectivity of the mechanism for formation of nanorod. *(From Panikkanvalappil R. Sajanlal, Theruvakkattil S. Sreeprasad, Akshaya K. Samal, Thalappil Pradeep, Anisotropic nanomaterials: structure, growth, assembly, and functions, Nano Rev. 2011 2, doi:10.3402/nano.v2i0.5883)*

a solution is prepared consisting of $HAuCl_4$, ascorbic acid, silver nitrate, and CTAB. The seed solution is then added to the growth solution. The nanoparticles in the seed solution provide nucleation sites for growing nanorods. There is a one well-known hypothesis that may be explained with the aid of Fig. 8.23. It is believed that CTAB preferentially attach to the {100} facets of the seeds. This acts to retard growth in that direction and to permit elongation on the {111} facets at the ends of the rods. It should be emphasized that without the organic component only zero-dimensional nanostructures, such as nanocubes, prisms and spheres would form. But when the shape-control or directing surfactant is added, such as CTAB and DCAB, they form a rod-shaped organic template that is filled with gold atoms.

Cartoon representation of (A) 3-D morphology showing (111) end-faces and (100) side-faces. (B) Illustration of "zipping" mechanism for the formation of the bilayer of CTAB (squiggles) on the NR (black rectangle) surface may assist NR formation as more gold ion (black dots) is introduced. Reproduced with permission from Reference (170). American Chemical Society, Copyright (2005 and 2003).

The procedure is not fully selective on the synthesis of gold nanorods. The yield will contain both nanoparticles and nanorods. Separation of gold nanorods can be performed using conventicle method if the application requires high purity gold nanorods. Fig. 8.24 shows some of the structures produced by this method. Since for most practical applications monodispersed structures are required, there is a need to separate, even crudely, the nanoparticles based on shape.

- Silver ion additives have been demonstrated to provide some control on the size and aspect ratio of obtained nanostructures.
- Multistep protocols of centrifuging or/and filtering were developed that are capable of separating the nanorods from the smaller structures with 90% purity.
- Shape separation-based on centrifugation have been introduced. Shape separation centrifugation techniques are based on different settling time of nanoparticles with different shapes.

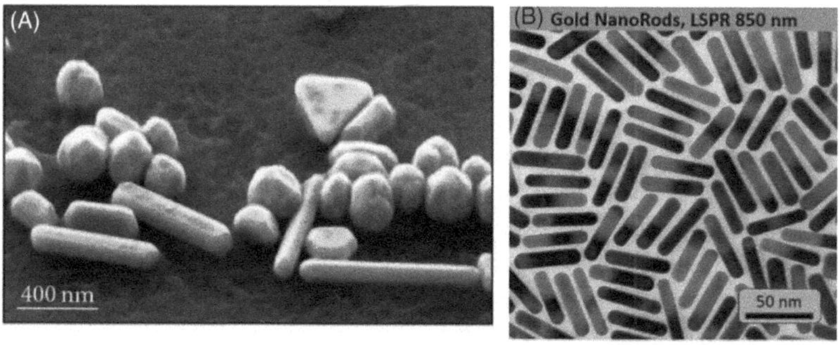

FIGURE 8.24 Examples of gold nanostructures produced by the method in Fig. 8.23 (particles and rods). *((A), From https://www.hindawi.com/journals/jnm/2011/515049/fig8/; (B), From https://nanohybrids.net/products/gold-nanorods-1.Reprinted with permission from NanoHybrids, Inc.)*

It should be noted that settling velocity is shape and size related. Therefore, different nanoparticles with different shapes behave differently under the same centrifugal force. The separation purity may be confirmed/monitored by different characterization techniques including UV-V is absorbance. Finally, unlike the silicon-based procedure, the organic-based procedure is potentially susceptible to organic and metal contamination.

A seedless method has been developed for the synthesis of high aspect ratio gold nanorods. Gold nanorods with a longitudinal surface plasmon resonance of larger than 1400 nm were synthesized in one step [30]. The yield is high, and no purification step is needed. Normal aspect ratio gold nanorods are synthesized by a seedless method. The conditions for the synthesis of uniform gold nanorods with different width and aspect ratio by a seedless method are demonstrated. A better understanding of the seedless method enables the facile synthesis of nanomaterials with a broader size tenability and better reproducibility. Another seedless procedure that has been recently developed for the synthesis of mainly gold nanorods with other residual nanostructures relies on the use of ultrasmall H-terminated silicon nanoparticles as the reducing agent. The procedure is described in more detail in Chapter 11 [31].

8.7.3 Refining and Harvesting Gold

There is a number of methods for refining gold, depending upon the quantity of gold one is working with and the desired level of purity. The two most common methods for refining gold are the use of high-temperature flame and the use of chemicals.

Flame: Refining with flame is one of the oldest methods. Refining by fire is the preferable method for larger quantities of gold [32]. In this procedure molten gold in a crucible is stirred and skimmed to remove the impurities. The flames may reach temperatures in excess of 1000°C. The tradition remains largely untouched today with the exception of a few advancements in safety and precision.

Wet chemistry: The second method involves the use of chemicals. Strong acids, such as nitric acid and hydrochloric acid are used to dissolve the impurities in the gold ore [33]. Once this is accomplished the acid is neutralized and washed away, taking the impurities with them. The resulting product has 99.999% or 24 K purity. This substance is dried until it is a powdered residue and then heated with a torch or other source of heat to melt the gold powder into useable gold.

Hydrometallurgy: This procedure refines ores, concentrates, and recycled or residual materials using aqueous chemistry. The harvested gold has 10^{-3} residual metal impurity (metallurgical grade). The process consists of leaching, solution concentration and purification and metal or metal compound recovery [34].

Electroparting is a recent gold refining procedure where gold is separated from other metals in electrolytic cell in a way that dissolves all metals except gold and platinum. This procedure refines gold directly, producing up to 99.5% purity gold powder in a single step [35–41].

For higher purity, refining steps using dissolution and precipitation of the metallurgical grade are used. A frequently used dissolution process involves the use of Aqua regia acid, which is a mixture of nitric and hydrochloric acids ($Au + HNO_3 + 4\,HCl \rightarrow HAuCl_4 + NO + 2\,H_2O$) to dissolve the low grade gold, or scrap gold used in jewelry manufacturing, or recycled gold from consumer's integrated circuit IC into hydrochloric acid solution. Hydroauric acid is widely used as a precursor gold ion to reduce and precipitate high purity gold powders as discussed in earlier.

8.7.4 Toxicity and Safety Issues of Metal-Based Nanoparticles [42]

Metal-based nanoparticles are classified into four different categories: (i) metallic nanoparticles, (ii) bimetallic or alloy nanoparticles, (iii) metal oxide nanoparticles and (iv) magnetic nanoparticles. Metallic nanoparticles like gold (Au), platinum (Pt), and silver (Ag), bimetallic nanoparticles like iron–cobalt (Fe–Co) and iron–platinum (Fe–Pt), and metal oxides, including titanium dioxide (TiO_2) cerium dioxide (CeO_2), silica (SiO_2), and zinc oxide (ZnO) with a focus on the lesser studied nanoparticles, such as silver (Ag), iron–platinum (Fe-Pt) and titanium dioxide (TiO_2) and how their unique properties allow for their potential use in various biomedical applications.

We discuss in this section the safety issues of metal-based nanoparticles, both pure metal and metal oxide. Common routes of introduction include skin absorption, ingestion, and inhalation. As there are many common exposure routes, the toxicity of these materials to several types of cell lines must be considered. Pure nanoparticles are presently being produced in large quantities for use as catalysts, drug delivery vehicles, and anti-bacterial agents. Some of those provide high degree of biocompatibility; hence they are safe for biological applications. Others are reactive and could pose a serious threat to health.

A common route of exposure for metal nanoparticles is skin absorption making tests on human skin cell lines vital to determine acute effects of nanoparticles.

In general, metal nanoparticles have cytotoxicology properties. This is not surprising since they are known to have catalytic properties and have reactive metal ion species. In fact, their toxicity is due in part to bulk properties of the metals. Having nanoparticles with stable shapes can limit ion exposure effects. Stability against dissolution or having self-limiting interaction in biological material renders such nanoparticles safe.

Gold nanoparticles show low toxicity consistent with the bulk biocompatibility of gold. Silver nanoparticles show cytotoxic effects within certain concentration regimes; but below a certain concentration they may lack toxicity. Each regime has a dominant death mechanism. Iron nanoparticles could be very cytotoxic by reacting on the surface of cells to form more stable iron oxides.

Oxide nanoparticles consisting of metal and semiconductor oxides include (CeO_2) ($Ca_3(PO_4)_2$) ZnO nanoparticles, and Fe_2O_3 nanoparticles. Silica, ZrO_2, Zinc oxide and titanium dioxide particles are both used in sunscreen to absorb UV radiation. Silica nanoparticles are generated in many industrial activities to usefulness in glass, food additives, etc. Iron oxide particles are useful due to their superparamagnetic properties.

Generally speaking, metal oxide nanoparticles are cytotoxic in at least some concentration regime. The commercially used TiO_2 and ZnO were found to be disturbingly toxic given their use in common sunscreen. Silica and $Ca_3(PO_4)_2$ nanoparticles were found to be non-toxic much like their bulk phases similar to gold nanoparticles. CeO_2 and ZrO_2 showed that for some nanoparticles, cytotoxicity is not always permanent and cell recovery can occur. For toxic oxide nanoparticles, cell death mechanisms varied widely between material, concentration, and exposed cell line. The studies presented prove that due to widely varying properties and effects of oxide nanoparticles, hazard potential must be studied with parameters specific to the nanoparticles involved as well as the environment (cell line, environmental factor, etc.) of exposure.

There remains a lot of investigating to be carried out. As exposure to these substances increases, toxicity must be investigated to prevent unnecessary health hazards.

REFERENCES

[1] Pines D, Bohm D. A collective description of electron interactions: II. Collective vs individual particle aspects of the interactions. Phys. Rev. 1952;85:338.

[2] Messiah A. Quantum Mechanics. USA: Dover Publications; 1999.

[3] Maier SA. Plasmonics: Fundamentals and Applications. Berlin, Germany: Springer Science + Business Media LLC; 2007.

[4] Margapoti E. Plasmonics: Fundamentals and Applications. Munich, Germany: Technical University München; 2012. Summer semester.

[5] Sólyom J. Fundamentals of the Physics of Solids: Volume II: Electronic Properties. Berlin, Germany: Springer Science & Business Media; 2008.

[6] Ditlbacher H, Hohenau A, Wagner D, Kreibig U, Rogers M, Hofer F, Aussenegg FR, Krenn JR. Silver nano wires as surface plasmon resonators. Phys. Rev. Lett. 2005;95:257403.
[7] X. Fan, W. Zheng, D.J. Singh, http://uw.physics.wisc.edu/~himpsel/551/Lectures/Plasmonics.pdf. Light scattering and surface plasmons on small spherical particles. http://www.nature.com/lsa/journal/v3/n6/full/lsa201460a.html
[8] Nanophotonics, http://www.hamamatsu.com/eu/en/technology/innovation/nanophotonics/index.html
[9] Baou G. Mie theory for metal nanoparticles. J. Phys. Chem. 2015;C119:28586–96. http://www.guillaume.baffou.com/pdf/Mie_Theory.pdf.
[10] https://nanocomposix.com/pages/tools
[11] Stockman MI. Nanofocusing of optical energy in tapered plasmonic waveguides. Phys. Rev. Lett. 2004;93:137404.
[12] Evlyukhin AB, Bozhevolnyi SI, Stepanov AL, Kiyan R, Reinhardt C, Passinger S, Chichkov BN. Focusing and directing of surface plasmon polaritons by curved chains of nanoparticles. Opt. Exp. 2007;15:16667–80.
[13] Yin L, Vlasko-Vlasov VK, Pearson J, Hiller JM, Hua J, Welp U, Brown DE, Kimball CW. Sub wavelength focusing and guiding of surface plasmons. Nano Lett. 2005;5:1399–402.
[14] Verslegers L, Catrysse PB, Yu Z, White JS, Barnard ES, Brongersma ML, Fan S. Planar lenses based on nanoscale slit arrays in a metallic film. Nano Lett. 2009;9:235–8.
[15] E. Yasun, H. Kang, H. Erdal, S. Cansiz, I. Ocsoy, Y.-F. Huang, We. Tan, Cancer cell sensing and therapy using affinity tag-conjugated gold nanorods, Interface Focus, 3 (2013) 0006. http://dx.doi.org/10.1098/rsfs.2013.0006.
[16] Veselago VG. The electrodynamics of substances with simultaneously negative values of ε and μ. Sov. Phys. Usp. 1968;10:509–14. (Russian text 1967).
[17] P. West, S. Ishii, G. Naik, N. Emani, A. Boltasseva, Identifying low-loss plasmonicmaterials, SPIE Newsroom: http://www.academia.edu/1037839/Identifying_low-loss_plasmonic_materials
[18] F. Rossi, F. Ratto, R. Pini, Laser activated gold nanorods for the photothermal treatment of cancer, in: Proceeding of the COMSOL Conference, Milan (2012).
[19] Chapter 2 surface plasmon polaritons at metal, Springer. www.springer.com/cda/content/document/cda.../9780387331508-c2.pdf?...0-0-45. A. Stefan, Plasmonics: Fundamentals and Applications, Maier, 2007, Springer Science.
[20] Fan X, Zheng W, Singh DJ. Light scattering and surface plasmons on small spherical particles. Light 2014;3:e179. Ultra-directional scattering by plasmonic particles: Forward vs back. doi:10.1038/lsa.2014.60; http://www.nature.com/lsa/journal/v3/n6/full/lsa201460a.html?foxtrotcallback=true#bib14.
[21] Fan X, Zheng W, J Singh D. Light scattering and surface plasmons on small spherical particles. Light Sci. Appl. 2014;3:e179. (nature.comNanotweezers jets); http://www.nature.com/lsa/journal/v3/n6/full/lsa201460a.html?foxtrotcallback(true#bib14.
[22] Tapan KS, Andrey LR. Complex-shaped metal nanoparticles. Bottom-Up Syntheses and Applications. Hoboken, NJ: John Wiley & Sons; 2012.
[23] Turkevich J, Stevenson PC, Hillier J. A study of the nucleation and growth processes in the synthesis of colloidal gold. Discuss. Faraday Soc. 1951;11:55–75. Kimling J, Maier M, Okenve B, Kotaidis V, Ballot H, Plech A. Turkevich Method for Gold Nanoparticle Synthesis Revisited. J. Phys. Chem. B 2006;110:15700–7.
[24] Martin MN, Basham JI, Chando P, Eah S-K. Charged gold nanoparticles in non-polar solvents 10-min synthesis and 2D self-assembly. Langmuir 2010;26(10):7410–7. A 3-min demonstration video for the Martin synthesis method is available at YouTube.

[25] Brust M, Walker M, Bethell D, Schiffrin DJ, Whyman R. Synthesis of thiol-derivatised gold nanoparticles in a two-phase liquid-liquid system. Chem. Commun. 1994;7:801–2.
[26] Hou Y, David JS. Magnetic nanomaterials: fundamentals. Synthesis and Applications. Hoboken, NJ: John Wiley & Sons; 2017.
[27] Khanna PK, Gaikwad S, Adhyapak PV, Singh N, Marimuthu R. Synthesis and characterization of copper nanoparticles. Mater. Lett. 2007;61(25):4711–4. Mott D, Galkowski J, Wang L, Luo J, Zhong C. Synthesis of Size-Controlled and Shaped Copper Nanoparticles. Langmuir 2007;23(10):5740–5. doi: 10.1021/la0635092. Umer A, Naveed S, Ramzan N, Rafique MS, Imran M. A green method for the synthesis of Copper Nanoparticles using L-ascorbic acid. Matéria. 2014;19(3):197–203. doi: 10.1590/S1517-70762014000300002, 2007.
[28] Lohse SE, Murphy CJ. The quest for shape control: a history of gold nanorod synthesis. Chem. Mater. 2013;25:1250–61.
[29] Babak BN, El-Sayed MA. Preparation and growth mechanism of gold nanorods (NRs) using seed-mediated growth method. Chem. Mater. 2003;15:1957–62.
[30] Xu X, Zhao Y, Xue X, Huo S, Chen F, Zou G, Liang X-J. Seedless synthesis of high aspect ratio gold nanorods with high yield. J. Mater. Chem. A 2014;10:3528–35.
[31] N. Elhalawany, B. Enders, E. Bahceci, A. Kocyigit, A. Nayfeh, H. Ates, M.H. Nayfeh (to be published) Synthesis of gold nanorods and filaments by reduction/growth on silicon nanoparticles.
[32] The World's Dominant Gold Refineries, BullionStar : http://www.zerohedge.com/news/2016-08-19/worlds-dominant-gold-refineries
[33] How to Refine Gold by the Aqua Regia Acid Method, Shor International Corporation, https://www.ishor.com/refine-gold-by-the-acid-method
[34] Andrews D, Raychaudhuri A, Frias C. Environmentally sound technologies for recycling secondary lead. J. Power Sources 2000;88:124–9. (Hydromettalurgy).
[35] Syed S. Recovery of gold from secondary, sources: a review. Hydrometallurgy 2012;30(51):115–6.
[36] Sheng PP, Etsell TH. Recovery of gold from computer circuit board scrap using aqua regia. Waste Manage. Res. 2007;25(4):380–3.
[37] Xiong L, Shen S, Liu L, Zhang L. Selective, effective reduction of gold ions from acidic HCl–NaClO3 leachate with oxalic acid. Ind. Eng. Chem. Res 2014;53(43):16672–7.
[38] https://www.youtube.com/watch?v(0yOBdkxk1IE
[39] Aitimbetov T, White DM, Seth I. Biological gold recovery from gold–cyanide solutions. Int. J. Miner. Process 2005;76:33–42.
[40] Ishikawa SI, Suyama K, Arihara K, Itoh M. Uptake and recovery of gold ions from electroplating wastes using eggshell membrane. Bioresour. Technol. 2002;81:201–6.
[41] Pethkar AV, Paknikar KM. Recovery of gold from solutions using Cladosporium cladosporioides biomass beads. J. Biotechnol. 1998;63:121–36.
[42] Schrand AM, Rahman MF, Hussain SM, Schlager JJ, Smith DA, Syed AF. Metal-based nanoparticles and their toxicity assessment. Nanomed. Nanobiotechnol. 2010;2(5).

Chapter 9

Nanosilicon

9.1 BANDGAP AND EXCITONS

Unlike metals, semiconductors do not electrically conduct, that is, they have no free electrons in the conduction band. All of the electrons are bound to atoms, that is, they are in the valence electronic bands. This is because in metals the conduction and valence bands overlap but in semiconductors the edges of bands are separated by an energy gap called band gap as shown in Fig. 9.1 [1].

In semiconductors an external agent, such as external light is needed to impart enough energy to a bound electron to overcome the band gap energy. When this takes place the electron is placed in the conduction band accompanied with its separation from the positive charge. Separation of the electron from the positive charge leaves behind in the valence band what is called a positive hole. The electron, as well as the positive hole is free to move away from their original common site; but they move together as an electron–hole pair or exciton, with the electron orbiting around the hole at an average distance or exciton's Bohr radius. In silicon for example, the Bohr radius is of 4.2 nm. The exciton structure is actually an "atom" which in some respect resembles the structure of a hydrogen atom, but 100–200-fold larger. Because of the large size, excitons are less bound and more fragile.

9.2 DIRECT AND INDIRECT BANDGAP MATERIALS

The interaction (absorption or emission) of light with matter must satisfy the conservation of energy and momentum. The ability of meeting these conservations laws depends on the type of material. Two classes of semiconductors are interesting in this regard: indirect bandgap materials, such as silicon and direct bandgap semiconductor such as CdSe (see Fig. 9.2). The band gap is called "direct" if the momentum of electrons and holes is the same in both the conduction band and the valence band. In this case an electron can directly emit a photon, which conserves energy while momentum is automatically conserved as shown in Fig. 9.2A . In an "indirect" gap material, the momentum of electron and hole are not the same; hence a photon cannot be readily emitted because, to conserve momentum, the electron must pass

Fundamentals and Applications of Nano Silicon in Plasmonics and Fullerines.
http://dx.doi.org/10.1016/B978-0-323-48057-4.00009-8

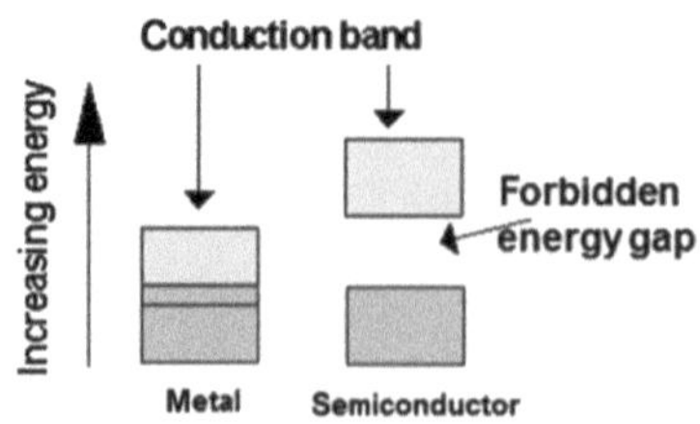

FIGURE 9.1 Conduction and valence bands for metal and semiconductor. *(Adapted from: http://webs.mn.catholic.edu.au/physics/emery/hsc_ideas_implementation.htm; X. Fan, W. Zheng, D.J. Singh, Light scattering and surface plasmons on small spherical particles, Open 3 (2014) e179, http://www.nature.com/lsa/journal/v3/n6/full/lsa201460a.html.)*

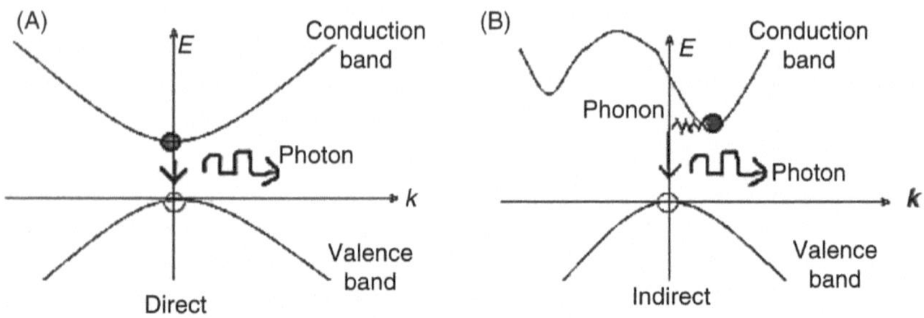

FIGURE 9.2 Schematic of the band structure showing the conduction and valence bands for (A) direct band gap material and (B) indirect band gasp material.

through an intermediate state and transfer momentum in the form of phonon quanta to the crystal lattice as shown in Fig. 9.2B. The probability of coincidence of three particles electron, hole, and phonon with specific properties is significantly lower than the coincidence of only two such particles (electron and hole). Thus, the probability of emission of a photon is much lower in indirect bandgap semiconductors than in direct bandgap ones. Bulk silicon is therefore a very poor light emitter, while in most cases, the direct bandgap semiconductors are good light emitters. Direct bandgap material is therefore used in light emitting devices, while indirect bandgap material is used in electronic devices.

9.2.1 Effective Mass

In a solid, the motion of a particle is governed by external forces, as well as forces exerted by the atoms in the solid. One procedure for handling the forces due to the atoms in the solid is the *effective mass* concept of the particle (often denoted m^*). In this picture all the atomic forces are lumped in term of a simple change of the mass of the particle when it is in free space. Formally, this is one of the results from the band theory of solids describing the movement of particles in a periodic potential, over long distances larger than the lattice spacing. Therefore, the effective mass is a quantity that is used to simplify band structures by constructing an analogy to the behavior of a free particle with that mass.

The effective mass can be considered to be a simple constant of a material. In general, the effective mass depends on the purpose for which it is used, however; and can vary depending on a number of factors. For electrons or electron holes in a solid, the effective mass is usually stated in units of the true mass of an electron, m_e (9.11×10^{-31} kg). In these units it is usually in the range 0.01–10, but can also be lower or higher—for example, reaching 1,000 in exotic heavy fermion materials, or anywhere from zero to infinity (depending on definition) in graphene. As it simplifies the more general band theory, the electronic effective mass can be seen as an important basic parameter that influences measurable properties of a solid, including everything from the efficiency of a solar cell to the speed of an integrated circuit.

9.3 ENHANCING AND BLUE SHIFTING OF LUMINESCENCE BY QUANTUM CONFINEMENT

As we shrink the size of direct semiconductor material, novel optical properties begin to emerge as we approach a certain characteristic size scale. The characteristic scale is the electron–hole distance (the Bohr radius) of the material. For example, for CdSe the Bohr radius is 5.6 nm. In this regime, the interaction of light with the material gets modified because quantum quantization of the energy levels of the electron as well as that of the hole according to Pauli's exclusion principle becomes important. A simplified treatment considers the energy of the electron and the hole as the energy of the charge in an infinite well (see Fig. 9.3A). The top well is for an electron and the bottom inverted well is for a hole.

This model results in three effects. First the energy levels in the model are discrete according to: $E_n = (h^2 n^2)/(8\ m_c\ R^2)$, where n is an integer designating energy levels, h is Planck's constant, m_c is the effective mass of the electron and hole system, and R is the radius of the quantum dot (particle). In fact those levels are embedded in the continuum states of the conduction and valence band as shown in Fig. 9.3B. In this simplified model, the bandgap of the material widens to $E_g = E_{g0} + (h^2)/(8\ m_c\ R^2)$. Second, this result shows that as the particle decreases in size, the band gap energy increases as shown in Fig. 9.4 [1]. More energy is then needed to excite the dot, and concurrently, more energy is released when the crystal returns to its ground state, resulting in a color shift toward blue in the emitted light. Third, in addition to the shift towards the blue the emission becomes stronger due to the discrete nature of the levels, making the nanoparticles much brighter than bulk [2,3]. As a result

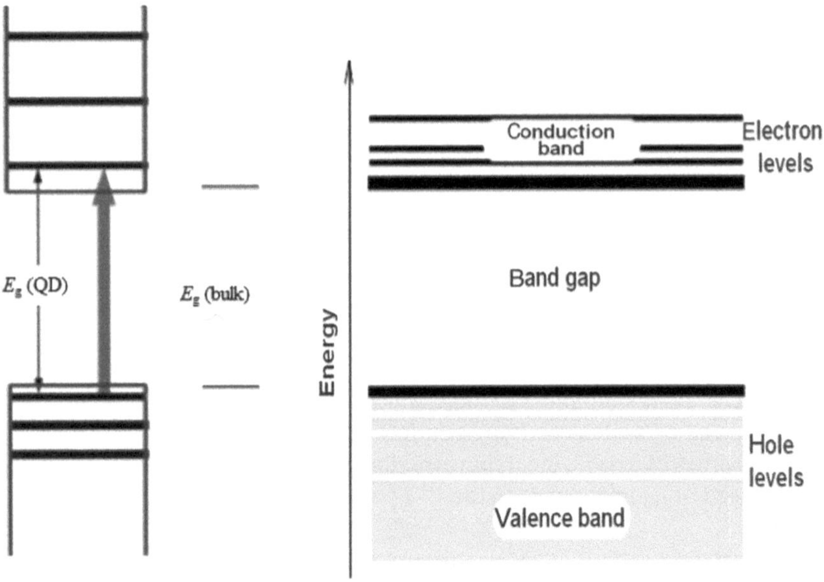

FIGURE 9.3 Two sketches of simplified infinite well model of the energy of the electron and the hole in a quantum dot. The top well is for an electron and the bottom inverted well is for a hole.

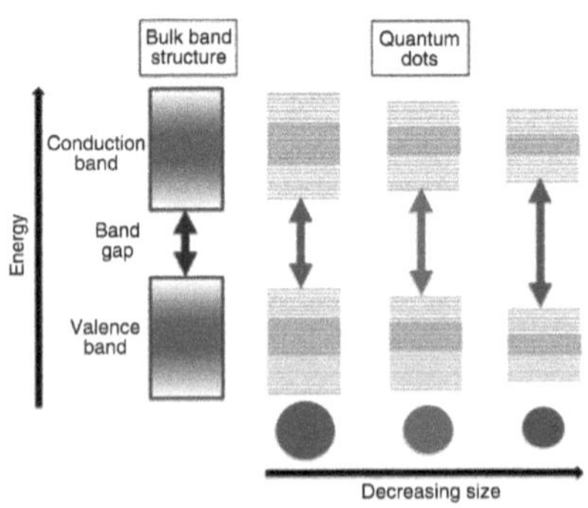

FIGURE 9.4 The conduction and valence bands with decreasing size of a semiconductor nanoparticle (quantum dot), showing splitting of energy levels due to the quantum confinement effect. The band gap increases with decrease in size of the nanocrystal. *(From https://www.sigmaaldrich.com/technical-documents/articles/materials-science/nanomaterials/quantum-dots.html)*

of this phenomenon, quantum dots can emit any color of light from the same material simply by changing the dot size. Additionally, with control over the size of the nanocrystals, quantum dots can be tuned during manufacturing to emit any color of light as shown in Fig. 9.4.

Example 9.1 Size Dependence Of Quantum Confinement In Semiconductor Nanoparticle [3]

In Fig. 9.3, a sketch of the energy level diagram in a direct semiconductors nanoparticle of rectangular shape is given. The hole levels are confined just like the electrons. Inside semiconductors the electron has a different mass than that of free electrons due to the interaction of the electrons with all of the atoms in the material. Thus the new mass is called the effective masses m^*e. Similarly we have an effective mass for holes and m^*h. Using the simple particle in a box model theory we get the following expressions for the energy levels of the electrons and holes in a nanoparticle box of length a.

$$\Delta E = E_g + \frac{\pi^2\hbar^2}{2\mu_{e,h}} \cdot \frac{1}{a^2} - \frac{1.8 \cdot e^2}{\varepsilon} \cdot \frac{1}{a} \text{ where } E_{coulomb} = \frac{1.8 \cdot e^2}{\varepsilon} \cdot \frac{1}{a}$$

Where n is an integer. Let us give an actual example which occurs in CdSe nanoparticles or quantum dots. The bulk bandgap energy is known to be $E_g = 1.74$ eV. In fact the band gap is not really constant when the size of the sample drops. The change is due to strain induced by miniaturization. Also the effective masses of holes and electrons are $m^*h = 0.45$ me and $m^*e = 0.13$ me, where *me* is the mass of free electrons. In Fig. 9.5, the measured absorption of nanoparticles as a function of the incident photon energy (hence their size) is plotted. The overall effect changes the light absorption maximum from about 1.9 eV to almost 2.6 eV with changing nanoparticle size. This displays the effect of nano technology.

9.4 MAKING SILICON GLOW: QUANTUM CONFINEMENT

Silicon is the eighth most common element in the universe by mass, but very rarely occurs as the pure free element in nature. Monocrystalline silicon, manufactured from sand using sophisticated technology is the backbone of the microelectronics industry. However silicon is the dullest material with regard to its optical activity because it belongs to the class of indirect bandgap semiconductors.

Fig. 9.7 shows the energy–momentum diagram for silicon. Light emission in silicon is highly improbable because it requires both conservation of energy, which is satisfied by the emission of the appropriate photon energy at 1.1 eV, and conservation of momentum, which is satisfied by emission of a specific phonon (vibration of the crystal) of 55 meV energy. Processes that require simultaneous emission of light and vibrations are highly unlikely and take a long time to happen, providing emission lifetimes on the order of milli seconds.

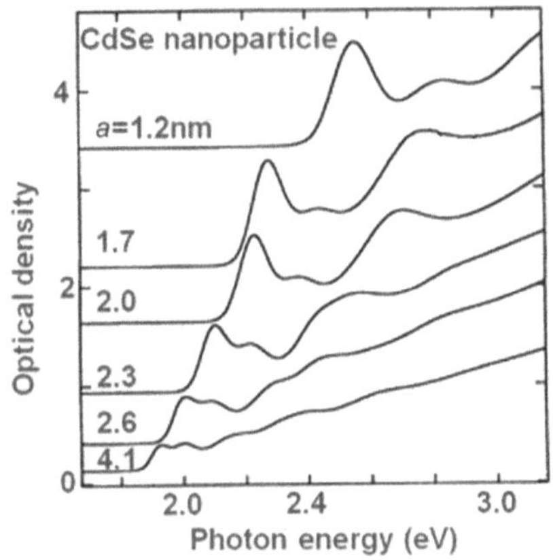

FIGURE 9.5 Measured absorption of CdSe nanoparticles as a function of the incident photon energy (hence their size) is plotted. *(From D.J. Norris, M.G. Bawendi, Measurement and assignment of the size-dependent optical spectrum in CdSe quantum dots, Phys. Rev. B, 53, (24), Adapted from https://journals.aps.org/prb/pdf/10.1103/PhysRevB.53.16338)*

Example 9.2 Binding Energy Of Exciton In Nanoparticles

Taking an infinite potential well as a simple model for a quantum dot, one calculates by solving Schrödinger's equation the energy levels to

$$E = E_{bulk} + \frac{\hbar^2 \pi^2}{2a^2}\left(\frac{1}{m_e^*} + \frac{1}{m_h^*}\right) - \frac{1.8e^2}{4\pi\varepsilon\varepsilon_0 a}$$

$$E_n = \left(h^2 n^2\right) / \left(8 mc R^2\right)$$

Where n is an integer designating energy levels, h is Planck's constant, mc is the effective mass of the electron and hole system, and R is the radius of the quantum dot (particle).

$m_0 = 9.11 \times 10^{-31}$ kg is the free electron rest mass.

Example 9.3 Shape Dependence Of Quantum Confinement (Rectangular Vs Spherical) In Metal Nanoparticle

Fig. 9.6 bottom sketches three types of confinement architectures, one dimensional (1D) quantum well, two-dimensional (2D) quantum wire, and three dimensional (3D) quantum dot or nanoparticle. We will derive the energy levels of an electron trapped inside the quantum well using quantum mechanical treatment, as well as those of the 3D case.

One-dimensional: Consider an electron in a one-dimensional infinite quantum well of width a. We are interested in the lowest confinement energy of an electron. To get an analytical solution, consider the infinite square well potential, that is, the limit V ~ infinity. Then states outside the box will be completely forbidden, and we are left with the solution inside the box.

$$\psi(x) = A\sin(kx) + B\cos(kx)$$

By requiring that

$$\psi(x) = 0 \text{ when } x = \pm a/2$$

one can easily show that the exact solutions of the Schrödinger equation are simple sine and function, and the lowest of these gives

$$\psi_0(x) = \sqrt{\frac{2}{a}}\cos\frac{\pi x}{a}$$

which has the ground state energy

$$E_0 = \frac{\pi^2 \hbar^2}{2ma^2}$$

The full solution for all energy levels is

$$E_n = n^2 \pi^2 \frac{\hbar^2}{2ma^2}$$

where

$$n = 1, 2, 3, \ldots$$

As the electron levels in a quantum well are discrete (rather than continuous as in a bulk metal), then only allowed transitions of the outermost electrons are those between the discrete levels. Substituting n = 1 and a = 1.55 nm gives

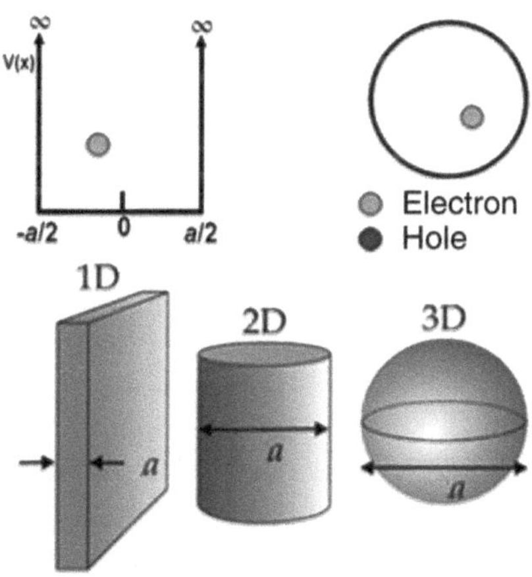

FIGURE 9.6 Sketches of three types of confinement architectures, one dimensional (1D) quantum well, two-dimensional (2D) quantum wire, and three dimensional (3D) quantum dot or nanoparticle.

E0 = 0.156 eV. The first transition energy is between $n = 2$ and $n = 1$ or $3 \times$ 0.156 eV = 0.468 eV

Three dimensional spherical nanoparticle.: Unlike the rectangular case in the 1D case above, the quantum mechanical solution involves much more complicated functions of Bessel functions instead of sines and cosines. Therefore we will leave the details out. Interested readers may find the expressions in other reference specialized books. Here we give only the expression for the energy levels. The three-dimensional (spherical symmetry) corresponding result is

$$E_{nl} = \beta_{nl}^2 \frac{\hbar^2}{2ma^2}$$

where l is the angular quantum number, and n is the energy quantum number and βnl is the nth zero of the lth spherical Bessel function jl(x). These are tabulated in Abramovitz and Stegun p. 467 (nowadays available in the web: [http://www.convertit.com/Go/Maps/Reference/AMS55.ASP?Res=150&Page=467)

- The first few zeroes are: 3.141593; 4.493409, 5.762459, 6.283185, 7.72525.
- Comparison with the 1D solution shows that the first energy value is exactly the same, but above that the 3D case has many more solutions.

In direct semiconductors, light emission proceeds readily with a lifetime of nanoseconds as the momentum is automatically satisfied without vibration and emission of phonons.

Recent developments enabled by nanotechnology are starting to change the picture. Some silicon nanostructures provide interactions with light now approach or even exceed the performance of equivalent direct-bandgap materials, which promise to take silicon into the realm of optics. A significant deviation from bulk properties was found in 1990 when L.T. Canham noticed visible photoluminescence in porous silicon he produced via chemical etching of bulk silicon in HF acid [5].

But porous silicon is just an interconnected nanoscale network of silicon skeletal (sponge-like structure). The first report of micro- and nanoparticles prepared from porous Si came from the Sailor group at University of California at San Diego [6]. Strong ultrasound was used to shatter porous silicon into micro- and nanoparticles. But the particles do not have specific configuration or uniformity because silicon is a hard material and the particles are basically a result of mechanical stress and shattering of the interconnected skeletal backbone. In 1997, a new self-limiting etching procedure was developed in the Nayfeh's group at the University of Illinois which produced on the silicon wafer disconnected individual spherical nanoparticles of preferred or magic configuration or sizes, which can be softly retrieved and stored in a liquid of choice [7–11]. The particles are protected by mono-hydride coating and can be produced in commercial amounts and stored for later use for many years. The smallest of these particles is 1 nm and fluoresces in the blue. Other sizes include 1.7 nm, fluorescing in the green; 2.15 nm, fluorescing in the yellow-orange; and 2.9 nm,

Example 9.4 Effective Mass m* in Bulk And Nanoparticles

In a solid, the motion of a particle is governed by external forces as well as forces exerted by the atoms in the solid. One procedure for handling the forces due to the atoms in the solid is the *effective mass* concept (often denoted m^*). In this picture all the atomic forces are lumped in term of a simple change of the mass of the particle. Formally, this is one of the results from the band theory of solids describing the movement of particles in a periodic potential, over long distances larger than the lattice spacing. Therefore, the effective mass is a quantity that is used to simplify band structures by constructing an analogy to the behavior of a free particle with that mass.

The effective mass can be considered to be a simple constant of a material. In general, it depends on the purpose for which it is used, however; and can vary depending on a number of factors. For electrons or electron holes in a solid, the effective mass is usually stated in units of the true mass of an electron, me (9.11×10^{-31} kg). In these units it is usually in the range 0.01–10. There are environments in which it can also be lower or higher. For example, it can reach 1,000 in exotic heavy fermion materials. In graphene, it can assume a value anywhere from zero to infinity (depending on definition).

Example 9.5 Silicon Parameters

Crystal structure	**Diamond**
Group of symmetry	Oh^7-Fd3m
Number of atoms in 1 cm^3	51022
Density	2.329 g cm^{-3}
Dielectric constant	11.7
Effective electron masses, ml	0.98 mo
Effective electron masses, mt	0.19 mo
Effective hole masses, mh	0.49 mo
Effective hole masses, mlp	0.16 mo
Electron affinity	4.05 eV
Lattice constant	5.431 A
Optical phonon energy	0.063 eV

Number of Si atoms

Consider 1 cm^3 of silicon. How many atoms does this contain? Solution: The atomic mass of silicon is 28.1 g, which contains Avagadro's number of atoms.

Avagadro's number N is 6.02 × 1023 atoms/mol. The density of silicon: 2.3 × 103 kg/m3, therefore, 1 cm3 of silicon weighs 2.3 gram and contains 4.93 × 1022 atoms.
The number of silicon atoms in 1 nm diameter particle. Take 5 × 1022 for the number in 1 cm3. The volume of 1 nm particle is (4ϖ/3) r3, which gives 0.523 × 10–21, hence we get 26.2 atom. In fact if we actually carve a sphere of 1 nm from silicon crystalline structure we find 29 silicon atoms.

Example 9.6 Germanium vs Silicon

Silicon is the backbone of the microelectronics industry, but attention has recently shifted to Ge for nanodevices in the sub 100 nm regime, as well as for Ge-based optoelectronics applications. Because its electron and hole mobilities are two- and four-fold larger than those of Si, germanium is emerging as a viable candidate for replacing silicon CMOS. In addition, successful growth of Ge on Si allows subsequent growth of optically active materials, such as GaAs, because Ge and GaAs have the same lattice constant.
Ge-based nanomaterials have also attracted a lot of attention in recent years due to their potential applications in electronics and optoelectronics. For instance, germanium has a smaller absorption coefficient, which makes it attractive for integration of monolithic photodetectors that can be used in optical interconnects. Moreover, because the bandgaps of bulk Ge are smaller than those of silicon, being 0.66 versus 1.1 eV for the indirect bandgap, and 0.9 versus 3.2 eV for the direct bandgap, Ge-based nanomaterials are more promising to be suitable for infrared applications? However the Ge-based nanomaterials pose some disadvantages. Unlike Si, the Ge chemistry is complex as a result of the fact that the Ge oxide is unstable in water. There is also an economic disadvantage since the cost of bulk Ge wafers is considerably higher than that of bulk silicon wafers.
Silicon–germanium (Si–Ge) alloys are becoming an important semiconductor material, for use in high-speed integrated circuits. Circuits utilizing the properties of Si–SiGe junctions can be much faster than those using silicon alone. Silicon–germanium is beginning to replace gallium arsenide (GaAs) in wireless communications devices. The SiGe chips, with high-speed properties, can be made with low-cost, well-established production techniques of the silicon chip industry.
The recent rise in energy cost has improved the economics of solar panels, a potential major new use of germanium. Germanium is the substrate of the wafers for high-efficiency multijunction photovoltaic cells for space applications. Because germanium and gallium arsenide have very similar lattice constants, germanium substrates can be used to make gallium arsenide solar cells. The Mars Exploration Rovers and several satellites use triple junction gallium arsenide on germanium cells.
Germanium-on-insulator substrates are seen as a potential replacement for silicon on miniaturized chips. Other uses in electronics include phosphors in fluorescent lamps, and germanium-base solid-state light-emitting diodes (LEDs). Germanium transistors are still used in some effects pedals by musicians who

wish to reproduce the distinctive tonal character of the "fuzz"-tone from the early rock and roll era, most notably the Dallas Arbiter Fuzz Face.
The chemistry of Ge is different from that of silicon. For example, whereas Si oxide (glass) is stable in water, germanium oxide dissolves in water.

Example 9.7 Types Of Confinement (Strong Vs Weak) [4]

Consider a nanoparticle of radius a. An electron–hole pair is photo-excited in the particle. We define a Bohr radius of the exciton aB, There are two cases with regard to the size of the exciton compared to the size of the nanoparticle as sketched in.

1. Exicton much smaller than the nanoparticle (Weak confinement regime):

In this regime, the exciton fits very well inside the nanoparticle, that is, particle radius is much larger than the Bohr exciton (a >> aB. In this case the electron and hole will be free to quantum mechanically bind to each other. In other words the exciton acts as a "hydrogen atom" with the hole acting as a "nucleus." The exciton binding energy is designated by Eex. We can also say that this condition means that binding energy is larger than the quantization energy of the electron and the hole. The lowest exciton energy is given by (μeh is the reduced mass of the electron and hole):
where Eg is the band gap energy of the bulk. In this size regime, the electron and hole states originate from quantum confinement of the exciton canter of mass. For example, this treatment has found successful applications in CuCl QDs.

2. Strong confinement regime:

For small nanocrystal the exciton size is larger than the particle (a << aB) as sketched in; hence it cannot fit in it. The interaction with the wall of the nanoparticle is strong enough to foul their binding and as a result the electron and hole are forced not to bind to each other. They then are treated independently. In this regime, the band gap state is given by :

$$\Delta E = E_g + \frac{\pi^2\hbar^2}{2\mu_{e,h}} \cdot \frac{1}{a^2} - \frac{1.8 \cdot e^2}{\varepsilon} \cdot \frac{1}{a} \text{ where } E_{coulomb} = \frac{1.8 \cdot e^2}{\varepsilon} \cdot \frac{1}{a}$$

is the Coulomb energy correction for electrons in the 1 Se level.

3. Intermediate confinement regime:

In the case of the intermediate confinement regime, the particle radius $a \asymp aB$. That is the case for large particles of CdS or CuBr, for which $aB =$ 2.8 and 1.8 nm, respectively. The large difference between the electron and hole effective masses ($me < mh$) permits adiabatic separation of their movement. The electron wave function is size quantized; whereas the hole moves in the coulomb potential caused by the fast electron and is pushed to the center of the particle.
http://infoscience.epfl.ch/record/100036/files/EPFL_TH3767.pdf

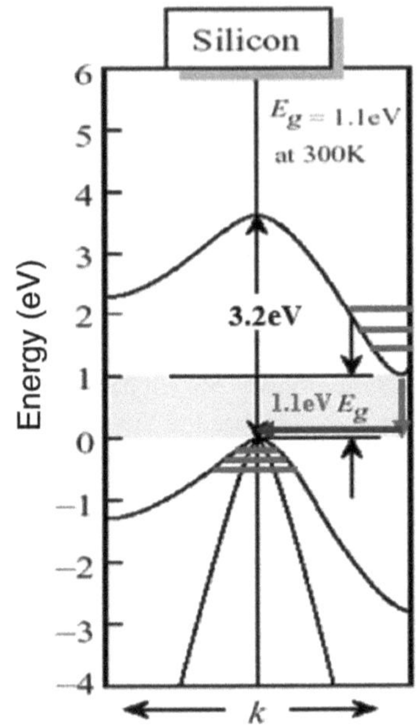

FIGURE 9.7 **The energy–momentum diagram of the band structure of silicon.** Emission of light requires both conservation of energy, which is satisfied by the emission of the appropriate photon energy at 1.1 eV (red vertical arrow), and conservation of momentum, which is satisfied by emission of a specific phonon (vibration of the crystal) of 55 meV energy (horizontal blue arrow). The upper set of green lines are discrete levels due to confinement of an electron and the lower set of green lines are due to the confinement of a hole.

fluorescing in the orange-red. Fig. 9.8A shows the luminous of colloids of 1 and 2.9 nm particles in alcohol using incoherent light excitation at a wavelength of 365 nm. The figure gives also the emission using laser beam excitation at 353 nm and 442 nm from a Nd–Yag and He–Cd lasers, respectively. The 1-nm Si nanoclusters are amenable to testing and accurate first principle simulations because they consist of a manageable number of Si and H atoms and are produced in macroscopic amounts. Lubos Mitas (then at the University of Illinois) of the University of North Carolina developed a computer prototype based on the experimental characteristics and those theoretical simulations. He showed that the 1-nm particle has a configuration of a super molecule $Si_{29}H_{24}$ as shown in Fig. 9.8B [12–15].

One important characteristic of these particles is that the energy and momentum conservation rules governing light interaction get modified. In bulk silicon crystals or large particles, photo produced excitons move freely in all directions. Because it is not easy to conserve momentum in indirect gap material when light is emitted, they recombine again before light is emitted with energy turning into vibrations and heat imparted to the crystal as sketched in Fig. 9.9.

Example 9.8 E and M scattering silicon nanoparticle [22]

Strong magnetic response of submicron Silicon particles in the infrared.

In this example we show that silicon nanoparticles can have light scattering due to induce magnetic dipole response. Consider a silicon nanoparticle with a radius of $a = 230$ nm, and refractive index $n_p = 3.5$. The particle is embedded in air (hence $m = n_p/n_h = 3.5$). We focus on the wavelength range 1000–2000 nm because the refractive index of silicon in this range is real. We apply the Mie scattering theory. The cross section for scattering normalized to the geometrical cross section (a^2) is plotted in Fig. 9.13 as a function of frequency (wavelength) of the incident beam.

The contribution of the first few terms in the Mie expansion to the scattering cross-section σ_S are calculated and shown individually. We included the a_1, b_1, a_2, and b_2 terms. The red line corresponds to a_1, the electric dipole contribution while green line corresponds to b_1, the magnetic dipole contribution. In addition we have the term a2 corresponding to the electric quadrupole as well as b2 corresponding to the magnetic quadrupole. The black curve represents the sum or the total cross section. For wavelengths larger than $\lambda = 1200$ nm, the cross section is completely determined by the first b1 and a1 coefficients. In other words, in this regime Si particles can be treated as non-Rayleigh dipolar particles. The magnetic contribution b1 dominates in this region. High-permittivity dielectric particles with resonant magnetic properties are being explored as constitutive elements of new metamaterials and devices in the microwave regime. Magnetic properties of low-loss dielectric nanoparticles in the visible or infrared are not expected due to intrinsic low refractive index of optical materials in these regimes.

When the particle is reduced to a size smaller than the exciton radius, especially in the sub 3-nm regime, the particles become less rigid allowing the silicon atoms to move, relax or adjust especially those on the surface. Moreover, excitons get more strongly confined spatially, causing the momentum of the electron and hole to spread out appreciably according to the Heisenberg uncertainty principle, such that their momentum distributions can overlap. In the overlapping region, the momentum conservation may be readily satisfied; thus recombination of the electron and hole to turn the excess energy into light becomes stronger and faster than producing vibration and heat. In this case a photon (light) is produced, with the energy of the emitted photon (color of the light emitted) depending on the size of the particle.

For a nanoparticle size of 1nm for example, emission of blue photons dominates the production of vibration and heat with a performance that matches particles of direct-band gap material as in Fig. 9.10A [16]. Fig. 9.10B is a computer-simulation image of 1-nm nanoparticle carried out by researchers at Livermore National Laboratory [14]. Turning silicon the dullest material in the universe to a material as bright as the brightest material warranted Nature magazine to report it in terms of the theme "Let there be light" as depicted in

FIGURE 9.8 (A) Blue and red luminescence of colloids of 1 and 2.9 nm particles in alcohol respectively with excitation at a wavelength of 365 nm (top row). Blue and red luminescence of colloids of 1 and 2,9 nm particles with excitation at a wavelength of 353 and 442 nm respectively (bottom row). (B) A prototype structure of 1-nm particle has a configuration of $Si_{29}H_{24}$. Silicon atoms in orange. Hydrogen atoms in white.

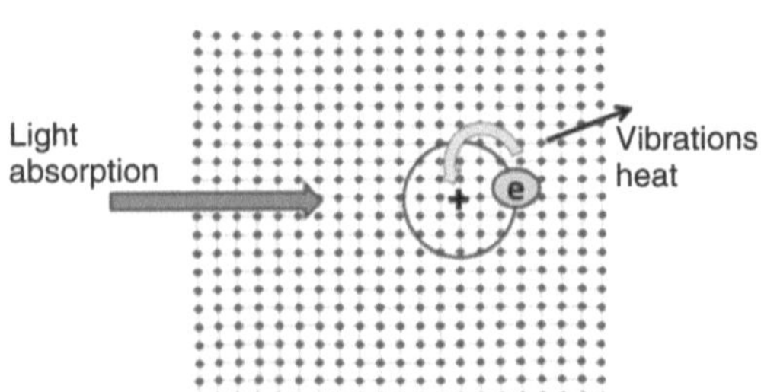

FIGURE 9.9 **Response of bulk silicon to light.** Exciton forms followed by recombination to produce vibrations and heat before it can emit a photon.

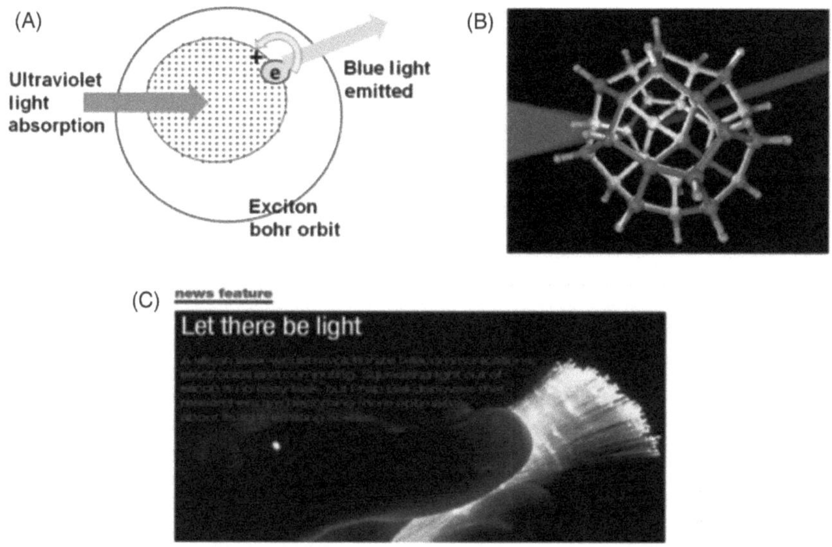

FIGURE 9.10 (A) Schematic of response of 1-nm silicon nanoparticle to light. In 1-nm nanoparticle, much smaller than the Bohr radius, recombination produces a blue photon very fast much before vibrations and heat is produced. *(Private drawing by author).* (B) Computer simulation image of 1-nm nanoparticle [14]. *(Private drawing by Andrew Williamson of Livermore National Laboratory and presently with Cambridge Innovation Capital).* (C) Nature magazine report on the brightness of the silicon nanoparticles in terms of a biblical theme "Let there be light". *(From http://research.physics.illinois.edu/nanosi/news.htm).*

Fig. 9.10C; it is a biblical theme of creation, signaling the transformation of the universe from the dark state to the bright state [17].

9.5 LARGE NONLUMINESCENT PARTICLES: LIGHT SCATTERING

Large particles in general and silicon particles in particular are not luminescent but they exhibit very interesting light scattering phenomenon; light scattering from molecules to nanoparticles and micro-particles [18].

9.5.1 Rayleigh Scattering

Rayleigh scattering [19] refers to the scattering of light off of small molecules, and can be extended to scattering from particles up to about a tenth of the wavelength of the light (~50 nm). It is this kind of scattering off the molecules of the air which gives us the blue sky. Rayleigh calculated the scattered intensity from dipoles much smaller than the wavelength to be:

$$I = I_0 \frac{8\pi^4 N\alpha^2}{\lambda^4 R^2}\left(1+\cos^2\theta\right)$$

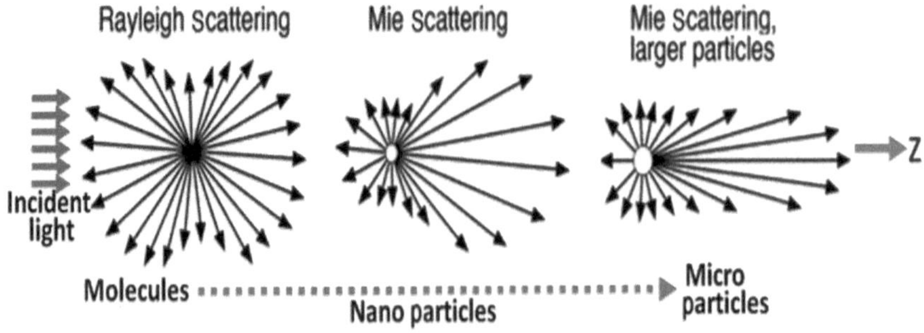

FIGURE 9.11 The angular distribution Rayleigh scattering of light off of small molecules, Mie scattering-off of nanoparticles and Mie scattering off microparticles.

where I_0 is the incident light intensity, N is the number of scatters, α is the polarizability of the scatters, λ is the wavelength of light, θ is the angle of scattering from the forward direction, and R is the distance to the observer. The angular distribution is plotted in Fig. 9.11A. This angular distribution is interesting as one can see it is symmetric around the direction 90 from the forward direction. It has the same intensity distribution for forward and backward directions. The process is an elastic scattering; that is the photon energies of the scattered photons is not changed. For comparison scattering in which the scattered photons have either higher or lower photon energy is Raman scattering. Usually this kind of scattering involves exciting some vibrational mode of the molecules, giving a lower scattered photon energy, or scattering off an excited vibrational state of a molecule which adds its vibrational energy to the incident photon.

9.5.2 Mie Scattering

As the size increases the Rayleigh theory will not applicable and more accurate analysis must be employed. It is based on an exact formalism developed by Mie [20], which accounts for all multipole expansion of the charge distribution in the particle (see Examples 8.5, 8.6, and 8.7 *for details of the Mie theory*). One major feature of the theory is that unlike molecules and very tiny particles (<1/10 wavelength) scattering produces a pattern like an antenna lobe, with a sharper and more intense forward lobe for larger particles as shown in Fig. 9.11B-C. In general scattering is characterized by a non-dimensional size parameter, $x = 2\pi a/\lambda$, where a is the radius of the spherical particle, and λ is the wavelength of incoming light. The angular distribution (phase function) derived from the Mie theory can be sketched for different sizes of particles or specifically in terms of the parameter x as sketched in Fig. 9.12.

9.5.3 Directional Light Scattering

Directional light scattering [21] by spherical silicon nanoparticles in the visible spectral range is experimentally demonstrated. These unique scattering

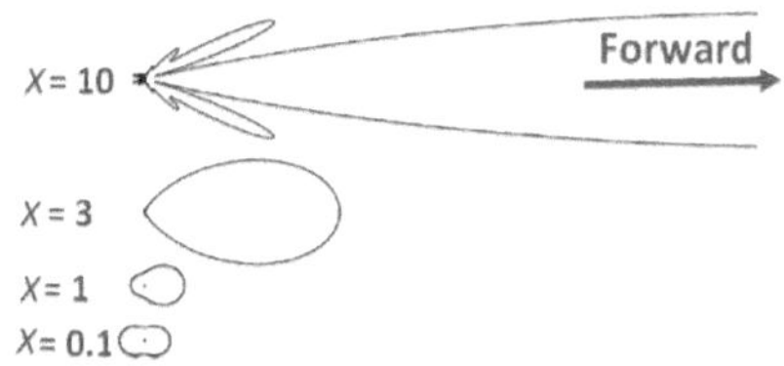

FIGURE 9.12 The angular distribution (phase function) derived from the Mie theory can be sketched for different sizes of particles or specifically in terms of the parameter $x = 2\pi a/\lambda$, where a is the radius of the spherical particle, and λ is the wavelength of incoming light.

properties arise due to simultaneous excitation and mutual interference of magnetic and electric dipole resonances inside a single nanosphere. Directivity of the far-field radiation pattern can be controlled by changing light wavelength and the nanoparticle size. Forward-to-backward scattering ratio above 6 can be experimentally obtained at visible wavelengths as shown in Fig. 9.13, as explained in Example 9.9. These unique properties of silicon nanoparticles make them promising for design of novel low loss visible- and telecom-range nano-antenna devices Figs. 9.14–9.16.

A core shell configuration of metal and silicon or silica has been used to fine tune the process. Tuning the size of the core as well as the shell thickness and the material composition of the nanoparticles along with the incident wavelength can be used to bring simultaneous resonance of magnetic

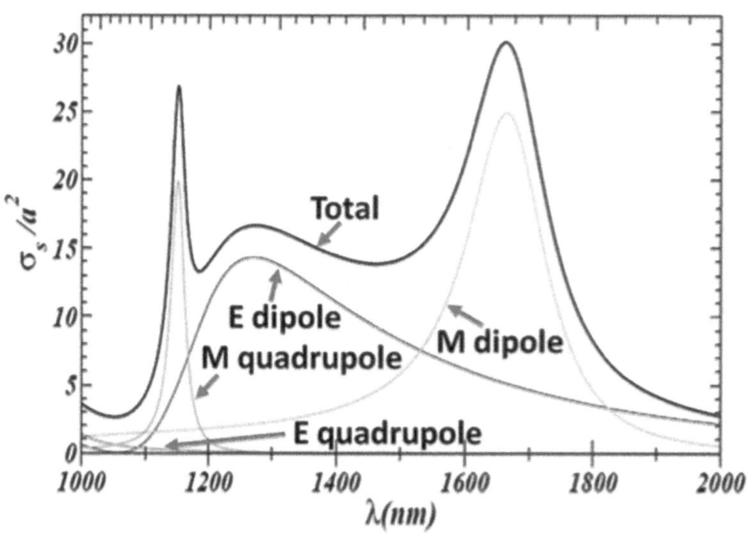

FIGURE 9.13 The cross section for scattering normalized to the geometrical cross section (a^2) according to Mie scattering theory is plotted as a function of frequency (wavelength) of the incident beam.

and electric response as well as match their strength. Detailed examples are given in Figs. 9.17 and 9.18 in Example 9.11. This is found to enhance he directionality in the forward direction. We study the angular scattering properties of individual core–shell nanoparticles that support simultaneously both electric and optically induced magnetic resonances of different orders. In contrast to the approach to suppress the backward scattering and enhance the forward scattering relying on overlapping electric and magnetic dipoles, we reveal that the directionality of the forward scattering can be further improved through the interferences of higher order electric and magnetic modes. As the major contributing electric and magnetic responses can be tuned to close magnitudes, ultra-directional forward scattering can be achieved by single nanoparticles without compromising the feature of backward scattering suppression, which may offer new opportunities for nano-antennas, photovoltaic devices, bio-sensing and many other interdisciplinary researches Fig. 9.19.

The effect of the conductivity of the material on light scattering and plasmon resonance are covered in Example 9.10. In the example we compare the responses of Si and gold.

9.6 OPTICAL NONLINEARITY IN NANO SILICON

In the presence of an electric field *E* of an incident laser radiation in a medium there is an induced polarization *P* in the medium at the frequency ω *of the* electric field *E* of an incident laser radiation. At low laser intensities *P* depends linearly on *E*. The linear optical properties of the material at that frequency are fully described by the refractive index n or the relative dielectric constant or the susceptibility χ.

At sufficiently high laser intensities, *P* deviates from its linear dependence on *E*. The expression for *P* involves higher-order terms in *E*. In electric dipole approximation, *P* can be mathematically expressed in a Taylor series as

$$P=\varepsilon_0\left[\chi^{(1)}E+\chi^{(2)}EE+\chi^{(3)}EEE+\ldots\right]$$

where ε_0, χ and *E* are the dielectric permittivity of free space, the nth order susceptibility tensor and the electric field in the medium respectively. It is to be noted that this expression is applicable only for non-resonant optical frequency and *E* is not too large so that the series expansion converges rapidly.

Nonlinear optical phenomena are exploited in developing various devices for frequency conversion, optical switching, optical limiting and optical logic circuits, etc. These susceptibilities are solely the properties of the material and depend upon the electronic structure of the atoms/molecules as well as their geometrical arrangement in the bulk. The second term *Â*(2)*EE* has a quadratic

Example 9.9 Scattering Of Si Vs Gold Nanoparticles

We compare light scattering of gold nanoparticles with that of silicon nanoparticles of the same diameter size (Fig. 9.14). The nanoparticles are of 180 nm diameter.

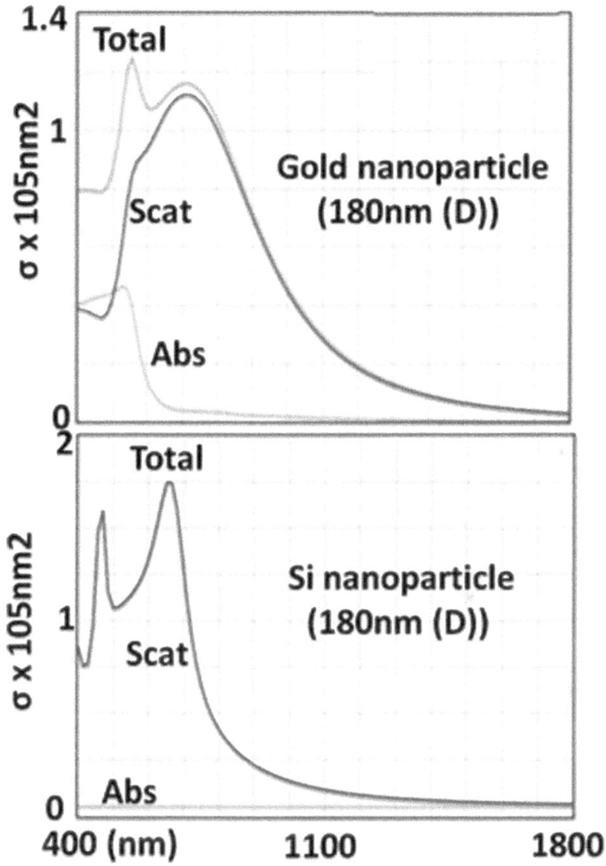

FIGURE 9.14 Calculated interaction of light with nanoparticles of 180-nm diameter using Mie's theory giving the scattering, absorption and total contributions (top) gold nanoparticles (bottom) silicon nanoparticles.

dependence on the electric field E. This second order polarization gives rise to the electro-optic effect, second harmonic generation, sum frequency generation and other second-order processes. The third-order polarization denoted by $\chi(3)EEE$ is responsible for third harmonic generation, optical phase conjugation and intensity dependent absorption and refraction in the medium.

Example 9.10 Near Field Scattering From Ultrasmall Dielectric Nanoparticles: Nano Jet And Tweezer [23]

Comparison of light scattering of gold nanoparticles with that of silicon nanoparticles of the same diameter size (180 nm diameter).

Near field distribution of the intensity (square of the electric field) for a nanoparticle of radius R = 1.8 nm (D = 3.6 nm). We treat three cases of dielectric of refractive index 1.33, 2.5, and 4. The incident light has a wavelength of 496 nm with an electric field strength of E0. The light scattering by dielectric particles is calculated using the Mie theory.

Fig. 9.15A–C shows the squares of the electric field (intensity) in the near field for the different refractive indices respectively. The electric field intensity increases in the near field region with increasing refractive index according to $E_{max}^2 = 1.90E_0^2, 5.15E_0^2, 7.10E_0^2$, respectively. As the nanoparticle radius increases, the Mie scattering shows some directionality as shown in Fig. 9.12. This has been also called a nanojet (directionally concentrated electromagnetic radiation). Localized regions of high intensity, such as those near the particle, can trap dielectric particles due to gradient forces (these are forces that arise because a particle with a dielectric constant higher than medium. If the focus of the beam just is above the object, a force will be generated to lift the object up towards the focus as depicted in Fig. 9.16.

It is to be noted that these nanojets are technologically important as they can be used to enhance Raman signals. The particle and its nanojet can be controllably moved and used as a nanoscale Raman probe using optical tweeze. Finally, we use 3.6 nm diameter of silicon and compare it to 3.6 nm diameter of gold as shown in Fig. 9.17. The silicon case there are no resonances found. Analysis in the infrared shows that there is no activity in the infrared beyond 800 nm. It is based on solely scattering with very little absorption. The gold particle shows a resonance at 500 nm solely based on absorption with very little scattering. There is no activity beyond 700 nm.

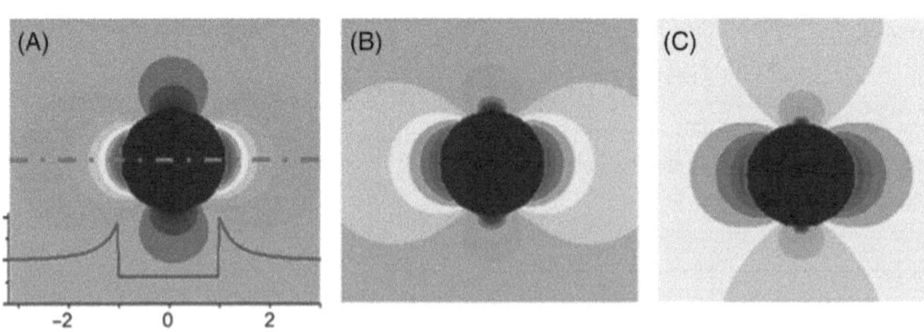

FIGURE 9.15 Near field scattering from ultrasmall dielectric nanoparticles of different refractive index. The squares of the electric field (intensity) in the near field for (A) 1.33 refractive index (B) 2.5 refractive index and (C) 4 refractive index. The intensity increases in the near field region with increasing refractive index.

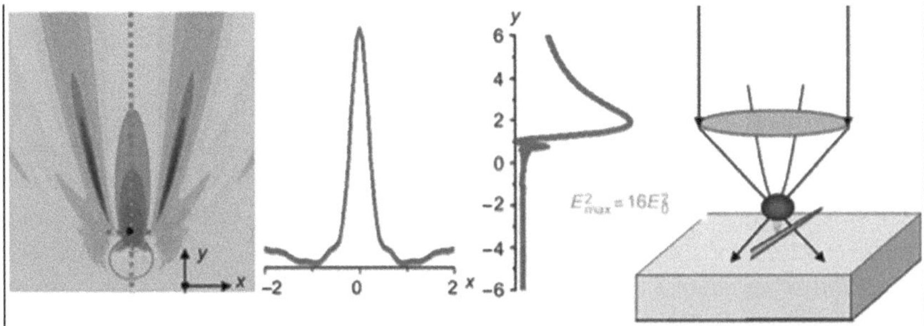

FIGURE 9.16 Plasmon nanotweezer: sketch showing focused light on a nano particle (right). Localized regions of high intensity, such as those near the particle, can trap dielectric particles due to gradient forces light distribution near the particle along the x- and y-axes (middle). Corresponding intensity distribution in the x-y plane (Left).

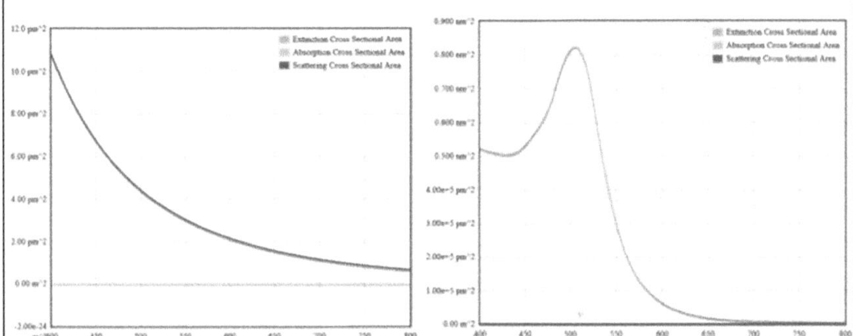

FIGURE 9.17 Light scattering from 3.6 nm diameter of silicon (A) and 3.6 nm diameter of gold (B).

For example, $\chi(2)$ vanishes for materials with inversion symmetry (centrosymmetry), such as bulk silicon. The structure of silicon crystal has cubic symmetry which exhibit inversion symmetry (centrosymmetry). But when the crystal becomes very small, it loses its rigidity and atoms especially the ones on the surface re-adjust, which exerts strain across the entire particle that alters the cubic structure and breaks the symmetry. Breaking the symmetry produces fundamental effects on its interaction with light. Fig. 9.20A displays one theoretical process that takes place in 1-nm particle in which some of the surface atoms lose their hydrogen atoms, move as much as 1.5 Å each towards each other followed by reconstruction/connection to form Si–Si dimer-like systems.

Blue luminescent 1-nm nanoparticles were used to test this mechanism. Particles prepared in water solvent were dried on device-quality silicon, which allowed them to form microcrystallites films as shown in Fig. 9.20B. The micro crystallites were then excited by high-intensity near-infrared femtosecond laser at 800-nm wavelength. Fig. 9.20C displays several frames of the emission from different crystallites. It shows strong wide blue luminescent band as well as competing sharp radiation at half the wavelength of the incident beam (second harmonic). The intensity of the second harmonic anti correlates with the wide band luminescence as seen in the figure. Both the sharp and wide band response indicates very strong coupling with light and the emergence of second-order nonlinear phenomena. The excitation of the blue luminescence band indicates a two photon excitation process, while the second harmonic indicates the breaking of the centrosymmetry. Other theoretical simulations and measurements, in collaboration with Charles Marsh and Ghassan Alchaar of the US Army Construction Engineering Research Laboratory (CERL) showed that the mechanical strain exerted is extremely high, corresponding to a pressure of several GPa [25]. Because atoms on the surface are under high strain, they can undergo large-amplitude molecular-like vibrations to relieve the strain [26]; and hence can also couple efficiently to thermal and mechanical stimuli. Second-order nonlinearity in silicon opens up other prospective applications in optics, including modulation, amplification, gain and laser action, and signal processing [27].

Example 9.11 Core-shell structures (Control of angular distribution) [24]

We study the angular distribution of light scattering of core-shell nanoparticles. The architecture promises to provide simultaneous support of both electric and optically induced magnetic resonances of different orders.

Dipole resonance versus quadrupole resonance: The structure shown in Fig. 9.18A consists of a silver core of radius 250 nm and refractive index equal to 2.5. Mie theory gives the response for different plasmon resonances including the E-dipole and E-Qudrupole. Also it gives the M-dipole and M-quadrupole. The figure plots the scattering cross section Q as a function of the wavelength of the incident light. The configuration shows resonance at point A between the E-dipole and the M-dipole. The two responses of the same amplitude. On the other hand when the size of the silver core is increased and the thickness of the shell is thinned such that the total size is kept the same, the resonance shifts to point B with equal amplitudes between the E- and M- quadrupoles. The interference at point A produces a forward nanojet. But the interference at point B gives a sharper forward scattering.

In contrast to the approach to suppress the backward scattering and enhance the forward scattering relying on overlapping electric and magnetic dipoles (covered in Examples 8.9), this result demonstrates that the directionality of the forward scattering can be further improved through the interferences of higher order electric and magnetic modes. As the major contributing electric and magnetic responses can be tuned to close magnitudes, ultra-directional forward scattering can be achieved by single nanoparticles without compromising the feature of backward scattering suppression, which may offer new opportunities for nanoantennas, photovoltaic devices, bio-sensing and many other interdisciplinary researches.

For comarionn we present the results for a homogenius nanoparticles made of pure silicon (refractive index 3.4) of radius 250 nm.Although ther eis resonance at point C, but there is not enough parameters to control to match intensities. There fore the interfernce process is unable to produce good forward scattering. We can now test anther core–shell structure in which the core is the dielectric and the shell is the metal as shown in Fig. 9.19. The matched resonance (wavelength and cross section) give forward scattering. This compares very favorably with the pure silicon case shown in the second row of Fig. 9.19.

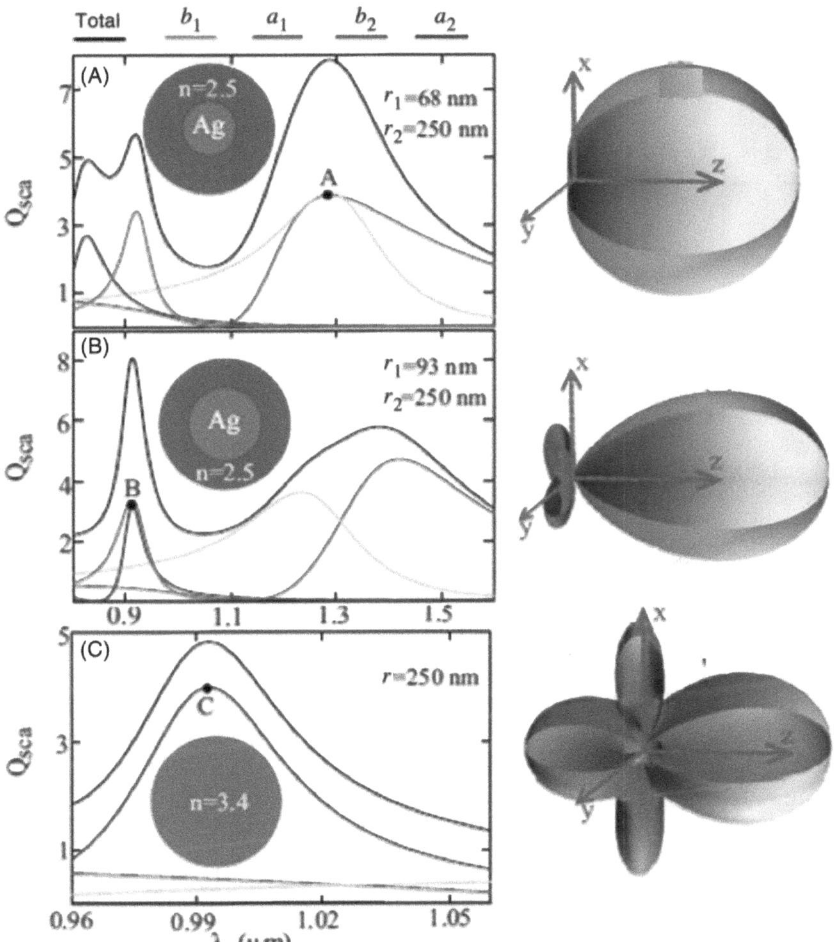

FIGURE 9.18 Control of angular distribution of scattering using core–shell structures of metal and dielectric. (A) Silver core and dielectric shell ($n = 2.5$). (B) Larger silver core and thinner dielectric shell ($n = 2.5$). (C) Pure dielectric material ($n = 3.4$).

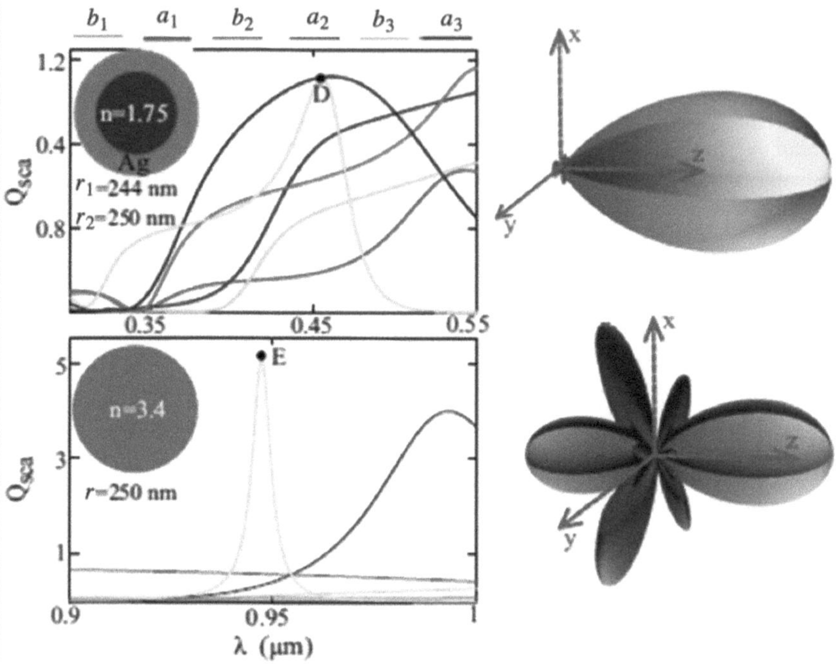

FIGURE 9.19 Control of angular distribution of scattering using core–shell structures of metal and dielectric. The calculated scattering cross section is plotted as a function of wavelength of the incident light (left column). It gives several multi-poles including first order (dipole), second order (quadrupole) and third order. Dielectric core (n = 1.75) and silver shell of 244 and 250 nm radii (top). Pure dielectric material (n = 3.4) of 250-nm radius (bottom). The corresponding angular distribution of the scattering (right column).

The third-order nonlinear susceptibility, hence the nonlinear refractive index has been studied using two methods: Z-scan and optical fringe pattern methods. The latter method was little explored for Si nanoparticles; but the Z-scan method has been used extensively by many research groups to obtain the contribution of quantum confinement in Si nanostructures. In this technique, the transmittance $T(x)$ of the sample is measured.

Fig. 9.21 gives a schematic of a general scheme of the setup of the Z-scan technique. In the conventional Z-scan experiment, the nonlinear sample is placed in the path of a focused laser beam. The sample is scanned through (Z-axis) the focused laser beam. The transmittance of the medium is recorded by a photodetector. The normalized transmittance is plotted as a function of sample position (z) measured with respect to the focal plane. The result gives the losses due to the linear and nonlinear absorptions of the sample. Using this plot, one can obtain the nonlinear absorption coefficient. If the transmitted light

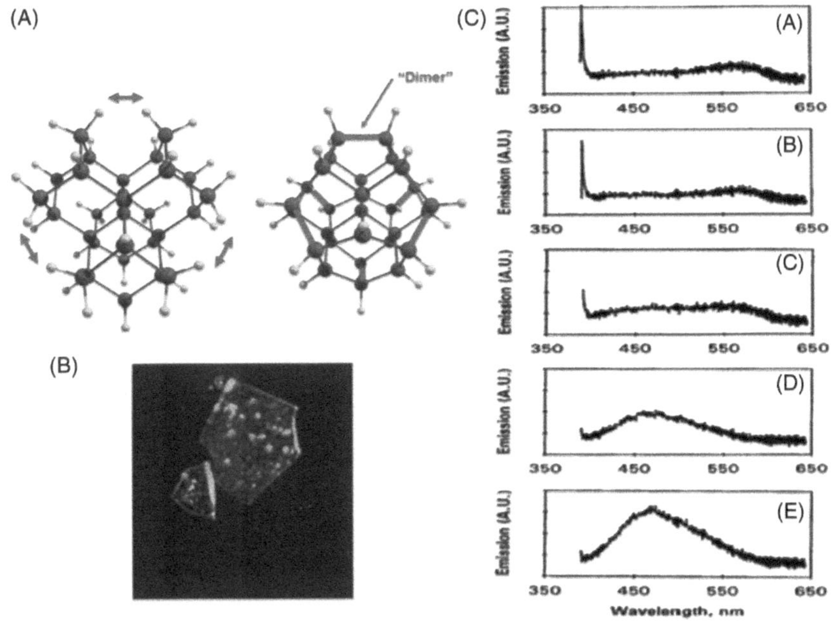

FIGURE 9.20 Response of micro-crystallites of 1-nm silicon particles to 780-nm near-infrared femtosecond excitation. (A) Model of surface reconstruction that breaks the crystal symmetry of silicon. (B) Image of reconstituted microcrystals. (C) Strong blue luminescent band indicating two-photon excitation, as well as competing sharp radiation at half the wavelength of the incident beam. The intensity of the second harmonic anti correlates with the wide band luminescence.

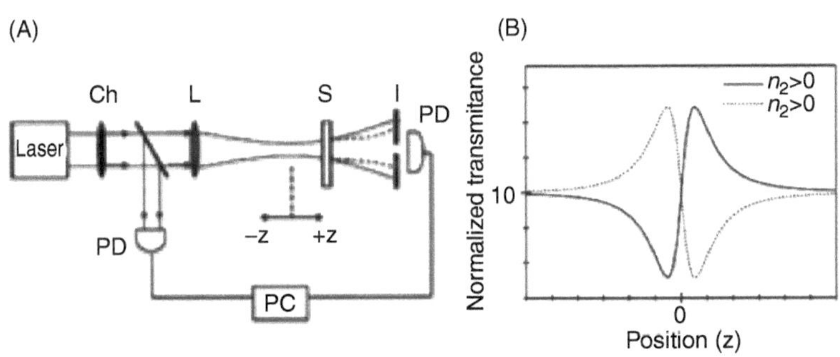

FIGURE 9.21 Measurement of the nonlinearity of large silicon nanoparticles. (A) Schematic of a general scheme of the setup of the Z-scan technique. (B) Normalized transmittance as a function of position of sample for both negative and positive refractive index.

is measured through a finite aperture in the far field, it is called a closed aperture Z-scan; otherwise it is called an open aperture Z-scan.

There several contributions to the nonlinearity. The electronic contribution originates from optical transitions involving bound electrons (second- and third-harmonic generation for example). The time scale of this contribution is of the order of 10^{-15} s. Other contributions are of non-electronic nature. Those are nonradiative interactions resulting from vibrational, electrostrictive and thermal processes. The vibrational response time is of the order of 10^{-13} s, while the electrostrictive response time is of the order of 10^{-8} s. Thermal non-electronic nonlinearities, such as those based on temperature and density changes exhibit longer time scales.

As the sample moves through the beam focus (at $z = 0$), self-focusing or defocusing modifies the wave front phase, thereby modifying the detected beam intensity. A pre-focal transmittance maximum (peak), followed by a post-focal transmittance minimum (valley) is a Z-scan signature of a negative nonlinearity. An inverse Z-scan curve (i.e., a valley followed by a peak) characterizes a positive nonlinearity. Fig. 9.21B depicts these two situations.

In this technique, the experimentally measured closed aperture transmittance $T(x)$ is first fitted with the following expressions to determine the phase shift $\Delta\phi$ and hence n_2 [28]

$$T(x) = 1 + \frac{4\Delta\phi}{(x^2+9)(x^2+1)}.$$

$$n_2 = \Delta\phi\lambda\alpha / \left[2\sqrt{2}\pi I_0 \left(1 - e^{-\alpha l}\right)\right]$$

where $x = z/z_0$, is defined as the reduced distance from the focal point, z being the absolute longitudinal distance and z_0 the Rayleigh range of the beam and $\Delta\phi$ is the phase change. Second, as the phase change $\Delta\phi$ is related to the nonlinear coefficient n_2 (in SI units) through the second equation the nonlinear coefficient may be calculated. In these expressions, α, I_0, and l are the linear absorption coefficient at the used wavelength, the peak intensity at the focus position, and the thickness of the sample, respectively.

The nonlinear coefficient of the refractive index has been analyzed at different wavelengths (or excitation energies) and time scales (from micro-second to femto-second) for Si nanoparticles prepared by various fabrication methods. The following general features have been concluded. (i) Optical nonlinearities of the Si nanoparticles are relatively larger than that in bulk Si and about 200 times larger than that for silica at fiber-optic wavelengths, making this system a good candidate for applications in nonlinear optics. (ii) It has been shown that there is a strong correlation of the third-order optical nonlinearity and hence the nonlinear refractive index of the Si nanoparticles with the electronic quantum confinement effect. (iii) The electronic susceptibility associated with intraband transitions is found to be highly enhanced for Si nanoparticles with a size below 2 nm due to the appearance of electronic transitions between discrete levels induced by quantum confinement. In fact, recently, an all-optical logic gate operating in the gigabits per second regime exploiting the nonlinear optical properties of Si nanoparticles has been demonstrated, as both the absorption coefficient and refractive index can be easily controlled by the application of external optical fluences.

Munir Nayfeh received his BSc/MSc from AUB and PhD in atomic physics from Stanford University. He was a postdoc at the Oak Ridge National Laboratory, a lecturer at the Yale University, and a consultant at the Argonne National Laboratory, and is currently a professor at the University of Illinois. He coauthored Electricity and Magnetism, coedited three books on lasers, and is the author of this book on silicon nanotechnology. He presents science fiction, using the trademark "Dr. Nano," to simplify nanotechnology for children and the layman. He codeveloped with colleagues from Oak Ridge National Laboratory procedures for detection of single atoms, and STM-based procedures for nanofabrication, and procedures for dispersion of crystalline silicon into sub-3-nm ultra-bright nanoparticles, with potential to enable advanced devices with diverse applications in renewable energy harvest and storage and lighting, nanoelectronics as well as biomedicine. He founded a number of companies to promote Si nanomaterial and its applications: NanoSi Advanced Technologies, Nano Silicon Solar, Parasat-Nanosi, and Reach-NanoSi, which are supported by 23 US patents in nanosilicon.

9.7 OPTICAL GAIN IN NANOSILICON-BASED MATERIAL

Not only miniaturization afforded silicon strong luminescence and optical nonlinearity, but there are some preliminary reports indicating optical gain and laser action in films of silicon nanoparticles [29]. In one report by Lorenze Pavesi of University of Trento, Italy [30], silicon nanoparticles were created inside a glass slab (Fig. 9.22) by silicon ion implantation followed by high-temperature annealing that allows the silicon atoms to condense into silicon nanoparticles of ~3-nm diameter. The slab acted as an optical amplifier of weak light beams. The authors who originally reported the optical gain propose that the excitons would recombine at a Si=O double bond at the interface between the silicon nanocrystals and the surrounding SiO_2 matrix. If the latter model proves true,

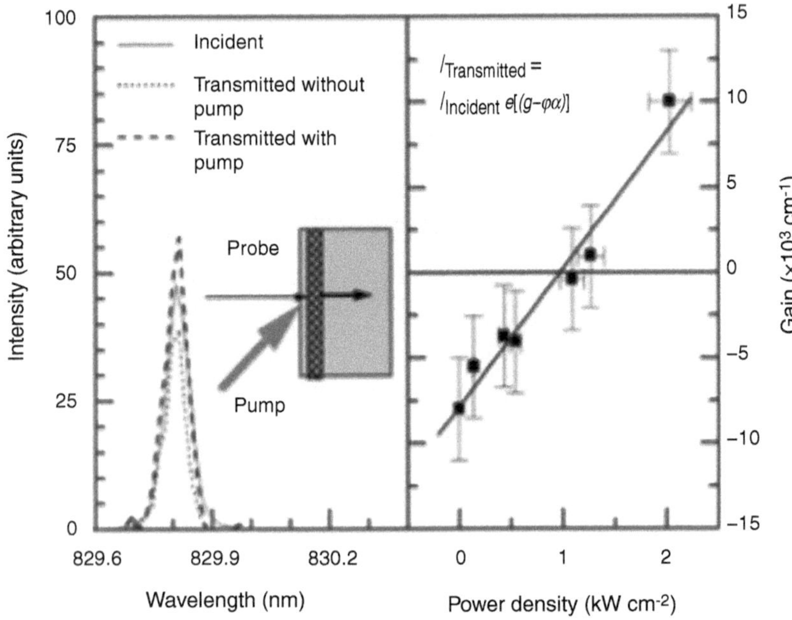

FIGURE 9.22 Optical gain in nano silicon. (Left) Layout of the measurement using pump-probe configuration of a film created inside a glass slab by silicon ion implantation followed by high temperature annealing that allows the silicon atoms to condense into silicon nanoparticles of ~3-nm diameter. (Right) Gain measurements in the silicon nanoparticles film. The slab acted as an optical amplifier of weak light beams.

atomic-scale engineering of the Si–SiO_2 interface will be the key to achieving efficient light emission from silicon nanocrystals. As a light source, silicon nanocrystals have an advantage over bulk silicon in that the electrical carriers are confined in a region in which no defects are present. This is because the SiO_2 matrix provides perfect passivation of the surface of these particles. However, the insulating effect of SiO_2 makes it difficult to inject electrical current into Si nanocrystals, and thus efficient light-emitting diodes are troublesome to make. Also, when large currents are passed through an oxide film it will eventually fail, resulting in a short circuit. Recent advances in using silicon-rich SiO_2 rather than pure SiO_2 as a host material indicate that more progress can be expected (S. Coffa, ST Microelectronics, Catania, Italy).

Some on the other hand believe that the light from silicon nanoparticles emission is caused by a "classical" model of the recombination of quantum-confined excitons inside the nanocrystals (F. Huisken, Max Planck Institute, Göttingen, Germany). For many applications of silicon nanocrystals, large quantities of isolated, size-selected nanocrystals are required. They can be made by using pulsed-laser-induced pyrolysis of SiH_4 in a gas flow reactor (F. Huisken, Max Planck Institute, Göttingen, Germany), resulting in emission quantum efficiencies—which measure the number of photons emitted per absorbed photon—as high as 100%.

In fact, in reports coming out of the University of Illinois, aggregates of ex situ prepared H-terminated silicon nanoparticles of 1-nm diameter showed optical gain and stimulated emission. Those measurements and others were conducted at the Florescence Dynamics Laboratory of the University of Illinois, in collaboration with the group of Enrico Gratton (then at the University of Illinois) of the University of California, Irvine. First, the power dependence of the emission of dispersed particles in a colloid was examined. The dependence shows quadratic dependence as in a two-photon process [31]. Next aggregates of particles were examined. Aggregates are found on the solid precursor before sonification and dispersion. Fig. 9.23 shows the emission intensity under femtosecond pulsed excitation as a function of the average incident power. The inset is the log–log of the data. At a threshold intensity there is a dramatic increase in the slope of the power pumping curve from ~1.6 to ~11–12, similar to stimulated emission in inverted systems [32].

When silicon nanoparticles were harvested and reconstituted into microcrystallites [33]. Those crystallites were irradiated by infrared femtosecond pulses of a laser beam at a wavelength of 780 nm with high peak power. Microscopic directed blue emission was observed. Microscopic red emission was also observed from clusters of 2.9-nm nanoparticles under excitation of strong incoherent green light [34].

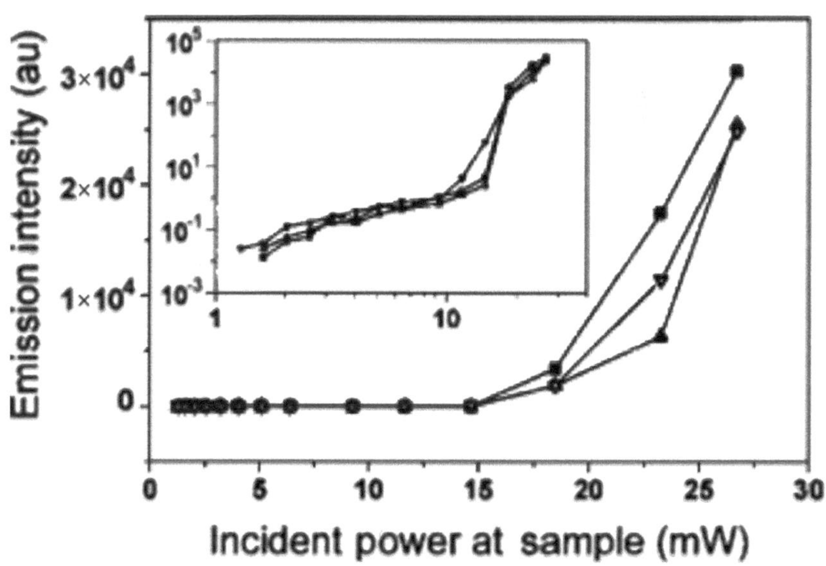

FIGURE 9.23 The emission intensity from a random nanoparticle distribution on the surface of an etched silicon substrate under femtosecond pulsed excitation as a function of the average incident power. The inset is the log–log of the data.

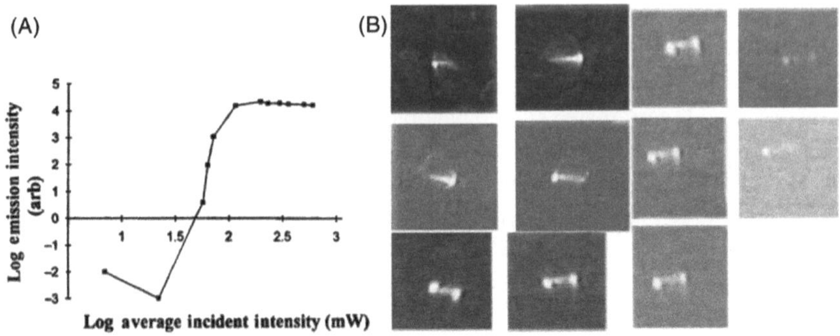

FIGURE 9.24 Response of thin microcrystallites reconstituted from 1 nm silicon nanoparticles to two-photon high intensity infrared femto second laser beam at 780 nm. (A) Blue emission intensity from such films as a function of the average incident intensity. (B) Variety of directed blue beams emerge, propagating in the interior of the crystallite, in the plane of the crystallite, locally normal to the closest side (normally incident beam).

Then large, thick, and uniform layers of the ultrasmall material were created by reconstituting the dispersed particles. Fig. 9.24A gives the emission intensity from such films as a function of the average incident intensity. For low intensity, the emission is small, but at an average intensity of ~10^6W/cm^2 (~20–25 mW, ~μm spot), the emission exhibits a sharp threshold, rising by several orders of magnitude. Beyond threshold, there sets in a low order power dependence that saturates at the highest intensity [33]. Fig. 9.24B displays the interaction of the incident beam with the micro crystallites. It is normally incident (normal to the plane of the figure). The four examples in Fig. 9.24B show that directed blue beams emerge, propagating in the interior of the crystallite, in the plane of the sample, locally normal to the closest side.

In addition to experiments with 1 nm particles, experiments were conducted with the 3-nm particles by the Nayfeh's group in collaboration with Sahraoui Chaieb (then at the University of Illinois). Laser behavior was observed at ~610 nm in aggregates of ultrasmall elemental 3-nm silicon nanoparticles [34]. The aggregates are excited by radiation at 560 nm from a mercury lamp. The power, 1–15 mW, focuses to a spot ~ 0.5 mm diameter, using an objective of 0.6 NA, giving an intensity of 0.13–2 W/cm^2. Intense directed beams, with a threshold, manifest the emission (see Fig. 9.25). We observe line narrowing, Gaussian beams, and speckle patterns, indicating spatial coherence. The directionality, line narrowing, coherence, and high-intensity point to laser oscillation. The aggregate and the face of the substrate may define a laser cavity.

The interface provides a reflectivity of 1–4% and the acrylic bubble provides ~4% reflectivity. Taking the radius of the bubble R ~ 4 μm and a cavity of length L ~ 1 μm, the thickness of the aggregates, gives a stable semi concentric cavity. Taking 4 μm for the radius of the aperture (interaction region), and 0.6 μm for λ gives a Fresnel number $N = a2/L\lambda$ ~ 26.7. Gaussian beams in a resonator require a stable cavity and a large Fresnel number (small diffraction

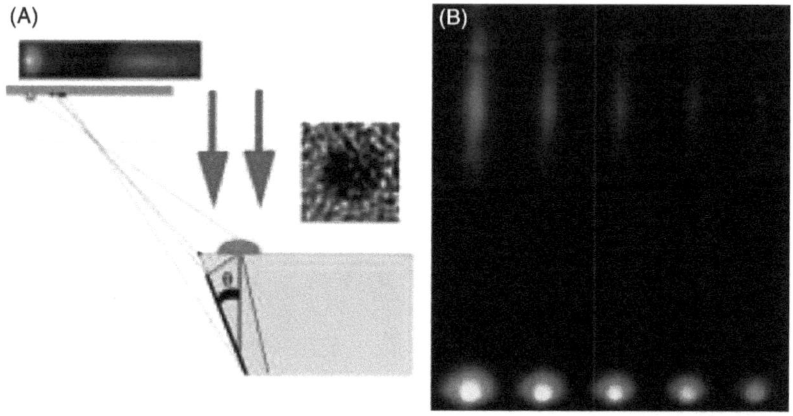

FIGURE 9.25 Response of small aggregates of 3-nm silicon nanoparticles wrapped in a PEG polymer under intense green light at 560 nm. (A) Layout of the experiment showing the excitation and emission collection configuration. (B) Intense directed beams, with a threshold, manifest line narrowing, Gaussian beams, and speckle patterns, indicating spatial coherence and laser action.

loss). This microlasing constitutes an important step towards the realization of a laser on a chip.

9.8 SPECTRAL BANDWIDTH AND LIFETIME OF LUMINESCENCE

The nanoparticles are deposited from a colloid suspension on a graphite grid. It is imaged by an HRTEM. Fig. 9.26A shows that the particles are nearly identical in diameter (~1 nm). Fig. 9.26B shows the emission spectrum of the dispersed particles and illustrates its development with the excitation wavelength at 330, 350, 365, and 400 nm. Each of the spectra encompasses the visible spectrum but is dominated by a strong blue band covering the range 400–550 nm with a peak at 450 nm and a structureless weak red/orange tail. Those spectra were recorded on a photon counting spectro-fluorometer with a Xe arc lamp light source and 4 nm bandpass excitation and emission monochrometers, with suitable filtering and correction for spectral response. The emission detector has a cut off wavelength at 950 nm. We also note that when the colloid is excited by the third harmonic of an Nd–Yag laser at 355 nm and 30 ns pulse duration, the blue emission is observable with the naked eye, in room light. For illustration, a photograph of the emission under these conditions is shown in Fig. 9.26C. Excitation spectra, which consist of recording the luminescence intensity as a function of excitation energy for a given emission wavelength, essentially measure a product of two factors: the efficiency of exciting electron–hole pairs at a given photon energy; and the probability of those pairs to relax into states from which emission at the chosen wavelength can occur. The excitation spectrum of the dispersed particles while monitoring the emission at 400 nm shows a band extending from 2.5 eV

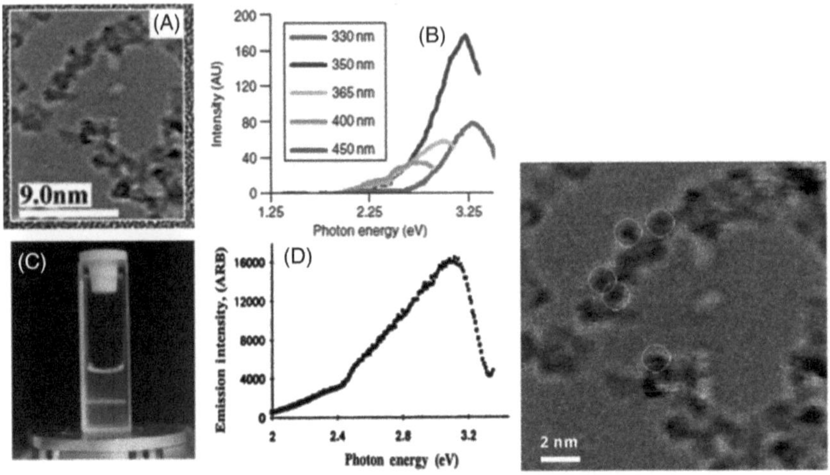

FIGURE 9.26 Characteristics of 1-nm silicon nanoparticle. (A) HRTEM of particles on a graphite grid shows that the particles are nearly identical in diameter (~1 nm). (B) Emission spectrum of the dispersed particles for excitation wavelength at 330, 350, 365, and 400 nm. (C) Observable blue emission with the naked eye, in room light colloid, excited by the third harmonic of an Nd–Yag laser at 355 nm. (D) The luminescence from a nanosolid that forms when nanoparticle colloid in water is slowly dried (excited at 355nm).

to ~4 eV with a maximum near 3.6 eV. A nanoparticle colloid in water is slowly dried giving thin nanosolid. The luminescence from the solid with excitation at 355 nm is given in Fig. 9.26D. Thus, these results show that the luminescence of 1 nm particle is wide band of FWHM ~80 nm. This is a surprising result, because in the classical picture of emission from excitons that are confined in a nanocrystal, the expected linewidth is more than an order of magnitude lower.

To check single nanoparticles, a very low density of nanoparticles was spread on a quartz plate and single particles were isolated. The spectrum was recorded as shown in Fig. 9.27A. Indeed the result show that the linewidth of the emission of individual nanoparticles can be as large as 80 nm.The fluorescence of an ensemble is shown in Fig. 9.27B for comparison. J. Linnros, Royal Institute of Technology, Sweden noted that the linewidth of the emission of individual nanoparticles can be as large as 120 meV [35].

Size (nm)	#
3.5	4
2.9	56
2.2	12
1.7	9
1	2

The table shows counts of 4, 56, 12, 9, and 2 at 3.5, 2.9, 2.2, 1.7, and 1 nm sizes, respectively. This is more of a normal than log-normal distribution.

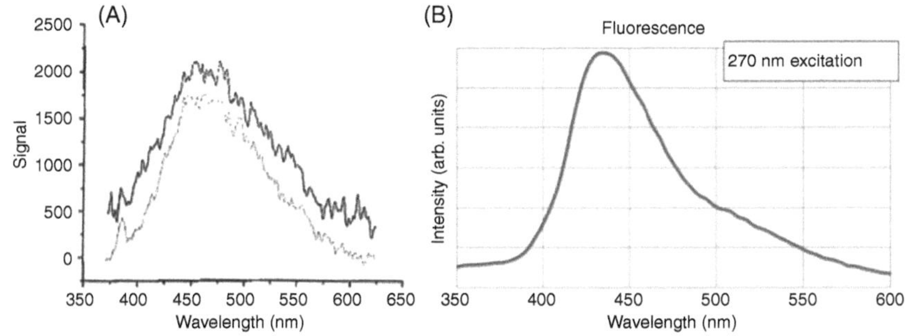

FIGURE 9.27 The luminescence spectrum of 1 nm silicon nanoparticles spread on a quartz plate. They are excited by a femto-secomd two-photon-process at 780 nm (A) single particle, (B) ensemble of particles.

The PL spectra taken from the precursor colloid is given in Fig. 9.28B. The linewidth is broad. But this not indicative of a very broad nanocrystals size distribution as can be seen in the TEM image. Thus the width and shape of the photoluminescence spectrum does not result from a size distribution; it is the spectrum of a single nano size, namely 3-nm. This is unlike the luminescence of nanoparticles of direct band gap material which give sharp lines. Unlike direct-bandgap nanoparticles of a given size, which emit sharp lines, indirect-bandgap nanoparticles may emit broadband. Emission (oscillator strength (x cross) from a single size of 1.67 nm diameter has been calculated to be wide. The width was attributed to structural changes on the surface upon emission. The bandhead position was predicted by several theories [13,36].

Time-resolved PL measurements on suspensions of silicon nanoparticles were also performed, in collaboration with Steve Granick group of the University of Illinois, using near-infrared two-photon femtosecond excitation as shown in Fig. 9.29 [37]. The excitation employs a two-photon process at 760–800 nm, corresponding to an effective single photon process at 380–400 nm. We used a

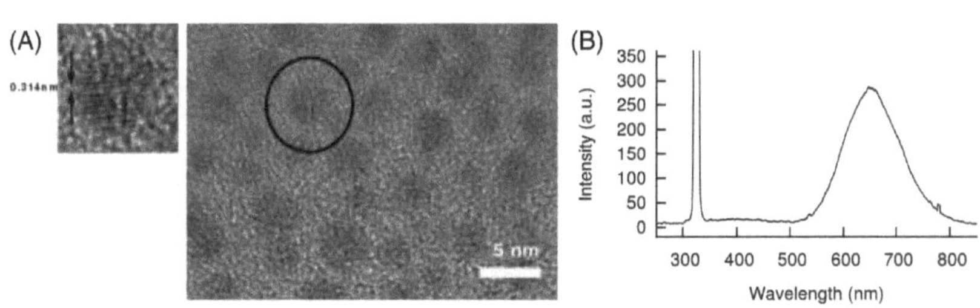

FIGURE 9.28 (A) TEM image of silicon nanoparticles distribution dominated by a 2,9-nm diameter particles (right). The particles are placed on a graphite grid. A zoom on a single particle showing resolution of silicon atomic planes (left). (B) The luminescence of the precursor colloid sample using at 365-nm wavelength excitation (table) Histogram of the size distribution in (A).

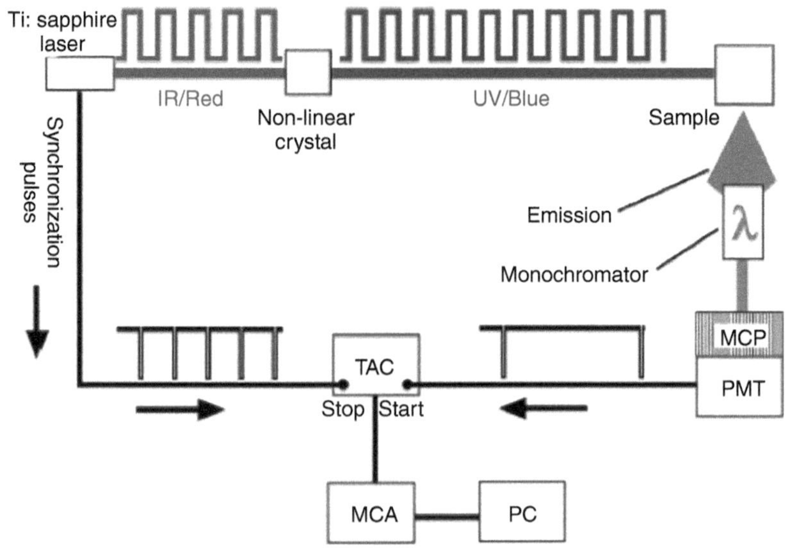

FIGURE 9.29 Schematic of the setup for performing time-resolved PL measurements on suspensions of silicon nanoparticles using near-infrared two-photon femtosecond excitation.

mode locked femtosecond Ti-Sapphire near-infrared laser system, generating pulses of <150 fs duration at a repetition rate of 80 MHz (12 ns duty cycle).

At the target, the average power, 20 mW, is focused to a beam waist of ~ 0.5 μm, giving an average intensity of 10^7 W/cm^2 (peak pulse intensity of 10^{12} W/cm^2). In the measurement of the time dynamics we did not employ wavelength resolution; all of the PLs in the blue band of the 1 nm particles or in the red band of the 2.85 nm particles were collected using a photomultiplier. The radiative lifetimes are measured as a function of both emission energy and temperature. All samples were illuminated with an excitation energy of 3.31 ± 0.05 eV (375 $\pm$ 5 nm). Suspensions of nanoparticles in several solvents were used including 2-propanol, benzene and 1-propanol.

Fig. 9.30 gives time-resolved PL of 1 and 3 nm particle colloids. It was taken under 780 nm excitation. The working range of the free decay is $\sim$2–11 ns within the 0–12 ns duty cycle. The flat steady-state background the free decay is riding on is due to accumulation resulting from decay rates with long characteristic time scale. For 1 nm, the background is $\sim 10\%$ of the maximum yield. This value of background shows that the 1 nm nanoparticle has predominantly short lifetimes <12 ns. For 3 nm, the fluorescence decays to near 50% level within the detection time window (12 ns). This implies that the nanoparticle has fast decay channels as well as considerably slower decay channels with times >12 ns. Our results for 1 nm particles show wide bandwidth but indicate full conversion to direct-like behavior, with a few nanosecond time characteristic (2.6 $\pm$ 0.3 ns) for the lifetime, corresponding to oscillator strength comparable to those in direct

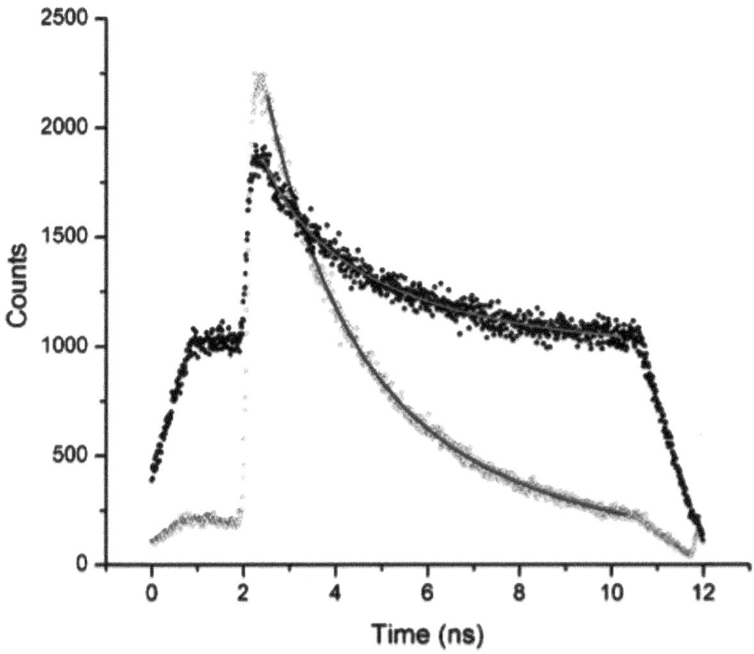

FIGURE 9.30 Time-resolved photoluminescence of 1 and 3 nm particle colloids, taken under 780 nm excitation using the facility described in Fig. 9.29.

semiconductors. In addition to fast nanosecond decay of 1.6 ns, the PL from 3 nm nanoparticle exhibits considerably slower decay of ~38 ns with amplitudes of 1:2, consistent with a transition regime to direct-like behavior.

Detailed room temperature lifetimes of 1nm silicon nanoparticles were taken at room temperature with the nanoparticles dispersed in a variety of solvents. The lifetimes are obtained from a fit of the decay histogram to a double exponential function. Both lifetime components are shown, one 3 ns, the other 0.5 ns. Nanoparticle lifetimes at several temperatures were also recorded. The room temperature (291 K) lifetimes (¤) are the results of a double exponential fit. Subsequent series were taken at 129 ± 7 K and 111 ± 3 K. Both lower temperature measurements exhibited single exponential decays.

9.9 ELECTRON TRANSPORT THROUGH SI NANOPARTICLES

A thin film of submonolayer of 1 nm (Si_{29}) silicon nanoparticles were prepared on device quality silicon wafer. The tungsten tip of a scanning tunneling microscope (STM) was placed above the film. Operationally, in the measurement, the tunneling current was set at 1 nA and the biasing voltage at 3 V in a constant current mode with the substrate being grounded, giving a tip height of several angstroms. The feedback loop is then disengaged allowing the tip to be held in

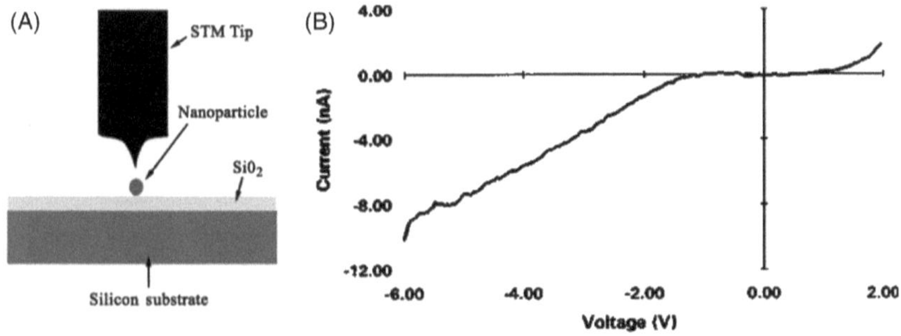

FIGURE 9.31 **Electron transport through Si nanoparticles** . (A) Schematic of the measurement using a scanning tunneling microscope. (B) An average of sixteen *I–V* individual spectra, showing very little staircase structure that may be due to charging, which indicates fast tunneling/transport across the particles.

a constant height mode as shown in Fig. 9.31A [38]. The voltage of the tip was then varied with respect to the grounded substrate from 10 to 13 V, while the tunneling current was recorded. Fig. 9.31B gives an average of sixteen *I–V* individual spectra. The figure shows very little staircase structure that may be due to charging. This indicates fast tunneling/transport across the particles. Averages over multiple spectra may show broadening and washing due to drift in the constant height mode. The measurement also shows that the source-drain current is blocked.

Fig. 9.32A shows the schematic with light from a mercury lamp directed onto the sample through pyrex glass window (~cutoff at 270 nm). The *I–V* response shown in Fig. 9.32B now shows quite visibly a regular step structure. The drop in the magnitude of the tunneling current, by a factor of fifty, is due to a change in the setting of the preamplifier. The series appears more like resonance, but in other spectra we see them closer to step-like. The derivative of the *I –V* curve, shown in Fig. 9.32C, clearly shows the progression of the resonance at large negative biasing up to a tip biasing of −2.2 eV with respect to the substrate. The light-induced series has a period of 1.0 eV. In addition, upon irradiation, the source blockade range extends now from −1.75 to 1.2 V, widening from 2 V under dark conditions to 3.75 V. This indicates that the source capacitance dropped upon irradiation.

Conductance resonance due to single-electron charging at room temperature was reported in a single silicon island 5 nm across and 70-nm long and single electron charging memory at room temperature were reported in 30 nm silicon particles, and in 4.5 nm nanoparticles. Though effects due to discrete charging effects in the transport of electrons in these structures are observable at room temperature, however, effects due to the quantum energy levels of the particles are basically washed out at room temperature and as such have not been observed so far [39]. Michael Crommie, http://research.physics.berkeley.edu/crommie/research:stm

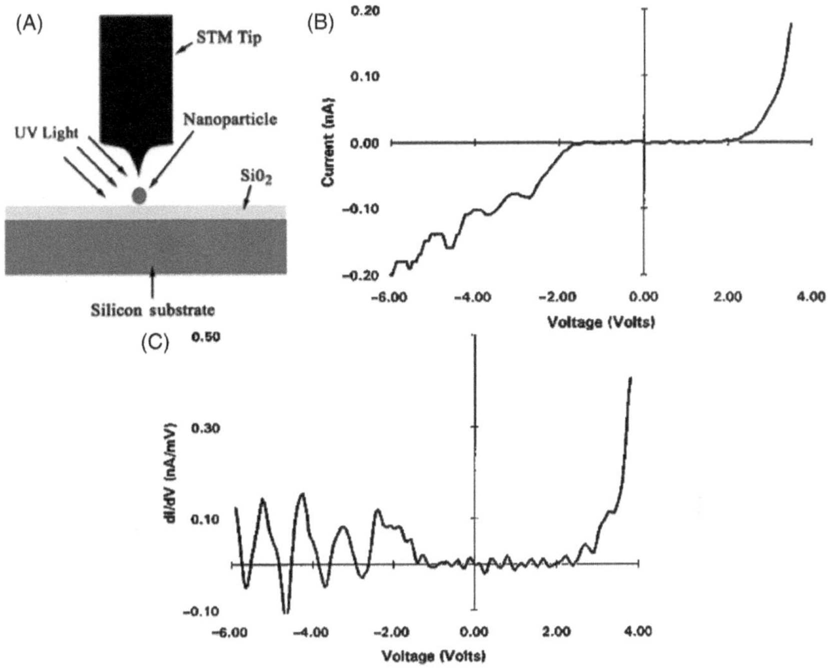

FIGURE 9.32 Electron transport through Si nanoparticles. (A) Schematic of the measurement using a scanning tunneling microscope with light from a mercury lamp directed onto the sample through Pyrex glass window (~cutoff at 270 nm). (B) I –V response shows quite visibly a regular step structure. (C) The derivative of I –V curve shows the progression of the resonance at large negative biasing up to a tip biasing of 2.2 eV with respect to the substrate.

9.10 MOLECULAR STRUCTURE

9.10.1 Prototype and Coordinates of Hydrogenated Particles (Supermolecule)

Fig. 9.33 shows a qualitative cartoon description of the formation of the 1 nm particle. The search for a realistic structural prototype started from single crystalline silicon wafer. A sphere of 1 nm diameter is carved out. The sphere will contain 29 silicon atoms and haves 36 broken bonds. The 36 broken bonds are terminated with hydrogen atoms. The number 29 is a magic number for the *T*d symmetry and spherical shape (see Fig. 9.33). However, the corresponding electronic energy gap was >6 eV, suggesting that the observed clusters possess smaller number of terminating hydrogen/oxygen, with a part of the dangling bonds saturated by a nanocrystal surface reconstruction. Pairs of nearby hydrogen atoms are stripped allowing the silicon atoms to reconstruct forming Si–Si bonds. This action lowers the bandgap of the particle. By eliminating six pairs incrementally (12H atoms total) we arrived at a structure of $Si_{29}H_{24}$ with six reconstructed surface Si—Si dimers similar to Si (001) surface 2 × 1

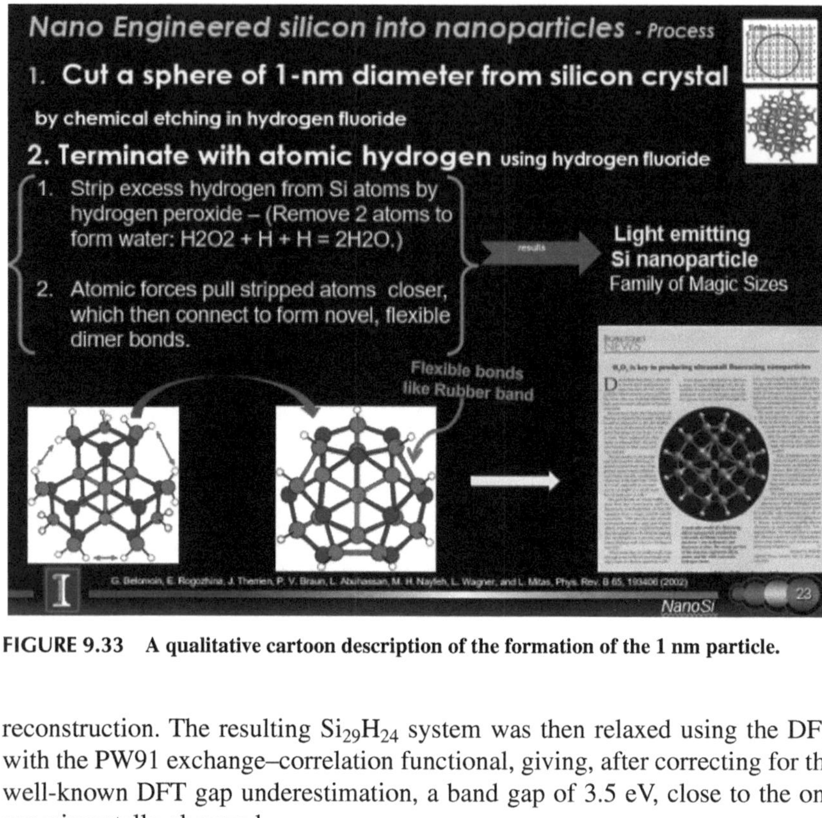

FIGURE 9.33 **A qualitative cartoon description of the formation of the 1 nm particle.**

reconstruction. The resulting $Si_{29}H_{24}$ system was then relaxed using the DFT with the PW91 exchange–correlation functional, giving, after correcting for the well-known DFT gap underestimation, a band gap of 3.5 eV, close to the one experimentally observed.

The formation of these reconstructions is favorable under our dispersion conditions. H_2O_2 in our etchant is critical for controlling the amount and configuration of hydrogen on the surface, hence the reconstruction. The reaction $Si_nH_m \rightarrow Si_nH_{m-2} + H_2$ has a barrier of 0.6 eV. However, the reaction $Si_nH_m + H_2O_2 \rightarrow Si_nH_{m-2} + 2H_2O$ proceeds followed by reconstruction to form novel radiative Si—Si dimer bonds while gaining an energy of 0.25 eV [36].

In the analysis H_2O_2 gets in the gap stripping two hydrogens and the bonds connect as shown in Fig. 9.33. Those reconstructions can take place more readily in ultrasmall particles for which the elasticity drops. This makes the atoms amenable to large movement. For instance, in the model, some surface Si atoms in 1 nm particles move by >0.65 Å from their bulk positions to reconstruct into novel radiative Si—Si dimers, with a binding energy of 0.75 eV, much larger than the thermal agitation energy of 0.025 eV.

If additional hydrogen below 24 total atoms and reconstruction takes place, sp^2 symmetry begins to set in, and the band gap of the particle begins to drop fast eventually collapsing resulting in a graphite-like structure (conduction system) [37]. Fig. 9.34A plots the bandgap of the nanoparticle with the number of stripped hydrogen pairs from the surface. When the number peeled increases

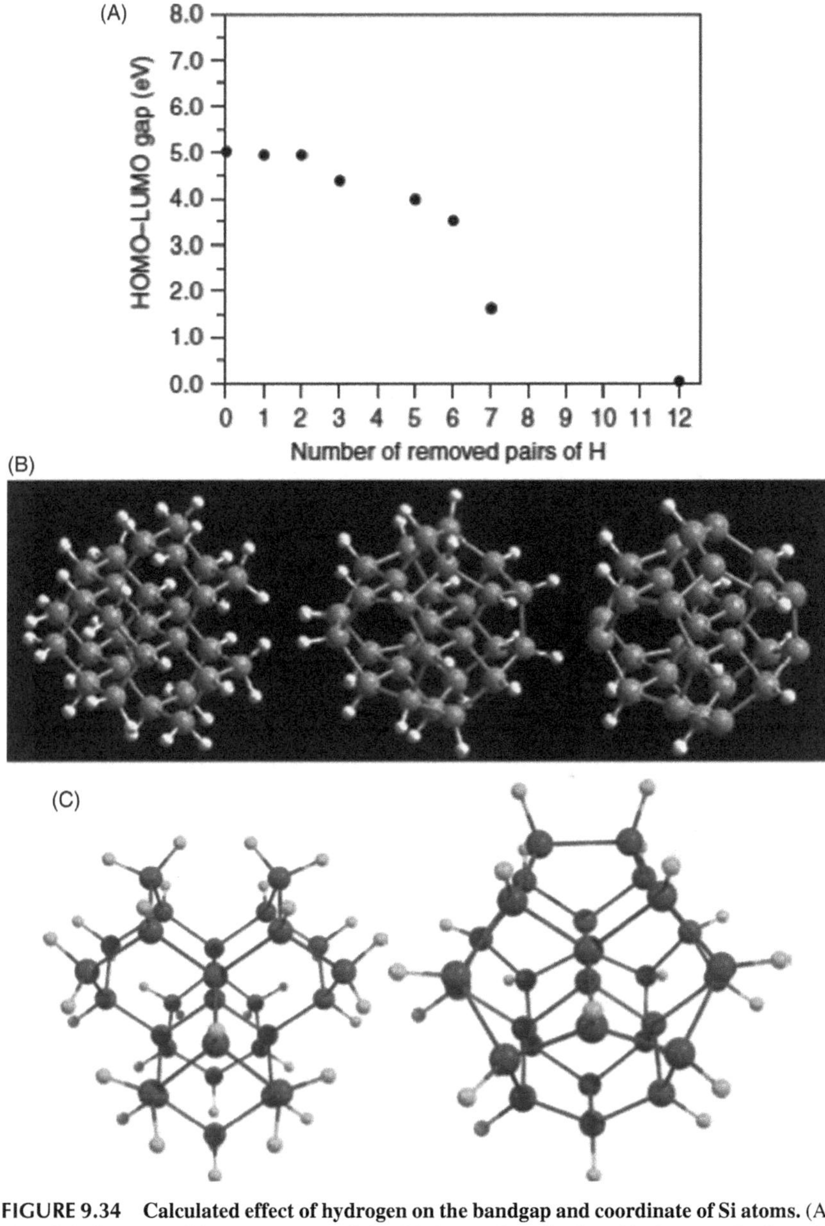

FIGURE 9.34 Calculated effect of hydrogen on the bandgap and coordinate of Si atoms. (A) Plot of the bandgap of the 1-nm nanoparticle with the number of stripped hydrogen pairs from the surface. (B) Three examples of the reconstructed configuration starting on the left from $Si_{29}H_{36}$ to $Si_{29}H_{24}$, and $Si_{29}H_{12}$. (C) A prototype of the bulk-like Si29H36 (left) and the reconstructed nanoparticle Si29H24 (right).

from zero, initially the bandgap drops slowly but as six pairs are peeled the band gap begins to collapse, resulting in a conducting state. Fig. 9.34B gives three examples of the reconstructed configuration starting on the left from $Si_{29}H_{36}$ to $Si_{29}H_{24}$, and $Si_{29}H_{12}$.

We should also note that the surface of a $Si_{29}H_{24}$ cluster can be represented as a 28-atom cage (12 pentagons, 4 hexagons) with a single silicon atom in the canter, bonded to four surface atoms (related by *T*d symmetry) to form a five-atom tetrahedral core. The remaining surface atoms are passivated with hydrogen.

The actual computational investigations started with a spherical piece of the bulk Si, which for a size of ~1 nm contains 29 atoms (magic number for the Td point group symmetry), which requires 36 H atoms to terminate the unsaturated bonds. The number of H atoms was varied due to reconstruction. The geometry optimizations were carried out using density functional generalized gradient approximations with a PW91 functional and 6-31G* and 6-31G** basis sets. The relaxed configuration shown in Fig. 9.34C, with five atoms constituting a single tetrahedral core and 24 constituting a H-terminated reconstructed surface ($Si_{29}H_{24}$, with six reconstructed dimers) provides the best agreement with the measurements [38]. The coordinates of the $Si_{29}H_{24}$ particle (*x, y, z*) in angstroms are summarized as follows. There are 12 Si atoms at (–2.720, –2.720, –0.0197) and their permutations. Another 12 atoms are at (–3.932, 0.886, –0.886) and their permutations. Four atoms are at (–1.394, –1.394, –1.394) and their permutations. Finally the center atom is at (0.000, 0.000, 0.000). The surface is highly wrinkled or puckered system of hexagon and pentagon rings. The prototype gives a diameter of 0.9 nm for the Si core and 1.066 nm, including the H-termination, and an absorption edge of 3.5 ± 0.3 eV. The surface Si atoms form four hexagonal rings. Three atoms from the surface with three of the internal tetrahedral unit (one is the canter atom) form a hexagonal ring. There are a total of eight such hexagons; those slice the particle as deep as the canter, creating ridges. There is also a total of eight pentagon rings, each is formed by four surface atoms and one of the tetrahedral internal atoms (excluding the center atom). The participation of the internal atoms in ring formation makes the particle highly wrinkled- or "puckered-ball."

We next calculated the optical absorption spectrum. The resulting spectrum with a Gaussian broadening of 0.14 eV, which qualitatively represent the temperature and size averaging, shows agreement with our measurement. We also evaluated the cluster polarizability, a uniquely defined property, giving ~793 a.u. A general definition of a dielectric constant in semiconductor nanoparticles is not possible since the energy levels are discrete. To arrive at an effective "dielectric constant" one has to define an appropriate "cluster interior volume." For an effective interior radius (~4.2 Å) we get a dielectric constant of ~5.7. This value is close to estimations done before by Allan et al.and Wang and Zunger [39]. The Penn's model-based analysis [39] gives, for a diameter of 0.8 nm, a more reduced static dielectric constant of 2.

The silicon–hydrogen prototype of the particle $Si_{29}H_{24}$ represent in some sense a new phase between solid and molecule. With a tetrahedral core and a strong molecule-like reconstruction of the surface, the new phase or supermolecule may exhibit solid-like behavior, as well as molecule-like behavior, which may account for the multitude of novel optical and electronic properties of our nanoparticles.

In principle other sizes of nanoparticles exhibiting magic sizes ~1, 1.67, 2.1, 2.9 nm in diameter may be theoretically constructed in a similar procedure. The calculation however may become out of reach of computational power.

9.10.2 Novel Si–Si Bonds (Molecular-Like Behavior)

As discussed in the previous section and with the aid of Figs. 9.33 and 9.34, there are six pairs of surface atoms, each of which has two broken bonds that are saturated with hydrogen. The atoms in each pair are at 4.84 Å interspacing. When each of those atoms gets stripped from one hydrogen, the atoms move closer from 3.84 Å interspacing to form a Si–Si molecule-like dimer bond at 2.35 Å, the tetrahedral distance. Allan et al. focused their attention on a 1 nm particle with only a single Si—Si molecule-like dimer (i.e., $Si_{29}H_{34}$ particle). Fig. 9.35 gives a schematic of the interatomic potential of the Si—Si reconstructed bond calculated by this group using LDA in the 1.03 nm diameter particle, as well as in 1.67 nm particle, and the various pathways for absorption and emission calculated by Nayfeh et al. [40]. The excited state is a double well with a potential barrier. The inner well (at 2.35 Å) is associated with the tetrahedral configuration and radiates on a long time scale of milliseconds. The outer well,

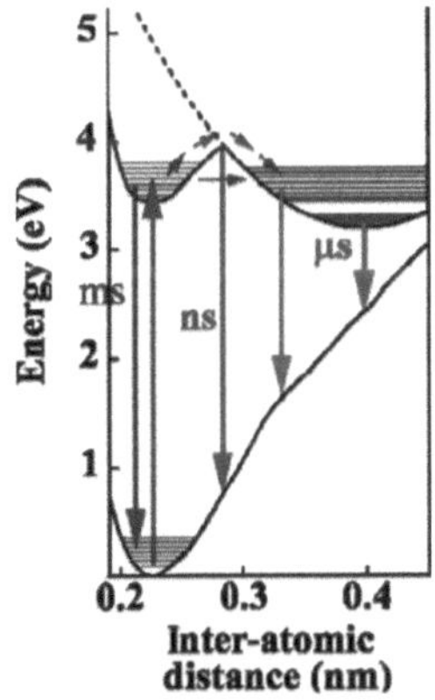

FIGURE 9.35 Schematic of the interatomic potential of the Si—Si reconstructed bond in Si29H34 from Ref. [13]. The ground state and excited state are calculated using LDA by Allan et al. focused their attention on a 1 nm particle with only a single Si—Si molecule-like dimer (i.e., $Si_{29}H_{34}$ particle) and the various pathways for absorption and emission calculated by Nayfeh et al.[44]. The excited state is a double well with a potential barrier. The inner well (at 2.35 Å) is associated with the tetrahedral configuration and radiates on a long time scale of milliseconds.

a new state found only in ultrasmall nanoparticles, is a trap well (at 3.85 Å) that radiates with lifetimes of 5 ns–100 μs. The calculations were performed with the silicon atoms terminated by hydrogen.

The state to which the outer well radiates vertically down is the ground electronic state. But, at extended bond lengths, the ground state is high lying and unpopulated, hence this system constitutes a stimulated emission/gain channel. Gain, stimulated emission and laser action have been reported by several groups (see Section 9.7). The blue emission proceeds at an interatomic distance at the top of the barrier (~3 Å), where the lifetime is in the nanosecond regime and mixing between the two states is most significant. Emission from near the bottom of the outer well is of longer time characteristic and of longer wavelength (in the red or near infrared). According to the Frank–Condon principle for molecular transitions, absorption proceeds vertically up into the inner well at a bond of 2.35 Å, followed by transfer into the outer by bond expansion via tunneling or thermal activation (i.e. self-trapping) as in Fig. 9.35. Also, above-barrier absorption, followed by relaxation populates the trapping well.

When more Si–Si dimers form, the band gap and the potential curves shift to lower energies. Moreover, the barrier between the inner well and the out well drops. For a total of six dimers produced [45]. The calculated life time as a function of the length of the Si–Si bond is also given in the figure. The $Si_{29}H_{24}$ forms (shown in Fig. 9.36A) and the potential curves transform to the ones shown in Fig. 9.36B.

9.10.3 Structural Stability of the Prototype

Theoretical and simulation studies were conducted at Lawrence Livermore National Laboratory (LLNL) [46] that employed a combination of first principles molecular dynamics (FPMD) and QMC calculations to determine the structural and optical properties of 1 nm silicon particle. As seen in Fig. 9.37A, the $Si_{29}H_{24}$ is a symmetrical with single-core configuration. The simulations found stable but asymmetric configurations consisting of a two-atom core configuration as depicted in Fig. 9.37B. The study sought to determine the stability of single-core nanoparticle configuration against hydrogen saturation. The number of hydrogens required to maintain a single-core crystalline structure in a 29-atom silicon cluster was examined. In the studies, FPMD simulations were performed in which a number of neighboring hydrogen atoms were removed from a tetrahedrally symmetric $Si_{29}H_{24}$ nanocrystal. The cluster was then heated to $T = 600$ and 1000 K. At $T = 1000$ K, removal of only six hydrogen atoms ($Si_{29}H_{18}$) resulted in a persistently stable, single-core structure. When 12 hydrogen atoms were removed ($Si_{29}H_{12}$), the silicon cluster formed a double core within 0.5 ps. When 18 hydrogen atoms were removed ($Si_{29}H_6$), the cluster again formed a double core within 0.5 ps, but also exhibited dynamics consistent with an amorphous cluster. The results showed that 14 neighboring hydrogen atoms ($Si_{29}H_{14}$) are required to maintain a single tetrahedral core.

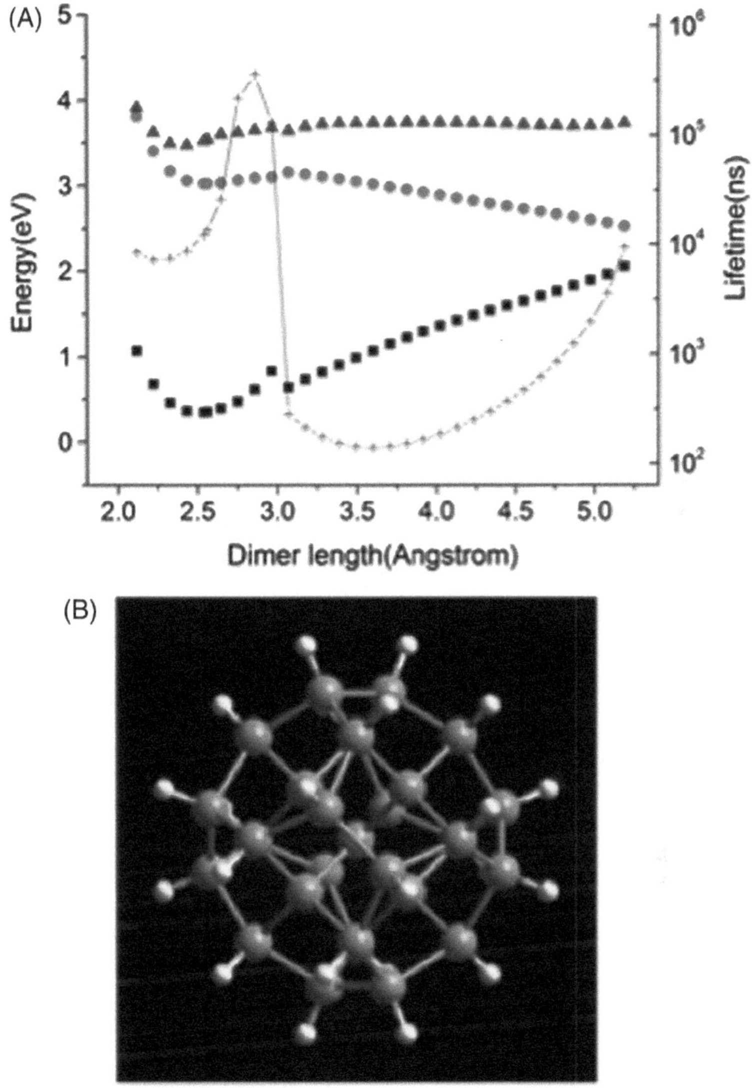

FIGURE 9.36 (A) Potential curves of the ground state and the first two excited states. When six Si–Si dimers form, showing that the band gap and the potential curves shift to lower energies. Moreover, the barrier between the inner well and the out well drops [45]. (B) The $Si_{29}H_{24}$ prototype, which forms for six dimers.

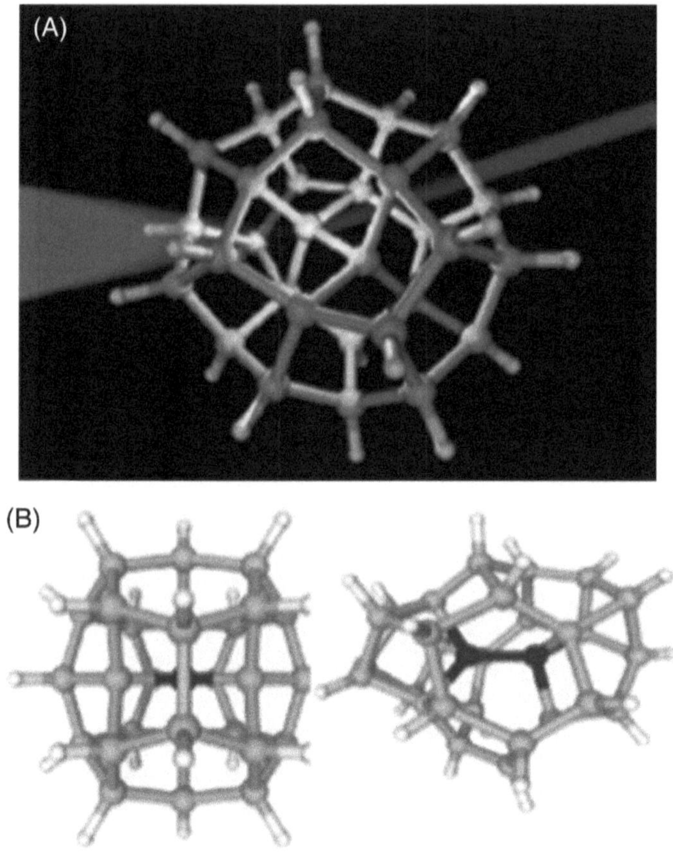

FIGURE 9.37 **Stability of the 1-nm silicon nanoparticles.** (A) The prototype $Si_{29}H_{24}$ is symmetrical with single-core configuration. (B) Asymmetric stable configurations consisting of a two-atom core configuration found by simulations conducted at Lawrence Livermore National Laboratory (LLNL) [46]. *(Private drawing by Andrew Williamson of Livermore National Laboratory and presently with Cambridge Innovation Capital, https://www.researchgate.net/figure/Different-reconstructions-of-crystalline-Si-29-ideal-nanocrystal-Si-29-H-36_fig1_8441694)*

At T = 600 K, similar behavior is observed, although the double core formed more slowly than at T = 1000 K. These findings suggest that the retention of a crystalline core depends on both the fraction of dangling bonds at the surface and the passivation time scale.

It is interesting to observe the $Si_{29}H_{24}$ particle from a different vantage point, namely the fullerene architecture. The surface of a $Si_{29}H_{24}$ cluster can be represented as a 28-atom cage (12 pentagons, 4 hexagons) with a single silicon atom in the center, bonded to four surface atoms (related by *T*d symmetry) to form a five-atom tetrahedral core. The remaining surface atoms are passivated with hydrogen.

The calculations also yielded two other configurations of the single-core particles $Si_{29}H_{24}$ with a different hydrogen distribution [46]. The first configuration is our bulk-like surface reconstruction of crystalline $Si_{29}H_{36}$ to $Si_{29}H_{24}$ and found to have an optical gap in agreement with the experimentally measured optical gap of 3.44 eV. The two other ways to connect the interior core atom to the surface and maintain a tetrahedral core are *unique to a curved nanostructure*, and have no bulk analogue. In these configurations, partial reconstruction takes place in which silicon atoms from the inner shell strip hydrogen atoms from their neighbors in the outer shell, which then bond to the central core atom, resulting in "isosize" distorted configurations. Exchange of one or two hydrogen atoms between the two shells is found to change the band gap from 3.5 to 4.5 or 4.1 eV, respectively.

9.10.4 Material Properties: Dielectric Constant and Effective Mass

Bulk properties of semiconductor systems rely on the periodicity of the atoms in the crystal, thus for smaller and smaller particles, there is a point where the particle can no longer support several unit cells of the crystal. This condition destroys the periodicity and translational symmetry, hence marks the commencement of the transition from bulk-like to molecule-like properties [47]. Moreover, as the size of the crystal decreases, the elasticity drops, especially in the surface layers, allowing appreciable restructuring to occur, accelerating the transition. The breakdown of bulk-based properties is not entirely unexpected, but the manner and the size at which it happens are unknown. Elucidating this transition is important since the structure governs the dielectric and optical constants, effective mass, and phonon structure, hence much of the electronic and optical behavior of the particle. The 1–3 nm regime may truly be a transition between the bulk-like and amorphous-like or molecule-like structure. Yet, there exists little systematic testing in this size regime.

Some insight may be seen when one revisit Figs. 9.31 and 9.32. In those measurements, particles are evaporated on *p-type* Si substrate. The tip of a STM is placed over the silicon particle film at a constant height for injection of charge. The figures give the *I–V* spectra taken at room temperature under dark condition and taken under light irradiation. Under light irradiation high visibility regular step structure are observed for negative tip biasing. Now it is to be noted that under standard doping of $\sim 10^{15}$ cm^{-3}, as in this sample, 1 nm particles contain less than one in a million holes, hence they behave as undoped, with the number of electrons in the conduction states extremely small. Moreover, laboratory temperatures are not sufficient to induce holes. Light irradiation, however, creates holes for the process to proceed. Measurements with size may yield the dielectric constant and the effective mass. Preliminary measurements showed that a transition at the size of 2.15 nm demarks a critical size, which separates particles with bulk-like properties from those with molecular-like properties. Most probably the effective mass approximation may fail for particles with sizes near this size.

9.10.5 Excited States (Molecular-Like Bands)

Richard Martin group at the University of Illinois, in collaboration with the Nayfeh's group conducted theoretical calculation and measurements to test this theory for each particle size [48]. All isosize reconstruction configurations of $Si_{29}H_{24}$ are analyzed. QMC and time-dependent density functional theory (TDDFT) were used. Fig. 9.38 gives the calculated absorption of the 1 nm particle. It shows overlapping bands. The lowest few bands occur at 350, 310, and 275 nm. However, observing the low-energy absorption spectrum below 3.6 eV in the TDDFT, we notice a small peak at 2.8 eV (443 nm). Measurements were conducted that map out all bands. The excitation wavelength was varied between 250 and 400 nm, while collecting the emission in the range 250–600 nm at each excitation wavelength. The results for the 1 nm particles are shown in Fig. 9.38. Distinct emission bands can be identified with peaks at 310, 360, and 450 nm, in agreement with the calculated TDDFT spectra. Moreover, the above calculations confirms that the particles behave in some sense also as "supermolecules," with excited electronic states from which higher-energy bands may originate.

The calculations give overlapping bands centered at 4.6, 4.0, and 3.6 eV, in fair agreement with the three bands seen in the measurements. The next higher excited bands or modulations in the absorption occur at 4.85, 5.15, and 5.75 eV. These lowest six bands are in reasonable agreement with the quantum Monte Carlo calculations of the Si_{29} configuration. The peak at 3.6 eV in the TDDFT calculation may be associated with the highest occupied–lowest unoccupied molecular orbital (HOMO–LUMO peak observed in the QMC calculations. However, observing the low-energy absorption spectrum below 3.6 eV in the TDDFT, we notice a small peak at 2.8 eV. The fact that the oscillator strength of this peak is so small signifies that it corresponds to a "forbidden" transition. Because of localization effects this transition is now possible but carries little oscillator strength.

The first principle atomistic theoretical simulations presented above using the TDDFT given in Fig. 9.38 show an absorption minimum at 10.2 eV. This is related to the atomic hydrogen termination layer on the nanoparticles and may have implication to astrophysics in term of resonance scattering of the abundant Lyman α near the sun (see Chapter 16 and Fig. 16.8).

9.10.6 Phonon Structure: Vibration Fingerprints (Raman Scattering)

The hypothesis of excitation and emission at single Si—Si dimers or collectively on all the dimers simultaneously draws evidence from luminescence activities. A direct finger print of the reconstructed Si–Si dimer network on the surface state can be achieved by conducting Raman light scattering spectroscopy. The process is very effective in providing information on vibrational modes of the particles, hence on the molecular composition. A block diagram of the measurement is given in Fig. 9.39A. In the measurements conducted by Satish Rao of

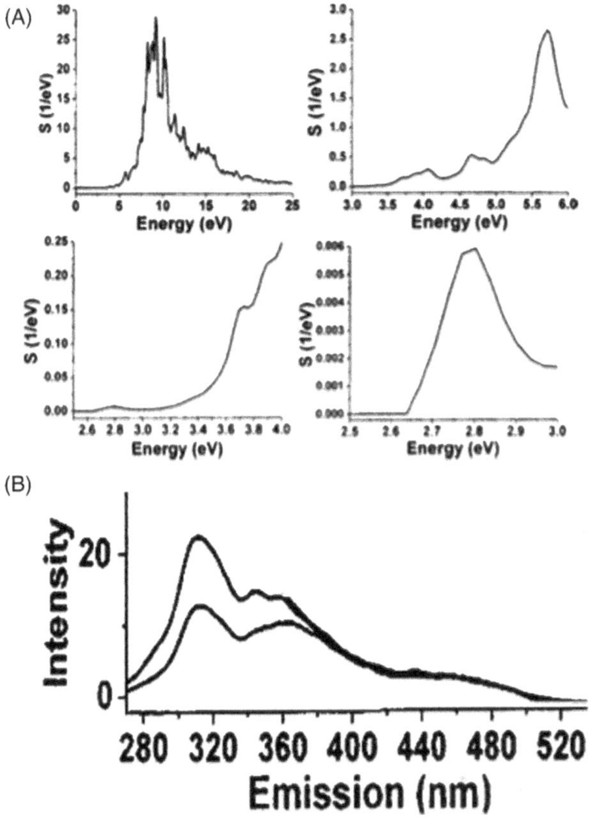

FIGURE 9.38 (A) Calculated absorption of 1-nm particle Si29H24 using QMC and time-dependent density functional theory (TDDFT) showing overlapping bands. The lowest few bands occur at 350, 310, and 275 nm. A minimum at 10.2 eV corresponding to the Lyman α. (B) Measurement of the spectrum of 1-nm particle.

the Nayfeh's group at the University of Illinois, the second harmonic of CW Nd-Yag laser at 532 nm is focused into a colloid sample. The 1 nm particle colloid solution was placed in a quartz cuvette with the detector of the scattered light at 90 degrees to the excitation. The scattered light is analyzed by a triple-grating monochrometer (SPEX 1000 Triplemate) for frequency and width; hence the frequency shift from the incident frequency can be obtained. The shift is related to the molecular vibration hence one can identify the vibrating bonds involved. By inserting a polarizer in the beam before it interacts with the sample and a polarizer in the scattered light before it gets analyzed information is obtained about the degree of polarization that is retained during scattering and the symtor is equipped with a charge coupled device (CCD). At the optimal settings for this experiment, the resolution of the system is $\sim$3 cm^{-1}.

Fig. 9.39B shows the measured spectrum [26]. Four distinct peaks are seen at 267, 292, 315, and 485 cm^{-1} with full width at half maximum (FWHM) values of 17, 7, 13, and 8 cm^{-1}, respectively. These values were found by fitting the peaks to a Lorentzian profile. The peak at 485 cm^{-1} is closest to the bulk Si TO mode (521 cm^{-1}) and is downshifted from it by 36 cm^{-1}. Fig. 9.39B presents also the Raman polarization results of the 1 nm particles. From the figures, the large peak at 315 cm^{-1} and the small peak at 292 cm^{-1} are greatly reduced in the perpendicular polarization, giving a nearly zero depolarization ratio. On the other hand, the peaks at 267 and 485 cm^{-1} exhibit depolarization ratios of 0.63 and 0.74, respectively.

To identify the vibrating component, detailed first principles calculation without an assumption of dispersion relations was carried out by Kevin Mantey of the Nayfeh's group using Hartree–Fock theory through the GAMESS quantum chemistry package, developed for molecules, to calculate the normal vibration modes and the Raman spectrum of the 1 nm particle [26]. We analyzed the modes further by the utilization of Chemcraft, a graphical interface for GAMESS that images the nanoparticle and animates its normal modes. We determined the atomic configuration and symmetry, frequency, reduced mass, and the Raman amplitude of the Si—H and Si-based modes. The calculated results are presented along with the measured results in Fig. 9.39C. The measurement agrees with calculated vibrations in the $Si_{29}H_{24}$ structural prototype and confirms the presence of surface dimer reconstruction, and high-symmetry collective vibrations localized on those dimers. This confirms electronic excitations on the dimers that were proposed as the source of the strong optical activity, and basis of the transformation of the particles into direct-like behavior. The peak positions and FWHMs, along with calculations of the normal modes, suggest that the nanoparticle exhibits strong molecular behavior despite its derivation from a bulk structure. The peak positions and full widths at half maximum, along with calculations of the normal modes, suggest that the nanoparticle exhibits strong molecular behavior despite its derivation from a bulk structure.

Another interesting feature was found in the simulations. The calculations indicate the presence of coherent vibration modes localized on pairs on opposite sides of the particle or collectively on all the surface dimers as shown in Fig. 9.40.

The measurement agrees with calculated vibrations in the $Si_{29}H_{24}$ structural prototype and confirms the presence of surface dimer reconstruction, and high symmetry collective vibrations localized on those dimers. This confirms electronic excitations on the dimers that were recently proposed as the source of the strong optical activity, and basis of the transformation of the particles into direct-like behavior. The peak positions and FWHMs, along with calculations of the normal modes, suggest that the nanoparticle exhibits strong molecular behavior despite its derivation from a bulk structure.

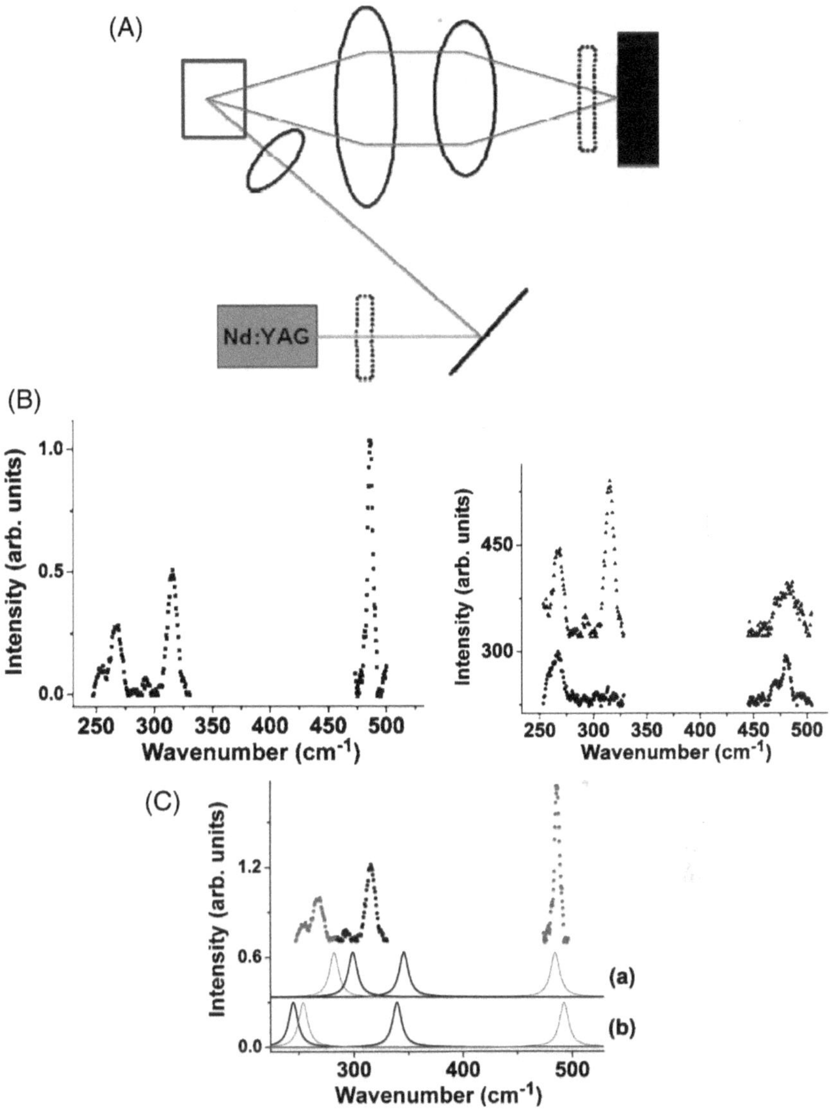

FIGURE 9.39 Raman light scattering spectroscopy. (A) Schematic of the block diagram of the Raman scattering measurement. (B) The measured Raman spectrum, as well as the Raman polarization spectrum of the 1 nm particles. (C) The calculated results are presented along with the measured results.

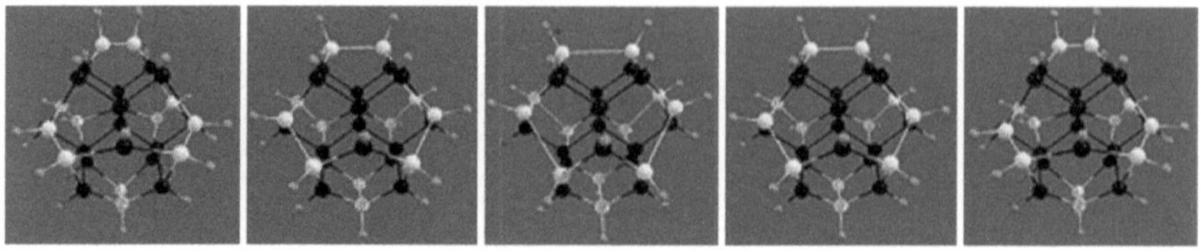

FIGURE 9.40 Coherent vibration modes of 1-nm particles localized on pairs on opposite sides of the particle or collectively on all the surface dimers.

9.10.7 Collective Molecular Surface

The particle may actually represent the transition between solid and molecule phases. Instead of forming only a single dimer on the particle, Mitas et al. examined the situation where the whole surface is reconstructed into the dimer molecule-like phase, as in the experimental synthesis. To calculate the molecular electronic structure of the silicon nanoparticle, the group used DFT for determining geometries of molecules, and the QMC method provides a significantly more accurate approach of calculating the energy differences such as the optically active excitations. Some of these calculations have been done in collaboration with a research group and supercomputers at LLNL led by Giulia Galli. One of the most interesting questions was whether the excited state was localized within the molecule and how much the excitation could affect the atomic geometries. Because QMC was capable of describing the excitonic effects very well, the team found that the excited electrons were delocalized across the whole structure. They also found that the excitation causes very small changes to the structure of the molecule. They also carried out calculations of the Stokes shift, or the difference between the energy absorbed by the molecule-like and the energy that the molecule emits afterward, for the excitation at the absorption edge of the molecule. The Stokes shift was in excellent agreement with the experimental data showing the actual difference in frequencies between absorbed and emitted energy. In particular, they were able to predict changes in the color of the emitted light upon attaching additional molecules such as methyl and ammonia groups to nanocrystals. The results of their QMC simulations of chemical doping were in excellent agreement with physical experiments.

9.10.8 X-Ray Scattering by Ultrasmall Si Nanoparticles: Form Factors

Consider N fixed atoms organized into an immobilized single nanoparticle as shown in Fig. 9.41. The form factor of the diffraction is a superposition of the scattering from the bound electrons, each with a phase shift **Q.R***mn*, where **Q**= Δ**k**= **k-ko**is the scattering wave vector transfer and **R***mn*is the atomic interspacing of pairs of atoms:

$$S=2\sum f_m f_n\left[1+\cos\left(\mathrm{Q}\cdot\mathrm{R}_{mn}\right)\right]$$

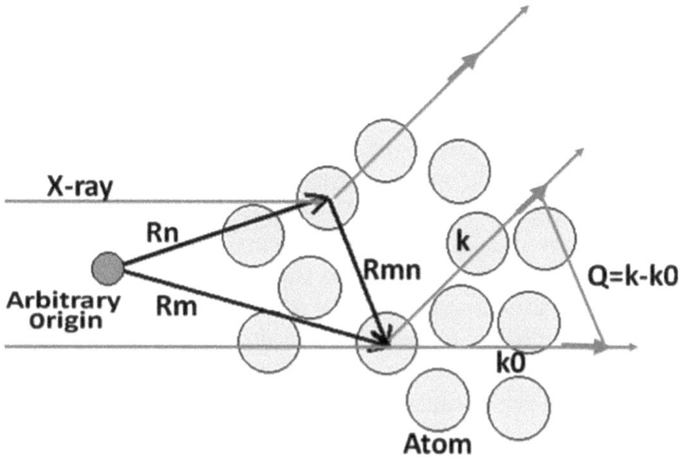

FIGURE 9.41 Sketch of X-ray scattering from N fixed Si atoms organized into an immobilized single nanoparticle.

where *m* and *n* integers designate the atoms in the system, and *fm* is the atomic form factor or the atomic scattering amplitude (may be described in terms the Thompson scattering length r_0). For many systems, such as molecules or aggregates in solution or deposited from a solution, the spatial distribution or orientation is randomly oriented with respect to **Q**. For scattering particles that are spherically symmetric, randomization gives

$$S=\sum f_m^2+2\sum f_m\, f_n \sin(\mathrm{Q}\cdot \mathrm{R}mn)/(\mathrm{Q}\cdot \mathrm{R}mn).$$

Averaging over random orientation decreases the degree of scattering compared to a crystallite aligned for a Bragg peak for larger scattering vector or scattering angle. The falloff of the atomic form factor with Q further damps the high Q response, and causes a shift in the diffraction pattern to smaller Q values. Numerical estimates show that this effect is negligible for the silicon system.

Fig. 9.42A gives the calculated molecular structure factor at 1.543 Å for 1 nm particles using the bulk-like interatomic spacing [49]. It consists of three broad peaks centered at 2θ= 49, 28, and 16 degrees. The scattering is similar to amorphous scattering. The peaks are broad, indicating that the particle has only a limited number of atomic planes. The band at 49° appears as a doublet and is associated with scattering from the tetrahedral bond lengths 2.3 Å. For reference, crystalline or polycrystalline have peaks at 56.25° and 47.5° due to scattering from the 220 and 331 planes. For amorphous silicon, a single broad peak replaces the two peaks at 49°. The band at 28 degrees, due to scattering from the next nearest neighbor bond length centered at 3.9 Å, is the precursor to scattering from the 111 crystalline plane. Additionally, there is high-order but weaker and somewhat unresolved broad resonance in the region ~70–90 degrees. In this angular region, crystalline or polycrystalline have peaks at 69.2, 76.4, and

88.1 degrees due to scattering from the 400, 331, and 422 planes. Finally, the scattering band in the low-angle region has a peak at 16 degrees; it is associated with pure surface atoms and has no crystalline or amorphous correspondence.

We next examined the fluorescent 1 nm particle ($Si_{29}H_{24}$), reconstructed from the bulk-like 1 nm cluster ($Si_{29}H_{36}$) as a result of surface relaxation. It has been proposed as a source of the optical activity. The calculated molecular structure factor of the reconstructed particle (Fig. 9.42A, bottom) consists of the same amorphous-like broad peaks of bulk-like clusters, however shifted, broadened, and less resolved. The 49 degrees complex shifts to 48 degrees, with the shoulder on the peek essentially disappearing. The substructure of the high-order complex of peaks is washed out further. The 27 degrees band shifts to 29.5 degrees. Finally, the strength of the large angle scattering diminishes with respect to the strength of the small angle scattering upon relaxation and reconstruction.

Fig. 9.42C displays the development of scattering with increasing bulk particle size 1, 1.67, 2.15, 2.9, 3.7, 5.0, and 6.0 nm. We do not show results for diameters >6 nm because we find very little change in the spectra beyond 6 nm. However, for particles >2.15 nm, scattering from all of the above planes is resolved. Thus the particle size of 2.15 nm among the above discrete set of sizes may represent a demarcation of the transition from amorphous/molecular-like to polycrystalline-like behavior. There is no experimental study on the size effects or the existence of critical size in silicon in this important size regime (1–4 nm).

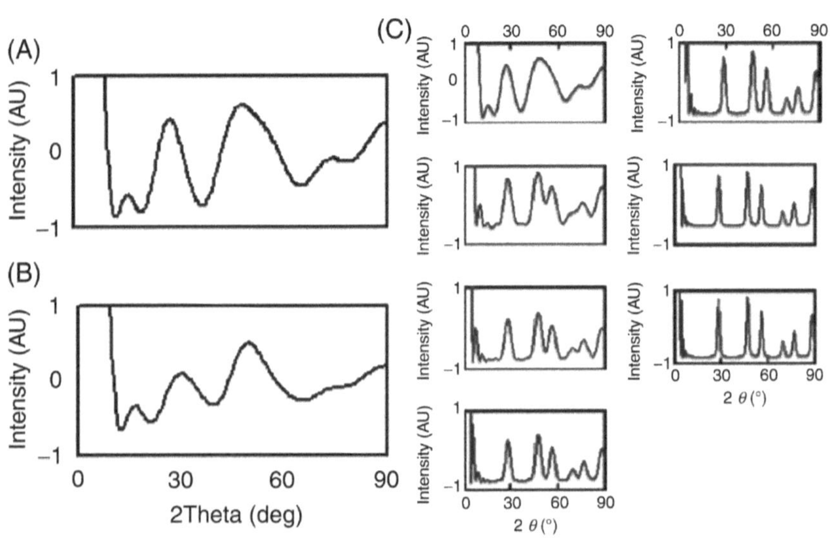

FIGURE 9.42 (A) The calculated molecular structure factor at 1.543 Å for 1 nm particles using the bulk-like interatomic spacing (Si29H36). (B) The calculated molecular structure factor of the reconstructed particle. (C) The development of scattering with increasing bulk particle size 1, 1.67, 2.15, 2.9, 3.7, 5.0, and 6.0 nm. For diameters >6 nm since there is very little change in the spectra beyond 6 nm.

The structure of the nanoparticles using X-ray diffraction may elucidate the transition from bulk-like to molecule-like properties. We numerically calculated the single particle molecular scattering of 1.543 Å X-rays for all members of the discrete family. The molecular form factor of the 1 nm particle shows molecular-like scattering at small angles, but amorphous-like at large angles. With size, the molecular-like structure shows an increase in the number of sub-peaks, narrowing, and shifting to forward scattering, but the amorphous-like structure transforms into polycrystalline scattering. The transition from amorphous/molecular-like to polycrystalline-like behavior occurs at 2.2 nm. Computed electron diffraction patterns of an array of the 1 nm particles show residual crystalline-like patterns for large scattering angle, but small angles show very little resemblance to crystalline scattering [49].

9.11 SYNTHESIS OF SILICON NANOPARTICLES

Before we discuss the synthesis of silicon nanoparticles, we briefly describe an etching process that polishes and shapes its surface [51]. Then we will describe synthesis of porous silicon, which exhibits luminescence [52,53]. The processes that result in a dispersion of silicon into luminescent silicon nanoparticles.

9.11.1 Synthesis of Porous Silicon

This electro-chemical etching of silicon was first studied by Arthur Uhlir Jr. and Ingeborg Uhlir at the Bell Labs in the 1950s [51]. The main purpose of their studies is polishing and shaping silicon surface. But occasionally underlower current, it was found that the process does not polish rather it produces an ugly looking thick black/brow/red film. A schematic of the electrochemical etching process for the synthesis of porous silicon is presented in Fig. 9.43A. The backside of a silicon wafer is contacted with a metal (copper for example). It is immersed in a solution of hydrofluoric acid (HF) in water (aqueous solution). The copper film is biased positively against an immersed platinum electrode. A voltages applied between the wafer backside contact and the platinum electrode, setting up a current. The current causes etching (dissolution) of the silicon wafer. The etching results in the progressive formation of a sponge-like structure with disordered pores. When the silicon wafer is highly doped (large conductivity) and the current is high, the current skims the surface resulting in very high current density. Under high-current density, the surface becomes mechanically weak and gets eaten away the top layer of silicon, which essentially electro polishes the surface. Another configuration which isolates the back metal from the etching solution is shown in Fig. 9.43B.

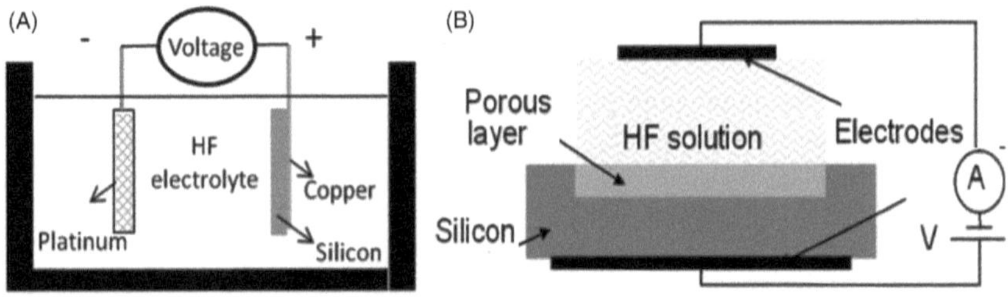

FIGURE 9.43 Synthesis of porous silicon. (A) Schematic of the electrochemical etching process for the synthesis of porous silicon. *(From http://cibaprocess.nus.edu.sg/process/index.html)* (B) Another configuration which isolates the back metal from the etching solution. *(From Elisabet Xifré-Pérez, Josep Ferré-Borrull, Josep Pallarés, Lluís F. Marsal, , Methods, Properties and Applications of Porous Silicon Electrochem. Eng. Nanopor. Mater. 37-63)*

Leigh Canham

Leigh Canham is a British scientist who has pioneered the optoelectronic and biomedical applications of porous silicon. Leigh Canham graduated from University College London in 1979 with a BSc in Physics and completed his PhD at King's College London in 1983. His early work in this area took place at the Royal Signals and Radar Establishment in Malvern, Worcestershire. Canham and his colleagues showed that electrochemically etched silicon could be made porous. This porous material could emit visible light when a current was passed through it (electroluminescence). Later the group demonstrated the biocompatibility of porous silicon. Canham now works as Chief Scientific Officer of psiMedica (part of pSiVida). According to the pSiVida web site, Canham is the most cited author on porous silicon. In a study of most cited physicists up to 1997 Canham ranked at 771.

Nearly 30 years later it was found that in some cases the etched silicon displayed photoluminescence under UV light [5]. Those studies were conducted by Leigh Canham at the Defense Agency in the UK. Close observations concluded that etching results in the progressive formation of a sponge-like structure with disordered pores (porous silicon layer). This caused the top layer of the wafer to exhibit red luminescence when exposed to UV light. It was then conjectured that the film consists of interconnected nanostructures, and quantum confinement effects was invoked where photoexcited electrons and holes, such as a particle in a box model, are behind the enhancement and blue-shift to the visible. The process is also called now anodic etching and the cell is called anodic cell.

There are several features worth noting with regard to the material choice, reaction equations and mechanism, electrode configuration, and material type, as well as other configurations:

(1) *Dopings:* Formation of fluorescent porous silicon could be obtained in all dopings from n + , n−, p−, and p+ if the etching conditions produced high density porosity.
(2) *Impurities:* The fluorescence was not observed in material with porous structure too large, nor was it observed in material made amorphous by defect concentration with ion implantation before anodization. Therefore if fluorescence is desired, a porous silicon fabrication technique can be largely judged by the crystallinity and size scale of the resulting structure.
(3) *Reaction equation*: for the etching process seems to be:

$$Si(surface) + 2HF + nh^+ \rightarrow SiF_2(surface) + 2H^+ + (2-n)e^-$$

The divalent silicon species formed on the surface is then oxidized to tetravalent silicon ions in the form of SiF_6^{2-} (solution) through a redox reaction in solution:

$$SiF_2(surface) + 4HF \rightarrow H_2SiF_6(solution) + H_2$$

with the partial reactions being:

$$SiF_2 + 4F^- \rightarrow SiF_6^{2-} \text{ and } 2H^+ + 2e^- \rightarrow H_2$$

1. *External current mechanism*: Holes come to the Si surface through the bulk where they weaken the Si–H passivating bonds allowing the fluorine ions to break them. The Si–F bonds weaken the Si–Si backbonds allowing them to be broken as well. Then a SiF4 molecule leaves the bulk which is now H passivated again, and the cycle repeats, which results in dissolution of silicon which is turned into porous silicon. In n-type silicon, holes need to be created by illumination of the sample, where electron–hole pairs are broken apart by the light to release more liberated holes. However, in p-type silicon, holes are majority charge carriers, which are easily driven to reach the wafer surface to participate in the reaction. Thus, p-type silicon is the preferred material for the process.

2. *Electrodes*: The backside of the p-type silicon wafer is usually connected to a metal plate or wire made of copper for example to achieve an Ohmic contact. The cathode is typically a platinum grid which is inert in HF. The back metal electrode may introduce metal impurities in the electrolyte which quenches luminescence. Recently, inert diamond cathodes are used instead of metal electrodes to avoid metallic impurities in the electrolyte; inert diamond anodes provide improved electrical back plate contact to the silicon wafers. Corrosion of the anode is produced by running electric current through the cell.
3. Both DC and pulsed current are used but the pulsed case allows formation of silicon wafers thicker than 50 μm [4].
4. *Solvent*: Ethanol is usually added to the HF electrolyte as a surfactant to prevent bubbles accumulating on the silicon surface. Usually, ethanol or methanol is added to the HF electrolyte to prevent bubbles from accumulating on the silicon surface and to facilitate infiltration of HF solution within the pores, as well as to improve the uniformity of pores and thickness. It is to be noted that during the formation of porous silicon, hydrogen gas forms. In the absence of ethanol or methanol hydrogen bubbles stick to the surface and induce lateral and in-depth inhomogeneity.

9.11.1.1 Stain Etching

Another method to obtain porous silicon is through what has been called stain-etching with HF acid, nitric acid, and water [54]. Porous silicon formation by stain-etching is particularly attractive because of its simplicity and the presence of readily available corrosive reagents; namely nitric acid (HNO_3) and hydrogen fluoride (HF). Furthermore, stain-etching is useful if one needs to produce a very thin porous Si films. R.J. Archer revealed that it is possible to create stain films as thin as 25 Å through stain-etching with HF–HNO_3 solution. Finally, porous silicon can be synthesized chemically from silicon tetrachloride, using self-forming salt byproducts as templates for pore formation. The salt templates are later removed with water.

9.11.2 Synthesis of Si Nanoparticles

Several procedures have been developed throughout the last decade for the synthesis of luminescent Si nanoparticles [2–40,22,23]. These include physical, physiochemical, chemical, and electrochemical procedures. We will discuss the procedures, size, uniformity in size, throughput, cost, and amenability to mass production, and recovery.

9.11.3 Physical Techniques

A variety of *physical techniques* were developed in the last decade.

First, Si nanoclusters (Si-nc) were formed, for example, in the matrices of glass and SiO_2. In these efforts, luminescent nanocrystals of ~3 nm across embedded in quartz by implanting high energy Si ions into quartz, followed by annealing at 1100 °C were created [55–58].

Second, silicon wafers were dispersed by ablation using a variety of agents to produce isolated Si particles. These agents included spark [59] and laser ablation [60]. In laser ablation a focused high power laser radiation is used as a brute force to blast the wafer, heat evaporates, or break chunks of the material. The broken pieces are transported downstream from the laser/material target spot by an inert gas jet, to be collected by filters. The particles are then recovered from the filter. These ablation techniques suffer from low throughput and cost, lack of control on the size and condition of the particles produced. For instance the generation of laser radiation consumes electrical power with less than 5% efficiency. The quality and efficiency of formation of nanoparticles depends on several parameter of the laser beam including the pulse duration, peak power, the number of photons involved (single- vs. multi-photon process) as well as on the gas environment and target material. Recently, effects of nanosecond pulses and femtosecond pulses were compared as shown in Fig. 9.44 [61]. The nanosecond pulse at left melts surface material before ablation, transferring heat to adjacent areas, which alters many materials and produce large droplets. Femtosecond pulse at right ablates material by multiphoton ionization, with very little heat transfer to adjacent material. Unlike nanosecond ablation, most of the ablated material formed nanoparticles in the case femtosecond pulses. Other studies show that size of the nanoparticles produced by femtosecond pulses depends on the laser flux, gas environment, and target materials.

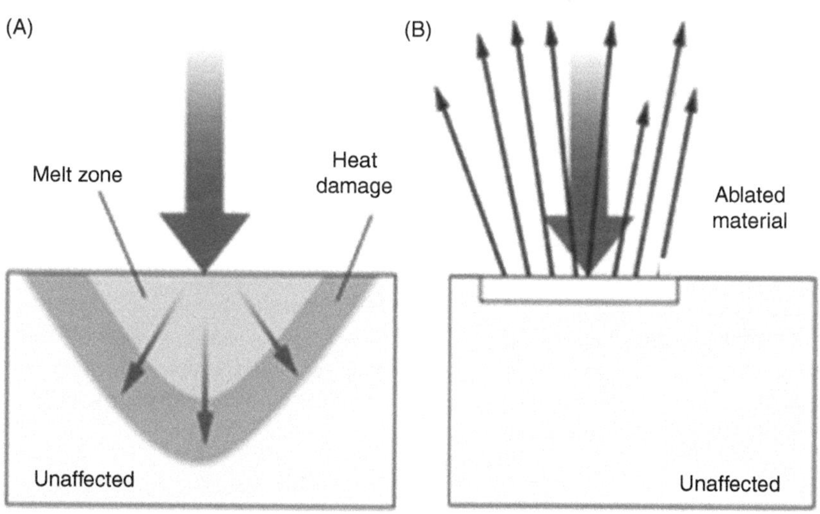

FIGURE 9.44 Laser ablation schematic of laser–material interaction. (A) Nanosecond pulses melts surface material before ablation, transferring heat to adjacent areas, which alters many materials and produce large droplets. (B) Femtosecond pulse ablates material by multiphoton ionization, with very little heat transfer to adjacent material, forming nanoparticles.

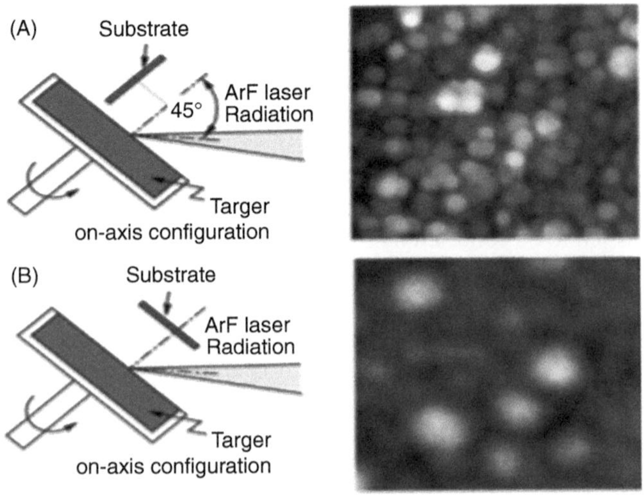

FIGURE 9.45 Schematic of collection of nanoparticles. (A) Off axis provides higher quality nanoparticles (smaller and more uniform) than (B) on-axis collection.

Fig. 9.45 shows that collection in off axis provides higher quality nanoparticles (smaller and more uniform) than on-axis collection.

For many applications of silicon nanocrystals, large quantities of isolated, size-selected nanocrystals are required. They can be made by using pulsed-laser-induced pyrolysis of SiH_4 in a gas flow reactor (F. Huisken, Max Planck Institute, Göttingen, Germany) [62].

9.11.4 Physicochemical Techniques

Isolated particles were obtained by the gas-phase preparation from silanes via slow combustion [63,64], thermal decomposition [65], microwave plasma [66], gas-evaporation [67], or chemical vapor deposition (CVD) [68,69]. Such methods may involve particle formation in a discharge of gas mixtures that include the highly toxic silane (SiH4), followed by collection in filters and recovery from filters. These techniques suffer from low throughput and cost, lack of control on the size and shape of the particles. Although the decomposition of silanes produces nanoparticles with a relatively small size distribution, it does not lend itself to the easy manipulation of the surface of the particles, covered with SiO_2, or to their large-scale manufacturing.

9.11.5 Chemical Techniques

Recently Si nanoclusters in the range 2–10 nm were synthesized *chemically* via a reduction of anhydrous ionic salts SiX_4 (X = Cl, Br), dispersed in water-free reverse-micelles solutions, with $LiAlH_4$ under Ar atmosphere. A certain degree of control over the Si cluster size was achieved by variation of the micelles size, intermicellar interaction and reaction chemistry. Unlike the *physical preparations*, which produce impure Si crystallites that contain a large amount of SiO_2 on the surface, this method produces Si particles with the surface terminated by hydrogen from metal hydride. However the formed Si particles are stable for long time (6 month to 1 year) only in a glove box under Ar. They would undergo degradation upon exposure to oxygen, presumably due to surface oxidation.

Finally, in some cases, organic capping of nanoparticles may be used as a means to achieve some control over the particle size, directly during the synthesis step, especially during solution synthesis procedure like for molecular silicon compounds. Methods of chemical reduction of Si(IV) compounds, such as SiR_nCl_{4-n} (R = H, Alkyl) to Si(0), directly produce organic-monolayer stabilized Si particles.

9.11.6 Electrochemical Techniques

The ultrasonic fracturing of PS in toluene and acetonitrile resulting in colloid suspension of Si nanoclusters was first reported in 1992 [70]. PS was prepared from n-type (P-doped) and p-type (B-doped) single crystal polished Si(100) wafers by galvanostatical etching at current densities between 0.1 and 5 mA/cm^2, adjusted by illuminating with a 300-W tungsten lamp, in a solution of aqueous HF and ethanol. The IR spectra of anodized wafers, rinsed with water and air-dried, displayed strong Si–H stretches in the 2050–2150 cm^{-1} region and lacked discernible peaks for surface Si–O, indicative of the clean, H-terminated Si surface characteristic of freshly etched PS.

Ultrasonication was carried out under dry N_2 atmosphere, using purified, dried solvents, added to the flasks with anodized Si wafers. Typically, after sonication the solution was allowed to settle for 1h, and the supernatant was removed from the settled solids. TEM images of Si colloidal suspensions clearly showed crystalline, overlapping Si particles. The particles were irregularly shaped and ranged in size from many micrometers down to the resolution limit of the TEM.

HRTEM studies of obtained colloidal suspensions showed that a size of the particles ranges from 50 nm and smaller up to 2 nm. As it was expected the addition of ethanol to electrolyte solution and the increase in anodization time result in producing more uniform crystallites in the size range of 10 nm or less. It was found that the bigger size objects are not single crystalline and in reality are agglomerate of several small crystallites approximately 7–11 nm in diameter. The smallest crystallites in agglomeration are about 2 nm. In addition to these relatively big aggregates of crystallites, small individual crystallites were also observed.

Recent investigations of Si particles obtained as toluene colloid from PS also confirmed the presence of SiO_2 on the surface of nanoclusters. TEM investigation of obtained colloid suspensions showed similar size distribution as it was reported before: the particles, predominantly with diameters in the range 1–10 nm, and the some larger (up to 80 nm) agglomerates.

A variety of investigations of porous silicon showed that the resulting porous layer has a highly complex nanoscale architecture consisting of one-dimensional crystalline nanowires and zero-dimensional nanocrystallites. It is to be noted that these layers are hydride terminated (containing very little SiO_2), with mono- ($\equiv$SiH), di- ($=SiH_2$) and tri- ($-SiH_3$) hydride groups in a variety of different local orientations and environments, owing to the complex porous nature of the material.

9.11.7 Discreetly-Sized Ultrasmall Si Nanoparticles

Significant progress in the dispersion of single-crystal silicon wafers into ultrasmall Si nanoparticles with uniform size and large throughput was recently developed in the Nayfeh's group [7–11, 71–75]. Ex-members of the group, David Andsager, Zain Yamani (presently at the Center of Excellence in Nanotechnology (CENT), King Fahd University for Petroleum and Minerals, Saudi Arabia), Joel Therrien, and GennadeyBelomoin were among the early codevelopers of the processes.

- First the electrode configuration is changed to increase the current density by limiting/concentrating the current in the top skin of the wafer. This is accomplished by removing the back metal placed on the wafer.
- Second, the etchant composition is modified to incorporate hydrogen peroxide H_2O_2. H_2O_2 is a highly oxidative, catalytic, and cleansing agent.

The effect of the incorporation of both elements include

- Increasing the etching rate appreciably, allowing the production of ultrasmall silicon nanostructures, down to 1 nm consisting of only one tetrahedral unit.
- Controling hydrogen termination and surface re-construction which allows the production of uniform-size with complete shell (magic size)

The silicon samples used were (100) oriented, 1–10 ohm-cm resistivity, p-type boron doped silicon wafers. With the silicon substrate acting as the anode, and a platinum wire facing it acting as a cathode, the anodizing current density was fixed at 5 mA/cm^2. The method involves gradually advancing bulk silicon, for example, a wafer into a chemical etch bath at a low speed while etching is taking place. A meniscus forms as a very thin slice of the silicon that is at the interface of etching solution and air. The slow advance of the silicon creates a large section, which is meniscus like. The process enriches the ultra-small substructure of the material. Moreover, it makes the top skin material extremely fragile. Larger particles are less amenable to dispersion due to stronger interconnections. We use a post HF treatment to weaken those. The Si wafer is then immersed in an ultra sound acetone (ethanol, methanol, or any other solvent) bath for a brief period of time. Under the ultra sound treatment, the very top layer of the luminescent film, a weakly interconnected nanostructure network crumbles into ultra small particles which may then be separated from the material and recovered.

The silicon samples used were formed from boron doped Si (100) wafers with a resistivity of 1–10 Ωcm from ISS. The wafers were cleaved into ~0.5 × 1.5 cm pieces which were cleaned ultrasonically in a methanol bath for five minutes and immersed in a solution of HF:methanol: H_2O_2 for 5 min to remove the native oxide layer and to passivate the surface. The samples were then laterally anodized in an electrochemical cell containing the same solution of HF: methanol: H_2O_2. Anodization was performed in a HF resistant Teflon beaker. The cathode is a platinum wire while the samples were held by a Pt mesh pad and were mounted parallel to the cathode ~0.5 cm away, as shown in Fig. 9.46. The samples were typically anodized for 60 min and then placed in a vacuum desiccator until needed. Lateral anodization produces samples that have thicker porous layers near the meniscus, where the impedance path is shorter, than at the bottom. Current densities ranged from 4–40 mA/cm^2 and were provided by a constant voltage source that could be sent through a constant current circuit when necessary. The solution was continuously agitated by a magnetic stirrer. That and the methanol in the solution helped prevent the formation of bubbles on the sample surface that impede the etching process.

Without the H_2O_2 in the etching solution, the samples must be exposed to UV radiation to become photoluminescent. The UV creates electron-hole pairs that enable further dissolution of the Si by HF. The H_2O_2 increases the etching process by turning Si into SiO_2 which dissolves readily in HF. This produces

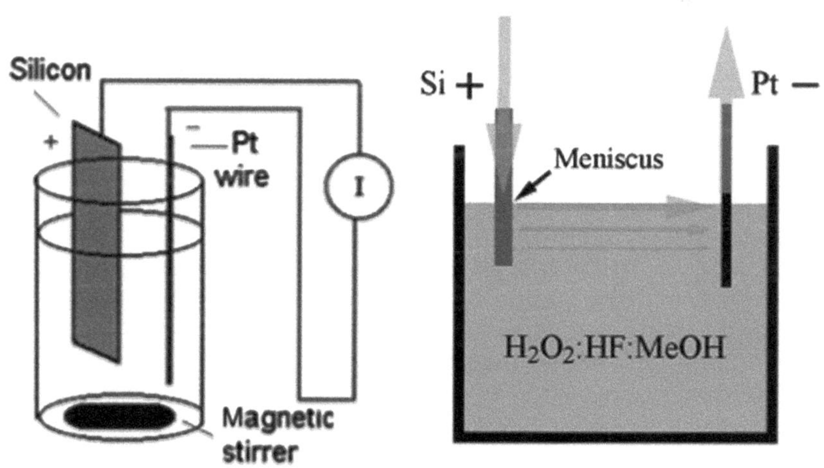

FIGURE 9.46 Experimental setup for lateral chemical etching of silicon (anodization) in HF/ H_2O_2 that produces a discrete set of particle size.

much smaller crystallites and removes the need for UV exposure to produce luminescent samples.

Ultrasonic dispersion of electrochemically etched Si wafers in a variety of solvents results in a colloidal suspension of Si particles. The size distribution depends on the anodization time and the composition of the electrolysis bath solution. Unlike the gas-phase preparation of colloidal particles, this technique is more convenient and starts from high-purity semiconductor-grade substrates. More importantly, the resulting H-passivation of the surface minimizes its contamination by SiO_2 and other impurities. Due to dispersion and fracturing in the ultrasonification step, some of the bare silicon will be exposed, resulting in some degree of hydrocarbon contamination and oxidation as it was demonstrated by recent FTIR investigation.

Commercial gel permeation chromatography was used to separate the particles. The particles are separated into four vials each contains particles of uniform size, with near 90–100 % efficiency. A thin graphite grid is immersed in the colloid and imaged by high-resolution TEM. Fig. 9.47A shows that particles are nearly spherical and classify into a small number of sizes. These include 1.0 (see inset) 1.67, 2.15, 2.9, and 3.7 nm. Fig. 9.47C shows the atomic planes in close ups of the 1.67, 2.15, 2.9 and 3.7 nm particles. This is in contrast to what has been known that the abundance spectrum self-terminated clusters Sin with n >20 exhibit neither special features nor discrete magic numbers. The smallest four are ultrabright blue, green, yellow, and red luminescent particles respectively as shown in Fig. 9.48. Fabricating size-, shape-, and orientation-controlled fluorescent Si nanoparticles in the range 1–3 nm (30–1000 atoms), with reproducibility would be critical to the understanding of nanostructures and would be of significant interest to the microelectronics, optoelectonics and biomedical industries. The availability of discrete sizes and distinct emission in the red, green and blue range is useful for biomedical tagging, RGB displays, and flash memories. Moreover, size uniformity is important for applications that require superlattices, high-quality films, or single nanoparticle-based devices, such as fluorescent imaging and tagging.

The incorporation of H_2O_2 in the anodization mixture brings several features and advantages. It produces electronic and chemical high-quality particles a homogeneous mono-hydride stretching phase on the surface, with no defect like tri-hydrides to di-hydrides and di-hydrides. The process also reduces the crystallite size, and reduces impurities. Moreover, it produces substructure that is very loose, requiring very brief period of time for sonication with very little fracturing, hence subsequent oxidation.

The size of the smallest particle was also confirmed by two-photon auto correlation fluctuation spectroscopy (FCS) in the colloid in which we determined the auto correlation function of the luminescence and compared it to a fluorescein dye molecule standard with known density, diffusion coefficient, and size. Previous differential excitation measurements and transmission electron microscopy of the solid before sonication showed structures of $\sim$1nm.

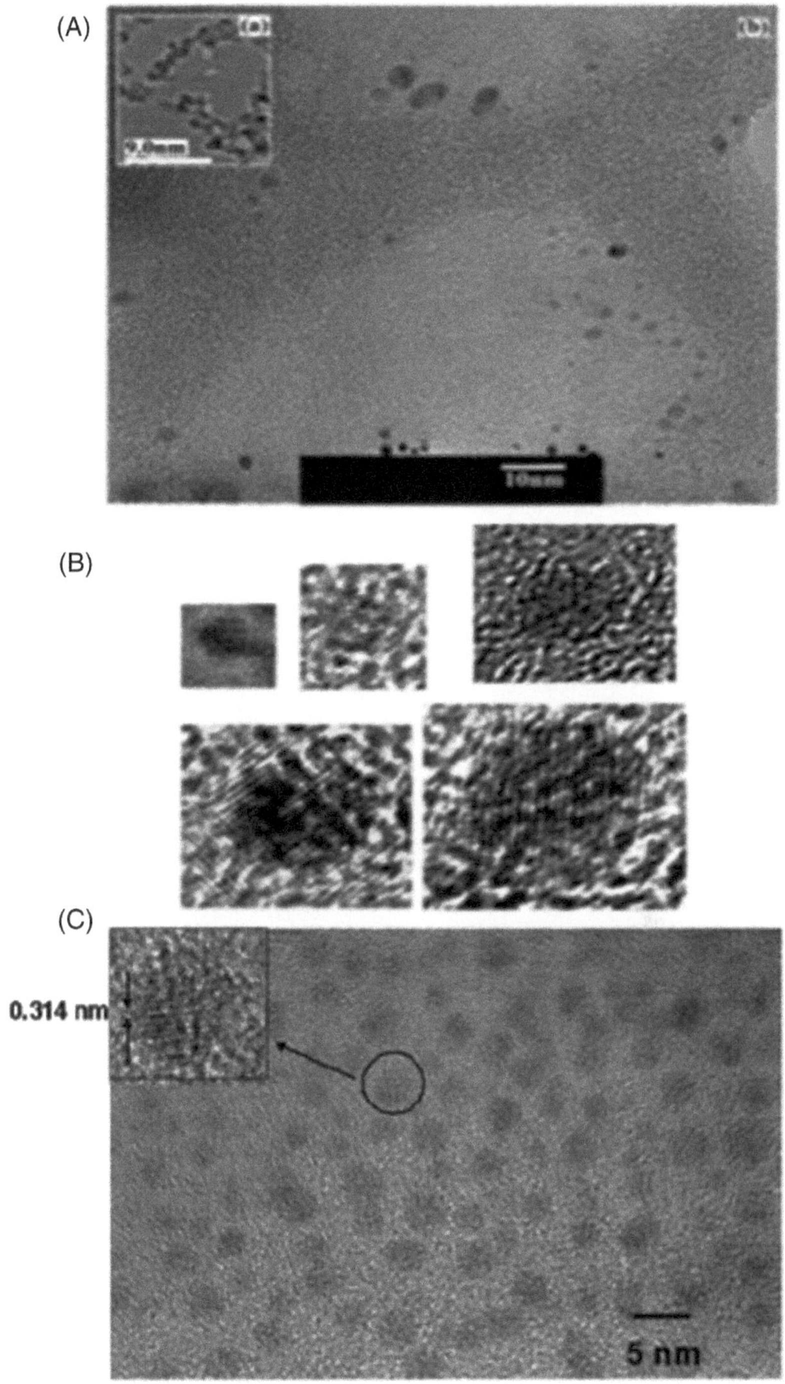

FIGURE 9.47 TEM images of a graphite grid when was dipped into colloids of the nanoparticle. (A) Particles are nearly spherical and classify into a small number of sizes: 1.0 (see inset), 1.67, 2.15, 2.9, and 3.7 nm. (B) Close ups of the 1, 1.67, 2.15, 2.9 and 3.7 nm particles. (C) Sample predominantly of 3-nm particles with a close up showing atomic planes spacing of 3.1 Å, characteristic of silicon.

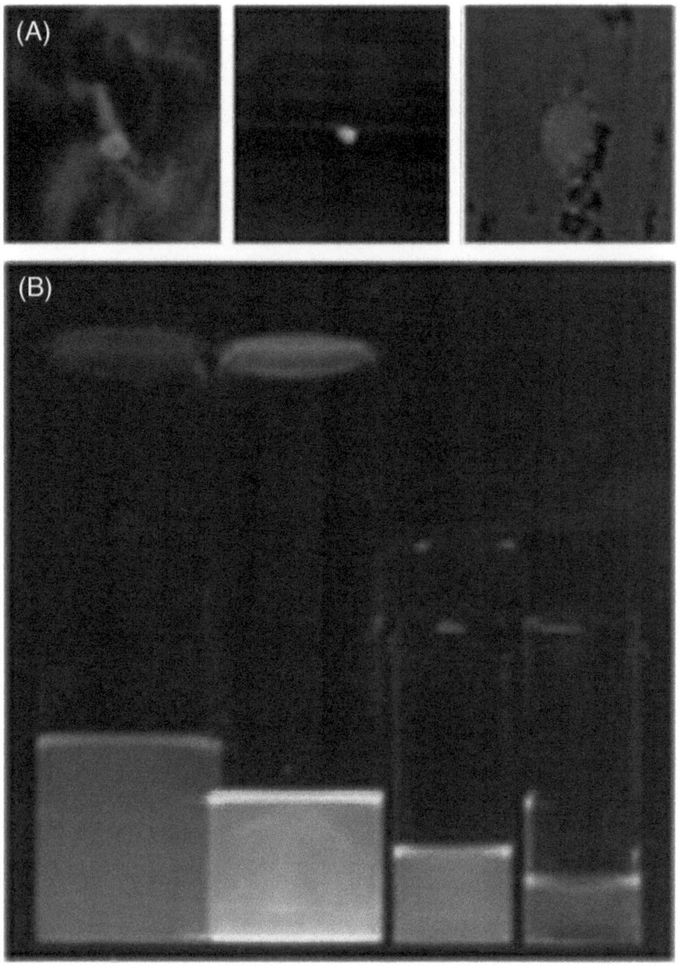

FIGURE 9.48 (A) The smallest four *particle are purified in separate vials,* blue, green, yellow, and red luminescent particles. (B) Colloidal crystals luminescent in the red, green, and blue.

Moreover, material analysis using electron photoluminescence shows that the particles are composed of silicon with less than 10% oxygen.

In contrast to the physical methods for preparation, the anodization procedures and ultrasound are straightforward and inexpensive. Unfortunately, the effort, required for control of the size distribution and quantitative characterization, is prohibitive for large-scale synthesis and surface manipulation is limited.

9.11.8 Metal-Assisted Synthesis of Si Nanoparticles

The external current used in the above electrochemical etching procedure is intended to charge the wafer with holes that catalyzes the HF attack to dissolution silicon. Another procedure that has been used eliminates and replaces the external current by a local source of charge. The technique uses metal-assist electroless chemical etching[11,45]. This procedure was developed for anodization of silicon wafers, crumbs or powder by the Nayfeh's group in collaboration with Laila Abuhassan (University of Jordan), Mouna Alchihabi (Allepo), Abdulrahman Al-muhanna (KACST) and Jon Host (Hemlock Semiconductor). Instead of 1-step etching step like the above electrochemical-etching method, this method requires two steps. Fig. 9.49A shows the schematic illustration: This method is fast and robust, which can be easily scaled for mass production.

- Deposition/plating platinum (Pt) particles onto the Si surface
- The plating of Pt onto silicon surface is done by immersing a piece of a silicon wafers into the plating solution, which is a 3:1 mixture of hexachloroplatinic acid and HF. In this process the metal ions are reduced and form nuclei on the surface of the silicon. With increasing plating time, these nuclei

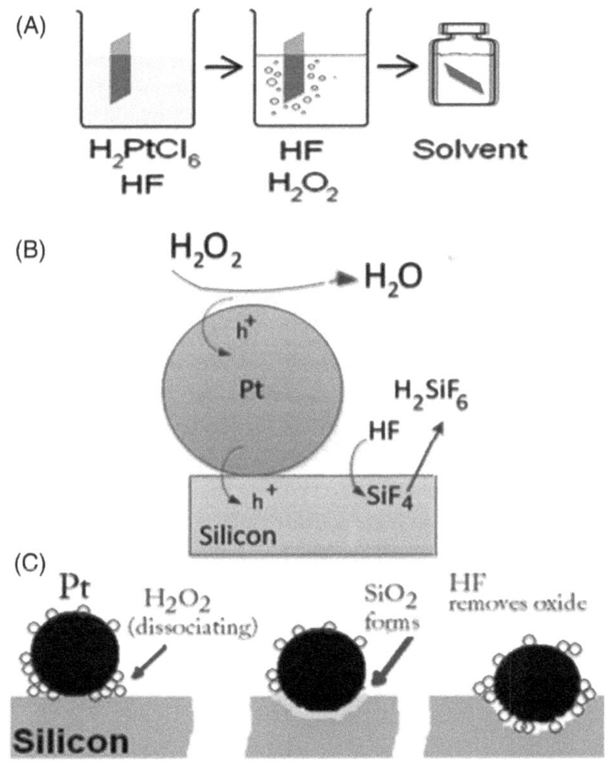

FIGURE 9.49 Platinum assisted etching. (A) Schematic illustration of the electroless setup consisting of a platinum deposition/nanoparticle formation bath and an etching bath for formation of silicon nanoparticles. (B) Schematic of the reduction of H_2O_2 at the Pt nanoparticles injecting holes into the Si substrate to initiate etching. (C) Schematic of platinum particles burrowing into the silicon substrate.

will grow into Pt nanoparticles. Good coverage happens after around 10 minutes of plating. A typical plating time is about 7–8 min.

- Etching: After plating, the wafers are rinsed and transferred to the etching bath HF/H_2O_2 for about 30 s.
- Sonication and recovery: The wafer is then removed, rinsed and dried and sonicated to recover the nanoparticles.

Fig. 9.49B gives the schematic illustration of the reduction of H_2O_2 at the Pt electrode injecting holes into the Si substrate to initiate etching in the Pt-catalyzed etching method. In the etchant, the Pt particle acts as a local cathode to mediate the reduction of H_2O_2 and subsequent hole injection. The mechanism was first proposed by Li and Bohn [74]. At the cathode we have the two reactions: the reduction of H_2O_2 and the reduction of protons into hydrogen:

$$H_2O_2 + 2H+ = 2H_2O + 2h+;$$
$$2H+ = H_2 + 2h+;$$

The generated holes are transferred inside the Pt particle toward the Si in contact with the particle. At the Pt–Si contact point, which is where the local anode is in this scheme, the holes are injected into the Si and oxidize that Si atom as a result:

$$Si + 4h+ + 4HF = SiF4 + 4H+:$$

The oxidized Si atom is then pulled off the Si surface through the reaction:

$$SiF4 + 2HF = H2SiF6:$$

The etching mechanism of the Pt catalyst is summarized as follows: peroxide dissociates on the surface of platinum particles, which oxidizes the surrounded silicon, and the oxide is in turn removed by the hydrouoricacid. This process allows the platinum particles to effectively "eat" into the silicon. The process of the platinum burrowing into the silicon can be seen in SEM images taken of the surface of a sample which was etched for only a couple seconds (Fig. 9.50A) gives the schematic illustration). Many bright platinum deposits can be seen sitting in rounded pits which they have already started etching into the silicon surface. The platinum deposits vary greatly in size from below 0.2 microns to about 1.3 microns. In Fig. 9.50B a higher magnification is used to reveal the smallest platinum deposits are about 100 nm in size. After allowing etching for 30 s the silicon surface becomes riddled with etching pits (Fig. 9.50C). Porous silicon forms on the walls of the pits, where the dissociated peroxide and HF, caused rapid oxidation and etching. In essence, the platinum deposits create intense focused etching, and in this sense play the role of the current focused on the meniscus in the anodization method.

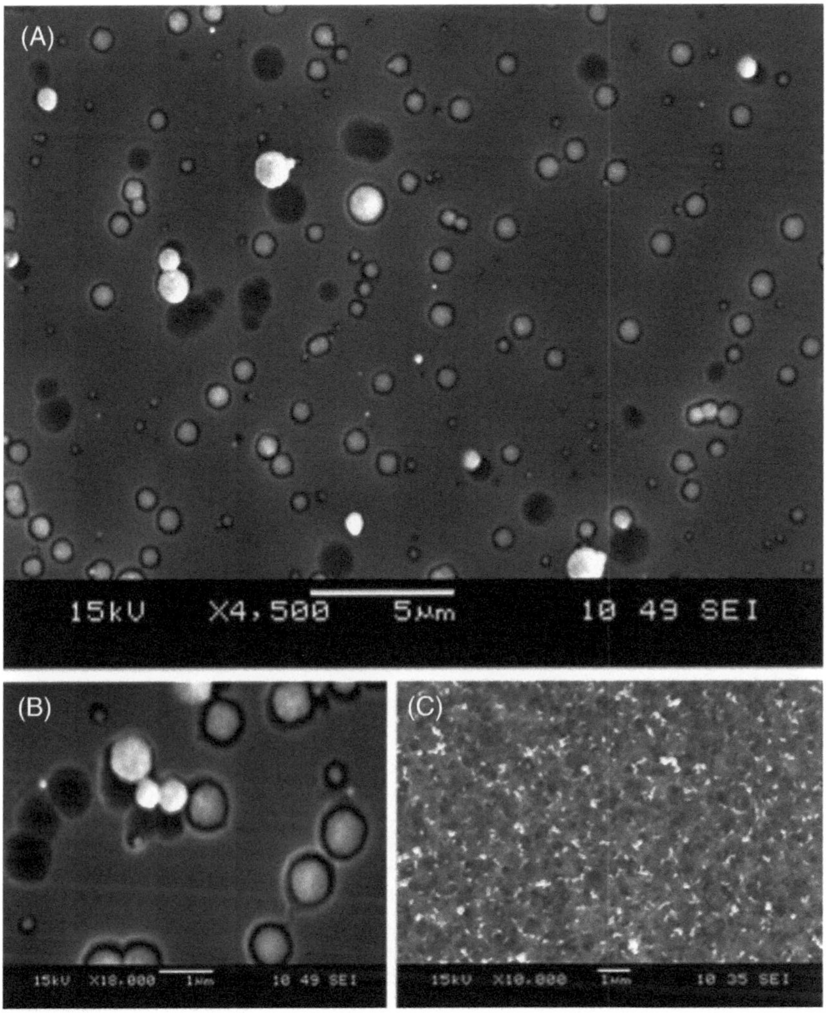

FIGURE 9.50 The process of the platinum burrowing. (A) SEM images taken of the surface of a sample which was etched for only a couple seconds, showing platinum, nanoparticles sine below 0.2 microns to about 1.3 microns. (B) higher magnification reveals the smallest platinum deposits are about 100 nm in size. (C) SEM imaging after etching for 30 s show the silicon surface being riddled with etching pits.

There are several other models describing the dissolution process of the Si at the anode. The first group of these models is direct dissolution of Si in the tetravalent state, which is essentially what was described above. The second group is direct dissolution of Si in divalent state. Instead we have:

$$Si + 4HF_2^- = SiF_6^{2-} + 2HF + H2 + 2e^-$$

The last group of models proposes Si oxide formation followed by dissolution of oxide:

$$Si + 2H_2O = SiO_2 + 4H + +4e^-;$$
$$SiO_2 + 6HF = H_2SiF_6 + 2H_2O:$$

The location of the hydrogen gas formation is also a matter of debate. Li and Bohn [74]suggested that it occurred at the cathode, Chartier [75] attributed such reaction to the anode end of the redox reaction. When they attached Ag particles to a Si substrate in solution with a high concentration of H_2O_2 in the absence of HF, they did not see any gas formation. Regardless where the gas is formed, the fact that the Pt particles catalyze the etching process of the Si surface remains the same. The collection of the particles is the same as the electrochemical etching method.

The use of platinum in this procedure was extended to other metals. Instead of the use of platinum, gold is used. Just like platinum, gold diffuse in silicon to form novel Bi-alloys but it poses some differences and difficulties (see Section 11.7 for details) [76].

9.12 METAL-ASSISTED SYNTHESIS OF SI NANOWIRES OR PILLARS

Metal catalyst may be used for synthesis of silicon nano-wire or pillar. This topic will be covered in detail in Chapter 11. For completeness we give a very brief introduction. In one procedure wet chemistry is used to etch a wafer in a bath of HF solution, which contains silver nitrate. The solution is heated to a temperature of 50°C. This results in etching away some silicon leaving behind a wire-like arrays of pillars attached to the substrate (see Chapter 11, Section 11.6). The array may be under etched at the base by using an SOI wafer. In another procedure hot gold nanoparticles are used as a catalyst to grow silicon nanowires from a silicon bearing gas, such as silane. The wire may grow in any direction (see Chapter 11, Section 11.5).

9.13 LOW-COST METALLURGICAL ROUTE FOR SYNTHESIS OF SILICON NANOPARTICLES

Currently, silicon nanoparticles are produced from device-quality mono-crystalline silicon. Those are expensive single crystal wafers or electronic grade Si (EGSi). To lower the cost of synthesis of silicon nanoparticles, recent work focused on the use solar grade silicon (SoGSi) or other lower grades, which are less pure (boron/phosphor) than the electronic grade, down to a limit on the maximum impurity content of ~1ppmw, nearly a factor of 1000 or less. In fact, silicon from low-grade silicon wafers manufactured from waste silicate slags has been tested in collaboration between the Nayfeh's group and the group of

Serekbol Tokmoldin of the Institute of Physics and Technology in Kazakhstan and Ersin Baceci of Iskandaron Univeristy, Turkey. The low cost of this grade is due to the fact that the grade is manufactured by short cutting chlorosilination purifying steps. As a result, the grade exhibit much higher residual impurity levels (boron/phosphor, as well as metal) and shorter carrier lifetimes than those required for low-grade (solar grade) silicon photovoltaic operation.

Example 9.12 Nanoparticle Size From X-Ray Diffraction Peak Broadening Analysis [50]

The structure of the nanoparticles using X-ray diffraction may elucidate the transition from bulk-like to molecule-like properties [100]. We numerically calculated the single particle molecular scattering of 1.543 Å X-rays for all members of the discrete family. The molecular form factor of the 1 nm particle shows molecular-like scattering at small angles, but amorphous-like at large angles. With size, the molecular-like structure shows an increase in the number of sub-peaks, narrowing, and shifting to forward scattering, but the amorphous-like structure transforms into polycrystalline scattering. The transition from amorphous/molecular-like to polycrystalline-like behavior occurs at 2.2 nm. Computed electron diffraction patterns of an array of the 1 nm particles show residual crystalline-like patterns for large scattering angle, but small angles show very little resemblance to crystalline scattering

As a primary characterization tool for obtaining critical features, such as crystal structure, crystallite size, and strain, X-ray diffraction patterns have been widely used in nanoparticle research. The randomly oriented crystals in nanocrystalline materials cause broadening of diffraction peaks. This has been attributed to the absence of total constructive and destructive interferences of X-rays in a finite sized lattice. Moreover, inhomogeneous lattice strain and structural faults lead to broadening of peaks in the diffraction patterns. The size calculated from X-ray diffraction peak broadening is a measure of the smallest unfaulted regions or coherently scattering domains of the material. In fact, this is the size of regions bounded by defects and grain boundaries and separated from surrounding by a small misorientation, typically one or two degrees. Three methods of Williamson and Hall, Warren and Averbach and Scherrer can be used to calculate crystallite size, and strain. The simplest and most widely used method for estimating the average crystallite size is from the FWHM of a diffraction peak using the Scherrer equation as follow:

$$d_{XRD} = K\lambda / BCos\theta$$

Where d is the crystallite size, λ the diffraction wavelength, B is the corrected FWHM, θ is the diffraction angle and K is a constant is close to unity. The major assumption is that the sample is free of the strain. B can be obtained from observed FWHM by convoluting Gaussian profile, which models the specimen broadening Br, as follows:

$$B_r^2 = B_0^2 - B_i^2$$

Where B0 is observed broadening and Bi is instrumental broadening.
Williamson and Hall is a simplified integral breadth method for deconvolution size and strain contributions to line broadening as a function of 2. They lead to a Williamson–Hall in which they intercept can be used to calculate crystallite size, while strain can be calculated from the
http://ijmse.iust.ac.ir/files/site1/user_files_4qu804/ijmse-A-10-3-149-c629680.pdf
PARTICLE SIZE CHARACTERIZATION OF NANOPARTICLES – A PRACTICALAP-PROACH B. Akbari, M. PirhadiTavandashti, and M. Zandrahimi, Iranian Journal of Materials Science & Engineering Vol. 8, Number 2, Spring 2011

Example 9.13 Attributes of Nanosilicon for Industry

A breakthrough technology was developed at the University of Illinois and Nano-Si Advanced Technologies that provides silicon nanoparticles of the highest technical and commercial quality available today. We believe it meets all the conditions necessary for a successful commercial product:

- Uniformity of size and shape: uniformity and reproducibility: uniformity provides sharp, constant threshold across a full wafer, which is critical for fast switching and circuit design.
- ultrasmall sizes of 1–3 nm in diameter allowing extreme down scaling, smaller flash, larger density that can be accommodated, and faster operation
- Ex-situ production: This allows CMOS compatibility using straightforward spin-coating at room temperature spread the particles uniformly on top of the tunnel oxide, which provides constant or even tunneling distance and constant control oxide thickness to the gate across a full wafer, without changes to the standard CMOS fabrication process.
- Easy to manufacture: The manufacturing proceeds utilizing common raw materials and chemicals. It uses silicon, the most abundant, cheapest and least toxic semiconductor material as well as common chemicals of acids and hydrogen peroxide and low power electricity in table-top setups. We start with same silicon that is used in silicon solar cells or computer chips. We ground the silicon to a consistency of flour through a special process that has been protected by 26 US patents (23 issued)).
- Mass production: The process is very efficient, scalable and can be automated. It provides commercial low-cost yield
- Easy to use: Differently sized silicon nanoprticles can be excited with a single source to obtain various colors, making them easier to use
- Quick process: The process takes 30–45 min to completion
- Reaction temperature: Room temperature
- Commercial low-cost yield: The particles are produced by novel dispersion of crystalline silicon using low cost electrochemical etching

The silicon used comes from a metrological process developed by the Institute of Physics and Technology (IPT), Kazakhstan (IPT process) [77]. In the process, phosphorous industry waste and synthetic slag are used as a silicate slag. The slag consists of more than 90% silicon and calcium oxides and less than 10% other elemental oxides. Alumothermic reduction of silica from this silicate slag is used to obtain silicon-containing Alloy-I and Alloy-II (metallurgical silicon). Silicon-containing Alloy-I was upgraded by acid leaching to silicon of a fine powder structure. Using this powder, poly- and mono-crystalline p-type silicon, with resistivity of 0.24 Ohm-cm, are grown by the Czochralski method. Silicon-containing Alloy-II was used for obtaining monosilane by aqueous treatment with hydrochloric acid under atmospheric conditions and without any catalyst. There was no trace of diborane, which is a common source for boron contamination in crude silane. This process constitutes a metallurgical refining lower cost short-cut technology for production of metal silicon. The short cut eliminated the expensive purification step of chlorosilane processing. The total amount of silicon produced by IPT was quite limited (50 kg). The IPT wafers were characterized for solar cell applications (solar grade silicon) at the US National Renewable Energy Laboratory (NREL) and the Solar Power Industries (SPI) [78]. The tests showed metal levels of ~ 1ppbmwt, which is suitable for poor solar grade silicon applications. However, impurity measurement of boron (B) and phosphor (P) showed high unacceptable levels. For instance it found the boron, aluminum (Al) and phorphor levels at 7.7–24 ppmwt, 1.8–22 ppmwt, and 1.3–44 ppmwt, respectively. Resistivity measurements of the wafers gave 0.1–0.3 Ω-cm, with a corresponding lifetime of charge carriers of 0.1–0.3 μs. Those values are outside the SPI's minimum requirement of Upgraded Metallurgical Grade Silicon for photovoltaic device operation, which is Boron <0.5 ppmwt, Phosphorus <2.0 ppmwt.

Our results show we can produce high-quality, high-added value red luminescent silicon nanoparticles, comparable to those obtained by the mono crystalline rout. After the etching is complete the strips are removed, rinsed in de-ionized water, then isopropanol and dried. Inspecting the strip with UV radiation after the completion of the treatment shows orange luminescence. The wafer's luminesce is red under UV irradiation as shown in Fig. 9.51A. The strips are then placed in a solvent of choice and sonicated for 5 min to dislodge the nanoparticles in the liquid. The nanoparticle concentration is enriched by recycling of the wafer five times. The enriched nanoparticle colloid is then centrifuged to filter out larger chip pieces that may have been broken due to the fragility of the wafer. The liquid appears nearly clear in visible light but luminesce red under UV irradiation as shown in Fig. 9.51B. The luminescence spectrum of the colloid is given Fig. 9.51C.

We isolated the silicon nanoparticles from the solvent by making a thin film of the colloid on a device-quality wafer using drop-drying of some of

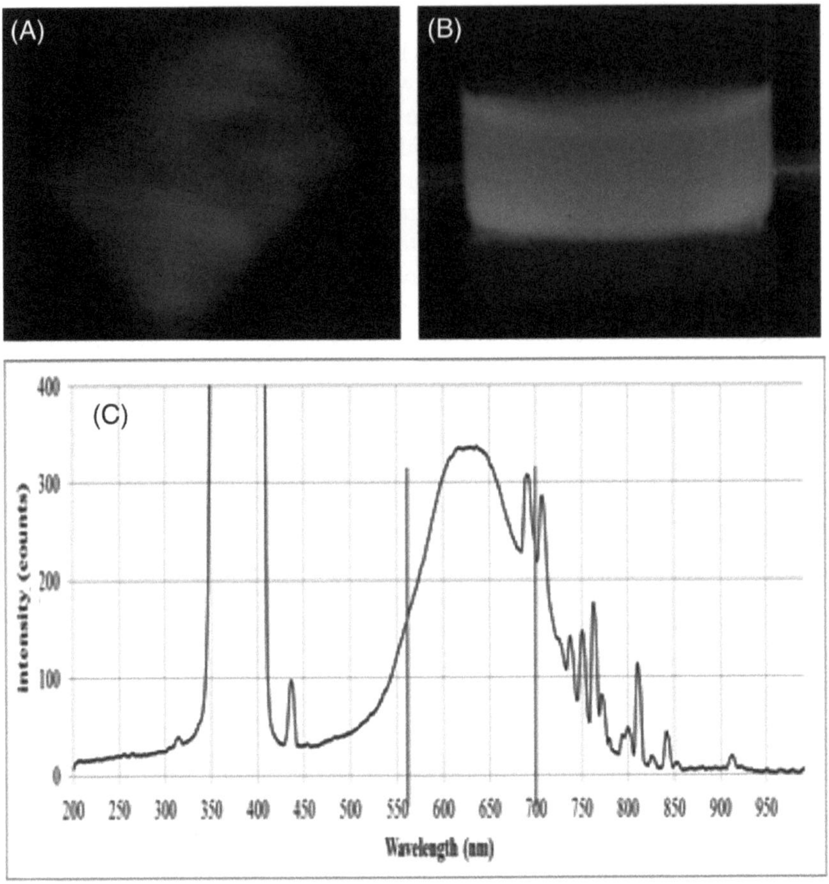

FIGURE 9.51 **Synthesis using low grade wafers.** Chemically etched grade lower than the solar grade silicon wafer. (A) Luminescent photo of the etched wafer with UV excitation at a wavel4ngth of 365 nm. (B) Luminescent photo of a dispersion of nanoparticles harvested from the etched sample. (C) The luminescence spectrum of the colloid.

the colloid. The film is transparent in room light, but when observed with a fluorescence microscope using UV 365-nm excitation it shows clusters that are red luminescent. Fig. 9.52 presents photos from three different sections of the wafer.

The results are understood in terms of impurity-induced charge confinement to a nano sub surface layer, shielding injected holes from bulk impurities and effectively prolonging their lifetime. Using low-grade metallurgical silicon route allows significant cost reduction, an important step for commercialization.

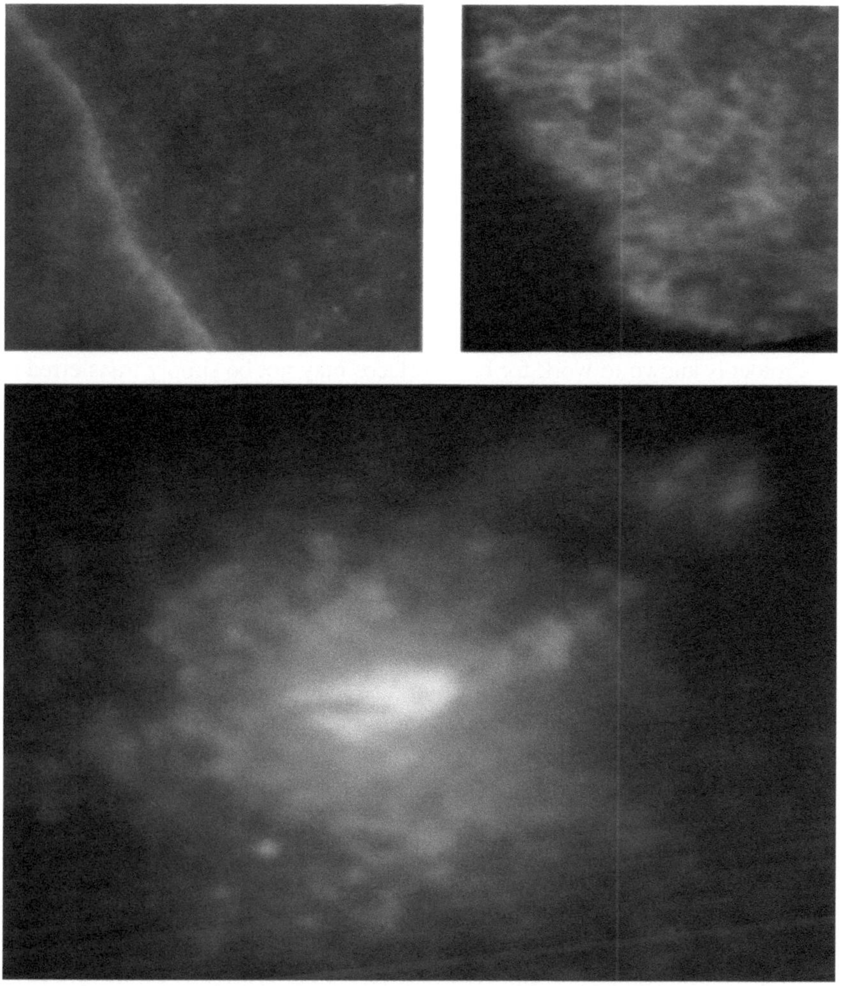

FIGURE 9.52 **Three examples of thin films of nanoparticles on a device-quality wafer prepared using drop-drying under excitation with UV at a wavelength of 365 nm.**

9.14 MAKING SILICON SMART (FUNCTIONALIZATION) AND ITS EFFECT ON THE BANDGAP

Organic groups on the surface of the particles provide several features, including an increase in the resistance towards oxidation, an opportunity to control particle's interaction and hence particle's arrangement on surfaces, the stabilization of the electronic and/or optical properties of the particles, and improved solubility (preventing bulk aggregation). The linkage facilitates the attachment of organic functional groups capable of recognizing specific analytes, such as proteins, DNA, or viruses. The exceptional optical properties, ultrasmall size (approximately small molecules), uniformity of size, low cost and photostability provide the ultimate sensitivity and spatial resolution in imaging capabilities.

As to the initial surface condition, electrochemical-based synthesis of silicon nanoparticles results in nanosurfaces that are to a large extent H-passivated. Because of its unique and simple structure, the hydride passivation opens a unique opportunity for manipulation of the surface of particles [79]. On the other hand, in general, SiO_2 overlayer completely inhibits the chemical activity of silicon. Si—H and Si—Si bonds on the particle's surface can be reactive. Indirectly in ultrasmall species, Si—O bonds may increase the reaction ability of silicon nanosurface because those bonds tend to polarize Si—Si and Si—H bonds, rendering them more amenable to functionalization.

Protocols known to work for bulk surfaces may not be simply transferred to the case of nanosurfaces. Those have several particularities, which require the adjustment of the modification strategies applied to bulk Si surfaces. The reaction ability of the nanosurface can be higher than the surface of the bulk material due to a higher surface curvature and consequently more defected surface composed of stressed Si—Si bonds. For instance, ultrasmall (~1 nm) Si nanoparticles have highly radiative surface reconstructed Si—Si network, found only in ultrasmall nanoparticles, on which excited excitons are self-trapped. Preserving the optical activity is the key to selection of strategies for semiconductor surface modification. The chemistry chosen for the modification must therefore involve only replacement of hydrogen of Si—H groups and not the cleavage of Si—Si bonds.

Silicon nanoparticles have been functionalized with alcohol, butylamine, 1-pentene, carboxyl acid, thiol groups, etc. Protocols were designed and tested by Elena Rogozhina (then at the University of Illinois) in a collaboration between the Paul Braun's group of the University of Illinois the Nayfeh's group. Fig. 9.53 gives an example for illustration [80]. In the process Si—H is reacted with Cl_2 gas in a wet environment, which replaces Si—H termination with Si—Cl termination. The product is then reacted with butyl amine $H_2N–C_4H_9$ which results in $Si–NH–C_4H_9$. The solvent used in the wet reactions is chlorobenzene and HCl is a byproduct. The effect of a variety of surface terminations including this aminization termination on the band gap of ultrabright Si29 nanoparticles was carried out both experimently and using computational models[81].

9.14.1 Aggregation and Solubility [79]

Alkylation of the particles significantly increases their solubility in organic solvents and the stability of colloid dispersions. The ammonized and alkylated Si nanocrystals in organic solvents (heptane, THF, CH_2Cl_2) were found to remain stable for months, indicating an increase in solubility and reduction of bulk aggregation. The colloid dispersions of H-terminated silicon nanoparticles in several organic solvents (Cl–benzene, xylenes, heptane, THF) were not stable and rapidly formed amorphous precipitates. A resonication restored their transparency only for short time. The colloids of the alkylated particles dispersed

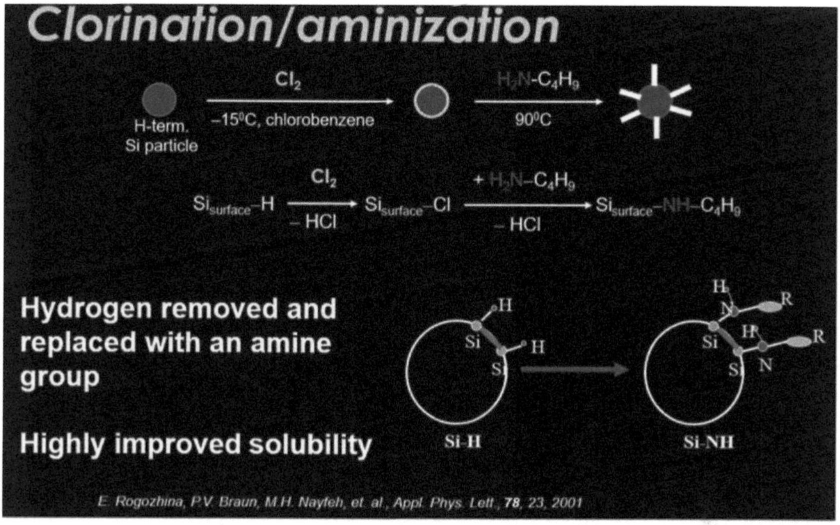

FIGURE 9.53 An example of functionalization of the silicon nanoparticles with butyl amine, producing Si–NH–C_4H_9.

in the same organic solvents were stable for months. In contrast to particles alkylated with butylamine and 1-pentene, the acid-functionalized particles were not dissolvable in heptane. However, they completely dissolved in CH_2Cl_2 and THF and more polar methanol, forming colloids that are stable for months. The higher thermal stability of Me ester of PA compared to t-Bu ester has diminished cross-linking, but not completely suppressing it.

The aggregation of *the particles* was characterized by gel permission chromatography (GPC) in THF at 365 nm. Prior to the modification GPC showed a single monodispersion but there is observed aggregation which may be explained by interconnection of the particles through H-bonding of Si—OH species formed because of partial oxidation of the surface, which results in Si—O—Si covalent bridges between particles in the dried sample. Contrary to the H-terminated particles, the GPC plot of particles functionalized with butylamine (Si—NH—C_4H_9) (see Fig. 9.53) showed a much narrower peak. This narrowing points to the absence of bulk aggregates (even after vacuum drying). The GPC of 1-pentene treated particles showed a weak broad shoulder to a sharp peak. Low intensity of this shoulder is evidence of low possibility of cross-linking for the hydrosilylation of 1-alkenes. The low polydispersity of silicon nanoparticles after modification may have resulted from an increased resistance of alkylated Si surface of nanoparticles to oxidation, which likely is the main reason of aggregation of H-terminated silicon nanoparticle. The GPC of the acid-functionalized particles dispersed in THF showed monodispersion but contains some cross-linked particle's aggregates. The aggregation is believed to be due to a side reaction of the Si particles with the –COOH groups that are liberated due to the partial thermal decomposition of the t-Bu ester groups. Moreover, free radicals on the different particles may recombine, resulting in interconnected particles.

9.14.2 Stability in Acid

Because of the presence of acid groups on the surface, the solubility of the modified particles was sensitive to pH. For example, aggregation was observed upon lowering of the pH of the water colloid to pH < 5, but increasing the pH to the original value restored its transparency. HCl interacts with carboxylated particles in basic water solution if there is residual Si–H (incomplete functionalization) groups. After keeping the aqueous sample with pH $= 2$ for 2 months, its PL was completely quenched. The sample, hydrolyzed under softer acidic conditions (starting pH equal to 4.9 and self-increased up to 7 at the end of the experiment), lost the PL after 5 months.

9.15 SAFETY OF NANOSILICON: SILICIC ACID (MONOTOXIC WASTE PRODUCT)

There are a number of positive attribute of porous silicon in general and silicon nanoparticles with regard to health issues.

- Degradation of those in the human body into monomeric silicic acid ($SiOH_4$). Silicic acid is known to be the most natural form of element in the environment and is readily removed by kidneys.
- Silicic acid is needed for the functioning of muscles in the human body
- Monomeric silicic acid is already found in the human blood plasma (at levels of less than 1 mg Si/l), corresponding to the average dietary intake of 20–50 mg/day.
- Porous silicon was not only biocompatible but could even provide a health benefit. When these silicon nanoparticles dissolve, they become silicic acid, which is a vital substance for strong bones and healthy connective tissues.

Recent research conducted at Lomonosov Moscow State University and the Leibniz Institute of Photonic Technology in Germany [82] used Raman spectroscopy to deduce information about the fate of silicon nanoparticles when they enter a cell and interact with the sub-celluar components and the fluid. The method gives information about material composition, location of the nanoparticles in the cell and tracing their disintegration. Watching disintegration is possible because the Raman spectrum of silicon nanoparticles changes with size reduction. Size reductions causes broadening and down shifting of the spectrum.

Silicon nanoparticles are found to localize on the cells of the cancer cell membranes of breast cancer within few hours then penetrate into the cells during the next few hours. Soon after penetrating a cell they start to biodegrade. This biodegrading is indicated by a decrease in signal amplitude, spectral broadening, and the appearance of the peak of the amorphous silicon phase.

Finally, the experiments showed that by the 13th day after being introduced, the nanoparticles had dissolved completely and the signal had disappeared.

These results have demonstrated that silicon nanoparticles can be applied to the diagnosis and treatment of cancer. Si nanoparticles penetrate into diseased cells and completely dissolve after delivering their cancer treatment drug payload. In contrast metal nanoparticles, for similar functions, present problems and biocompatibility issues. Despite delivering drugs accurately to a group of cancer cells and providing fast action, there are negative side effects such as kidney and liver damage. The damage is directly related to the fact that gold, silver, titanium oxide, cadmium selenide, and plenty of other nanoparticles are almost not excreted.

Other studies made by Fujioka et al. have studied silicon NP cytotoxicity to determine viability of silicon for biological staining. Si QDs were found to be much safer than CdSe QDs with and without UV light exposure. Si QDs appear to be cytotoxic at high concentrations by inducing oxidative stress as determined by MDA measurement, lipid peroxidation, and superoxide detection. The data presented gives credible evidence that Si QDs are non-toxic in the concentration regimes used in other studies. Comparing the IC_{50} to the density of Si, the volumetric ratio of Si to solvent at the IC_{50} is on the order of 1:1000. This ratio is huge compared to common accidental exposure amounts, showing that Si is extremely non-toxic at commonly used concentrations.

Although UV exposure increases the toxicity of CdSe QDs, Si exposed cells appear to be unaffected by radiation. The lack of cytotoxicity during UV exposure with Si QDs suggest that not only does Si not become toxic with UV radiation, the QDs absorb photons that might otherwise cause damage to cell tissue. Si QDs appear to protect cells from dangerous UV radiation while providing fluorescence for tracking. Si QDs with sizes less than 10–20 nm have strong absorption spectra in UV wavelengths, usually from 250 to 375 nm, supporting the idea of QDs as cell protectors from UV radiation.

REFERENCES

[1] http://whitenoise.kinja.com/dimensions-in-semiconductors-can-something-be-zero-dim-1543310309 (fig)

[2] Trwoga PF, Kenyon AJ, Pitt CW. Modeling the contribution of quantum confinement to luminescence from silicon nanoclusters. J. Appl. Phys. 1998;83:3791.

[3] Quantum Mechanics and Electronic Properties of Nanostructures, http://www.acclab.helsinki.fi/~knordlun/nanotiede/nanosc5nc.pdf

[4] A.S. Awos, Spectroscopy and kinetic studies of electron-hole recombination in CdSe nanoparticles: effect of size, shape, and lattice structure, École Polytechnique Fédérale De Lausanne, http://infoscience.epfl.ch/record/100036/files/EPFL_TH3767.pdf.

[5] Canham LT. Silicon quantum wire array fabrication by electrochemical and chemical dissolution of wafers. Appl. Phys. Lett. 1990;57:1046.

[6] Heinrich JL, Curtis CL, Credo GM, Kavanagh KL, Sailor MJ. Luminescent colloidal Si suspensions from porous Si. Science 1992;255:66.

[7] Yamani Z, Thompson H, AbuHassan L, Nayfeh MH. Ideal anodization of silicon. Appl. Phys. Lett. 1997;70:3404.
[8] Yamani Z, Ashhab S, Nayfeh A, Nayfeh MH. Red to green rainbow photoluminescence from unoxidized silicon nanocrystallites. J. Appl. Phys. 1998;83:3929.
[9] Ackakir O, Therrien J, Belomoin G, Barry N, Muller J, Gratton E, Nayfeh M. Detection of luminescent single ultrasmall silicon nanoparticle using fluctuation spectroscopy. Appl. Phys. Lett. 2000;76:1857–9.
[10] Belomoin G, Therrien J, Smith A, Rao S, Chaieb S, Nayfeh MH. Observation of a magic discrete family of ultrabright Si nanoparticles. Appl. Phys. Lett. 2002;80:841.
[11] Nielsen D, Abuhassan L, Alchihabi M, Al-Muhanna A, Host J, Nayfeh MH. Current-less anodization of intrinsic silicon powder grains: formation of fluorescent Si nanoparticles. JAP 2007;101:114302.
[12] Mitas L, Therrien J, Twesten R, Belomoin G, Nayfeh MH. Effect of surface reconstruction on the structural prototypes of ultrasmall ultrabright Si_{29} nanoparticles. Appl. Phys. Lett. 2001;78:1918.
[13] Allan G, Delerue C, Lannoo M. Nature of luminescent surface states of semiconductor nanocrystallites. Phys. Rev. Lett. 1996;76:2961.
[14] Draeger EW, Grossman JC, Williamson AJ, Galli G. Influence of synthesis conditions on the structural and optical properties of passivated silicon nanoclusters. Phys. Rev. Lett. 2003;90:167402.
[15] Sundholm D (2003) First principles calculations of the absorption spectrum of Si29H36. Nano Lett. 3: 847; O. Lehtonen, and D. Sundholm (2005) Density-functional studies of excited states of silicon nanoclusters. Phys Rev B 72: 085424
[16] Smith A, Yamani Z, Turner J, Habbal S, Granick S, Nayfeh MH. Observation of strong direct-like oscillator strength in the photoluminescence of 1nm silicon nanoparticles. PRB 2005;72:205307.
[17] Ball P. Let there be light. Nature 2001;409:974.
[18] http://hyperphysics.phy-astr.gsu.edu/hbase/atmos/blusky.html
[19] McCartney EJ. Raleigh Optics of the Atmosphere: Scattering by Molecules and Particles. Hoboken, NJ: Wiley; 1976.
[20] Mie G. Beiträgezur optiktrüber medien, speziellkolloidaler metallösungen. Ann. Phys. 1908;25:377–452.
[21] Yuan FH, Kuznetsov AI, Miroshnichenko AE, Ye YF, Boris L. Angular distribution: directional visible light scattering by silicon nanoparticles. Nat. Commun. 2013;4:1527.
[22] García-Etxarri A, Gómez-Medina R, Froufe-Pérez LS, López C, Chantada L, Scheffold F, Aizpurua J, Nieto-Vesperinas M, Sáenz1 JJ. Strong magnetic response of submicron silicon particles in the infrared. Opt. Express 2011;19(6):4815.
[23] Fan X, Zheng W, Singh DJ. Light scattering and surface plasmons on small spherical particles 2014;. Open 3 (2014) e179 http://www.nature.com/lsa/journal/v3/n6/full/lsa201460a.html.
[24] Liu W, Zhang J, Lei B, Ma H, Xie W, Hu H. Ultra-directional forward scattering by individual core-shell nanoparticles. Opt. Express 2014;22:16178–87. doi: 10.1364/OE.22.016178.
[25] Mantey K, Zhu A, Boparai J, Nayfeh M, Marsh C, Alchaar G. Observation of linear solid-solid phase transformation in silicon nanoparticles. Phys. Rev. B 2012;85:085417.
[26] Rao S, Mantey K, Therrien J, Smith A, Nayfeh M. Molecular behavior in the vibronic and excitonic properties of hydrogenated silicon nanoparticles. Phys. Rev. B 2007;76:155316.
[27] Nayfeh M, Akcakir O, Belomoin G, Barry N, Therrien J, Gratton E. Second harmonic generation in microcrystallite films of ultrasmall Si nanoparticles. Appl. Phys. Lett. 2000;77:4086.
[28] Sudakshina P. Nonlinear refractive index of silicon nanoparticles: a short review. J. At. Mol. Condens. Nano Phys. 2014;1(2):87–97.

[29] Kanemitsu Y. In: Pavesi L, editor. Luminescence From Si/Si02 Nanostructures, in Towards the First Silicon Laser. Norwell, MA: Kluwer Academic Publishers; 2003. p. 109–22.
[30] Pavesi L, Dal Negro L, Mazzoleni C, Franzò G, Priolo F. Optical gain in silicon nanocrystals. Nature 2000;408:440–4.
[31] Highly nonlinear photoluminescence threshold in porous silicon, M. Nayfeh, O. Akcakir, J. Therrien, Z. Yamani, N. Barry, W. Yu, and E. Gratton, Appl. Phys. Lett.75, 4112-4114 (1999); Nayfeh MH, Rogozhina E, and Mitas L, Silicon nanoparticles: Next generation of ultrasensitive fluorescent markers, in Synthesis, in Functionalization, and Surface Treatment of Nanoparticles, M.-Isabelle Baratron (ed.) American Scientific Publishers, (2002)
[32] Nayfeh MH, Mitas L. Silicon nanoparticles: New photonic and electronic material at the transition between solid and molecule. In: Kumar V, editor. Nanosilicon. Amsterdam, Netherlands: Elsevier; 2007.
[33] Nayfeh MH, Barry N, Therrien J, Akcakir O, Gratton E, Belomoin G. Stimulated blue emission in reconstituted films of ultrasmall silicon nanoparticles. Appl. Phys. Lett. 2001;78:1131.
[34] Nayfeh MH, Chaieb S, Rao S, Barry N, Therrien J, Belomoin G, Smith A. Observation of laser oscillation in aggregates of ultrasmall silicon nanoparticles. Appl. Phys. Lett. 2002;80:121.
[35] Valenta J, Lalic N, Linnros J. Electroluminescence of single silicon nanocrystals. Appl. Phys. Lett. 2004;84:1459.
[36] Cazzanelli M, Navarro-Urriós D, Riboli F, Daldosso N, Pavesi L, Heitmann J, Yi LX, Scholz R, Zacharias M, Gösele U. Optical gain in monodispersed silicon nanocrystals. J. Appl. Phys. 2004;96:3164.
[37] Smith A, Yamani Z, Turner J, Habbal S, Granick S, Nayfeh MH. Observation of strong direct-like oscillator strength in the photoluminescence of 1 nm silicon nanoparticles. PRB 2005;72:205307.
[38] Therrien J, Belomoin G, Nayfeh MH. Light-induced conductance resonance in ultrasmall Si nanoparticles. Appl. Phys. Lett. 2000;77:1668.
[39] Michael Crommie, http://research.physics.berkeley.edu/crommie/research:stm
[40] (a) Puzder P, Williamson A, Reboredo F, Galli G. Structural stability and optical properties of nanomaterials with reconstructed surfaces. Phys. Rev. Lett. 2003;91:157405. (b) Draeger E, Grossman AJ, Williamson A, Galli G. Synthesis dynamics of passivated silicon nanoclusters. Phys. Stat. Sol. B 2003;239(11). Influence of synthesis conditions on the structural and optical properties of passivated silicon nanoclusters, Phys. Rev. Lett. 90 16740.
[41] Mitas L, Therrien J, Belomoin G, Nayfeh MH. Effect of surface reconstruction on the structural prototypes of ultrasmall ultrabright nanoparticles. Appl. Phys. Lett. 2001;78:1918.
[42] Belomoin G, Rogozhina E, Therrien J, Braun PV, Abuhassan L, Nayfeh MH, Wagner L, Mitas L. Effects of surface termination on the band gap of ultrabright nanoparticles: Experiments and computational models. Phys. Rev. B 2002;65:193406.
[43] (a) Wang LW, Zunger A. Dielectric constants of silicon quantum dots. Phys. Rev. Lett. 1994;73:1039. (b) Tsu R, Ioriatti L, Harvey J, Shen H, Lux R. Quantum confinement effects on the dielectric constant of porous silicon. Mater. Res. Soc. Symp. Proc. 1993;283–437. (c) Allan G, Delerue C, Lannoo M, Martin E. Hydrogenic impurity levels, dielectric constant, and Coulomb charging effects in silicon crystallites. Phys. Rev. 1995;. B 52 11982. (d)Penn DR. Wave-number-dependent dielectric function of semiconductors. Phys. Rev. 1962;128. 2093.
[44] Nayfeh M, Rigakis N, Yamani Z. Photoexcitation of Si-Si surface states in nanocrystallites. Phys. Rev. B 1997;56:2079. (MRS, 486 (1998) 243).
[45] Kevin Mantey, Structure, electronic levels, and ionic interactions of 1 nanometer silicon particles phd thesis, UIuc 2011

[46] (a) Puzder P, Williamson A, Reboredo F, Galli G. Structural stability and optical properties of nanomaterials with reconstructed surfaces. Phys. Rev. Lett. 2003;91:157405. (b) Draeger E, Grossman AJ, Williamson A, Galli G. Synthesis dynamics of passivated silicon nanoclusters. Phys. Stat. 2003;11. Sol. B, 239, Influence of synthesis conditions on the structural and optical properties of passivated silicon nanoclusters, Phys. Rev. Lett. 90 167402 (2003).
[47] Joel Therrien, Size dependence of the electrical characteristics of silicon nanoparticles, PhD thesis, UIuc 2002.
[48] (a) Rao S, Sutin J, Clegg R, Gratton E, Habbal S, Nayfeh MH, Tsolakidis A, Martin R. Excited states of tetrahedral single-core nanoparticles. Phys. Rev. B 2004;69:205319. (b) Tsolakidis A, Martin R. Comparison of the optical response of hydrogen-passivated germanium and silicon clusters. Phys. Rev. B 2005;71. 125319.
[49] Belomoin G, Alsalhi M, Al Aql A, Nayfeh MH. X-ray structure factors for Si nanoparticles. J. Appl. Phys. 2004;95:5019.
[50] Akbari B, Pirhadi MT, Zandrahimi M. Particle size characterization of nanoparticles-a practical approach, Iran. J. Mater. Sci. Eng. 2011;8(2):48–56. http://ijmse.iust.ac.ir/files/ site1/user_files_4qu804/ijmse -A-10-3-149- c629680.pdf, http://coecs.ou.edu/ Brian.P.Grady/ saxs,html.
[51] Uhlir A. *Electrolytic shaping of germanium and silicon*. Bell. Syst.Technol. J. 1956;35:333.
[52] Lehmann V. Electrochemistry of Silicon: Instrumentation, Science, Materials and Applications. New York, NY: Wiley-VCH; 2002.
[53] Sailor MJ. Porous Silicon in Practice: Preparation, Characterization and Applications. Weinheim: Wiley-VCH; 2011.
[54] (a) Archer R. Stain films on silicon, Stain films on silicon. J. Phys. Chem. Solids 1960;14:104. (b) Bibcode:1960JPCS.14.104A. 10.1016/0022-3697(60)90215-8. Coffer J. L. 1997, "Porous silicon formation by stain etching", in: L.T. Canham, Properties of Porous Silicon, Institution of Engineering and Technology, London, ISBN 0-85296-932-5, pp. 23–28
[55] Risbud SH, Liu L-C, Shackelford JF. Synthesis and luminescence of silicon remnants formed by truncated glassmelt-particle reaction. Appl. Phys. Lett. 1993;63:1648.
[56] Fujii M, Hayashi S, Yamamoto K. Growth of Ge microcrystals in SiO2 thin film matrices: a Raman and electron microscopic study. Jpn. J. Appl. Phys. 1991;30:687.
[57] Zhu JG, White CW, Budal JD, Withrow SP, Chen Y. Growth of Ge, Si, and SiGe nanocrystals in SiO2 matrices. J. Appl. Phys. 1995;78:4386.
[58] (a) Pavasi L, Del Negro L, Mazzoleni C, Franzo G, Priolo F. Optical gain in silicon nanocrystals. Nature 2000;408:440. (b) Borgensi A, Sasella A, Pivas B, Pavesi L. Characterization of porous silicon inhomogeneities by high spatial resolution infrared spectroscopy. Solid State Commun. 1993;87(1):1–4.
[59] Saunders WA, Sercel PC, Lee RB, Atwater HA, Vahala KJ, Flagan RC, Escorcia-Aparcio EJ. Synthesis of luminescent silicon clusters by spark ablation. Appl. Phys. Lett. 1993;63:1549.
[60] Yoshida T, Takeyama T, Yamada Y, Mutoh K. Nanometer-sized silicon crystallites prepared by excimer laser ablation in constant pressure inert gas. Appl. Phys. Lett. 1996;68:1772.
[61] J. Hecht (contributing editor), Photonic frontiers: ultrafast laser processing: Ultrafast lasers make ultraprecise tools, , Laser FocusWorld, http://www.laserfocusworld.com/articles/print/volume-48/issue-03/features/ultrafast-lasers-make-ultraprecise-tools.html; Preparations of Monodispersed Nanoparticles by Pulsed Laser Ablation, https://staff.aist.go.jp/takeshi.sasaki/NPs_PLD_HTML/NPS.htm
[62] Huisken F. Max Planck Institute Albert Polman, Photonic materials: Teaching silicon new tricks. Nat. Mater. 2002;11:10–2.
[63] Fojtik A, Weller H, Fiechter S, Henglein A. Preparation of colloidal silicon and preliminary photochemical experiments. Chem. Phys. Lett. 1987;134:477.

[64] Fojtik A, Henglein A. Luminescent colloidal silicon particles. Chem. Phys. Lett. 1994;221:363.

[65] Wu JJ, Flagan RC. Onset of runaway nucleation in aerosol reactors. J. Appl. Phys. 1993;73:407.

[66] Takagi H, Ogawa H, Yamazaki Y, Ishizaki A, Nakagiri T. Quantum size effects on photoluminescence in ultrafine Si particles. Appl. Phys. Lett. 1990;56:2379.

[67] Okada R, Iijima S. Oxidation property of silicon small particles. Appl. Phys. Lett. 1991;58:1662.

[68] DiMaria DJ<ETAL>. Formation of 2-D Arrays of Silicon Nanocrystals in Thin SiO2 Films by Very-Low Energy Si+ Ion Implantation. J. Appl. Phys. 1998;56:401.

[69] Berger S, Schächter L, Tamir S. Photoluminescence as a surface-effect in nanostructures. Nanostruct. Mater. 1997;8:231.

[70] Heinrich JL, Curtis CL, Credo GM, Kavanagh KL, Saylor MJ. Luminescent colloidal silicon suspensions from porous silicon. Science 1992;255:66.

[71] Therrien JM, Belomoin GA, Nayfeh MH. Synthesis of ultra-small Si nano particle colloids and thin films-high temperature single electronics. Mater. Res. Soc. Symp. Proc 1999;582. H11.4 . https://doi.org/10.1557/PROC-582-H11.4.

[72] Belomoin G, Therrien J, Nayfeh MH. Oxide and hydrogen capped ultrasmall blue luminescent silicon nanoparticles. Appl. Phys. Lett 2000;77:778.

[73] Smith A, Rao S, Twesten R, Chaieb S, Nayfeh MH, Wagner L, Mitas L. Mat. Res. Soc. Symp. Proc. 703 2001;11.4:703.

[74] Li X, Bohn PW. Metal-assisted chemical etching in produces porous silicon. Appl. Phys. Lett. 2000;77:2572–4.

[75] Chartier C, Bastide S. Metal-assisted chemical etching of silicon in hf-h2o2. Electrochim. Acta 2008;53:5509. L_evy-Cl_ement.

[76] Auric acid assisted silicon nanoparticle formation method, Munir Nayfeh, United States Patent, 9,096,432.

[77] Development of Photovoltaics and Energy-Efficient Technologies in Kazakhstan, Serekbol Tokmoldin, Kazakhstan Energy Forum November 1, 2011, Almaty Kazakhstan http://documents. mx/documents/development-of-photovoltaics-and-energy-efficient-technologies-in-kazakhstan-serekbol-tokmoldin-institute-of-physics-and-technology-kazakhstan-national. html

[78] Evaluation of solar grade silicon produced by the Institute of Physics and Technology, cooperative research and development final report, CRADA Number: CRD-07-211, NREL Technical Contact: Matthew Page, CRADA Report, NREL/TP-7A10-53847 (February 2013) Contract No. DE-AC36-08GO28308.

[79] Nayfeh MH, Rogozhina E, Mitas L. Silicon nanoparticles: Next generation of ultrasensitive fluorescent markers. In: Isabelle Baratron M, editor. Synthesis in Functionalization Surface Treatment of Nanoparticles. Valencia, CA: American Scientific Publishers; 2002.

[80] Rogozhina E, Belomoin G, Smith A, Abuhassan L, Barry N, Akcakir O, Braun PV, Nayfeh MH. Nitrogen Si-N linkage in ultrabright, ultrasmall Si nanoparticles. Appl. Phys. Lett. 2001;78:3711.

[81] Belomoin G, Rogozhina E, Therrien J, Braun PV, Abuhassan L, Nayfeh MH, Wagner L, Mitas L. Effect of surface termination on the band gap of ultrabright Si29 nanoparticles: experiments and computational models. Phys. Rev. B 2002;65:193406.

[82] Lubov Osminkina, Researchers show porous silicon nanoparticles could be harmless to diagnose and treat cancer, July, 2016, https://www.news-medical.net/news/20160721/Researchers-show-porous-silicon-nanoparticles-could-be-harmless-to-diagnose-and-treat-cancer.aspx

Chapter 10

Nanocarbon

10.1 BULK DIAMOND AND GRAPHITE

Carbon is the fourth most abundant element in the universe by mass after hydrogen, helium, and oxygen. On Earth, carbon is the 15th most abundant element, because unlike silicon, which ranks second, it is mostly abundant in the biosphere of the surface. Being in silicates (compound of silicon and oxygen), Oxygen ranks number 1. Solid carbon exists as diamond and graphite. Graphite is one of the softest known substances, while diamond is the hardest naturally occurring substance [1]. Fig. 10.1 gives sketches of the diamond and graphite structures.

The crystalline structure of diamond is tetrahedral, having sp3 hybridization with each carbon atom linked through four covalences by other four carbon atoms. The C–C bonds in diamond, are very strong, which makes it the hardest known mineral. It is a wide-bandgap material, having clear, *almost* crystal lattice. The structure of the crystal has a cubic elementary cell, looking like a "Giant molecule."

Graphite, on the other hand is made up of many carbon ring layers (graphene – see later) stacked together, with the elementary cell being of hexagonal type. Unlike diamond, graphite has zero bandgap, which conducts heat and electricity very well. Moreover, unlike diamond, graphite is opaque gray, soft, and slippery; but it is the most stable carbon structure. The flat layers making graphite are made up of carbon atoms that are arranged in hexagons, as combs. The parallel planes are linked by Van der Waals type weak bonds. Carbon atoms are Sp2 hybridized. Those three links made with the three hybrid orbitals are strong. But the third 2p orbital of each atom of the plane remains non-hybridized and contains a single electron. All electrons of non-hybridized orbitals of atoms in a plane are common and mobile, and they belong to all the atoms in the plane. The delocalization of those electrons affords graphite good stability.

Diamond is not used as a semiconductor because it a wideband gap, having a high bandgap of 5.5 eV. In comparison, silicon has a much lower band gap of 1.11 eV. At room temperature, which provides thermal agitation of 0.03 eV, the thermal release across the bandgap from carbon is insignificant compared to that in silicon. As a result at room temperature this makes silicon a semiconductor but diamond an insulator. Because of the high band gap of diamond, it would only be useful as a semiconductor at elevated temperatures >300–400°C.

Fundamentals and Applications of Nano Silicon in Plasmonics and Fullerines.
http://dx.doi.org/10.1016/B978-0-323-48057-4.00010-4

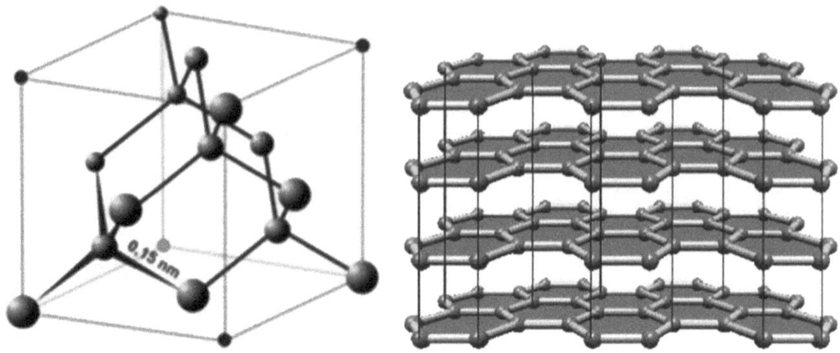

FIGURE 10.1 **Sketches of the diamond and graphite structures.**

10.2 FULLERENES (NANOCOMPONENTS)

In addition to diamond and graphite, ultrasmall pure carbon structures have been discovered. Several of the structures are presented in Fig. 10.2 [1]. One of those is spherical particles that consist of 60 carbon atoms (60-carbon spheres). They were called buckminsterfullerene (or *Bucky balls* for short), after Buckminster Fuller, an architect who created geodesic dome structures. Other shapes have also been discovered, including ellipsoids, and tubes. Another structure which was recently discovered is graphene, an open structure. It is a one atom thick layer of hexagonal carbon rings. The 2010 Nobel Prize in Physics was awarded to Andre Gem and Konstantin Novoselov for their discovery of this two-dimensional structure.

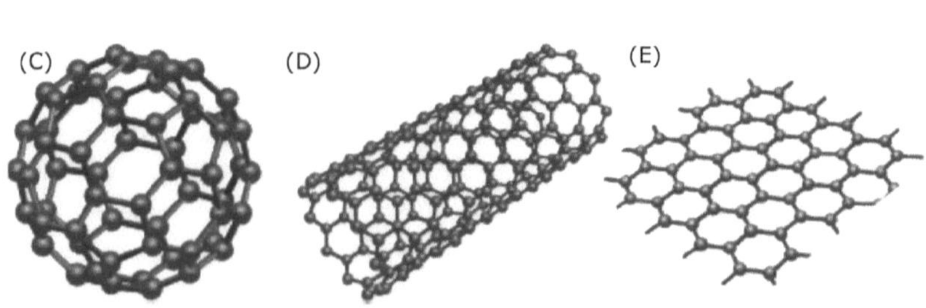

FIGURE 10.2 **Carbon structures.** (A) Graphite, (B) diamond (C) spherical particles that consist of 60 carbon atoms (60-carbon spheres). (D) Carbon nanotube and (E) graphene.

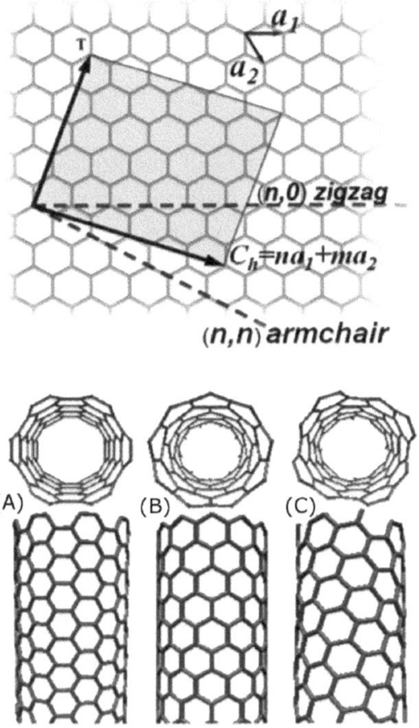

FIGURE 10.3 Three types of CNTs. (A) zigzag, (B) armchair, and (C) chiral.

It is interesting to note that graphene sheets can be seen by the naked eye despite the fact they are only one-atom thick. This is attributed to the fact that it absorbs 2.3% of white light. Carbon *nanotube* (CNT) is another form of carbon; it is a semi open structure in the form of a single layer of graphene which is rolled into a seamless hollow tube. The simplest of these structures is the single-walled nanotubes (SWNTs) but there are multiwall CNTs. SWNTs have a diameter close to 1 nm, and can be as long as few millimeters.

There are several ways for wrapping graphene sheets into SWNTs. The rolling is represented by a pair of indices (n,m). The integer n and m denote the number of unit vectors along two directions in the honeycomb crystal lattice of graphene. There are three cases that are shown in Fig. 10.3. The nanotubes are called (A) *zigzag* nanotubes if the two integers are equal ($n = m$). *If* one of the integers is zero ($m = 0$ for example), they are called (B) *armchair* nanotubes. Otherwise, they are called (C) *chiral*. In terms of the integer indices and $a = 0.246$ nm, the diameter of an ideal nanotube can be written as follows

$$d=\frac{a}{\pi}\sqrt{\left(n^{1}+nm+m^{2}\right)}=78.3\sqrt{\left((n+m)^{2}\;nm\right)}\,\text{pm}$$

The band gap of SWNTs can vary from zero to about 2 eV and their electrical conductivity can show metallic or semiconducting behavior, depending on the values of the indices or on the diameter and chiral angle. From a practical point of view this feature is problematic as it not straightforward to separate CNTs based on their bandgap.

Nanotubes come also as multi-walled nanotubes (MWNT). Those consist of multiple rolled layers (concentric tubes) of graphite. Sheets of graphite are arranged in concentric cylinders. Also a single sheet of graphite is rolled in around itself, resembling a scroll of parchment or a rolled newspaper. By now, diameters of SWNTs and MWNTs fall typically in the range 0.8–2 nm and 5 –20 nm, respectively. However, MWNT diameters can exceed 100 nm. As to lengths, CNT lengths range from less than 100 nm–0.5 m.

B. Fuller, R. Curl, H. Kroto, R. Smalley, and S. Iijima

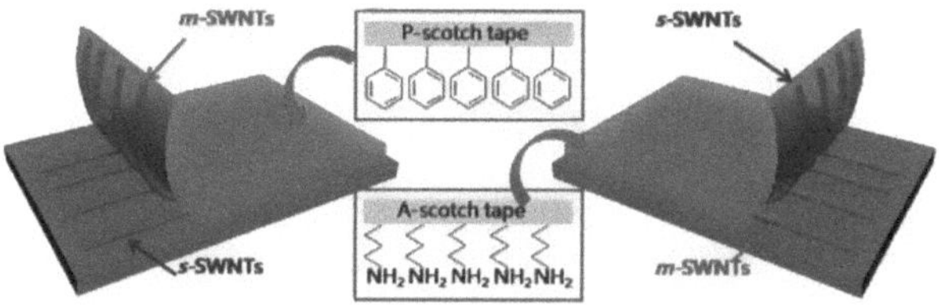

FIGURE 10.4 **Schematics of a nondestructive separation of the different types carbon nanotypes based on adhesion forces.** P- and A-scotch tapes are used to selectively remove m- and s- SWCNTs respectively, and leave their counterparts on the sapphire substrates.

10.3 SORTING METALLIC AND SEMICONDUCTING SINGLE-WALLED CARBON NANOTUBES

Current production methods for CNTs result in units with different diameter, length, chirality and electronic properties, all packed together in bundles, and often blended with some amount of amorphous carbon. Often, these mixtures are of little practical use since many advanced applications, especially for nanoelectronics, are sensitively dependent on the structures. Consequently, the separation of metallic (m) and semiconducting (s) CNTs is essential for future applications and studies. Separation of nanotubes according to desired properties is still proving to be a challenging task, especially single-walled carbon nanotube (SWCNT) sorting. The composition of SWCNTs of different types is very similar, but the chemical properties are not. Conventional separation techniques are efficient in separation of CNTs, but there are many drawbacks. Fig. 10.4 shows the schematics of a nondestructive separation based on adhesion forces [2]. PDMS thin-films are used as supporting material. Aminopropyltriethoxysilane (APTES – the "A-scotch tape") and triethoxyphenylsilane (PTEOS – the "P-scotch tape") are used as bonding material. Those are used to attach amine and phenyl functional groups, respectively. This is followed by subjecting the structures to air plasma. The Si–O–Si bonds on PDMS surface were oxidized to Si–OH groups, which could further react with APTES/PTEOS to form a strong interaction between the supporting PDMS layer and functionalized bonding layer (see Fig. 10.4).

The A-scotch tape selectively removed s-SWCNTs, while the P-scotch tape adhered to metallic ones, leaving their counterparts on the substrate. The tape-based procedure eliminate wet environment and is damage-free but the throughput and cost are limited for industrial use. Designing adhesive tape-based procedure to be able to separate single-chirality SWCNTs remains.

10.4 DEFECTS IN CARBON NANOSTRUCTURES

The majority of electronic applications require structurally perfect CNTs to maximize the electron mobility. Thus the quality of nanotubes is an issue. Quality is expected to impact performance and would have a great influence on the physical properties.

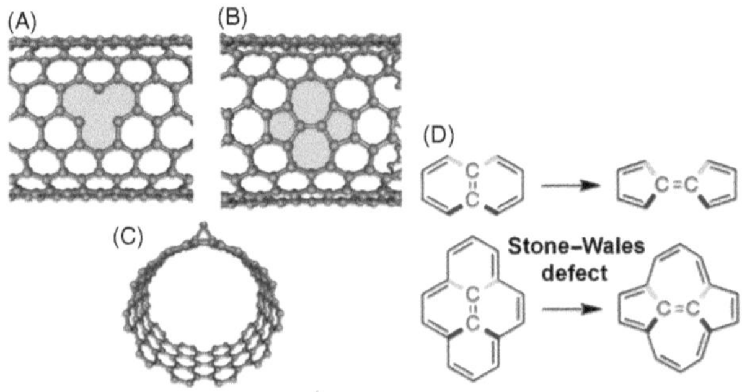

FIGURE 10.5 Schematic of typical defects, such as (A) vacancy, (B) Stone–Wales defect (bond rotation), and (C) adatom in SWCNTs. (D) the Stone–Wales defect showing the resulting rotation of the C=C bond by 90 degrees.

10.4.1 Types of Defects

In real SWCNTs, various types of defects are found [3–5]. Vacancies, Stone–Wales defects, adatoms, or H–C complex are among the defects. Such defects can be introduced at the stage of SWCNT growth or later on during device or composite production. They can be also created deliberately by chemical treatment or by irradiation with electron, ion, or laser light. Understanding the properties of such defects in SWCNTs is important for improving SWCNT growth methods, tailoring their physical properties, and controlling the irradiation-induced damages.

Despite the fact that CNTs are rarely perfect, they still show some defects. The strong sp2 bonding network of graphitic carbons is expected to energetically preclude defect formation, making atomic vacancies quite unfavorable. Still many topological perturbations are possible as they are governed by their Boltzman factor exp $(-Ed/kbT)$. But line and screw defects, interstitials, and bond rotation defects are all observed in graphites in considerable numbers. Fig. 10.5 shows a schematic of typical defects, such as (a) vacancy, (b) Stone–Wales defect (bond rotation), and (c) adatom in SWCNTs. Fig. 10.5 shows more details about the S, showing the resulting rotation of the C = C bond by 90 degrees.

10.4.2 Synthesis-Induced Defects

It is to be noted that SWCNTs are actually rugged against formation of defects during growth despite the fact that synthesis processes involve very high temperatures. This is due to the fact that the C = C bond in these systems is very strong. For example, a defect involving a single graphene vacancy breaks three short, strong C = C bonds as can be seen in Fig. 10.5A and costs nearly 7.8 eV. Even though CNT synthesis is a high-temperature process (up to 3000°C), the likelihood of incorporating such sites during crystal growth remains extremely low. Furthermore, the carbon-rich synthetic environment excludes the likelihood of vacancy survival, beyond mere thermodynamic considerations.

Electron irradiation and strain might induce structural transformations in CNTs; these consist mainly of vacancies and interstitials which can change the helicity of the nanotube and modify its electronic structure. The structural defects in CNTs are mainly atomic vacancies and interstitials. A vacancy is a deviation from the perfect hexagonal lattice arrangement due to the absence of one or more carbon atoms. Other defects like pentagons, heptagons, Stone–Wales reconstruction, adatoms, or substitutional impurities have been proposed theoretically and observed experimentally. CNT walls may contain other types of defects, such as chemical impurities. The chemical impurities are generally by-products, such as amorphous carbon or fullerenes, as well as the remains of the primary materials used in the synthesis process, such as pieces of graphite from arc-discharge or catalyst metal nanoparticles from CVD. Hence, purification is part of the synthesis process; the most problematic step is to get rid of the undesired by-products without damaging the CNT structure. While vacancies and interstitials are highly disfavored, bond rotations are not, and these constitute the most prevalent type of defect in high quality graphites. A single bond rotation can be incorporated into graphene at a cost of approximately 3.5 eV without disturbing the sheet's topology or sp2 conjugation. The rotation only affects four adjacent hexagons, converting two into pentagons and two into seven-sided heptagons.

10.4.3 Postsynthesis Defects and Reconstruction

Irradiation with high-energy electron, ion, or neutron radiation can dislodge or fully remove carbon atoms [5]. In addition to irradiation, strain might induce similar structural transformations in CNTs. Fig. 10.6A shows cartoons of CNTs with two (2)-atom vacancy, five (5)-atom vacancy, and eight (8)-atom vacancy. This type of a single vacancy initially results in three dangling (broken) bonds that will immediately rehybridize or react with surrounding molecules. Small vacancy defects in the graphene system. Fig. 10.6B illustrates how reconstruction and saturation with hydrogen occurs after vacancies (removal of carbon atoms) are induced; gives mono-vacancy (a) before and (b) after reconstruction and H-termination of the remaining dangling bond; di-vacancy (c) before and (d) after reconstruction. One rearrangement amenable to theoretical study is the so-called 5-1DB vacancy defect, in which two of the dangling bonds bridge to form a strained, pentagonal 5-membered ring, leaving only a single dangling bond "1DB" (Ajayan et al.), confirmed by Lu and Pan [5]. The resulting structure costs only 5 – 6 eV, though substantial disagreement continues over whether this rearrangement occurs spontaneously and whether it is even the lowest energy configuration. Thus vacancies and interstitials can change the helicity of the nanotube and modify its electronic structure. A vacancy is a deviation from the perfect hexagonal lattice arrangement due to the absence

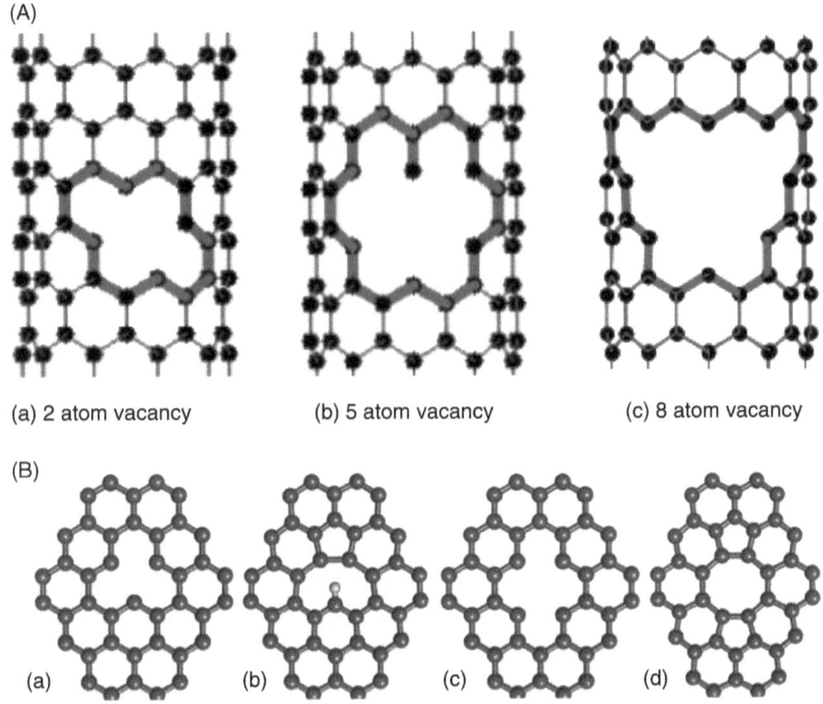

FIGURE 10.6 (A) A cartoons of carbon nanotubes with two (2)-atom vacancy, five (5)-atom vacancy, and eight (8)-atom vacancy. (B) reconstruction and saturation with hydrogen after vacancies (removal of carbon atoms) are induced. Gives Mono-vacancy (a) before and (b) after reconstruction and H-termination of the remaining dangling bond. Di-vacancy (c) before and (d) after reconstruction.

of one or more carbon atoms. Other defects like pentagons, heptagons, Stone–Wales reconstruction, adatoms, or substitutional impurities have been proposed theoretically and observed experimentally.

10.4.4 Confinement Effects

Two special differences between graphites and CNTs arise. First, the allowed categories of defects are restricted by dimensionality. SWNTs obviously cannot contain higher dimensional defects like line and screw dislocations. Isolated SWNTs also cannot support many of the common interstitial defects, which are stabilized by intersheet bonding in the graphites. On the other hand, the point defects found in graphites are more complicated in CNTs because of circumferential, curvature-induced strain. As carbon bonds exist at various angles to this strain, defect stabilities depend precisely on position within the lattice, as well as CNT diameter and helicity.

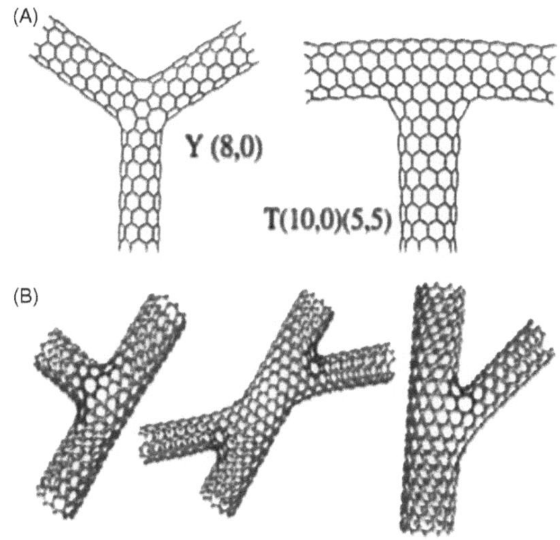

FIGURE 10.7 Welding of CNT after irradiation with electrons that remove atoms. (A) Y(n,m) = Y(8,0) is a semiconductor nanotube connected by six pentagons and six octagons. T(10,0)(5,5) connected by metal nanotube (5,5) and one semiconductor (10,0) nanotube. (B) Variations to the junctions (denoted as T, X, and Y).

10.4.5 Uses of Defects

Defects can be useful for other applications. For example they are good anchors for foreign species such as single atoms, organic or inorganic molecules. The influence of defects induced by electron and ion irradiation in pristine CNTs has been studied experimentally and theoretically.

Dangling bonds associated with one or more vacancies (missing carbon atoms) can provide chemical sensitivity to serve as sites for interconnecting nanotubes. Moreover, such vacancies may be used for incorporating dopants. In the process SWCNTs are irradiated with electrons to remove atoms. This induces local covalent welding between tubes of similar diameters. Examples are shown in Fig. 10.7A–B. The one designated by Y(n,m) = Y(8,0) is a semiconductor nano tube and it is connected by six pentagons and six octagons. On the other hand, T(10,0)(5,5) is connected by metal nanotube (5,5) and one semiconductor (10,0) nanotube. At the T-junction there are four heptagonal rings. Fig. 10.6B gives variations to the junctions (denoted as T, X, and Y).

10.5 PURIFICATION/CLEAN-UP OF CNTS

As discussed above, as-prepared single-walled carbon-nanotubes contain a large quantity of impurities, such as amorphous carbon and residual catalyst particles, usually encased in graphitic carbon shells or within the tubes themselves, catalytic particles and encapsulating graphitic shells. There are many purification methods and can be broadly categorized as follows:

- *Physical separation approaches*
  Those employ selective adsorption, often in combination with size-exclusion chromatography and/or density gradient ultracentrifugation. Those approaches are reasonably useful for electronic/chirality/length separation. But they have low yield and low concentration of SWCNT dispersions.
- *Oxidative*
  Oxidation processes may rely on a number of oxidative agents, including acids, hydrogen peroxides, oxygen, air, water vapor, etc. Acidic purifications using HCL, HNO_3, $KMnO_4$, or H_2SO_4/HNO_3 dissolve exposed catalyst particles found within graphitic shells or the SWCNT interiors. However, the process may disruption the carbon framework, which reduces their mechanical and electronic properties, as well as introduces acidic debris and carboxylated carbonaceous impurities. Hydrogen peroxide oxidant is less damaging than oxidizing acids. Gaseous purifications preferentially destroying amorphous carbons and high curvature sp^2 carbons through competitive combustion. However, the SWCNT structure is damaged when a strong gaseous oxidant (e.g., oxygen) is used, however milder oxidants, notably water vapor, can be employed to minimize SWCNT damage. With all gaseous purification approaches, the temperature and duration of heating can be varied to influence the balance between SWCNT damage and extent of purification.
- *Electrochemical*
  Electrochemical charging of SWCNTs and their impurities in aprotic, polar solvents, notably amides has been used. Reductions in liquid ammonia, or with sodium naphthalide (NaNp) in *N,N*-dimethylacetamide (DMAc). If the charge is quenched carefully, the SWCNTs can be left undamaged by this process.
- *Annealing*
  Annealing of SWCNTs at high temperature in inert atmosphere or in vacuo can be performed after purification to remove sp^3 defects/functional groups introduced during purification to recover the SWCNT properties. However, routes which involve severe oxidation (or sonication) will introduce vacancy defects and/or become shortened irreparably; in this case, annealing to remove surface functionality will only retrieve a fraction of the pristine properties, analogously to the reduction of graphene oxide.

The choice of purification strategy is primarily determined by two factors: the end use of the nanotubes and the SWCNT type. When selecting a purification route for a particular application, it is important to balance the efficiency of removing metal catalyst and/or non-SWCNT carbons against the damage accumulated in the SWCNTs. In addition, the dimensions, crystallinity, chirality, and surface chemistry must be taken into account since they can affect the effectiveness of certain purifications, for example by influencing dissolution kinetics and resistance to oxidation.

10.6 STRIDES, BREAKTHROUGH, AND CHALLENGES TOWARD APPLICATIONS (ELECTRONIC AND MECHANICAL)

Since their discovery in 1993, over the next quarter century SWCNTs has seen continues progress as well as many strides. We mention the strides achieved in the fields of electronics, photonics, integration in CMOS, physical properties and a variety [7].

10.6.1 Electronics

several concepts has been demonstrated including electron emission source, single-electron transistors (operating at low temperature), optical antennas, field-effect transistors, fast oscillators (>50 GHz), 9 nm CNT transistors that outperforms silicon. Y-shaped nanotubes ready-made transistors and CNT computer.

10.6.2 Photonics

A prototype high-definition 10-cm flat screen, *diode, a photovoltaic effect* was also observed in the nanotube diode device that could lead to breakthroughs in *solar* cells, built an electronic circuit around a CNT. Nanotubes incorporated in virus battery.

10.6.3 Integration in CMOS

A single-walled CNT was grown by chemical vapor deposition (CVD) across a 10-micron gap in a silicon chip.

10.6.4 Physical (Mechanical/Thermal/Bio)

Nanotube-enhanced bike, scaffold for damaged nerve regeneration, wet-spun CNT fibers with enhanced tensile strength and electrical and thermal conductivity, demonstration proving that bending CNTs changes their resistance

10.6.5 Platform Developments

Development of a technique for separating semiconducting and metallic nanotubes, nanotube sheet synthesized with dimensions 5 × 100 cm, precision placement of nanotube, sorting nanotubes by size and electrical properties, developing a new wet-spun nanotech fiber. The new fiber is made with an industrial scalable process, 4 cm long SWNT. 3D all-carbon scaffolds/architectures for next generation of energy storage, supercapacitors, field emission transistors, high-performance catalysis, photovoltaics, and biomedical devices and implants, CNT forests, yarns and regular sheets were produced in much smaller volume, tether for a purported space elevator.

10.6.6 Challenges

As of 2013, CNT production exceeded several thousand tons per year, used for application in energy storage, automotive parts, boat hulls, sporting goods, water filters, thin-film electronics, coatings, actuators and electromagnetic shields. There are key technical challenges and research needs to enable the successful integration of CNTs into bulk materials or composites for commercial applications [8]. One major challenge is the poor yield, slow production rate, variations in material quality, and lack of real time process control.

10.6.7 CNT Entanglement and CNT-Paper

One challenge for the use of CNTs is the fact that they get easily entangled [9]. We have seen above that it is unlikely that defects develop in single-walled CNTs during growth because C = C bonds are very strong. Their tendency to exist in the form of bundles due to their very strong van der Waals interactions. Infact it is found that when they touch, CNTs exhibit very high-cohesive energies (on the order of 10^5 eV/μm). This causes nanotubes to assemble into bundles. Multi-walled CNT tend to form highly entangled meshes that are mechanically interwoven.

Pristine CNTs possess a small negative electrostatic potential. This potential is sufficient to prevent mesoscale aggregation in polymer chains that possess a permanent dipole moment. But it is insufficient for common solvents. Chemical oxidation increases the electrostatic potential of the CNTs, hence it reduces aggregate size. But care must be taken because oxidation processes may significantly damage the mechanical and electrical performance. Physical or chemical modification of either the CNTs or the polymer matrix is necessary to achieve a stable dispersion in non-polar polymers (see Section 10.5).

The entanglement property may be put to use [10]. The so called Bucky paper for example consists of entangled CNTs. It comes as a thin sheet made from an aggregate of CNTs or CNT grid paper. Fig. 10.8A shows an SEM image of Bucky paper, showing entangled CNTs. Fig. 10.8B shows the appurtenance of the paper to the naked eye. The sheets show potential applications as vehicle armor, personal armor, and next-generation electronics and displays.

10.7 TREATMENT AND FUNCTIONALIZATION

Most of the applications require the functionalization of CNTs, such as changing the surface properties to make nanotubes soluble in different media, or attaching functional groups or polymer chains for specific utilizations of modified nanotubes.

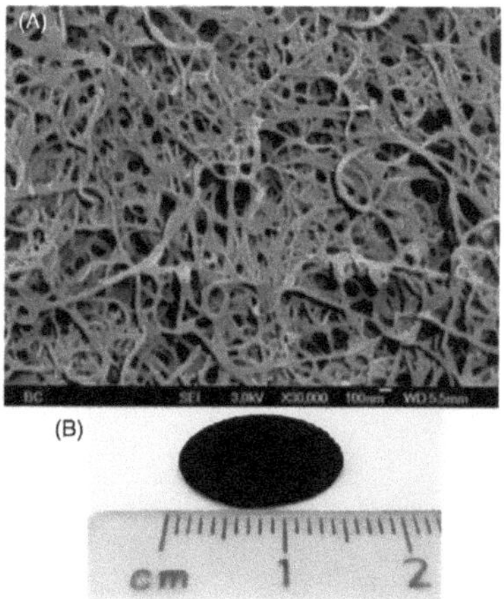

FIGURE 10.8 (A) SEM image of Bucky paper, showing entangled carbon nanotubes. (B) Appearance of the paper to the naked eye.

There are two types of functionalization: endohedral and exohedral functionalization [11]. In endohedral functionalization, CNTs are treated by filling their inner empty cavity with different molecules or nano particles. Schematic representation of a SWNT filled with C60 fullerenes. Exohedral functionalization involves grafting of molecules on the outer surface of nanotubes. For this purpose, several approaches have been developed and include defect functionalization covalent functionalization and noncovalent functionalization with surfactants or polymers. Shows a variety of sites (A, B, C, and D) on CNT (open end, bends, defects, rings) of the SWNT terminated with –COOH groups. Other terminal groups, such as –OH, –H and =O, are also possible. The noncovalent interaction is based on van der Waals forces. Two major groups of chemical functionalization of CNTs via covalent attachment can be distinguished, the end and "defect-group" chemistry and the sidewall functionalization.

In unintended interactions, CNT surfaces interact with adsorbed gases, moisture, supporting substrates, and nearby amorphous carbons, all of which provide spontaneous reaction pathways to saturate dangling bonds. In all but ultrahigh vacuum conditions, most intermediates are susceptible to nucleophilic attack by H_2O, making –OH terminated vacancies one of the most likely, and physically relevant, configurations.

10.8 GRAPHENE, PENCILS, AND CHICKEN WIRES

Graphite is the soft dark material used in pencils. It consists of layers of graphene as shown in Fig. 10.9. Thus graphene is a basic material found in nature. When graphite is taken apart its sheets become graphene.

The shape and structure of graphene are revealed using electron microscopy. The microscope image of graphene resembles the chicken wire on a farm (Fig. 10.10). The structure consists of rings of six sides or hexagons in the shape

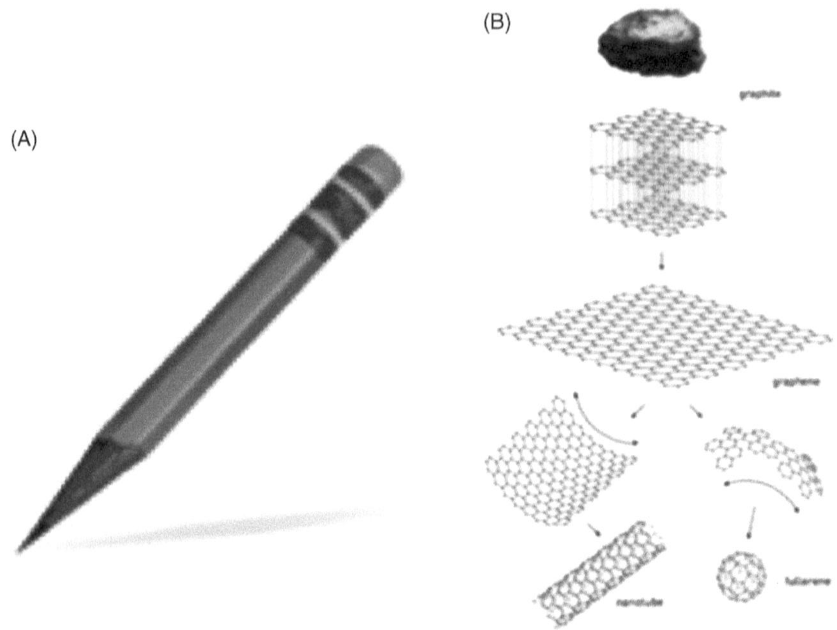

FIGURE 10.9 (A) Graphite pencil. (B) chunk of graphite, a soft dark material used in pencils, consists of layers of graphene.

honeycomb. On each corner of a ring there is a single carbon atom. Thus graphene is most easily visualized as an atomic-scale chicken wire made of carbon atoms and their bonds.

Graphene is the thinnest material in nature because it is just one-atom thick. Fig. 10.11 shows that the carbon–carbon bond length is about 0.142 nm (1.42 Å). It is to be noted that a rolled up layer of graphene forms a CNT shown in Fig. 10.2. On the other hand if folded up, it becomes a buck ball fullerene. It is to be noted that graphene is found in charcoal. Thus it is the basic structural element of all carbon allotropes including graphite, charcoal, CNTs and fullerenes, and may be considered as an indefinitely large aromatic molecule.

FIGURE 10.10 Shape and structure of graphene (A) are revealed using electron microscopy. The microscope resembling the chicken wire on a farm (B).

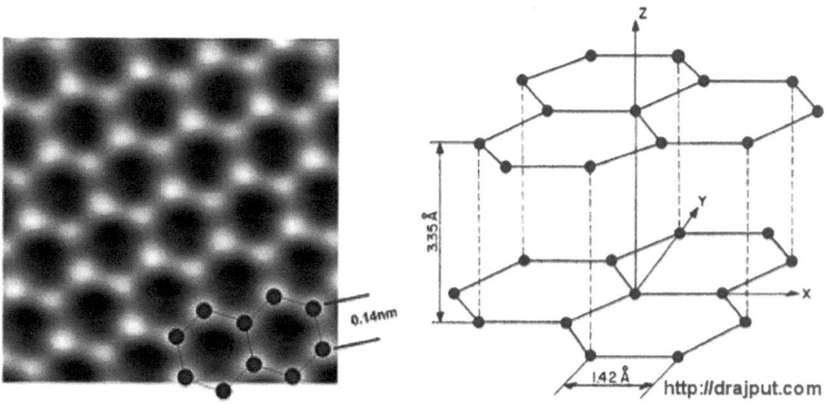

FIGURE 10.11 **Carbon–carbon bond length in graphene (left) (about 0.142 nm (1.42 Å)).**

Number of Graphene Sheets

Graphene sheets stack to form graphite. It is the thinnest material in nature because it is just one-atom thick. The distance between a sheet and another is only of 3.35 Å. This means that a stack of 3 million sheets would be only one millimeter thick.

Who named it and why? Hans-Peter Boehm gave it its name in 1962 a long time ago. He used half of the word graphite (graph) and added the suffix (ene) to it to indicate miniaturized and came up with graphene.

10.9 ELECTRONIC IMPACT

10.9.1 Synthesis of Graphene

10.9.1.1 Isolation of Graphene

As mentioned earlier graphene is already hound in nature as the building block of graphite. One crude way to isolate grapheme is to take graphite apart into its sheets using scotch tape. This method has been known early as the 1980s. Very thin layers, which consist of many graphene sheets can be easily removed from a graphite sample. Such samples were used to demonstrate the newly developed atomic resolution of scanning tunneling microscope (STM) imaging. The microscope can easily resolve individual carbon atoms and the honeycomb structure in the graphite sample or flakes as shown in Fig. 10.12.

In 2004, physicists Geim and Novoselov isolated single graphene planes. They extracted the graphene from a piece of graphite, such as is found in ordinary pencils using regular adhesive tape. They managed to obtain a flake of carbon with a thickness of just one atom for the first time. More importantly, they measured electronic properties of the obtained flakes and showed that the material has unique properties. This was important because many scientists believed it was impossible for such thin crystalline materials to be stable.

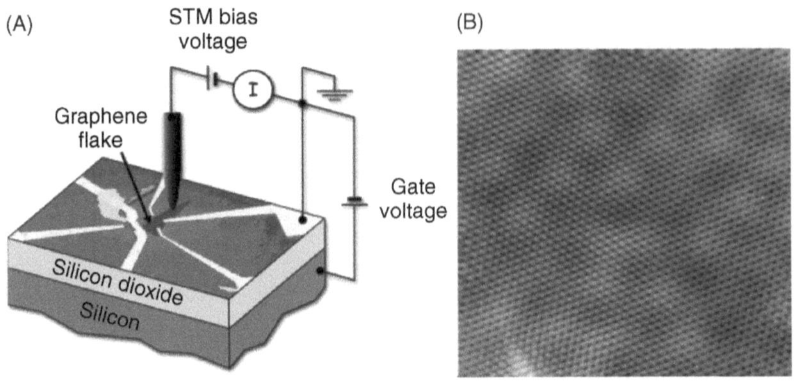

FIGURE 10.12 (A) STM imaging of graphene flakes. (B) Image of graphene individual carbon atoms and the honeycomb structure are resolved.

Andre Geim (left) and Dr. Konstantin Novoselov, two Russian expatriates working in Britain, were awarded the Nobel Prize for Physics for "groundbreaking experiments" with the thinnest, strongest material known to mankind – a carbon vital for the creation of faster computers and transparent touch screens.

10.9.1.2 Chemical Preparation of Large Graphene Sheets

Taking apart graphite by peeling into grapheme with scotch tape is not easy, clean or reproducible. Often one gets "flakes" consisting of many graphene sheets stacked together. A procedure for preparing graphene sheets [12] as large as 1 cm by 1cm has been developed as shown in Fig. 10.13. One starts with a thin copper or nickel foil of 1 cm by 1 cm. The foil is placed inside a box which is filled by methane gas. The copper or nickel foil is heated to 1000 degrees. When a methane gas molecule collides with the foil, heat breaks it up into hydrogen atoms and carbon atoms. The carbon atoms deposit on the surface while hydrogen atoms fly into the gas. The treatment is run for an hour to deposit enough carbon atoms. Because the surface is very hot, carbon atoms diffuse readily on the foil and when there are enough carbon atoms, they connect to each other to form graphene. Covering the whole area of the foil is a matter

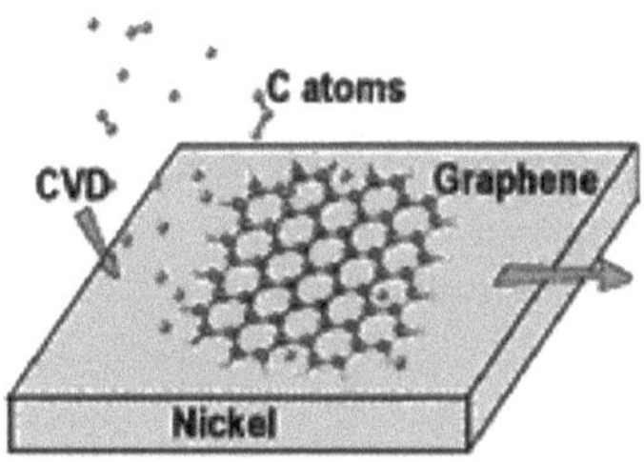

FIGURE 10.13 A chemical-based procedure for preparing graphene sheets as large as 1 cm by 1 cm.

of time, which results in a grapheme layer of 1 cm by 1 cm. The process is self-limiting since excess carbon atoms cannot diffuse freely over the first one and as such will not form a second grapheme sheet.

Now the graphene may be peeled off from the copper foil. It is not, however easy to peel it mechanically because it bonds to it due to the fact that the electrons of the carbon atoms are pulled into the copper foil. A protocol using chemical procedures has been developed for this purpose. The graphene side is first painted with a polymer to protect it. The copper foil is now dipped in a ferric chloride ($FeCl_3$) liquid, which dissolves the copper foil without affecting the graphene and polymer. The graphene then floats to the surface of the chemical, which can be readily scooped on a surface of choice. For example one can scoop it on a glass plate by simple contact. The polymer coating may finally be removed by treating the layer with acetone.

Graphene can be also prepared from sugar. In the procedure, methane gas in the above method is replaced by sugar. Sugar is placed on the copper or nickel plate. In addition, argon and hydrogen gas are fed. The treatment is fast, lasting about ten minutes and it is scalable for industrial production. The process is also amenable to doping the graphene with other atoms or molecules as impurities. These impurities may give it other electric properties (see later).

10.9.2 Graphene and Nanotube Transistors and Chips [13]

Fig. 10.14 shows a cartoon of what would be a graphene transistor. It is to be noted, however that a grapheme transistor has serious problems, which may not allow it to be realized like the silicon transistor as we will see later. First, a large area of a grapheme sheet is prepared using methane and copper as was explained above.

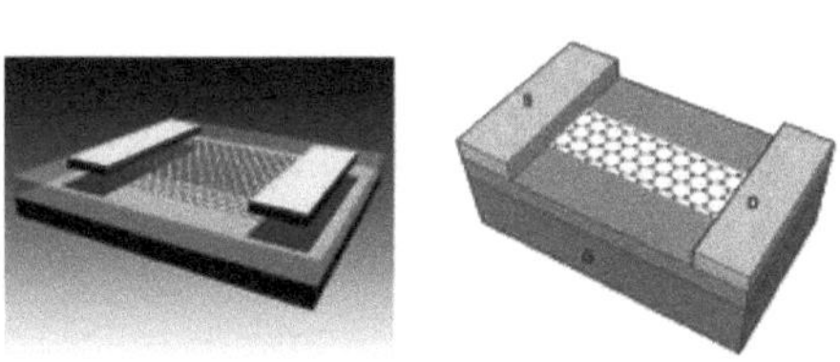

FIGURE 10.14 Cartoon of what would be a graphene transistor. *(Used with permission from Jian-Hao Chen, University of Maryland)*

Then a silicon wafer is coated with a thin silicon oxide insulator (glass). Then the graphene sheet is placed over the oxide layer. A metal strip is then deposited on one side to act as a source of electrons and another on the other side of the sheet to act as a drain for electrons. The source and drain are then wired to a battery. The back of the silicon wafer is connected to a second battery. The second battery acts as a back gate that can control the flow of electrons the source to the drain.

In more elaborate designs one can additionally deposit a thin film of glass (silicon oxide) over the graphene sheet to insulate graphene, and a thin layer of metal over the entire surface. The metal layer on top will be connected to a third battery to act as a top gate that can control the flow of electron from the source to the drain. A single gate, either top or back gate, however is sufficient to make the transistor work.

The problem with graphene is that it has a zero band gap, which makes it a conductor. It conducts electricity at any small voltage, which allows making resistors from it but one cannot use them to make rectifiers or transistors. Semiconductors have a non-zero bandgap, thus a certain voltage must be exceed to conduct, enabling building rectifiers and transistors. Having zero bandgap does not allow turning the transistor off, which greatly limits the uses of graphene in electronics. There are two promising designs to impart a bandgap to grapheme.

- Impurity induced gap: Some scientists at the University of California at Berkeley proposed to use two sheets on top of each other instead of one, and place some impurity like metal atoms over one of graphene sheet. This configuration can give it a bandgap, enabling transistor action with controllable switch off [14].
- Electric field induced gap: In the two sheet configuration, the impurity is replaced by an applied electric field on one side [15]. The electric field induces a gap. In this case the transistor essentially has two gates one from the top and one from the back. Scientists were able to use the back gate to give the two sheets a band gap of 0.25 eV, and used the top gate to control the transistor and turn it on and off. This bandgap is smaller than those of silicon, which is 1.1 eV or germanium, which is 0.6 eV, but it is for now good enough.

A CNT, on the other hand, has a band gap but depending on their structure it can vary from zero to about 2 eV and their electrical conductivity can show metallic (zero bandgap) or semiconducting (non-zero bandgap) behavior. Therefore it is possible to build transistors. But one needs to isolate purely semiconductor CNTs because residual metallic nanotubes short the electronic device, disrupting the semiconducting properties. Recently, a UW-Madison team used polymers to selectively sort out the semiconducting nanotubes, achieving a solution of ultra-high-purity semiconducting CNTs, which enabled them to successfully build transistors [16].

10.9.2.1 Graphene Chip

Graphene transistors are problematic but it affords other applications, such as superfast chips. A group at MIT built a microchip using graphene that operated at much higher speeds than is possible with the silicon chips [17]. As a first demonstration, the MIT researchers built a frequency multiplier chip which works much better than what is available. For example, it consumes less power and more efficient, which potentially would allow manufacturing cell phones and other communications systems that can transmit data much faster. Intel tested the first graphene wrapped CPUs and ramp up to 100 GHz.

10.9.3 Graphene Paper

In making the paper, one first dispenses oxidized and chemically processed graphite in water. Using paper-making techniques, the monolayer flakes may form a single sheet and bond very powerfully [18]. These sheets are called graphene oxide paper. They have a measured tensile modulus of 32 GPa, which is equivalent to 320,000 atm.

10.9.4 Graphene Supercapacitor Electrode

Due to the extremely high surface area to mass ratio of graphene, one potential application is in the conductive plates of super capacitor. A normal capacitor consists of two conducting plates with gas, liquid or insulating polymer between them. It is believed that graphene could be used to produce super capacitors with a greater energy storage density with over five times the energy density of commercial devices. Such devices are not yet enough to replace nickel metal hydride batteries and lithium-ion batteries but advances are coming all the time and it will not be long before this happens. We at the University of Illinois, in collaboration with Cleveland State University were able to improve the graphene super capacitors further by placing silicon nano particles on the surface [19]. We mixed silicon nano particles and graphene flakes in a conducting polymer and made a paste. We took a very thin plastic sheet like the saran brand plastic with which we wrap a sandwich. It weighs very little. We used a paint brush to "paint" both sides of the plastic sheet with the paste and left it to dry. The sheet now is simply a super capacitor. We then cut it into eight pieces each 1 cm × 1 cm and stacked them on top of each other and pressed them good. Then we charged the stack using a battery. This is now ready to act as a battery but it is much lighter than any battery. We used this super capacitor to light up a large panel of white LEDs as shown in Fig. 10.15.

10.9.5 Bendable Electronics

Because graphene is flexible, translucent and ultrathin, it can be used to make flexible electronics as shown in Fig. 10.16 [20]. Standard chip-making techniques are used to etch circuit designs into a nickel surface. A graphene sheet is then grown on the nickel in the same way it was done on copper as described earlier. The graphene formation takes on the circuit template that was etched

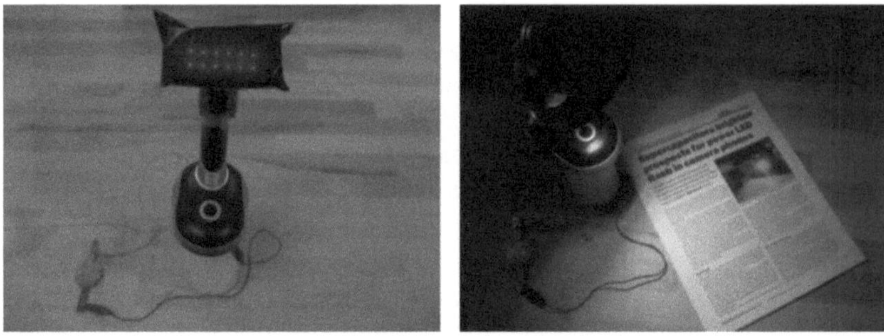

FIGURE 10.15 Graphene supercapacitor electrode application. A paste made from a mixure of silicon nanoparticles and graphene flakes in a conducting polymer is used as the active layer on very thin plastic sheet. It is used to light up a large panel of white LEDs.

FIGURE 10.16 Bendable electronics application: graphene used to make flexible electronics using standard chip-making techniques.

into the nickel. This procedure could find use in flexible computer displays, molecular electronics, and new wireless communications.

10.10 OTHER APPLICATIONS

There are other important applications. Graphene is practically transparent, absorbing less than 3% and a good conductor; therefore it is suitable for producing a number of devices, such as transparent touch screens, light panels, and solar cells. It can also be mixed into plastics, which affords them electric conductivity of electricity, heat resistivity, and mechanical robustness.

10.11 IS GRAPHENE THE NEW SILICON?

The intrinsic electron mobility in graphene would potentially be higher than any other known material, and more than 100 times higher than silicon. But this hinges on the elimination of other extrinsic factors that limit mobility in graphene, such as impurities and lattice vibrations in the underlying substrate. Because of its high electron mobility graphene promises transistors that can switch extremely fast, such as in the processing of extremely high frequency signals. But as discussed above the problem of switching on and off is still a pivotal problem, which governs the ultimate adoption of its transistors. Thus the question of "Will Graphene turn out to be the material that will replace silicon?" remains.

10.12 SYNTHESIS OF CARBON NANOTUBE

Fullerenes [21–23] in general and CNTs in particular can be found naturally in soot from indoor and outdoor air alike. Moreover, they can be found in mundane places, such as ordinary flames produced by burning carbon bearing substances, such as methane, ethylene, and benzene. However, those are not of high quality, that is, they are highly irregular in size and quality (lacking a high degree of uniformity) as they are produced under highly uncontrolled environments. More controllable procedures with high throughput and low-cost have been developed in recent years. The methods include arc discharge, laser ablation, and CVD.

Fig. 10.17 shows a schematic of the arc discharge housed in a vacuum chamber or under an inert gas [24]. In the process, a sharp carbon rod is biased positively to several thousand volts. The voltage is then discharged from the rod creating a current of 100 A, which rapidly heats the rod to temperatures above 1,700°C. The high temperature vaporizes/sublimates some of the carbon contained in the negative electrode. When the carbon vapor cools CNTs are produced. The yield can be ~30% by weight. The vaporization process produces both single- and MWNTs with lengths that can be as long as 50 micrometer. The tubes have few structural defects. The yield of SWCNTs may be enhanced when metal ions are introduced.

Other methods include laser ablation, which was demonstrated by Richard Smalley and co-workers at Rice University [25]. In the procedure a high intensity pulsed laser strikes a graphite target in a high-temperature reactor while an inert gas is circulated into the chamber. The laser vaporizes carbon. Nanotubes develop on the cooler surfaces (water cooled for example) of the reactor as

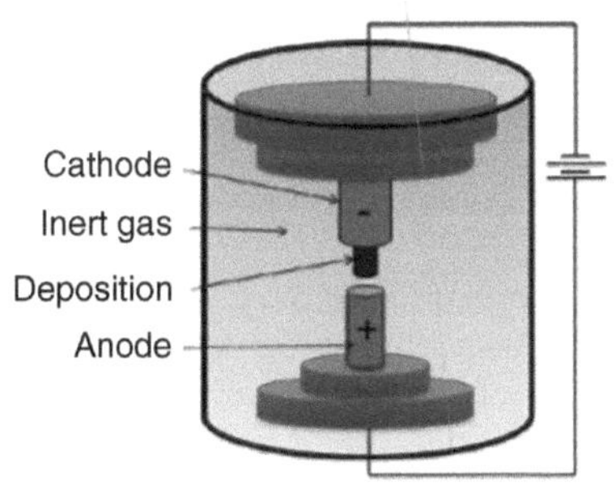

FIGURE 10.17 A schematic of the arc discharge housed in a vacuum chamber or under an inert gas for production of CNT.

the vaporized carbon condenses. Another method employing a plasma torch was also demonstrated. In this case, a carbon-bearing gas is used instead of graphite vapors to supply the carbon atoms. Finally, CVD-based procedure has been developed. Here metal catalyst particles, such as nickel, cobalt, iron, or a combination are deposited on a substrate. The substrate is heated to ~700°C. Hydrogen, nitrogen, or ammonium gasses are introduced. Also a carbon-bearing gas, such as ethanol or methanol is introduced. The carbon-containing gas is dissociated, releasing carbon at the surface of the catalyst particle. The generated carbon is transported to the edges of the particle, where it forms the nanotubes, that is, nanotubes grow at the sites of the metal catalyst. The diameters of the produced nanotubes are related to the size of the metal particles. Finally, the CVD process may be enhanced by introducing water into the CVD reactor. This enhances both the activity and lifetime of the metal catalyst. Dense millimeter-tall vertically aligned nanotube arrays (VANTAs) or "forests", aligned normal to the substrate, were produced.

REFERENCES

[1] O. Pierson, W. Andrew, Handbook of Carbon, Graphite, Diamonds and Fullerenes: Processing, Properties and Applications, 2012.

[2] Hong G, Zhou M, Zhang R, Hou S, Choi W, Woo YS, Choi J-Y, Liu Z, Zhang J. Separation of metallic and semiconducting single-walled carbon nanotube arrays by "Scotch Tape" 2011;50:6819–23.

[3] M. Tachibana, Characterization of laser-induced defects and modification in carbon nanotubes by Raman spectroscopy, Nanotechnol. Nanomater; S. Suzuki (Ed.), Physical and Chemical Properties of Carbon Nanotubes, ISBN 978-953-51-1002-6, Published: February 27, 2013. Chapter 2.

[4] Brayfindley E, Irace EE, Castro C, Karney WL. Stone–Wales rearrangements in polycyclic aromatic hydrocarbons: a computational stud. J. Org. Chem. 2015;80(8):3825–31. PMID 25843555.

[5] (a) Ajayan PM, Ravikumar V, Charlier JC. Surface reconstructions and dimensional changes in single-walled carbon nanotubes. Phys. Rev. Lett. 1998;81(1437). (b) Lu AJ, Pan BC. Nature of single vacancy in achiral carbon nanotubes. Phys. Rev. Lett. 2004;92:105504.

[6] Adam J, White CER, Hui HT, Milo HCY, Shaffer SP. Systematic comparison of conventional and reductive single-walled carbon nanotube purifications. Carbon 2016;108:423–32.

[7] Zhang Q. Carbon Nanotubes and Their Applications. Boca Raton, FL: CRC Press, Pan Stanford Publishing Pte. Ltd; 2012.

[8] Buckyballs, Nanotubes, Graphene, and Beyond. https://www.nano.gov/sites/default/files/pub_resource/2014_nni_cnt_tech_meeting_report.pdf.

[9] http://www.technovelgy.com/ct/Science-Fiction-News.asp? NewsNum (1939) https://en.wikipedia.org/wiki/Buckypaper.

[10] Future planes, cars may be made of, 'buckypaper' Yahoo! Tech, News 2008.

[11] https://www.slideshare.net/AttittudeBlogger/lini-42914063.

[12] (a) Hye JP, Meyer J, Roth S, Skakalova V. Growth and properties of few-layer graphene prepared by chemical vapor deposition. Carbon 2010;48(4):1088–94. (b) Castro MO, Liebold-Ribeiro Y, Barros EB, Salomao FCC, Mendes FJ, Filho AGS, Chesman C. Chesman Synthesis of graphene on nickel films by CVD method using methane. INIS 2011;43.

[13] Darren Quick. http://news.stanford.edu/news/2008/may28/ribbon-052808.html. New graphene-based single-transistor amplifiers are a triple threat, October 13, 2010. http://newatlas.com/triple-mode-graphene-based-transistor/16646/
[14] Berashevich J, Chakraborty T. Impurity-induced spin gap asymmetry in nanoscale graphene. Phys. Rev. B 2009;80:115430.
[15] Kin MF, Chun LH, Jie S, Tony FH. Observation of an electric-field-induced band gap in bilayer graphene by infrared spectroscopy. Phys. Rev. Lett. 2009;102:256405.
[16] http://news.wisc.edu/for-first-time-carbon-nanotube-transistors-outperform-silicon/.
[17] http://news.mit.edu/2009/graphene-palacios-0319.
[18] Zhang M, Hou C, Halder A, Wang H, Chi Q. Graphene papers: smart architecture and specific functionalization for biomimetics, electrocatalytic sensing and energy storage. Mater. Chem. Front. 2017;1:37–60.
[19] Liu Q, Nayfeh O, Nayfeh MH, Yau S-T. Flexible supercapacitor sheets based on hybrid nanocomposite materials. Nano Energy 2013;2:133–7.
[20] Georgiou T, Jalil R, Belle BD, Britnell L, Gorbachev RV, Morozov SV, Kim Y-J, Gholinia A, Haigh SJ, Makarovsky O, Eaves L, Ponomarenko LA, Geim AK, Kostya S, Novoselov A. Vertical field-effect transistor based on graphene–WS_2 heterostructures for flexible and transparent electronics. Nat. Nanotechnol. 2013;8:100–3.
[21] Takeuchi K, Hayashi T, Kim YA, Fujisawa K, Endo M. The state-of-the-art science and applications of carbon nanotubes. Nanosyst. Phys. Chem. Math. 2014;5(1):15.
[22] (a) Iijima S. Helical microtubules of graphitic carbon. Nature 1991;354(6348):56–8. (b) Bibcode. Nature 1991;354(56I)doi: 10.1038/354056a0.
[23] Ebbesen TW, Ajayan PM. Large-scale synthesis of carbon nanotubes. Nature 1992;358(6383):220–2. Bibcode, 1992, Nature, 358, 220E.
[24] Nanoscience Instruments http://www.nanoscience.com/applications/education/overview/cnt-technology-overview/cnt-synthesis/
[25] Guo T, Nikolaev P, Rinzler AG, Tomanek D, Colbert DT, Smalley RE. Self-assembly of tubular fullerenes (PDF). J. Phys. Chem. 1995;99(27):10694–7.

FURTHER READING

[26] Scarselli M, Castrucci P, De M, Crescenz J. Electronic and optoelectronic nano-devices based on carbon nanotubes. Phys. Condens. Matter 2012;24:313202.

Chapter 11

Material and Functionality Integration

11.1 BINARY SILICON-METAL SYSTEMS

Integration of silicon with metal opens a wide range of technological applications. There is current interest in integration of electronics and optics [1,2] by integrating silicon with noble metals, such as gold and silver, which promises to afford Si for example efficient visible photoluminescence or enhanced coupling to light. Integration of metal and Si transcends optoelectronics. This enables Si wafer bonding and connection of device elements (eutectic soldering or bonding) [3,4]. Further, in the nanoscale regime, the direct phase of the Au–Si binary system enables growth of silicon nanowires (SiNWs) and promises to open a means for growing a variety of gold nanostructures.

The integration is based on binary eutectic metal-silicon alloying. For example, the melting temperature of gold is 1064°C and the melting point of silicon is 1414°C. When gold and silicon are mixed a binary alloy (Au–Si) emerges, which exhibits novel thermal properties. Fig. 11.1 displays the phase diagram of the combined system as a function of the atomic percentage of silicon. For pure gold, the melting temperature is 1064°C. Above this temperature gold exists as a liquid. As we add silicon, the melting point drops but the system can exist in equilibrium as a solid and a liquid. As the percentage approaches 18.6%, melting temperature approaches 363°C. At this condition (18.6%, 363°C), the melting temperatures reaches its minim, and transformation between solid and liquid proceeds without passing two-phase equilibrium. With silicon percentage increasing further, the melting temperature increases and a two phase equilibrium emerges. For 100% silicon (no gold present), the melting temperatures reaches the 1414°C. The point (18.6%, 363°C) is called a eutectic point. At this point gold and silicon are mixed in the ratio of ~4/1.

The fundamental reason behind this feature is that the valence electrons of silicon and gold are compatible in this mixing ratio, allowing them to form a new type of joint crystal lattice. In general we have:

- Not all binary alloys have eutectic points because the valence electrons of the component species are not always compatible in any mixing ratio, to form a new type of joint crystal lattice. For example, in the silver-gold system the melt temperature (liquids) and freeze temperature (solidus) "meet at

Fundamentals and Applications of Nano Silicon in Plasmonics and Fullerines.
http://dx.doi.org/10.1016/B978-0-323-48057-4.00011-6

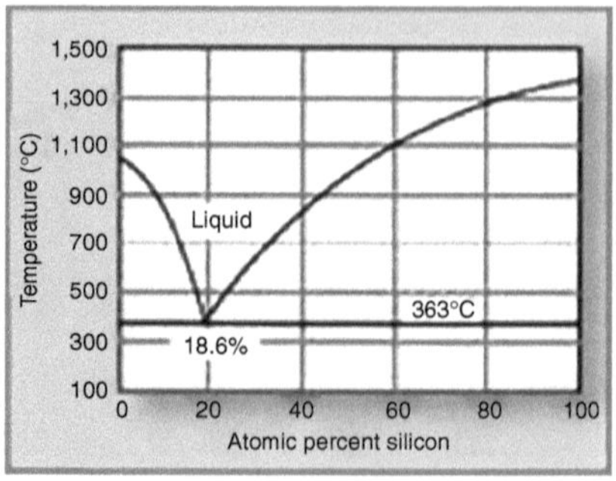

FIGURE 11.1 The phase diagram of the combined gold–silicon binary system as a function of the atomic percentage of silicon .

the pure element endpoints of the atomic ratio axis while slightly separating in the mixture region of this axis."

- All noble metals can diffuse into silicon and because of similarity between the crystal structures they can form stable novel binary alloys (eutectic alloys), which exhibit novel direct phase transition without passing a two-phase equilibrium.
- Gold's eutectic melting point falls at the conveniently low temperature of 363°C, but for silver and platinum it impractically rises by more than 500°C forming strong
- Au–Si bond is strong which enables low temperature Si wafer bonding (eutectic soldering or bonding); Pt, Ag–Si bonds are weak, which allows strong surface diffusion allowing formation of metal nanoparticles onto silicon.
- Schotcky potential barrier at metal-silicon interface, which can be used to generate, as well as inject charge across the interface.

11.2 WAFER BONDING

First the existing protective or native oxide layers are removed. This can be removed by several means including wet chemical etching (HF clean), dry chemical etching, chemical vapor deposition (CVD), H_2 plasma or CF_4 plasma. Second, gold is deposited on the substrate by either electroplating, physical vapor deposition or sputtering. Third the substrates are contacted. Fourth, the substrates may be pressed together mechanically and heated to temperature above the eutectic temperature. The last step tends to increase the diffusivity and solubility of gold into silicon. The reaction starts upon atomic contact; however rising bonding temperatures, especially temperatures higher than the eutectic temperature may result in the formation of a thicker Au–Si alloy layer, as well as stronger eutectic bond.

It is to be noted that removal of *thermal or native silicon oxides* on silicon are critical in achieving good silicides. Prior to the deposition of metal thin films, the thermal oxide must be etched. Secondly, a native oxide of 1.5-nm to

2-nm-thick SiO_2 layer usually develops at the silicon substrate surface following the removal of thc thermal oxide. For some metals an argon ion sputter-cleaning step of the native oxide is usually required. The cleanup is important because some metals cannot penetrate the thin oxide layer to react with the silicon to form silicides. For example, whereas titanium and nickel atoms are capable of penetrating through the thin oxide, cobalt atoms have difficulty forming silicide with silicon if a thin oxide layer is present at the interface. Because of other requirements, $CoSi_2$ is widely used in devices with line widths of 0.18 μm or smaller, a cleanup step is used, and sputtering at room temperature is used to deposit cobalt thin films, resulting in a mixture of Co_2Si and CoSi at 300°C. Moreover, $CoSi_2$ forms at 550°C.

The temperature is then decreased below the eutectic temperature or the concentration ratio changes (for Si–Au: $T < 370°C$) to solidify material mix. If successful, epitaxial growth of silicon and gold on top of the silicon substrate is achieved. Moreover, small silicon islands protruding from a polycrystalline gold alloy would have formed. In this case, the two wafers bond very well with bonding strengths around 70 MPa.

A variation to this procedure involves the use of additional layers of combinations of other metals including titanium (Ti), chromium (Cr), nickel (Ni), or a platinum (Pt) as shown in Fig. 11.2. Aluminum (Al) can also be used. The use of the extra intermediate layer(s) allows boding without the removal of the silicon oxide step. A typical wafer is composed of a silicon wafer with oxide, 30–200 nm Ti or Cr layer and Au layer of >500 nm thickness.

11.3 METAL SILICIDE

The interaction between Si and metal results in binary compounds of silicon and the metals (metal silicide) [5]. The binary compounds actually molecules, where the chemical bonds in silicides may exhibit primarily covalent or primarily ionic characteristics, depending on the electronegativity of participating elements. Examples include the noble metal silicides PtSi, AuSi, Pd_2Si, AgSi. Many more metals make compounds with silicon. Metal silicides play important roles in the field of advanced technology. Thin films of metal silicides are integral to the integrated circuits (ICs) of semiconductor and microelectronics devices, the most common including $CoSi_2$, $NiSi_2$, WSi_2, $MoSi_2$, $TaSi_2$, $TiSi_2$, and PtSi.

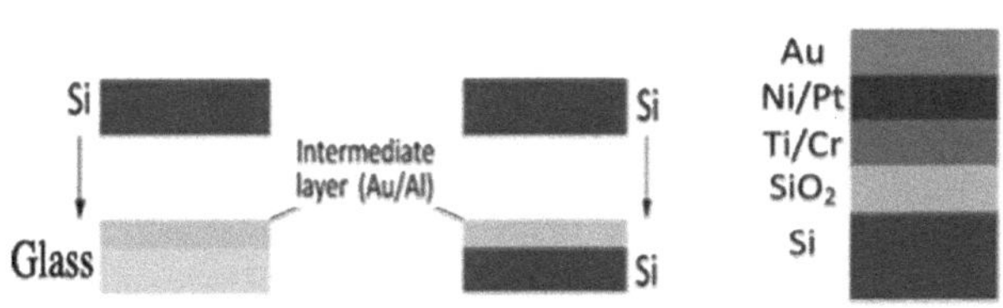

FIGURE 11.2 **Wafer bonding using the binary gold–silicon binary system .**

11.3.1 Silicide Films

The usual steps to form a silicide begin with the standard cleaning of the wafers consecutively by organic solution, dilute hydrochloric acid (HF), and deionized water. This is followed by sputter-cleaning by argon ions if necessary. Next, metal thin films are deposited using sputtering or ion implantation on silicon at room temperature or at a higher temperature, and finally, the wafers are subjected to standard annealing or by rapid thermal annealing to form silicides.

The nature of the film, especially the electronic nature depends on the metal, which makes metal silicide thin films integral parts of all microelectronics devices. They have been used as means to control

- Charge transport: Ohmic contacts, local interconnects, Schottky barrier contacts,
- Electric field distribution: gate electrodes,
- Material transport: diffusion barriers.

11.3.2 Nanostructured Silicide

Nanostructured silicide may be formed by depositing ultrathin (~1 nm) metal film on silicon surfaces. The metal may react with the silicon substrate to form silicide nanodots under appropriate annealing conditions. Ion implantation of metal ions, for example into SiNWs followed by annealing, may also produce silicide nanoparticles. Quantum dots or particles afford useful applications, such as single-electron transistors, high-density memories, light emission, semiconductor lasers, and tunnel diodes.

Silicide nanostructures may also be made on strained silicon substrates, such as $Si_{0.7}Ge_{0.3}$ on (001) Si. For example, epitaxial β-$FeSi_2$ nanodot arrays were fabricated on strained Si/ $Si_{0.8}Ge_{0.2}$ (001) substrate.

Another area which has been active is growing single-crystal silicide nanowires. For example, SiNWs were first synthesized. The wires were then coated with nickel, followed by annealing at 550°C. This resulted in the formation of the NiSi nanowires. In another development NiSi nanowires were coated with carbon using radio-frequency-induction heating CVD reactor, which resulted in carbon-coated NiSi nanowires.

11.4 PLASMONIC-SILICON INTEGRATION

11.4.1 Plasmonic Electric Stress

Understanding the silicon–gold system especially in the nanoscale regime is of great importance to integration of electronic and optics The interest stems from the fact that integration of thin films or nanoparticles of noble metal with silicon can be used to subject or inject intense electric flux into silicon, diffuse metal that forms novel bi-alloys, or extract or inject charge into silicon. Thin films or nanoparticles of metal focuses light to a few nanometers due to electron (plasmonic) resonances, which upon exposure to light, subjects underlying silicon to extremely high electric fields that changes the structure and crystal symmetry affording silicon novel strong optical activity. For gold, plasmonics resonances occur conveniently in the visible, but they blue shift inconveniently into UV and near UV for silver and platinum. Second, noble metals can diffuse into silicon and because of similarity between the crystal structures they can form stable novel binary alloys (eutectic alloys) which exhibit novel direct phase transition without passing a two-phase equilibrium. For gold, the eutectic melting point falls at the conveniently low temperature of 363 °C, forming strong Au-Si bond which enables low temperature Si wafer bonding (eutectic soldering or bonding); but it impractically rises by more than 500 °C for silver and platinum. Weak metal–Si bond for example in the case of Pt and Ag, however allows strong surface diffusion, which can result in the formation of metal nanoparticles upon delivery of the metal atoms onto the surface. Third, metal structures can be used to generate, as well as inject charge across the metal–silicon interface through a Schotcky potential barrier, which enhances the chemical etching of silicon in wet HF for example. If charging is spatially localized by localizing metal deposition, one can design specific nanostructure formations. In fact, charge induced etching is emerging as a viable technique for surface-texturization (antireflection) of Si photovoltaics, production of Si-nanowires, Si-hole arrays as well as formation of ultrasmall luminescent silicon nanostructures or particles.

11.4.2 Plasmonic Antireflection Effects

Over the last decade, researches have been extensively studying better antireflecting coatings to enhance light coupling hence improve the efficiency of various optical and optoelectronic devices such as lenses, solar cells, photodetectors, and others [6]. Nanostructuring of the front surface of the devices has been targeted. Nanostructuring may be achieved by roughening the surfaces or by depositing nanoparticles of various materials. Plasmonic nanoparticle coatings based on silver or on crystalline silicon nanoparticles have been proposed.

Numerical simulations of light coupling in the visible spectral range was carried out. Fig. 11.3 shows a silicon surface with nanoparticles decorating the surface. To simplify the calculations, periodic arrays of spherical nanoparticles

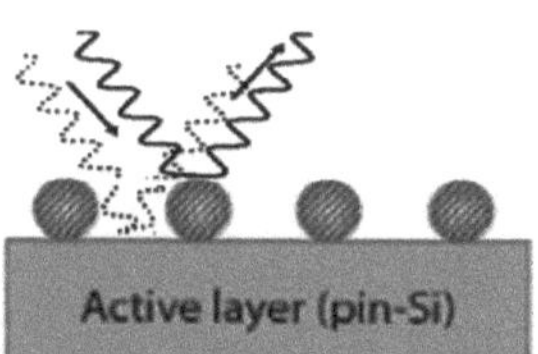

FIGURE 11.3 Sketch of reflection of visible light off a silicon surface decorated with metal or silicon nanoparticles, showing reflection from the nanoparticles and from the silicon surface.

were used. When an electromagnetic wave such as a light beam impinges on the nanostructured surface, there will be electric- (E-) and magnetic- (M-)dipole scattering by the nanoparticles or reflection. In addition there will be reflection from the surface of bulk. To contrast the effect, both silicon nanoparticles and silver nanoparticle coatings are analyzed.

Fig. 11.4A–C shows the absorption, transmission, and reflectance, respectively as a function of wavelength of silver and silicon nanoparticles arrays on the surface a function of wavelength. In the absorption measurements, one

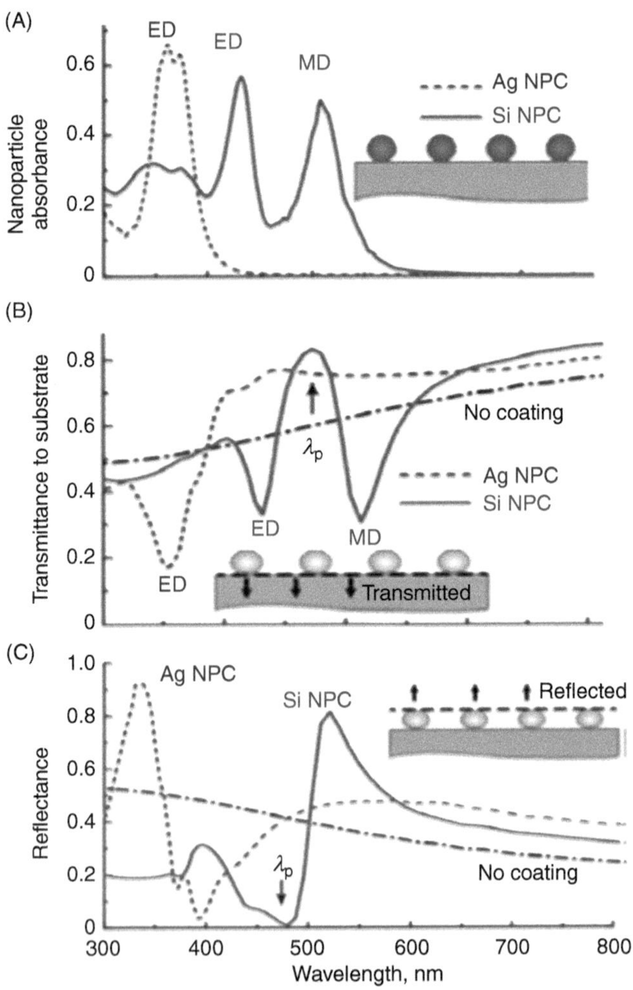

FIGURE 11.4 Numerical simulations of light coupling in the visible spectral range. With silicon surface, with nanoparticles decorating the surface. (A–C) The absorption, transmission, and reflectance, respectively, act as a function of wavelength of silver and silicon nanoparticles arrays on the surface.

can easily see the absorption of the electric-dipole and the magnetic dipole responses. The reflection from the underlying silicon substrate with a high refractive index (3.5) was also calculated. The interference of E- and M-dipole of Si nanoparticles and bulk silicon reflection was also calculated. Fig. 11.4 shows the corresponding wavelength dependence of the total transmission of the silver and silicon nanoparticle arrays and the substrate. Fig. 11.4 shows the corresponding wavelength dependence of reflectance. Comparing the absorption and transmission spectra one can see that both silver-nanoparticle coating (Ag-NPC) and silicon one (Si-NPC) enhance transmission to the substrate at the wavelengths longer than ED and MD resonances, respectively. It originates from the effect of constructive interference of scattered light with incident wave. Indeed, for wavelengths longer than the resonant, dipole oscillations occur in-phase with incident light, while below the resonant wavelength, dipole oscillates with phase-shift in respect to the driving field, which results in destructive interference observed for shorter wavelengths $\lambda \approx 300$–350 nm for Si-NPC (see Fig. 11.4B).

It is to be noted that silver in ambient conditions is sucesstible to severe degradation. This is unlike silicon, which has high chemical stability and lower optical losses, which makes them promising candidates for future nanophotonic devices. Moreover, recent advances in methods of their cost-effective fabrication by laser printing open up a possibility to vary nanoparticle size and crystallographic phase, making silicon nanostructures even more prospective for photovoltaic applications.

The transmittance spectra for Si-NPC with nanoparticles for different radii R are shown in Fig. 11.5. The positions of ED- and MD-resonances and transmittance enhancement are shown with the black dashed lines.

Fig. 11.5B presents the integral enhancement over all wavelengths Fl for Ag and Si nanoparticles (left axis). The peak enhancement Fp for Si-NPC at λp is also shown (right axis). The suppression of the reflectance results in the resonant increase of the transmittance with Si-NPC, and we define λp as the wavelength of maximum transmittance, so it is a function of nanoparticle radius R. The dipole resonances of silicon nanoparticles are defined by their sizes in accordance with Mie theory, and the transmission maximum λp shifts toward long wavelength region with increasing the nanoparticles sizes. One can notice that the position of λp shifts along with the positions of ED and MD resonances staying located between them. For larger nanoparticles, resonances are stronger, and the increase of transmission is higher than for small nanoparticles. Thus, we would also like to pay attention to the narrow-band increase of transmittance and introduce the peak transmittance enhancement of Si-NPC as the enhancement at λp which also can be plotted as a function of nanoparticle radius (see Fig. 11.4B; secondary y-axis). One can notice that both ED and MD resonances are vanished for nanoparticles smaller than 40 nm radius, and resonant enhancement is not observed (see Fig. 11.4A). The value of the resonant enhancement

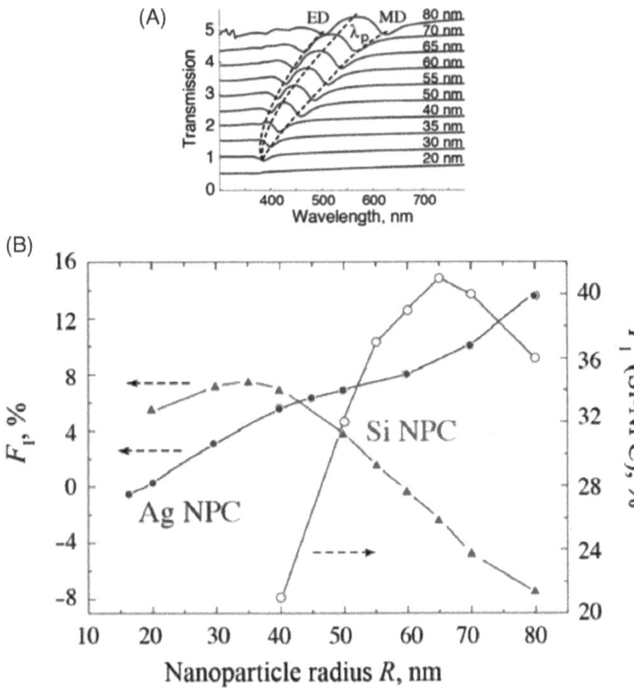

FIGURE 11.5 (A) The transmittance spectra for Si-NPC with nanoparticles for different radii R. The positions of ED- and MD-resonances and transmittance enhancement are shown with the black dashed lines. (B) The integral enhancement over all wavelengths Fl for Ag and Si nanoparticles (left axis). The peak enhancement Fp for Si-NPC at λ_p is also shown (right axis). The suppression of the reflectance results in the resonant increase of the transmittance with Si- NPC.

changes not monotonically and has maximum $Fp = 41\%$ for $R = 65$ nm (see Fig. 11.4B). Although the resonant transmission at λp increases with the increase of nanoparticle sizes, it also shifts to the longer wavelength region ($\lambda > 500$ nm). There, the transmission of the bare silicon substrate is higher, and thus, for nanoparticles with larger radiuses ($R > 70$ nm), the resulting transmission enhancement is decreased.

Overall, we observe that both narrow- and broad-band transmittance increases can be obtained with nanoparticle coatings. For the increase in the entire visible range, Ag-NPC with radiuses 70–80 nm give the best transmittance enhancement up to 15%, and Si-NPC with radiuses 30–40 nm causes up to 8% of enhancement. At the same time, at the narrow spectral range, Si-NPC enables up to 40% of transmittance increase. Thus, one can see that placing NPC on top of active layer of silicon photovoltaic element can efficiently increase absorbance inside it: up to 8–10% with Ag- or Si-NPC of small radius (30–40 nm) and up to 30% with Ag-NPC of radius 80 nm having advantages of multipole resonances of plasmonic localized modes.

11.5 SYNTHESIS OF SI NANOWIRES USING METAL (GOLD) NANOPARTICLE SEEDS

There are several methods for synthesis of SiNW. The methods used to form nanowires can be classified into top down and bottom up methods. Top down methods start with bulk silicon and remove material to arrive at the structure of the nanowire [7,8]. The method includes: laser beam ablation, ion beam etching, and thermal evaporation oxide-assisted growth (OAG).

The bottom-up synthesis starts form individual atoms to build up the structure of a nanowire. The method includes physical procedures such as molecular beam epitaxy, as well as metal-assisted CVD synthesis [9]. It works in a gas environment, as well as in a wet environment.

- Vapor liquid solid (VLS) growth: This is a type of catalyzed CVD, which often uses silane as Si precursor and gold nanoparticles as catalyst (or "seed").
- Precipitation from a solution: It uses a supercritical fluid (e.g., organosilane at high temperature and pressure) as Si precursor. The catalyst is a colloidal gold nanoparticles, and the SiNWs are grown in this solution.
- In the gas environment, the procedure utilizes the eutectic direct phase of the Au–Si binary system to grow SiNWs. In the technique the following steps are taken:
- *Gold deposition*: A gold (Au) film is deposited onto a silicon (Si) wafer substrate by sputter deposition or thermal evaporation. The thickness of the film can fall in the range 1–10 nm.
- *Annealing*: The wafer is then annealed at temperatures $T > 363°C$, higher than the Au–Si eutectic point. Heating creates Au–Si alloy droplets on the wafer surface (thicker Au films results in larger the droplets.
- <u>CVD:</u> A $SiCl_4{:}H_2$ gaseous mixture is introduced in the heated chamber. Si deposits on the Au–Si eutectic droplets at temperatures above 363°C and get adsorbed because Au can form a solid-solution with all Si concentrations up to 100%).
- *Nanowire growth*: When a supersaturation state of Si in Au is reached Si atoms precipitate out of the supersaturated liquid-alloy droplet at the liquid-alloy/solid-Si interface, and the droplet rises from the surface, leading to axial crystal growth, hence forming a wire. The diameter of the wire is limited to the diameter of the gold nanoparticle.

Fig. 11.6A–C is a schematic illustration of Si wire growth from the reaction of $SiCl_4$ and H_2 vapor phases. This reaction is catalyzed by gold–silicon droplet deposited on the wafer surface prior to whisker growth. The disadvantage of the growth mechanism is that the growth in not in mono direction. This feature is highlighted in Fig. 11.6D. Pre-synthesized gold nanoparticles were spin coated on the silicon surface. The resulting nanowires grow in all possible directions. There were attempts to grow vertical wires of a given length. In the procedure, another wafer is placed above the growth substrate. When the wire reaches the

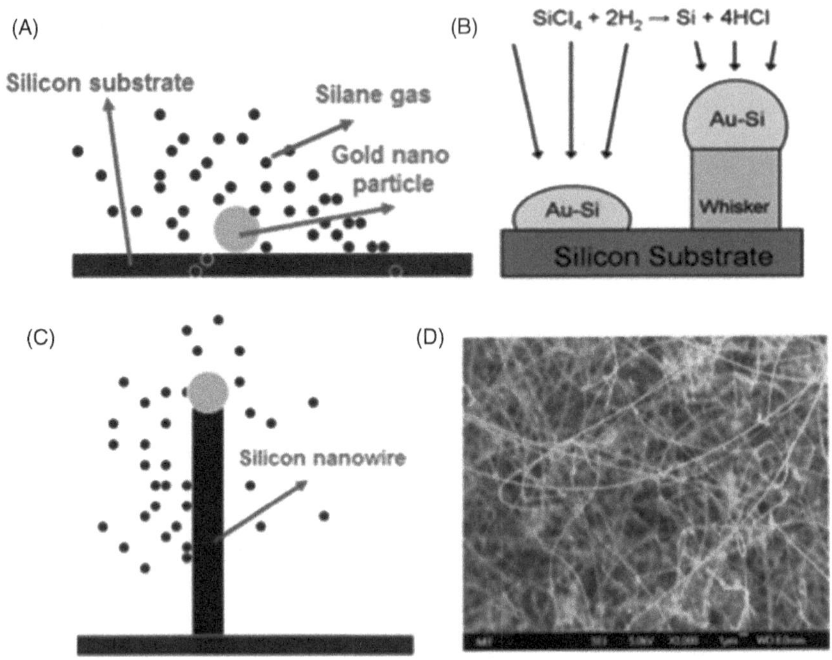

FIGURE 11.6 Si nanowire growth using Au seeds. (A–C) Schematic illustrations of Si wire growth from the reaction of $SiCl_4$ and H_2 vapor phases, catalyzed by gold–silicon droplet deposited on the wafer surface prior to whisker growth. (D) Si nanowires growth in not in mono direction.

top wafer the growth stops. Since the vertical wire has the shortest length, then stopping the process after a certain treatment time would leave a lot of wires that have not made contact with the top substrate. This procedure is useful when the lower and top substrates are actually the source and drain of a transistor. All other wires that do not terminate on both will not affect the electrical operation of the device.

Typical features of the VLS method

- Greatly lowered reaction energy compared to normal vapor-solid growth.
- Wires grow only in the areas activated by the metal catalysts and the size and position of the wires are determined by that of the metal catalysts.
- This growth mechanism can also produce highly anisotropic nanowire arrays from a variety of material.

Using noble metal such as Au, Ag, or Cu to grow Si nanowires can adversely affect the semiconductor properties of the nanowires as contamination from these metals may act as deep level traps in Si. Alternative seed metals, such as Al, Sb, and Sn are less parasitic but they are reactive and therefore care must be taken to protect them from oxidation during deposition and patterning [10]. Fig. 11.7A shows a schematic of the steps involved. The process starts

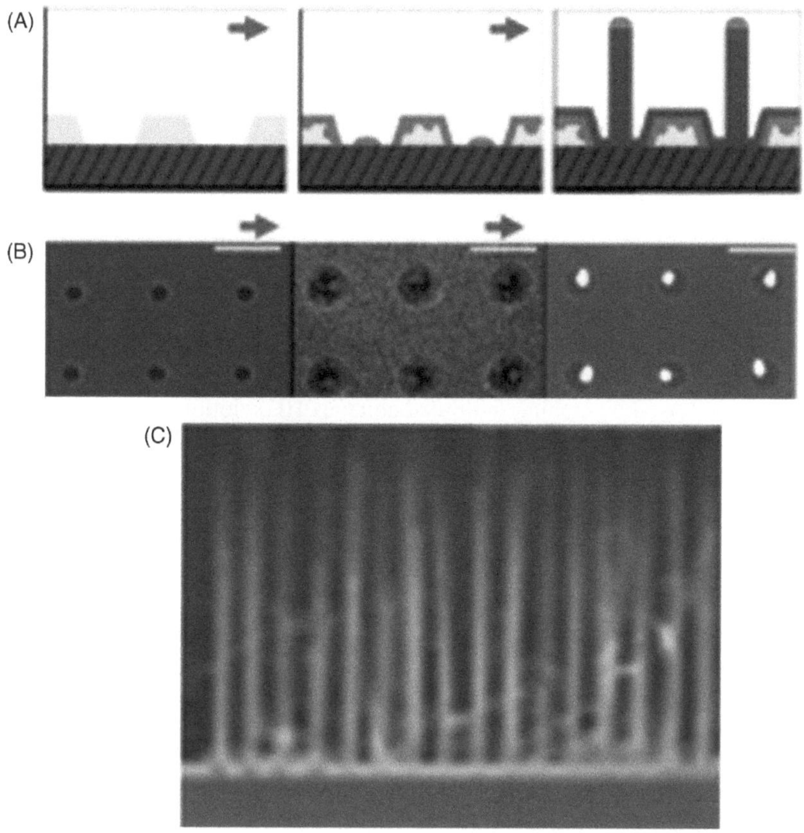

FIGURE 11.7 **Si nanowire growth using Al seeds.** (A) Schematic of the process. (B) Top view of the actual steps of processing. (C) Side view of the wires grown (Adapted from: M.M. Khayyat, B.A. Wacaser, M.C. Reuter, F.M. Ross, D.K. Sadana, T.-C. Chen, Nanoscale chemical templating of Si nanowires seeded with Al, Nanotechnology 24 (2013) 235301 (8pp)).

with a Si wafer substrate. A layer of SiO_2 (50-nm thick for example) is grown. A pattern of holes are created in the oxide. The bottom of the holes reaches the silicon substrate, using standard photolithography (see Section 12.3). All is then deposited as described previously. The sample was then transferred to the CVD chamber and annealed at 730°C. This annealing temperature is higher than the eutectic temperature (577°C) for Si and Al (higher than the Si–Au temperature); thus during annealing, Al that is in contact with Si at the bottom if the holes forms the Si:Al eutectic droplets. Those constitute the seeds from which the nanowires grow. Extra Al deposited on the SiO_2 is oxidized and they become unresponsive. The temperature is now lowered to ~490–500°C and pure silane gas (SiH_4) is flowed into the chamber with di-borane/helium gas for a certain period of time which results in the growth of the nanowires on the seeds. Fig. 11.7B shows top of view. Fig. 11.7C is a side view of a section of the nanowires grown. The growth is vertical which may be related to the fact that the seeds are inside of oxide holes.

11.6 SYNTHESIS OF SILICON NANOPILLARS USING SILVER ION ETCHING [7,8]

Silicon wire-like nanostructures can be made using silver ion etching. The process starts with a commercial boron-doped silicon on insulator wafer (SOI), <100> (4–8 Ω cm) SOI wafer. The process also works for n-type crystals. SOI wafers consist of three layers. The handle is a thick silicon layer of the order of half a millimeter, with a 300 nm thick silicon oxide layer. A 10 μm thick device-quality Si layer is on top of the oxide. The SOI wafer is first cleaned by acetone, followed by etching in diluted aqueous HF solution. The wafer is immersed in a horizontal configuration in a bath of 5.0 mol/L HF solution, which contains silver nitrate at a concentration of 0.02 mol/L. The solution is heated to a temperature of 50°C. It should be noted that the process is less reproducible at room temperature, requiring higher concentration of silver ions. After the etching process is completed, the silicon wafer is rinsed with deionized water and blown dry in air very gently. The sample is then immersed in nitric acid in a horizontal configuration as a simple selective chemical etching treatment to remove the residual silver.

Fig. 11.8A presents a 100 μm × 100 μm scanning electron microscopy (SEM) image of a typical section of the treated wafer. The image shows bundles of nanowire-like structures having the same length. The length is 10 μm is equal to the thickness of the device layer of the SOI wafer. The diameters of nanostructures are in the range of 30–200 nm. In the process, the wafers are treated for a period of time just short of what is needed for the wires to detach. Fig. 11.8B presents a 29 μm × 25.7 μm SEM image of a typical section of the treated wafer. It shows a crater region were some wires have edge of the crater in the SEM images shows also two dimensional wire-like formations with an average spacing of ~1 μm. From the interspacing, we estimate a density of $\sim 10^8$ /cm^2. On the sides of the crater we see a high density of disconnected wires. Fig. 11.8 shows a density of $\sim 5 \times 10^6$/cm^2, indicating some loss in the etchant upon detachment.

Chemical analysis using electron spectroscopy gives atomic concentration for silver of less than 1 in 1000. The treated wafers can be sonicated in a liquid of choice and deposited the released nanostructures on a fresh silicon wafer. Often wires taken from the bottom of the liquid are found to be still in bundles. Increasing the intensity of ultrasound disperses the bundles more effectively; however, harsh sonication shatters or chips some. Close-up of some of the wire-like structures show a cross section of 50 nm. Images show infrequent cases of distorted wires. The effect of crystal orientation on the wire formations is interesting. When <111> single crystal orientations is used, the wire-like structures do not form upright; rather they formed slanted. A silver sulfate

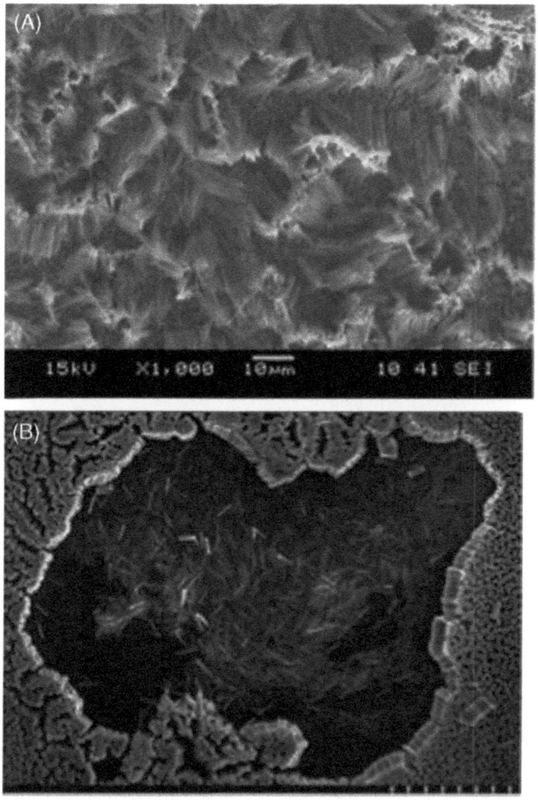

FIGURE 11.8 Synthesis of wire-like formations using metal catalyst on silicon in insulator (SOI) wafer. (A) 100 μm × 100 μm scanning electron microscopy (SEM) image of a typical section of the treated wafer. The image shows bundles of nanowire-like structures having the same length. (B) 29 μm × 25.7 μm SEM image of a typical section of the treated wafer. It shows a crater region were some wires have edge of the crater in the SEM images shows also two dimensional wire-like formations with an average spacing of ~1 μm.

group instead of silver nitrate can be used. Wire formation using the sulfate works just as well.

Silver sulfate (Ag_2SO_4) is an ionic compound of silver used in silver plating and as a nonstaining substitute to silver nitrate. This sulfate is stable under ordinary conditions of use and storage, though, unlike the nitrates, it darkens upon exposure to air or light. It is also unlike the nitrate minimally soluble in water.

A generic process for the preparation of curved silicon nanowires (SiNWs) with ribbon-like cross sections was developed as shown in Fig. 11.9 [11]. The present synthetic approach is based on chemical etching of (100)-oriented silicon wafers in mixture solutions of HF and H_2O_2 by using patterned thin gold films (Fig. 11.9A) as catalyst and provides a unique opportunity for the fabrication of extended arrays of zigzag SiNWs (Fig. 11.9B), ultrathin straight [11] SiNWs,

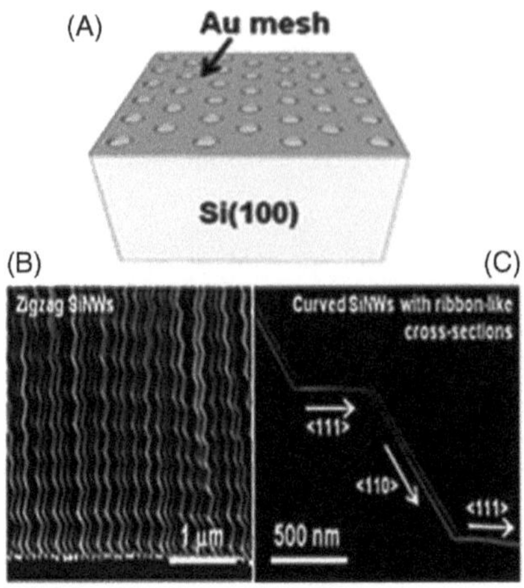

FIGURE 11.9 A generic process for the preparation of curved silicon nanowires (SiNWs) with ribbon-like cross sections. (A) Chemical etching of <100> Si in HF/H_2O_2 solution using patterned thin gold films. (B) Extended arrays of zigzag Si nanowires. (C) Controlled turning angles.

and curved SiNWs with controlled turning angles (Fig. 11.9C). On the basis of our experiments performed under various etching conditions, the factors governing the axial crystal orientation and morphology of SiNWs were systematically analyzed. We proposed a model that explains the formation of the present novel silicon nanostructures during chemical etching of silicon.

11.6.1 Control of Aspect Ratio and Orientation of Si Nanowires

Thermal oxidation of Si nanowires may provide control over their aspect ratio. The control is based on the fact that SiNWs exhibit a distinct self-limiting oxidation behavior whereby oxidation effectively ceases due to diffusion limitations. This phenomenon allows accurate control of their dimensions and aspect ratios. In fact, precise use of the effect has been used to obtain high aspect ratio with diameters below 5 nm. In addition to control of dimensions, external electric fields in polar alignment, electrophoresis, mircofluidic methods, and contact printing are among the means that may be used to control the orientation of the nanowires.

11.7 GOLD-ASSISTED SYNTHESIS OF SILICON NANOPARTICLES [12]

Chemical etching of silicon using platinum nanoparticles, which produces prerequisite platinum nanoparticles first, has been studied extensively and is emerging as a viable production technique of synthesis of luminescent silicon nanoparticles. While the procedure is capable of producing ultrasmall sizes of silicon (sub 3 nm) the size of the platinum nanoparticle produced is limited to few hundred nanometers due to the weak platinum eutectic bond and large platinum surface diffusion.

Moreover, the weak eutectic bond and the high eutectic melting point do not offer integration of nano structures of platinum and silicon.

In the case of gold, the low eutectic temperature point (363°C) and the strong eutectic Au–Si bond reduces diffusion of gold on silicon and as such may offer an opportunity for integration of the two materials at the nanoscale. But strong bonding and weak diffusion may hamper formation of pre requisite gold nanoparticles. Moreover, the higher Schottky potential barrier at the gold silicon interface (0.3 compared to 0.23 eV for Pt–Si), which slows down hole injection into silicon may allow holes to spread out, reducing hole density below levels needed to produce ultrasmall silicon features or nanoparticles. Moreover, the larger Schottcky barrier may cause charge to accumulate in the gold structures, which may cause charge breakdown and percolation between the gold structures.

In section we introduced platinum-assist electroless chemical etching of silicon to prepare silicon nanoparticles. In this method we first deposit (plate) Pt onto silicon surface by immersing silicon wafers into a hexachloroplatinic acid/HF mixture solution. Because the Si–Pt silicide is weak, allowing fast diffusion of platinum on the surface Pt nanoparticles form. The use of platinum in this procedure was extended to the use of gold instead of platinum. Just like platinum, gold diffuse in silicon to form novel Bi-alloys but it poses some differences and difficulties. Gold is deposited using chemical deposition from an acid carrying gold ($HAuCl_4$/HF).

11.7.1 Gold Deposition

Gold is deposited on silicon using a wet treatment. <001> boron doped single crystal wafers of 4-8 ohm-cm^{-1} resistivity are used. Hydro auric acid ($HAuCl_4$) is used to provide the gold Au^{3+} ions. HF is added to the auric acid solution to remove the native oxide from silicon as well as maintain the passivation of silicon with hydrogen, preventing oxidation in the aqueous environment. Deposition does not proceed on oxygen-terminated silicon. The underlying principle of deposition of gold is an electron exchange mechanism, where electrons or holes are efficiently drawn from or into the silicon surface because Si has lower reduction potential than that of gold (−0.85 V compared to 1.6 V).

SEM imaging of the silicon sample shows islands of gold as shown in Fig. 11.10A–B. AFM imaging shown also in Fig. 11.10C–D shows that the islands are actually columns or pillars, indicating the growth being predominantly vertical. As to the size or depth of the columns we used a combination of the SEM images as well as the AFM 2-d and phase images. Additional measurements show that with time of treatment (from a few seconds to 15 min) the deposition stays predominately vertical, with much less lateral growth such that most of the columns stay disconnected. Fig. 11.11 shows that the amount of gold deposited saturates with time, indicating a self-limiting deposition. This implies that as the top of the column gets further and further from the silicon surface the ability of the surface to reduce the gold ion becomes smaller and smaller, indicated a self-limiting process.

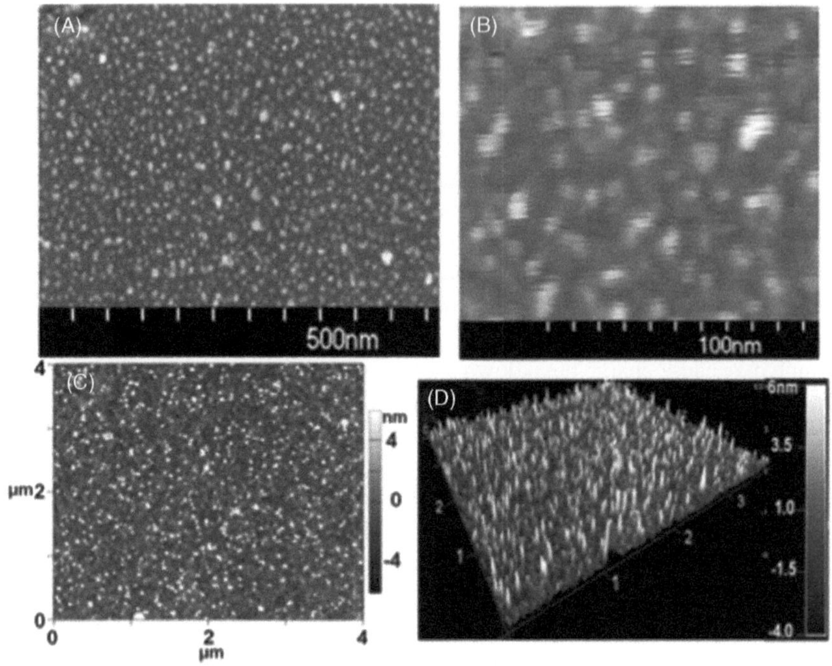

FIGURE 11.10 (A) SEM imaging of the silicon substrate after gold has been deposited electrochemically, showing islands of gold. (B) Close-up SEM image of the surface. (C) Two-dimensional AFM image of the surface. (D) AFM imaging shows that the islands are actually columns or pillars, indicating the growth being predominantly vertical.

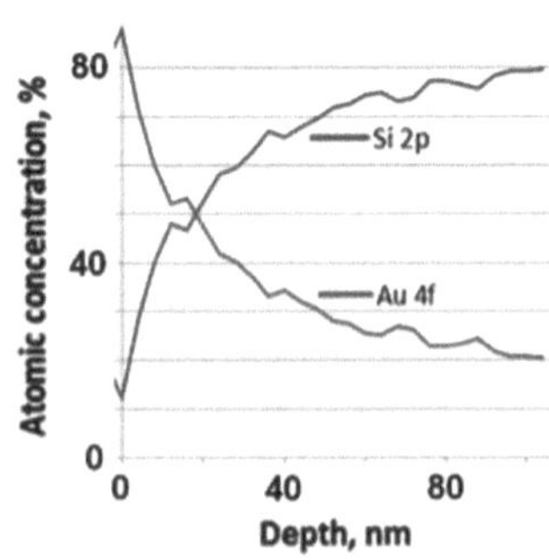

FIGURE 11.11 **The depth of gold column with time (from a few seconds to 15 min.) of deposition.** It shows that the amount of gold deposited saturates with time, indicating a self-limiting deposition.

11.7.2 Charge Injection

The gold coated wafer is now immersed in a 1/1 mixture of HF/H_2O_2, which in a couple of minutes turn from silver to yellow/orange stain. Fig. 11.12A–B shows, respectively, photos of the sample in room light, and under UV excitation of 365 nm wavelength, showing the red luminescence. The wafers are then observed with scanning electron spectroscopy as shown in Fig. 11.12C–D. The

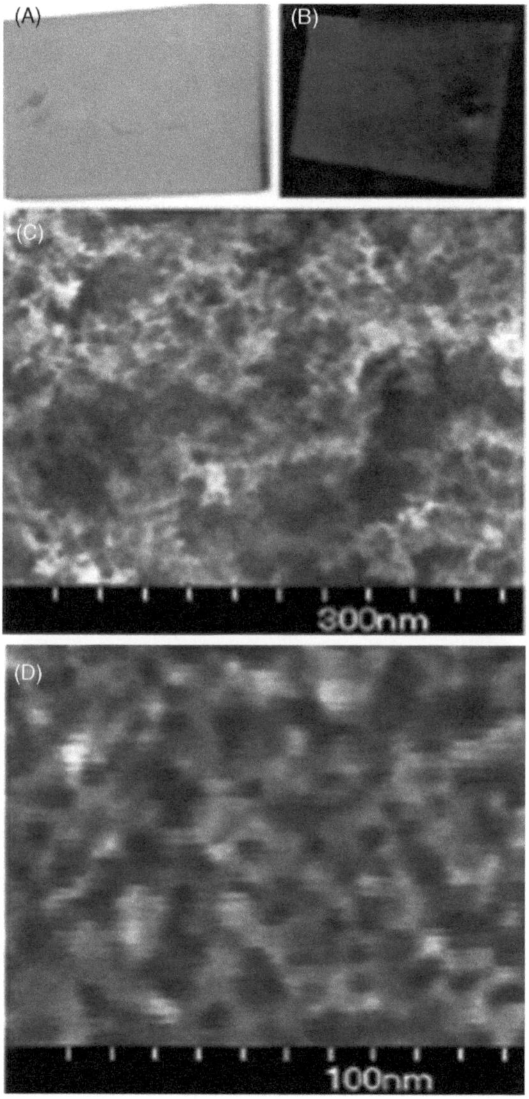

FIGURE 11.12 Chemical etching (A and B) photos of the sample in room light (showing yellow/orange staining), and under UV excitation of 365 nm wavelength (showing red luminescence), after the gold coated wafer (silver gray in room light) immersed in a 1/1 mixture of HF/H_2O_2. (C and D) Scanning electron spectroscopy images of the etched wafers show that the separated columns exhibit a long-range connectivity, as well as show a high density of pores.

images show that the separated columns exhibit a long-range connectivity, as well as show a high density of pores. Nearly each column has a pore down its middle of ~8–10 nm diameter. Less frequent are larger pores of ~40 nm across are seen.

The gold surface acts as a catalyst for reduction of H_2O_2 which generate hole charge. Hole charges are transferred inside the gold columns toward the Si surface in contact with gold. At the Au–Si contact point, the holes are injected into Si; however very slowly because of a high Schottky barrier which allows the charge to accumulate in the gold columns. In fact, the Schottky barrier at the gold–silicon interface where a thin film of gold silicide exits is sizable ~0.3 eV. Recent work has shown that the Schottky barrier of the Au–Si interface could increase with the thickness of the gold layer. Charge transport inside the gold columns to the Si substrate softens gold, creating a gold nanoparticle and a pore. Hole injection into the silicon substrate charges the local silicon surface; hence catalyzing its etching by HF, resulting in the formation of silicon nanoparticles. On the other hand, as the charge accumulates in the columns, charge breakdown could take place laterally.

The silicon wafer is now sonicated and centrifuged in isopropyl alcohol. This recovers some of the gold component and the luminescent component. Photo of the stable dispersion exhibits red luminescence under UV excitation (Fig. 11.13A. The luminescence spectrum shown in Fig. 11.13B is wide band over the range 550–750nm, characteristic of 2.9 nm red luminescent silicon nanoparticles. In room light, the suspension has a visible light green (red-green) tint Fig. 11.13C. We dispensed a drop of the liquid on a device quality silicon wafer and imaged it with SEM. The images are shown in Fig. 11.13. Half of the particles seen in the image are in the range 15–25nm while the rest of the particles fall in the range 30–35 nm. Those correlate with the sizes of the pores seen in the SEM imaging.

11.8 SYNTHESIS OF GOLD NANO WIRES/RODS USING SILICON NANOPARTICLE [13]

Si nanoparticles are used for growth of gold nanostructures. The measurements were conducted in the Nayfeh's group at the University of Illinois in collaboration with the Noha Elhalawany's group at the National Research Center in Cairo. In the procedure, H-terminated silicon nanoparticles, 2.9-nm in diameter, dissolved in isopropynol alcohol are added to $HAuCl_4$ water solution. The salt serves as a source of gold ions. Isopropanol alcohol, a weakly polar solvent, in which nanoparticles make stable dispersions, helps moderate the hydrophobic water forces to limit silicon's agglomerations.

Once gold ions are attracted to silicon nanoparticles by electrostatic polarization forces, they get reduced by extracting electrons. Diminishment of the luminescence indicates consumption of H on the Si nanoparticles. Gold, having electro negativity higher than that of dangling bonds of H-terminated silicon, has higher reduction potential (1.6 V compared to −0.45 V), undergoes electrochemical metal deposition by efficiently exchanging electrons. Once reduced and more and more of these gold atoms form, the solution becomes supersaturated, and gold gradually starts to nucleate in the form of nanometer structures.

The reaction proceeds instantaneously. The color in room light changes from yellow, characteristic of gold salt to pink, characteristic of gold nanostructures, in correlation with the diminishment of the red luminescence.

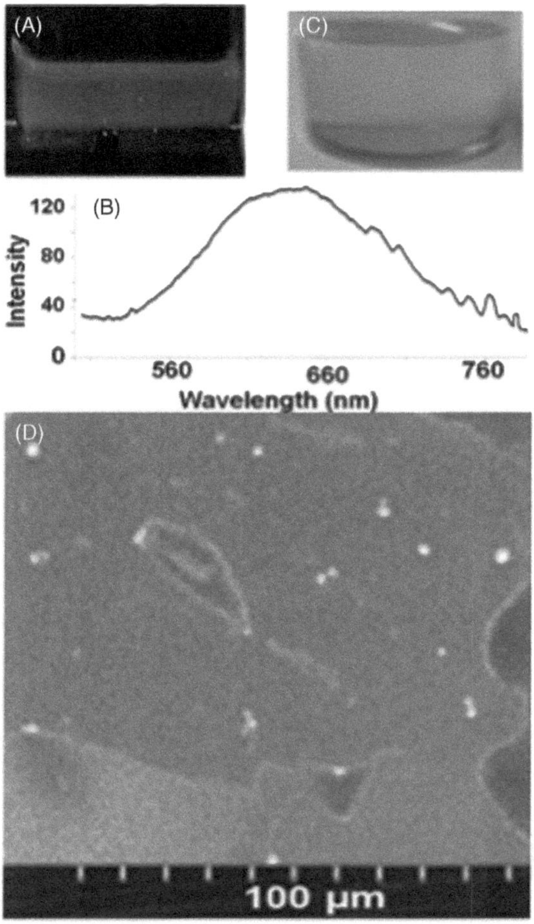

FIGURE 11.13 (A) Photo of the stable dispersion exhibiting red luminescence under UV excitation, after the silicon wafer is sonicated and centrifuged in isopropyl alcohol. (B) In room light, the suspension has a visible light green (red-green) tint. (C) The luminescence spectrumis wide band over the range 550–750 nm, characteristic of 2.9 nm red luminescent silicon nanoparticles. (D) SEM images after dispensing a drop of the liquid on a device quality silicon showing gold nanoparticles in the range 15–25 nm and others fall in the range 30–35 nm.

Thin films are prepared by dispensing a certain volume of the suspension on a device-quality Si wafer. Fig. 11.14A shows SEM images from the film, displaying nanorods from 100 nm to a few hundred µm long. Fig. 11.14B displays other nano structures that this procedure can yield, including sheets, plates (squares, rectangles, trapezoids, and hexagons), bipyramids, and nano particles. Nanoparticles in the rage 10–25 nm are shown on the top left of the figure. The bipyramids have two sharp tips at opposite ends, with the thickest diameter being at the waist. Owing to the sharp tips, they

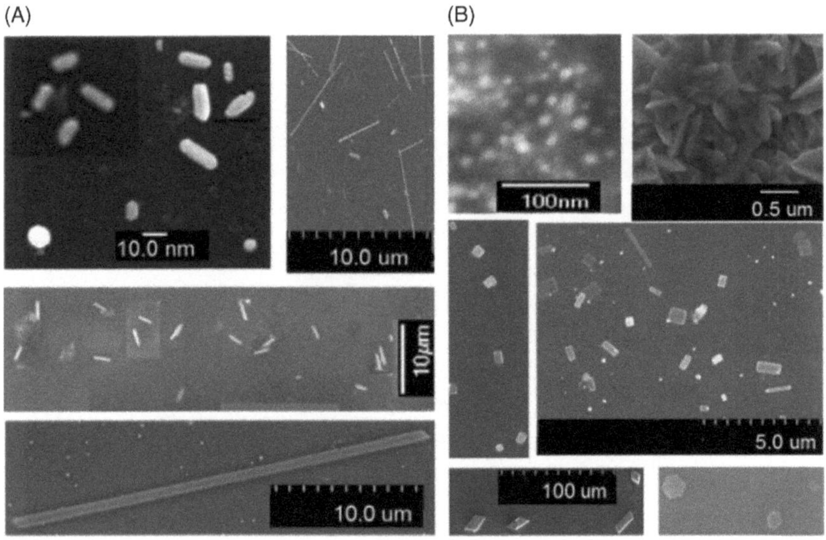

FIGURE 11.14 **Si nanoparticle-catalyzed synthesis of gold nanostructures.** Several SEM images of a wafer after dispensing some liquid (A) variety of gold nanorods, 100 nm to a few hundred μm-long. (B) Nano structures, including sheets, plates (squares, rectangles, trapezoids, and hexagons), bipyramids, and nano particles.

produce stronger local electric field than nanorods. Energy-dispersive X-ray spectroscopy (EDX) confirms a peak at the Mα shell of gold (88.9%), and a smaller peak at the Kα of Si (11.1%).

11.9 SILICON-METAL CORE SHELL [14]

Silicon nanoparticles were used in charge exchange reactions as a reducer of metal ions to harvest metals or their oxides. These studies were performed in the Nayfeh's group at the University of Illinois in collaboration with the Adem Kocyigit's group of Igdir University, Turkey and the Noha Elhalawany's group at the National Research Center in Cairo. For example, in the measurements Mg chloride salt in water is reacted with H-terminated Si nanoparticles of 2.9-nm diameter in isopropynol alcohol. The reaction of magnesium ions Mg^{2+} takes place immediately as evidenced by formation of luminescent clusters, which can be seen as shown in Fig. 11.15A. Mg^{2+} and clusters of nanoparticles are attracted to each other by polarization forces, followed by exchange of holes from Mg^{2+} to oxidize silicon. The reduced Mg simultaneously suffering combustion with water to produce $Mg(OH)_2$ and hydrogen: $Mg(s) + 2H_2O(g) \rightarrow Mg(OH)_2(aq) + H_2(g)$. This is made possible because, unlike bulk, Mg in powder or sub-micron form reacts vigorously with water exothermally. The produced $Mg(OH)_2$ is a low-solubility mineral brucite which forms solid precipitate coating on Si grains.

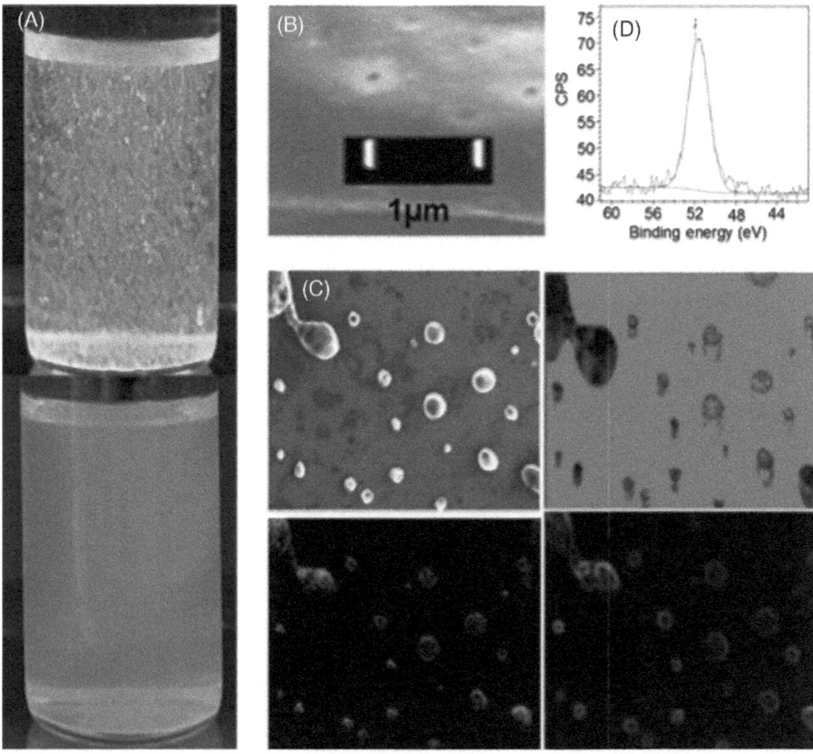

FIGURE 11.15 Silicon–metal core shell. (A) Luminescent photos of 2.9 nm silicon nanoparticles colloid in isopropyl with and without solution of Mg chloride salt in water.(B) SEM images of isolated core-shell structures after dispensing on a Si wafer. They consist of a dark core surrounded by a lighter lower density shell. (C) Elemental analysis and material map using energy-dispersive X-ray spectroscopy (EDX), showing that the core is Si, while the shell consists of magnesium and oxygen. (D) X-ray photoelectron spectrum showing a single peak having a binding energy indicative of the presence of Mg.

H-terminated Si nanoparticles are able to reduce Mg^{2+} to metal because it has electro negativity larger than that of Mg, that is, larger reduction potential (−0.45 V *compared to -2.37 V*), which favors exchange of holes.

A film was prepared by dispensing one drop of the suspension on a Si wafer. Fig. 11.15B shows SEM images of isolated core–shell structures. They consist of a dark core surrounded by a lighter lower density shell. Energy-dispersive X-ray spectroscopy (EDX) provided elemental analysis and material map. Fig. 11.15C shows that the core is Si, while the shell consists of magnesium and oxygen. The X-ray photoelectron spectrum shown in Fig. 11.15D showed a single peak having a binding energy indicative of the presence of Mg. As the phtolumiunescece showed stablity, Si(0) is not consumed while the shell may have provided protective cladding against luminescence quenching.

Similarly, H-terminated silicon nanoparticles, 2.9-nm in diameter dissolved in isopropynol alcohol are mixed with a solution of erbium salt ($ErCl_3$) in water. The salt is used as a source of metal Er^{3+}ions.

Upon mixing the reaction takes place immediately as evidenced by the generation of gas. In room light, the solution shows a pink tint. But with UV excitation at 302 nm, it shows the red luminescence of Si nanoparticles. The pink tint points to Er_2O_3 oxide with 3+ oxidation state. We dispensed several drops on a device-quality Si wafer. Fig. 11.16A. gives a photo luminescent image of the film with 302 nm excitation. Because water dries slowly, drop drying produces higher density of clusters at the edges. We imaged the films with a SEM. Fig. 11.16B shows a number of structures; they consist of a dark high

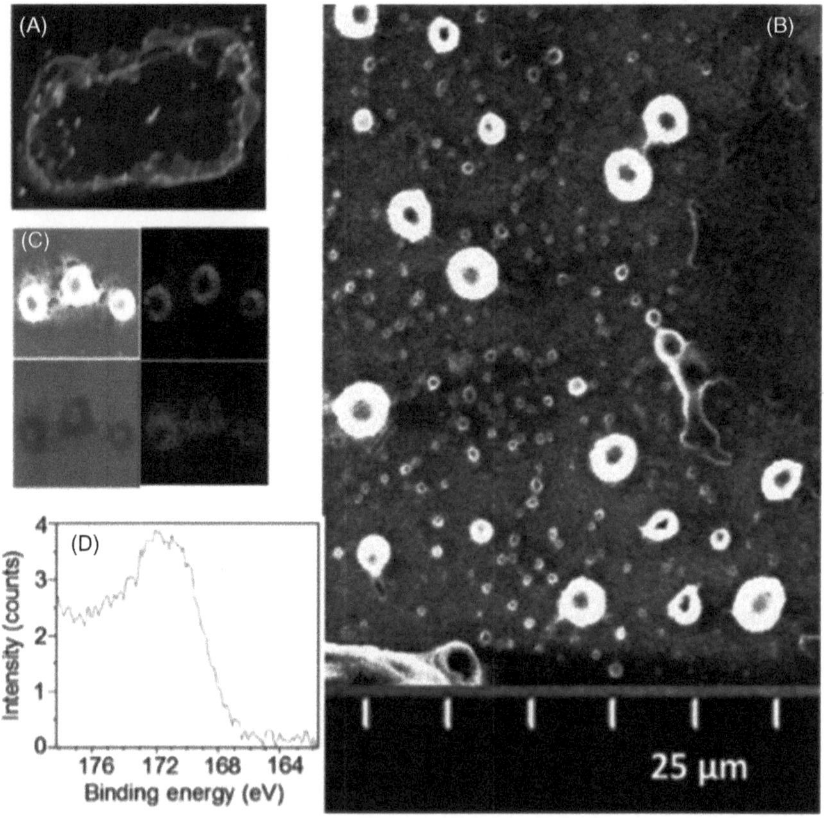

FIGURE 11.16 Silicon–metal core shell. (A) Luminescent photos (with 302 nm excitation) after dispensing some solution on a Si wafer of a mixture 2.9 nm silicon nanoparticles colloid in isopropyl and a solution of Er chloride salt in water. (B) SEM images of the wafer showing isolated core–shell structures. They consist of a dark core surrounded by a lighter lower density shell. (C) Elemental analysis and material map using energy-dispersive X-ray spectroscopy (EDX), showing that the core is Si, while the shell consists of erbium and oxygen. (D) X-ray photoelectron spectrum showing a single peak having a binding energy indicative of the presence of Er.

density core, surrounded by a lighter lower density shell. Material analysis using energy-dispersive X-ray spectroscopy (EDX) was used to provide the map of elemental analysis. The color coded images in Fig. 11.16C show that the core is Si, while the shell consists of erbium and oxygen. To identify the oxidation state, XPS measurements were taken as shown in Fig. 11.16D. The spectrum shows a peak at 170.4 eV, which is known to be associated with the Er4d in Er_2O_3. The sizes of the core–shell structures vary from 100 nm to several micrometers, with some of the larger ones are readily observable by a standard optical microscope. The size distribution is due to clustering of the hydrophobic silicon nanoparticles in water solution.

We believe, in the process, Er ions and clusters of nanoparticles, which are attracted to each other by polarization forces, first exchange holes, while the reduced Er simultaneously reacts with water to produce erbium hydroxide. Metal erbium especially in nanograin or atomic form reacts readily with water exothermally to form erbium hydroxide by releasing hydrogen, which can be dehydrated into Er_2O_3 when deposited on the Si wafer.

11.10 INTEGRATION OF CARBON IN SILICON [15]

There is a number of perquisites for integration of carbon nanotubes in conventional microelectronics as well as the development of novel functional devices. Those include:

- Dispersion of carbon nanotubes: So far limited progress has been accomplished in the dispersion and controlled/precision placement of carbon nanotubes on silicon surfaces. In one dispersion, optical tweezer cum microdissector system discussed in chapter of has been used to disperse carbon nanotubes. In the process, the ultra violet laser (UV) laser of the optical tweezers. The UV laser beam, with a wavelength of 337 nm was applied to bunch of multi-walled carbon nanotubes in water, 24 μJ. The laser had a pulse duration of 4 ns, and frequency of 33 Hz. So, the peak power per pulse is very high (around 75 kW out of which 10 kW is available at objective). A bunch of the multi-walled nanotubes was split and dispersed into smaller parts using the UV laser (UV) laser of the optical tweezers.
- Sorting according to size (bandgap): Development of techniques for separating semiconducting and metallic nanotubes is very important for our ability to integrate. Sorting nanotubes by size and electrical properties,
- Selective and precision positioning: There is a lack of techniques to arrange carbon nanotubes at desired location, which is a major hurdle. There is no proper way to manipulate the physical and chemical characteristics of carbon nanotubes in a controlled fashion. At the moment, insufficient knowledge of handling single nanotubes and performing measurements on them complicates the practical investigation of their physical properties. For investigation of only a single carbon nanotube, other nanotubes have to be well separated in the sample. Otherwise, different nanotubes will influence

each other's physical properties. It is not straightforward to isolate nanotubes since they try to bundle.

- Binding to silicon: There is need to bind carbon nanotubes to the silicon surface. This kind of binding will open up the path to applications like chemical sensors, scanning probe tips etc. based on carbon nanotubes. These carbon nanotubes were made to react with silicon surface. The silicon surface was –OH groups were created on the silicon surface. Carbon nanotubes were processed to form a carboxylic group at their ends.

11.10.1 Fullerenes and Nanotubes on Silicon Surfaces [16]

11.10.1.1 Dry Contact Transfer (DCT)

In the Joseph Lyding group at the University of Illinois, scanning tunneling microscopy/spectroscopy was used to control and investigate carbon nanotubes interfaced with silicon surfaces. First one starts atomically with flat Si (100)-2x1: H surface under UHV conditions. Individual SWNTs are isolated. The nanotube is transferred onto the surface in a dry contact transfer (DCT) manner. The process is carried through under UHV condition. Once, the SWNTs is located on the silicon surface, several characterizations are carried out including atomically resolved imaging, tunneling current–voltage spectroscopy.

Operational process can be performed on the nanotube itself or/and on the underlying silicon surface. Such processes may include lateral manipulation of the nanotube and cutting processes using the STM tip (under room-temperature UHV-STM conditions). On the other hand the, the state of surface passivation can be changed using the top to provide a means to affect Si–C interaction.

First those processes are accomplished with minimal disruption of the atomically flat Si(100)-2x1:H surface. To enhance the interaction between the tube and the surface, a process was carried out for the removal of hydrogen at the nanotube-silicon interface. The hydrogen is removed via UHV-STM electron-stimulated desorption. Unfortunately, the removal resulted in structural and electronic modifications to the SWNT. Moreover, the SWNTs are found to be initially unstable in the presence of the rastered STM tip. Once the hydrogen underneath the tube is removed, that is, depassivating the underlying H-Si(100) surface the tube could be stabilized. These local perturbations incurred by the tube were studied by first-principles simulations of the SWNT/Si(100) system.

Finally for some applications, it is advantageous to prepare finite-sized SWNTs. The UHV-STM system was used to cut a long extended SWNTs using a feedback-controlled cutting mode. Studies showed that the procedure is precise and reproducible.

11.10.1.2 Temperature Studies

Temperature induced deposition of carbon nanostructures on silicon, especially nanotubes is interesting since they may act as it is a means as a precursor to provide carbon atoms, with a purpose:

(i) Integration of carbon nanotubes with silicon-based electronics
(ii) Nanotube/semiconductor heterojunctions
(iii) Silicon carbide nanorods

Single-wall carbon nanotubes (SWNTs) were deposited on hydrogen passivated Si(111) surfaces, which were prepared under ambient atmosphere. Hydrogen is subsequently desorbed from beneath the nanotubes in an ultra-high vacuum (UHV) environment. The temperature of the system was raised. After a short anneal at a temperature of 650°C, defective carbon nanotubes decompose. The nanotube/silicon interface was studied by photoemission spectroscopy. It is to be noted C60 on Si(111)-7x7 is more stable. It decomposes at temperatures ~800 °C. Ex-situ atomic force microscopy (AFM) indicates nanotube decomposition results in silicon carbide islands. The morphology of the island is governed by initial nanotube geometry, which suggests a route for fabrication of nanometer scale silicon carbide structures.

11.10.2 C-Nanotubes in CMOS [17]

One example of monolithic integration of nanotube devices and MOS transistors is introduced in Fig. 11.17. The figure shows an IC. It shows an n-channel metal oxide semiconductor (NMOS) field effect transistor. A catalyst island located on an NMOS decoder circuit is created. Single-walled carbon nanotube (SWNT) devices were grown, using CVD, from the catalyst islands as shown. Each nanotube is addressed individually.

An IC combining SWNT devices with NMOS field effect transistors was fabricated. SWNTs were grown, using CVD, from catalyst islands located on an NMOS decoder circuit. Massive arrays of nanotube devices, each addressed individually using the NMOS circuit, were rapidly characterized. The nanotube devices may be used chemical sensor array, which is electronically addressable.

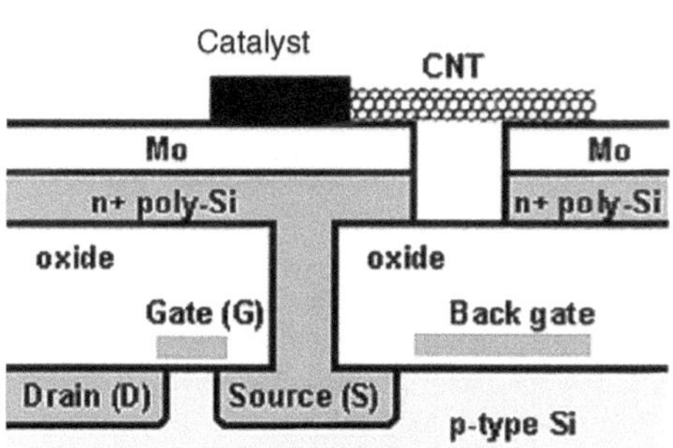

FIGURE 11.17 Monolithic integration of nanotube devices and MOS transistors.

11.10.3 Si Waveguides With Nanotubes [18]

Carbon-based material can emit, modulate and detect near-infrared light. This feature affords an alternative route towards the integration of optoelectronic devices in Si photonics. For example SWNTs may enable all active devices in the Si photonics platform. Those may be delivered straightforwardly using wet-based delivery, such as spin coating, circumventing complex epitaxial growth, or chip bonding approaches.

An optical silicon resonator (photonic waveguide) is built on silicon-on-insulator (SOI) platform (silicon oxide). A simplified model of the integrated system is depicted in Fig. 11.18. A technique based on a polymer-assisted extraction technique is used to select (pick) semiconducting SWNTs from a mixture of metallic and semiconducting SWNTs and impurities (HipCo SWNTs solution). The technique is based on use of polymers to allow diameter-selective dispersion of semiconducting SWNTs. In fact the technique and its mechanism was recently explained [19]. The result is a solution composed of polymer-wrapped semiconducting SWNTs and excess polymer. Second spin coating is used to deposit from this solution a few nanometer thick layer (10 nm) of polymer-wrapped semiconducting SWNTs on top of photonic waveguide. The nanotube-polymer component may be considered (modeled) as nearly homogeneous composite, with a refractive index of 1.5, dominated by the polymer. This modeling of the composite may be useful in conducting theoretical analysis of different geometries and configuration.

This procedure using carbon would be very convenient for integration considering the fact that current Si photonics require on-chip integration of several materials, including III–V for lasing, doped silicon for modulation and Ge for detection. The very different requirements of these materials result in complex fabrication processes that offset the cost-effectiveness of the Si photonics approach.

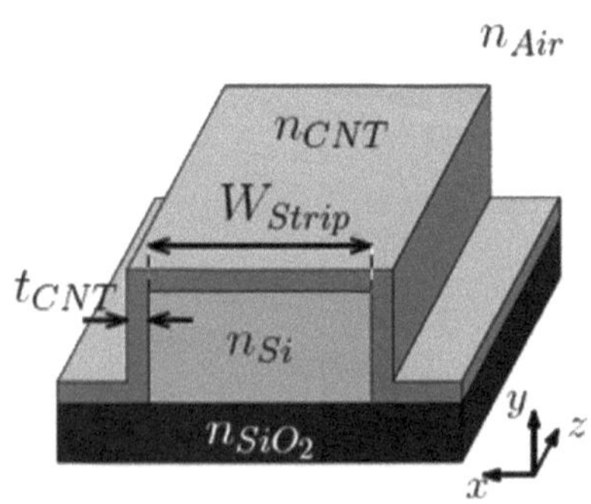

FIGURE 11.18 An optical silicon resonator (photonic waveguide) is built on silicon-on-insulator (SOI) platform (silicon oxide) with a few nanometer thick layer (10 nm) of polymer-wrapped semiconducting SWNTs on top of photonic waveguide. *(From E. Durán Valdeiglesias, W. Zhang, H.C. Hoang, C. Alonso-Ramos, A. Noury, S. Serna, X. Le Roux, E. Cassan, N. Izard, F. Sarti, U. Torrini, M. Balestrieri, A.-S. Keita, H. Yang, V. Bezugly, A. Vinattieri, G. Cuniberti, A. Filoramo, M. Gurioli, L. Vivien, Hybrid integration of carbon nanotubes into silicon slot photonic structures Proceedings Volume 9752, Silicon Photonics XI; 97520D (2016); doi: 10.1117/12.2211317)*

11.10.4 Carbon-Nanotube in Copper: Ultra Conductive [20]

Integrating carbon nanostructures and metal is also very useful. Fabrication techniques to produce metal–carbon composites that have been considered or studied include:

- Electrolytic co-deposition
- Die casting
- Hot-extrusion
- Laser formation and fiber infiltration

The aim is to develop composites with enhanced conductivity, and advanced physical and mechanical properties. One interesting metal in this regard is copper. Lowering copper's electrical resistivity by the incorporation of carbon nanotubes would be highly valuable. Initial attempts may have shown that the presence of carbon nanotubes increases the ambient-temperature electrical conductivity of copper by more than 40%. In particular the production of ultra conductive wires made up of copper/nanocarbon composites would be extremely useful for electric energy transport. In fact enhanced copper is being developed on a laboratory scale at several organizations around the world. But it is still many years away from mass production for commercial use.

Fig. 11.19A–B is morphology of ultraconductive wire based on a copper/CNT composite. Fig. 11.19A shows a longitudinally sectioned wire. It shows regions of high (gray, axial) and low (copper-colored near wire surface) electrical conductivity. Fig. 11.19B gives an electron-micrograph taken in the high conductivity area of Fig. 11.9A. It displays separated, dispersed and oriented CNTs (yellow arrows) embedded in the copper matrix (Source: Taysir Nayfeh, A. Wiederholt at Cleveland State University).

11.10.4.1 Integration of Graphene and Silicon Devices [21]

Hybrid integration of graphene and silicon devices has been recently demonstrated. The silicon device is a silicon complementary metal-oxide semiconductor field–effect transistor (CMOS FET) ring oscillator as shown in Fig. 11.20. The graphene device is an analog graphene FET multimode phase shifter. Fig. 11.20 shows the integrated system. In the process all CMOS components are first built on a silicon wafer. This is followed by the formation of a passivation layer and vertical interconnection. Then the graphene transistors are built on the passivation layer at low temperature. The output signal of the silicon CMOS ring oscillator is vertically linked to the graphene multimode phase shifter. In this scheme therefore, the silicon device is in the first level and the graphene device is in the second level. Thus the procedure constitutes a three-dimensional (3-D) integration in which the graphene device is fabricated on top of the silicon device rather than making them side-by-side. We mention here a number of features.

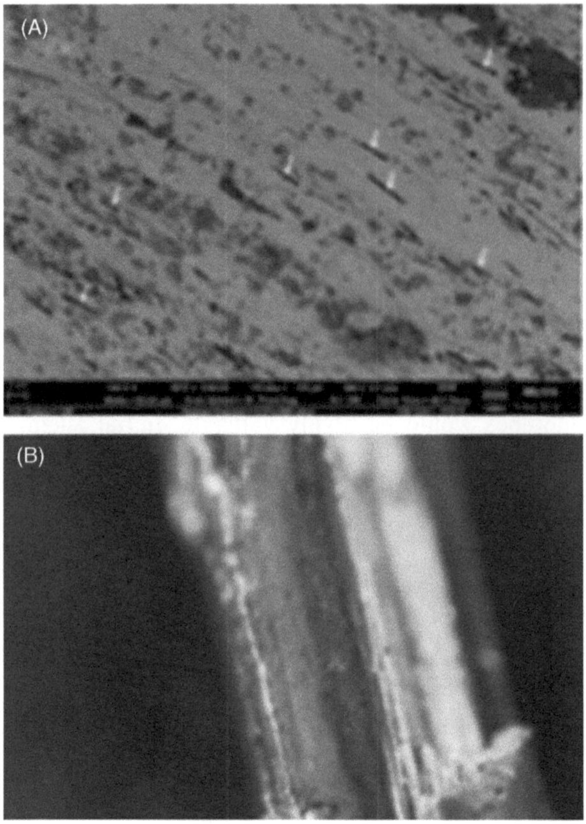

FIGURE 11.19 Morphology of ultraconductive wire based on a copper/CNT composite. (A) Longitudinally sectioned wire. (B) Electron-micrograph taken in the high conductivity area of (A), displaying separated, dispersed and oriented CNTs (yellow arrows) embedded in the copper matrix. *((Source: Taysir Nayfeh, A. Wiederholt at Cleveland State University))*

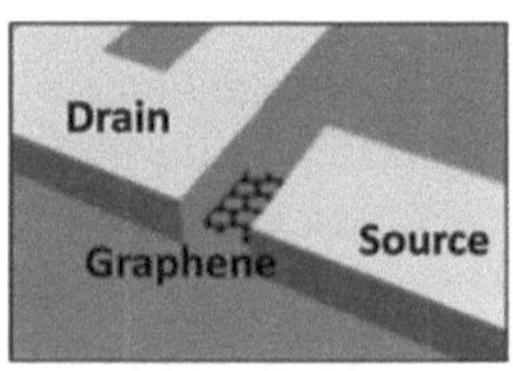

FIGURE 11.20 Hybrid integration of graphene in a silicon complementary metal – oxide – semiconductor field-effect transistor (CMOS FET) ring oscillator .

- Compatible with the conventional silicon CMOS process because of 3-D sequential fabrication flow
- Graphene FETs are not suitable for digital circuits, but suitable for analog devices, because of the absence of a bandgap in graphene (see Chapter 10).
- A hybrid integration a graphene-based analog circuit and a silicon-based digital circuit

- strong advantages in analog applications due to their high carrier mobility and high saturation velocity
- The hybrid device takes advantage of graphene's excellent inherent properties and the maturity of current silicon CMOS technology for future electronics.
- Reduced fabrication complexities, while also preserving the inherent electrical performances of each device.

REFERENCES

[1] Coulthard I, Degen S, Zhu YJ, Sham TK. Gold nanoclusters reductively deposited on porous silicon: morphology and electronic structures. Can. J. Chem. 2014;76(11):1707–16.

[2] Suzer O, Dag O. Reductive deposition of Au^{3+} (aq) on oxidized silicon surfaces. Can. J. Chem. 2000;78:516–9.

[3] Dziuban JA. Bonding in Microsystem Technology. Berlin, Germany: Springer Science & Business Media; 2007.

[4] L. Overmeyer, Y. Wang, T. Wolfer, Eutectic Bonding, CIRP Encyclopedia of Production Engineering, , pp. 488–492.

[5] L.J. Chen, TMS, JOM Metal Silicides: An Integral Part of Microelectronics, http://www.tms.org/pubs/journals/JOM/0509/Chen-0509.html.

[6] Baryshnikova KV, Petrov MI, Babicheva VE, Belov PA. Plasmonic and silicon spherical nanoparticle antireflective coatings. Scientific Reports 2016;6. Article number 22136.

[7] Mantey K, Shams S, Nayfeh MH, Nayfeh O, Alhoshan M, Alrokayan S. Synthesis of wire-like silicon nanostructures by dispersion of SOI using electroless etching. Appl. J. Phys. 2010;108:124321.

[8] (a) Qiu T, Wu XL, Mei YF, Wan GJ, Chu PK, Siu GG. From Si nanotubes to nanowires: Synthesis, characterization, and self-assembly. Cryst. Growth 2005;277:143. (b) Du XW, Lu YW, Kulinich SA, Sun J, Yao P. Microporous silicon connected with silicon wires. Semicond. Sci. Technol. 2006;21:498.

[9] Johnson RC, Xiong L, Shen S, Liu L, Zhang L. Selective and Effective Reduction of Gold Ions from Acidic HCl–$NaClO_3$ Leachate with Oxalic Acid. Ind. Eng. Chem. Res. 2014;53:16672. http://www.eetimes.com/document.asp?doc_id=1141299(2000).

[10] Khayyat MM, Wacaser BA, Reuter MC, Ross FM, Sadana DK, Chen T-C. Nanoscale chemical templating of Si nanowires seeded with Al. Nanotechnology 2013;24:235301. (8pp).

[11] Jungkil K, Kim YH, Choi S-H, Lee W. Curved silicon nanowires with ribbon-like cross sections by metal-assisted chemical etching. ACS Nano 2011;5(6):5242–8.

[12] M. Nayfeh, Auric acid assisted silicon nanoparticle formation method, United States Patent, 9,096,432.

[13] N. Elhalawany, B. Enders, E. Bahceci, A. Kocyigit, A. Nayfeh, H. Ates, M.H. Nayfeh, Synthesis of gold nanorods and filaments by reduction/growth on silicon nanoparticles (to be published).

[14] Hoang T, Elhalawany N, Enders B, Bahceci E, Laila A. Strong 1.54 μm cathode luminescence from core-shell structures of silicon nanoparticles and erbium. Appl. Phys. Lett. 2016;109:261103. Adem Kocyigit, Noha Elhalawany, Ersin Bahceci, Brian Enders, Krithik Puthalath, Laila Abuhassan, Zain Yamani, and Munir Nayfeh, **Wideband Luminescence From Bandgap-Matched Mg-based Si Core-Shell Nanostructures** (to be published).

[15] Park Hongsik, Afzali Ali, Han Shu-Jen, Tulevski George S, Franklin Aaron D, Tersoff Jerry, Hannon James B, Haensch Wilfried. High-density integration of carbon nanotubes via chemical self-assembly. Nature Nanotechnology 2012;7:787–91. doi:10.1038/nnano.2012.189.

Hong Guo, Zhou Matthew, Zhang Ruoxing, Hou Shimin, Choi Wonmook, Woo Yun Sung, Choi Jae-Young, Liu Zhongfan, Zhang Jin. . Separation of Metallic and Semiconducting Single-Walled Carbon Nanotube Arrays by "Scotch Tape" 2011;. doi: 10.1002/anie.201101700.

[16] Albrecht PM, Ruppalt LB, Lyding JW. UHV-STM nanofabrication on silicon. In: Kalinin S, Gruverman A, editors. Scanning Probe Microscopy. New York, NY: Springer; 2007.

[17] Tseng Y-C, Xuan P, Javey A, Malloy R, Wang Q, Bokor J, Dai H. Monolithic integration of carbon nanotube devices with silicon MOS technology. Nano Lett. 2004;4(1):123–7.

[18] Durán Valdeiglesias E, Zhang W, Hoang HC, Alonso-Ramos C, Noury A, Serna S, Le Roux X, Cassan E, Izard N, Sarti F, Torrini U, Balestrieri M, Keita A-S, Yang H, Bezugly V, Cuniberti G, Filoramo A, Gurioli M, Vivien L. Hybrid integration of carbon nanotubes into silicon slot photonic structures. SPIE OPTO, Silicon Photonics XI 2016;9752(97520D). San Francisco; doi: 10.1117/12.2211317.

[19] Star Alexander, Fraser Stoddart J, Steuerman David, Diehl Mike, Boukai Akram, Wong Eric W, Yang Xin, Chung Sung-Wook, Choi Hyeon, Heath James R. Preparation and Properties of Polymer-Wrapped Single-Walled Carbon Nanotubes. Angewandte Chemie International Edition 2001;40:1721–5.

[20] H. Stillman, M. Burwell, Ultraconductive Copper Wire: Overview Of Worldwide Research and Development, International Copper Association, https://www.slideshare.net/mcburwell/ultraconductive-copper-wire-overview-of-worldwide-research-and-development-dec-2013.

[21] Hong SK, Kim CS, Hwang WS, Cho BJ. Hybrid integration of graphene analog and silicon complementary metal-oxide-semiconductor digital circuits. ACS Nano 2016;10(7):7142–6.

Chapter 12

Delivery of Nanoparticles on Surfaces

12.1 MECHANICAL-BASED DELIVERY

This procedure includes dispense evaporation, spin coating, atomizer, incubation, and ink-jet printing and patterning.

12.1.1 Dispense Evaporation: Acetone, Water Crystals

In this procedure precipitation from a volatile solvent (acetone, alcohol, water, THF, etc.) is used to reconstitute nanoparticles into thin films. By gentle evaporation, microcrystallites can be formed on high-quality Si or silicon oxide, or in free standing mode. Slower growth rates using slower evaporation rates at reduced temperatures produces larger, flatter, clearer, and more uniform films. Careful regulation of the temperature allows precise adjustment of the destabilization. Fig. 12.1A shows evaporation from a colloid of 1-nm Si nanoparticles in acetone while Fig. 12.1B shows evaporation from a colloid of 1-nm Si nanoparticles in water. Optical imaging in initial experiments shows crystals of 5–50 μm [1]. Evaporation from water is slow and makes larger see-through nanofilms. Upon drying it cracks into smaller pieces to relive strain with the substrate, with some pieces flying away. On the other hand evaporation from acetone is faster which tend to make smaller, uglier clusters. The water-based structures are optically clear.

12.1.2 Guided Deposition/Assembly

A two-step maskless process to assemble gold nanoparticles from a water colloidal solution at specific sites of a silicon surface has been developed [2]. First, as depicted in Fig. 12.2A a silicon substrate covered by native oxide is exposed to a charged particle beam (ions or electrons). This step modifies the surface chemically by changing the type of surface-terminating groups at the selected areas exposed by the beam. Thus one can have a number of programmed terminations: bare, H-terminated, and O-terminated silicon. The sample is then immersed and incubated in an HF-modified solution of colloidal nanoparticles (Fig. 12.2B). Utilizing differences in the reactivity of gold with Si having these

Fundamentals and Applications of Nano Silicon in Plasmonics and Fullerines.
http://dx.doi.org/10.1016/B978-0-323-48057-4.00012-8

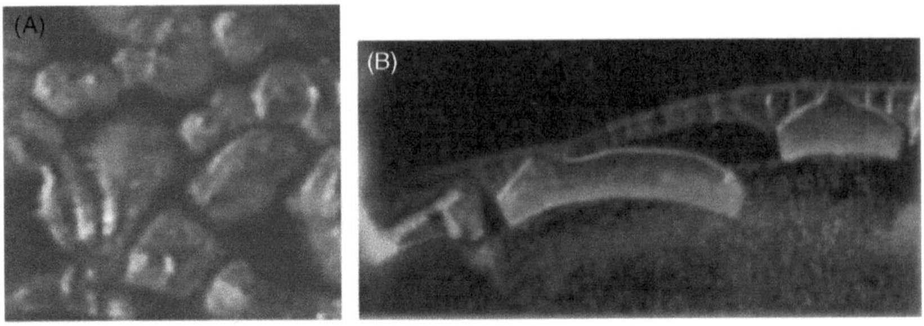

FIGURE 12.1 Crystal formation by drop drying. (A) Evaporation from a colloid of 1-nm Si nanoparticles in acetone. (B) Evaporation from a colloid of 1-nm Si nanoparticles in water.

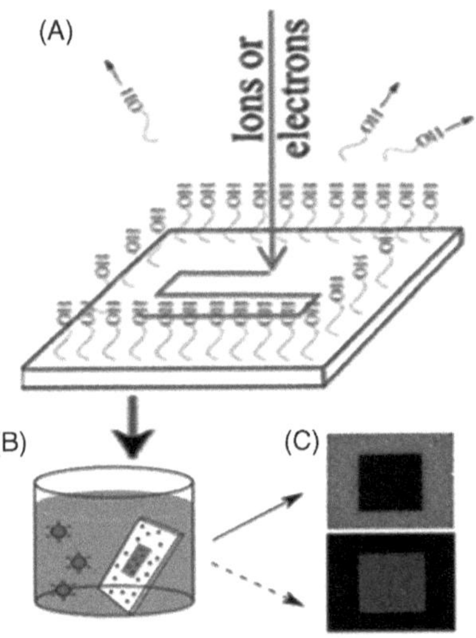

FIGURE 12.2 Mask-less assemble of gold nanoparticles. (A) A silicon substrate covered by native oxide is exposed to a charged particle beam (ions or electrons). (B) The exposed sample is incubated in an HF-modified solution of colloidal nanoparticles. (C) Deposition of colloidal gold nanoparticles.

types of silicon terminations, positive or negative deposition of the colloidal nanoparticles can be achieved (Fig. 12.2C).

When the intensity of the charge beam is high enough, physical removal of silicon may also take place hence creating physical holes in the substrate. This physical roughness may introduce another mechanism for targeted deposition. Thus a pre-programmed etching by an e-beam or ion beam may be used to create chemically surface modified patterns or physical-hole pattern. Nanoparticles of a choice of choice are then dispensed on the substrate from a colloid. A small angle ion beam etching to clean the pattern by remove nanoparticles on the surface while retaining the nanoparticles in the holes. This creates a pattern of clusters of nanoparticles. If the nanoparticles are luminescent, the

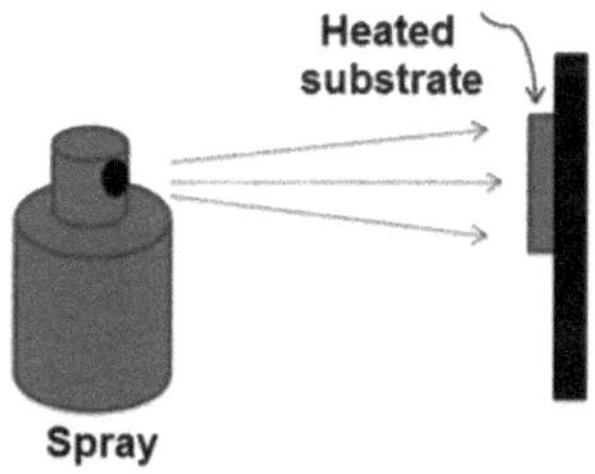

FIGURE 12.3 **Nanoparticles in an appropriate solvent (alcohol for example) is sprayed onto a substrate using differential pressure via a jet/nozzle of an atomizer.**

pattern may constitute a photonic crystal. So generally, one can use a bouquet of parameters, such as the initial substrate termination, the pH of the solution, and beam fluence one achieve either positive or negative deposition of the colloidal nanoparticles.

12.1.3 Atomizer (Spray Coating)

Nanoparticles in an appropriate solvent (alcohol for example) may be sprayed onto a substrate using differential pressure via a jet/nozzle (Fig. 12.3). The volatile liquid evaporates in flight allowing the particles to land on the substrate. If volatile solvent does not evaporate fully in flight, the target substrate may also be mildly heated, for example by a heat lamp so when the nanoparticles in alcohol hit the target, residual alcohol could evaporate right away, leaving nanoparticle spinned.

12.1.4 Spin Coating

In this demonstration, Si nanoparticles are spin coated directly on silicon dioxide layers. Colloids of as-synthesized H-terminated, that is, unfunctionalized, Si nanoparticles of predominately 2.9 nm diameter are used. Starting substrates were 150 mm *p*-Si wafers (10^{15} cm^{-3}). Samples were first cleaned using a Radio Corporation of America (RCA) clean protocol. This step was followed by a dry oxidation procedure at 800 °C. The oxidation procedure grew a 4.2 nm SiO_2 layer with a roughness of 0.3 nm [3].

The first step is to wet the surface. Isopropanol is first dispensed at 600 rpm in order to wet the substrate. The spin speed was then increased to 2000 rpm for 45 s to allow the isopropanol to spread. After wetting, the speed was reduced back to 600 rpm and the nanoparticle/isopropanol colloid was dispensed. The wafer was allowed to spin for another 45 s at 2000 RPM to allow distribution of the nanoparticles.

The uniformity of the film can be examined by spectroscopic ellipsometry. It is used to examine the uniformity of the film, as well as to calibrate the density of the starting colloid to approach sub-monolayer coverage. To estimate the film thickness, the optical constants were needed. In this analysis, the nanoparticle

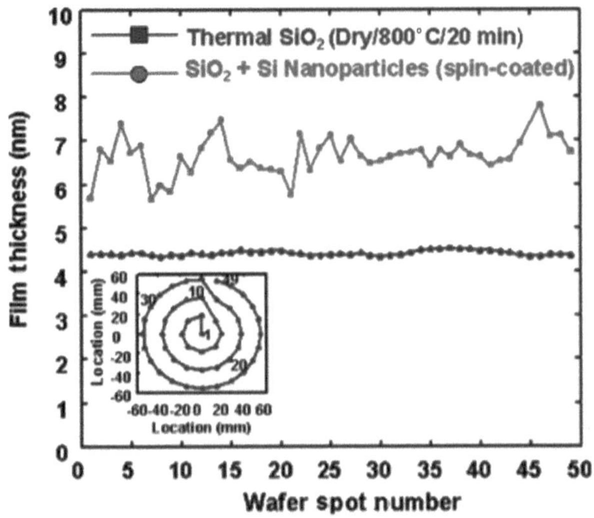

FIGURE 12.4 **Ellipsometry measurements of thickness of an oxide film that was grown on a device quality silicon substrate and the measured thickness of the film after Si nanoparticles were spin coated (SiO_2 + Si nanoparticles).**

film was treated as an extension of the SiO2 layer. This is a good approximation, as the refractive indices (real part) were previously shown to be similar. Moreover, inaccuracy in the absolute thickness due to uncertainties in the optical parameters may not significantly bear on the uniformity of the film derived from ellipsometry measurements. Fig. 12.4 shows the thickness of the oxide film that was grown on a device quality silicon substrate and the measured thickness of the film after the nanoparticles were spin coated (SiO2 + Si nanoparticles). The measurements give an oxide thickness of 4.2 ± .2 nm and a thickness of (2.5 ± .4 nm) for the nanoparticle film, indicating sub-monolayer coverage. The measurement indicates good global uniformity or coverage, an important criterion for integration of silicon nanoparticles across entire substrates for advanced device applications.

AFM was used on several samples to directly image the nanoparticle distribution. Fig. 12.5A–B shows tapping mode AFM images of a 50 × 50 nm^2 section. It shows a sub-monolayer with low levels of agglomeration and a regular interspacing of ~12 nm. The intervening void between the 2.9 nm nanoparticles is deduced to be ~9 nm. The corresponding nanoparticle density is estimated to be ~7 × $10^{11}/cm^2$. AFM was also used to measure the nanoparticle size. Fig. 12.5C–D shows a line profile through an individual nanoparticle. A vertical corrugation of ~3 nm is measured, however, in the lateral direction; a nanoparticle measures 10 nm, much larger than what is expected. It is known that AFM is only accurate vertically because laterally it is dominated by the tip-size effect, typically ~9–10 nm (see Section 4.3). AFM measurements are next compared to transmission electron microscopy

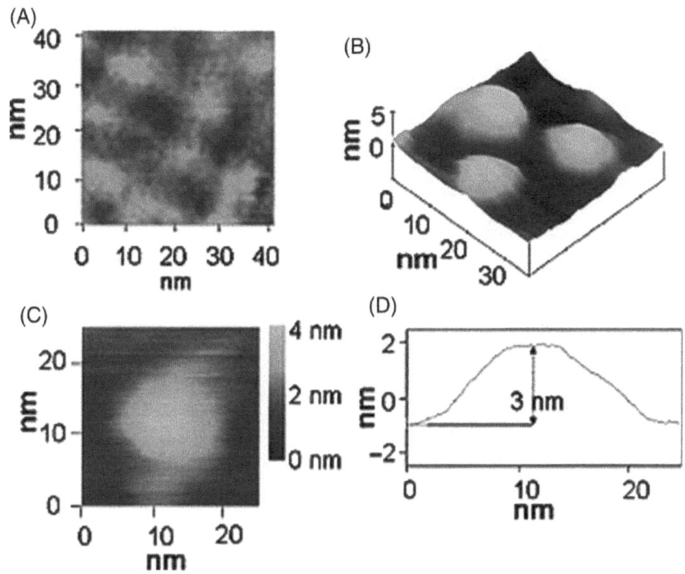

FIGURE 12.5 AFM direct imaging of Si nanoparticle distribution using tapping mode. (A) Two-dimensional AFM image of a 50 × 50 nm^2 section. (B) Three-dimensional image of three particles. (C) Three dimensional image of a single particle. (D) Line profile through an individual nanoparticle. A vertical corrugation of ~3 nm is measured.

(TEM) measurements. A carbon-coated TEM grid was immersed in the same colloid for 20 s and dried. Fig. 9.29A presented a typical high-resolution TEM (HRTEM) image that shows 3 nm nanoparticles, consistent with the AFM depth.

12.1.5 Incubation

Incubation is a convenient method to stain components. In the procedure, the object is simply submerged into the nanoparticle solution. As the solvent evaporates away, the particles in the solution are driven toward the surface, and attach themselves to the surface through some physical bond. Roughness of the surface increases the sticking factor. The procedure may be used to stain biological cells, fiber glass pads or salt crystals. For smooth surfaces like that of a Si wafer, the attachment might be rather weak. For rough surfaces such as that of glass fibers, the attachment is stronger.

As an example, a glass fiber pad is incubated in an isopropyl alcoholcolloid of the nanoparticles. Particles get drawn to and bind on the fibers. Fig. 12.6A–B shows images of the coated pad and of a control pad, respectively [4], both taken under excitation at a wavelength of 365 nm. The red/orange luminescent pad is very bright, indicating that the quantum efficiency of the particles have changed little if any upon surface contact. SEM imaging shows that the coating

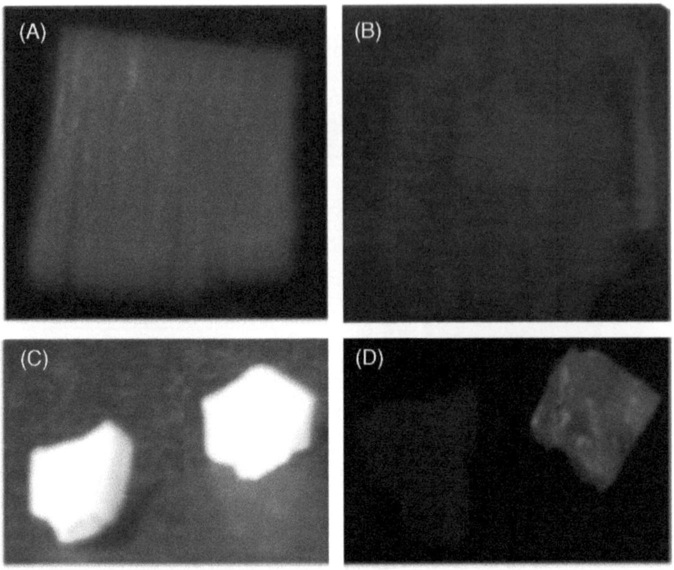

FIGURES 12.6 Coating a glass fiber pad by incubation. (A) Luminescent image/photo of a coated pad taken under 365 nm wavelength excitation. (B) Image of a control pad taken under excitation at a wavelength of 365 nm. (C) photos of coated NaCl crystal (left) and uncoated crystal (right) under room light condition. (D) photos of coated NaCl crystal (right) and uncoated NaCl crystal (left) under UV irradiation at 365-nm wavelength. The blue color of the control probe is caused by scattering of incident ultraviolet.

is not fully uniform but it is comprehensive. The blue color of the control probe is caused by scattering of incident ultraviolet.

Spectroscopic measurements of the spectrum from the colloid and from the coated pad show insignificant change. The spectrum is indeed characteristic of the nanoparticles when suspended in isopropyl alcohol; the spectral distribution shows no shift or broadening from that of the particle colloid, which indicates that the adsorption is physical rather than chemical in nature. This means that despite a variety of oxygen charge states on the surface, the nanoparticle–silica interface did not induce traps that could result in quenching or shifting of the luminescence. Monitoring the pad over a period of 1 year under ambient conditions showed that the optical activity is stable. The coating is found to be highly stable in aqueous solutions and in alcohols as it was incubated for extended periods in these liquids. Sonication of coated fibers in isopropanol results in a fluorescent solution, indicating that some of the particles detached from the fibers. However, the fibers are still fluorescent, indicating that there are still particles attached to the glass fibers.

The same procedure was used to stain NaCl crystals. Fig. 12.6C gives photos of two crystals (one coated and one uncoated) under room light condition. Fig. 12.6D contrasts coated (right) and uncoated (left) crystals under UV irradiation at 365 nm, showing the red luminescence from the coated crystal.

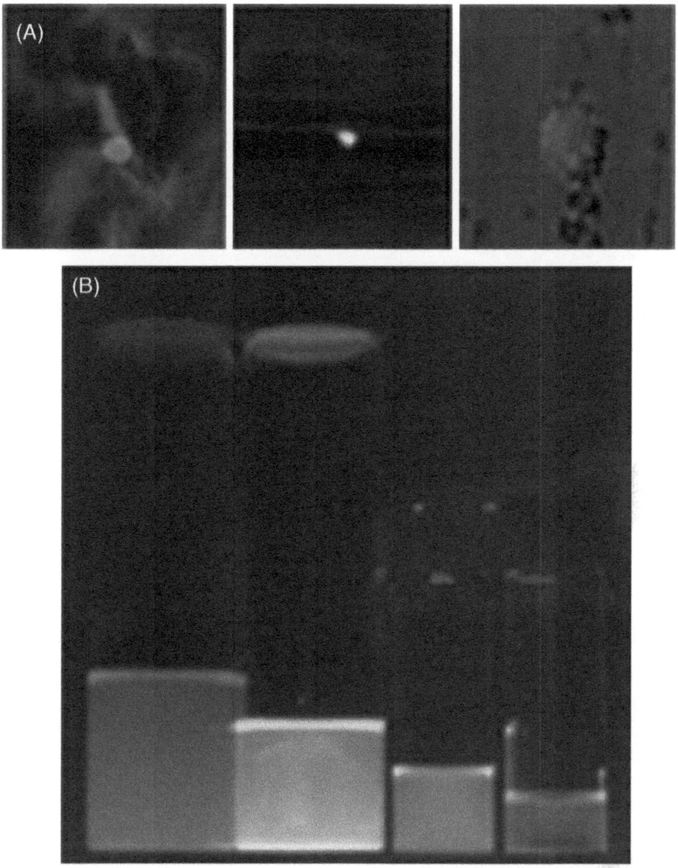

FIGURE 12.7 Colloidal self-assembly (bottom) (B) silicon nanoparticle colloid of 1, 1.67, 2.2, and 2.9 nm diameter stored in separate vials. (A) Upon drying of a soup (mixture) of the nanoparticles.

12.1.6 Colloidal Self Assembly

Silicon nanoparticles of a given size 1, 1.67, 2.2, and 2.9 nm diameters may be prepared and stored in separate vials as seen in Fig. 12.7B. A soup (mixture) of the nanoparticles of more than one size is prepared. Upon drying, larger colloidal crystals may form where same size nanoparticles aggregate together to form larger colloidal crystals (Fig. 12.7A) [5].

12.1.7 Ink-Jet

Fig. 12.8A depicts the setup of electro hydrodynamic jet (e-jet) printing developed by John Rogers at the University of Illinois [6]. It consists of a gold-coated nozzle (internal diameter: 1, 2, or 30 μm), which is located above a substrate that

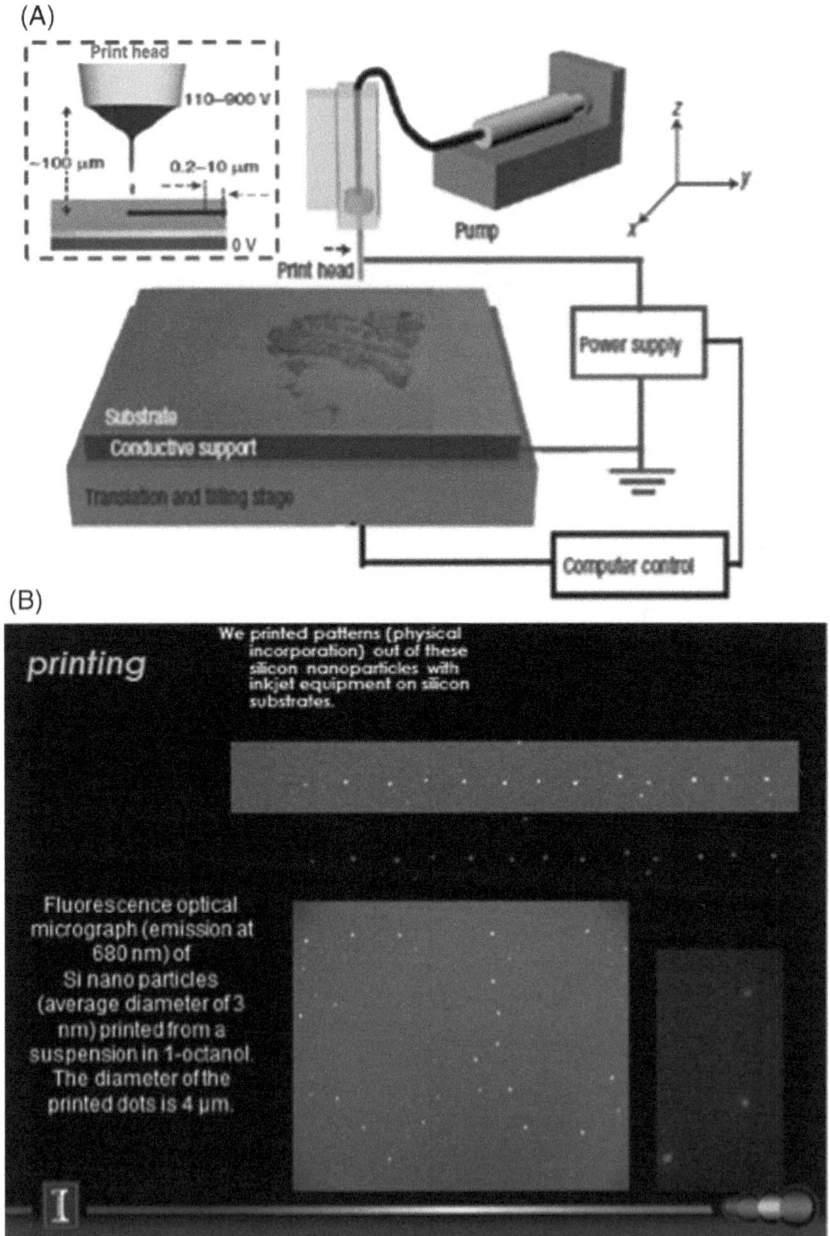

FIGURE 12.8 Electro hydrodynamic ink-jet printing. (A) Setup of ink-jet process (e-jet) printing developed by John Rogers at the University of Illinois. (B) Example of printing with a solution suspension of luminescent silicon nanoparticles on a silicon wafer.

rests on a grounded electrode with a separation (H) of ~100 μm. A power supply connects to the nozzle and the electrode under the substrate. The substrate/electrode combination mounts on a five-axis (x, y, z axes and two tilting axes in the x–y plane) stage for printing. The technique uses electric fields, rather than thermal or acoustic energy, to create the fluid flows necessary for delivering inks to a substrate. The system was used to print with resolution well within the sub-micrometer range. Fig. 12.8B shows an example of printing with a solution suspension of luminescent silicon nanoparticles on a silicon wafer. We printed patterns (physical incorporation) out of these silicon nanoparticles with ink-jet equipment on silicon substrates. A cartridge of an aqueous organic solvent in which the colloid of alkyl-functionalized particles is dissolved operates as a fluorescent nano-silicon organic ink. We can use this method to produce electronic and other devices such as displays based on ink-jet deposition of the silicon nanoparticles. Hybrid gold–silicon nanoparticle inks were developed using thiol linkages.

In fact, using integrated computer-controlled printer, printing has been achieved of complex patterns of a variety of inks, ranging from insulating and conducting polymers, to solution suspensions of silicon nanoparticles and rods, to single-walled carbon nanotubes. High-resolution printed metal interconnects electrodes and probing pads for representative circuit patterns and functional transistors with critical dimensions as small as 1 μm demonstrate potential applications in printed electronics.

This approach has been explored for modest-resolution applications (dot diameters ~20 μm using nozzle diameters (50 μm) in the graphic arts 26–29. This development provide high-resolution (that is, <10 μm) patterning or to fabricate devices in electronics or other areas of technology by use of functional or sacrificial inks.

Other inks commonly employed (dyes or pigments as colorants) in ink-jet technologies, can be used. Materials that may be dispensed include optical polymers, solders, thermoplastics, light-emitting polymers, semiconductor and conducting particles, biologically active fluids, and precursors for chemical synthesis.

12.2 ELECTRIC BASED: ELECTRIC DEPOSITION ELECTROSPRAY

12.2.1 Electrochemical Deposition

Electrodeposition or delivery of nanoparticles uses electrochemical-plating processes [7]. Those are analogous to metal plating. In the method which was developed in collaboration with Taysir Nayfeh's group of Cleveland State University, a conducting substrate is immersed in a solution in which the particles have been dissolved (Fig. 12.9A). Biasing the substrate positively against a

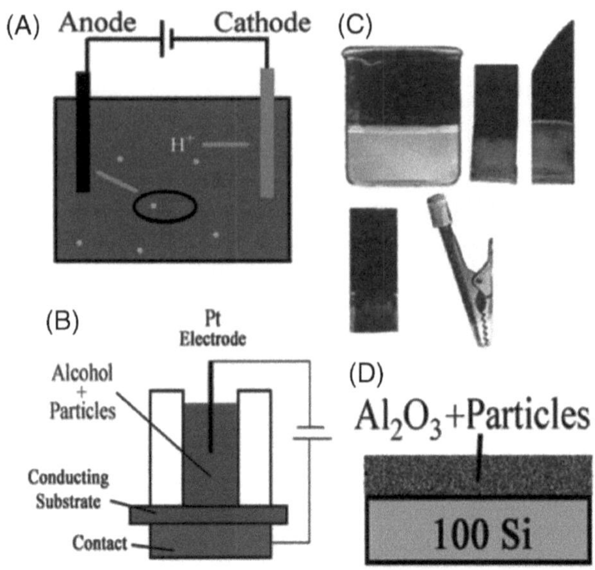

FIGURE 12.9 Electrodeposition or delivery of nanoparticles uses electrochemical-plating processes analogous to metal plating. (A) Schematic of the lateral geometry of electrodes in the plating bath. (B) Schematic of the vertical geometrical configurations of the electrodes. (C) Precursor colloid of red luminescent 2.9 nm particles, as well as a variety of plated surfaces metals, foils, or silicon substrates. (D) Deposition from a colloid of silicon nanoparticles and aluminum chloride salt, which produces thin films of aluminum oxide-silicon nanoparticle films.

negatively based platinum wire electrode draws the particles to the substrate, forming a film. One can use a lateral or vertical geometrical configuration of the electrodes as shown in Fig. 12.9B. A colloid of red luminescent 2.9 nm particles was used. Thin coatings of the particles on metals, foils, or silicon substrates are demonstrated (Fig. 12.9C).

The figure shows a series of images demonstrating this feature. A solution of (a) red fluorescing particles was used to electroplate (b) a stainless steel substrate, (c) silicon wafer, silicon wafer selectively deposited on. This was accomplished by modulation of the conductivity of the Si substrate using (d) oxide masking allowed selective area deposition, and (e) an alligator clip. In the process, the particles get attached to negative alcohol ions (ROHSi) and deposit on the positive electrode. Alcohol is known to ionize to produce the negative ion in the presence of even trace amount of water. We should note that adding aluminum metal salt ($AlCl_3$) to the solution deposits composite films of particles and metal (oxide). In this case the nanoparticles reverse direction and deposit on the negative electrode. The nanoparticles attach more strongly to the aluminum ion. This is depicted in Fig. 12.9D. In the presence of water, the process may produce thin films of silicon–oxygen networks. The film peels off and rolls into hollow cylinders or nanotube (see Fig. 5.35A).

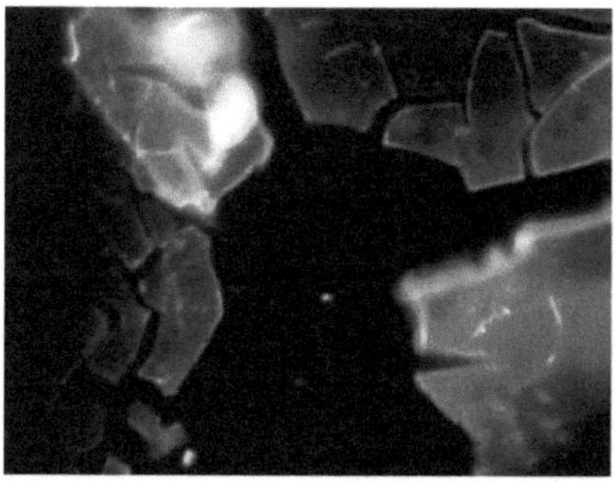

FIGURE 12.10 Electrodeposition described in Fig. 12.9 on a silicon substrate coated with a thin oxide. The deposition proceeds very slowly and mostly at the meniscus (liquid–air interface). Like size particles (same luminescence) assemble/segregate.

In another series of demonstrations, a soup (mixture) of nanoparticles of different sizes was used. A silicon substrate coated with a thin oxide is used. The deposition proceeds but very slowly and mostly at the meniscus (liquid–air interface). Inspecting the interface area shows a deposit where like size particles (same luminescence) assemble/aggregate as shown in the Fig. 12.10.

12.2.2 Field-Driven Assembly

A 300 nm thick thermal oxide was grown on a p-type (100) Si substrate. Patterns in the oxide were etched to expose conducting patterns hence current paths [8]. The substrate is immersed into the alcohol particle colloid, parallel to a platinum wire electrode. The substrate is biased at several hundred volts above the platinum electrode (reverse biasing). Under this bias an electric current of 100–500 μA flows and the particles get drawn on the substrate, mostly along the conducting paths. After the deposition, the substrate is dried. Scanning electron (SEM) and fluorescent microscopies were used to provide detailed information on selected areas. SEM is used for real-space imaging. Fig. 12.11A presents a wide-angle image of the deposit, showing a variety of formations. Fig. 12.11B shows close-ups to display the formations more closely. Upon drying, many of the deposits suffer some pile up as can be seen. However, from the few isolated formation left, one can see some features of tree-like self-assembly. In those, one can observe that the growth avoids the formation of closed loops, branching at an angle between 90 and 120 degrees, with most of them closer to 120 degrees. Fig. 12.11C presents a model of formation, which starts out with clustering and then assembly into branches.

It is interesting to note that the building block of tree formations are not individual nanoparticles, rather particle clusters of ~150 nm in diameter. This growth of uniform spherical aggregates is surprising. The network is illuminated with Hg UV lamp and detected in reverse direction. The image in Fig. 12.11D

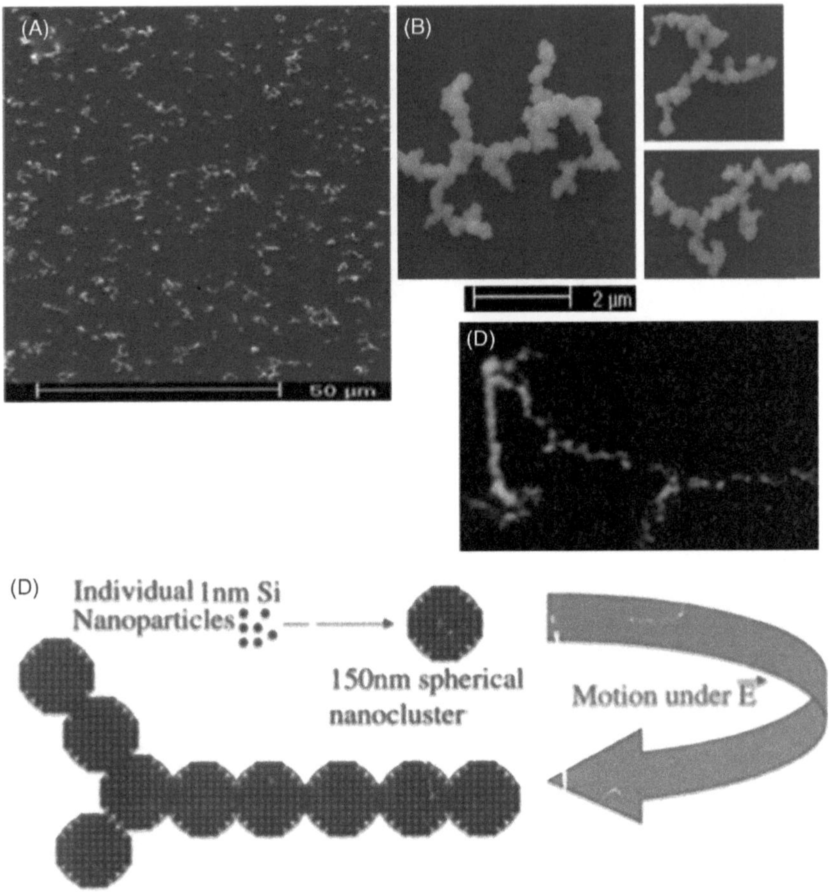

FIGURE 12.11 **Field-driven assembly/deposition of silicon nanoparticles.** (A) A wide-angle image of the deposit. (B) Close-ups show features of tree-like self-assembly, avoiding closed loops, while branching at an angle between 90 and 120 degrees. (D) Fluorescent image (C) presents a model of formation, which starts out with clustering and then assembly into branches.

shows that the clusters making up the trees have the characteristic blue/green emission of 1 nm particles.

12.2.3 Electrospray

This method relies on propelling nanoparticles using voltage [9]. The basic electrospray system (Fig. 12.12A) consists of three main components – a solution reservoir, a metal capillary tip, and a grounded target surface. The target surface is typically a 50 nm thick metallic film deposited on a Si substrate (the metal here is either titanium or ITO – a transparent metal). The capillary tip is biased with a voltage from 2 to 3 kV. A typical spray voltage is 2.2kV for

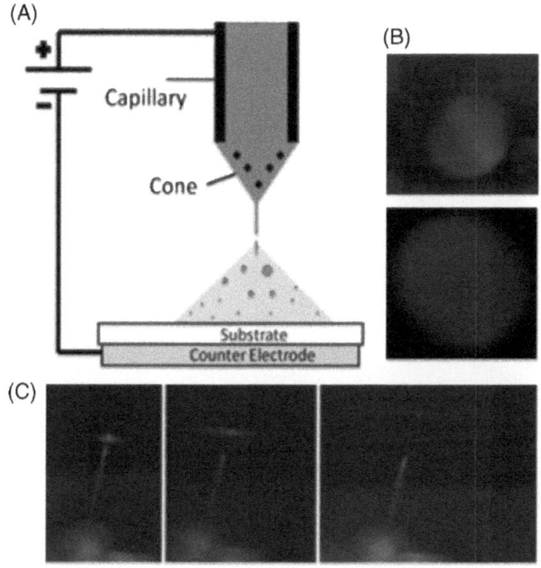

FIGURE 12.12 Electrospray of nanoparticles. (A) Schematic of the nozzle assembly consisting of a solution reservoir, a metal capillary tip, and a grounded target surface. (B) Luminescent image of the deposited film. (C) Light scattering of a red laser beam by the cloud crossing taken at different location between the nozzle and the substrate.

a 3-cm tip-target distance with isopropyl alcohol as solvent (for tip with opening of 100 μm in diameter). For a 50 μm tip, the spray voltage is typically lower, around 2 kV at a tip-target distance of 2 cm. The capillary tip could be biased with a positive or negative voltage, and in each mode, the sign of the charge is the same as that of the tip voltage.

A certain volume of the particle-ion solution is loaded in the nozzle's reservoir. At a positive threshold voltage bias of 2.26 kV with respect to the counter grounded conducting plate, the colloid is propelled and nanoparticles get pinned down on the substrate, resulting in the formation of a thin film. A nanoparticle–Er3+ solution was propelled from the reservoir and formed a film. Fig. 12.12B shows a luminescent image of the film, which displays the characteristic red luminescence of the Si nanoparticles, observable by the naked eye under exposure to UV radiation at a wavelength of 254 nm. The luminescence spectrum of the deposited film is recorded and confirms the presence of the nanoparticles. Fig. 12.12C shows a light scattering by the cloud of a red laser beam crossing at different location between the nozzle and the substrate.

One can add a dielectric mask in front of the target surface to shape, or distort, the electric field into desired patterns. The dielectric masks used here are made out of low density polymer sheets (transparency films) with thickness of 0.1 mm, which shield the electric field very effectively. Selective patterns of circular and square holes were punched into the masks. The masks were

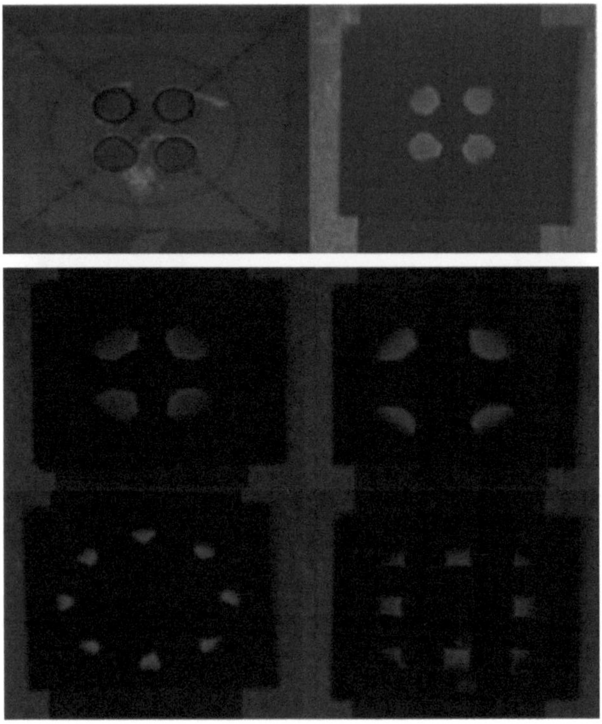

FIGURE 12.13 **Spatial selectivity of electrospray.** A plastic mask with patterns of holes covers the substrate. The fluorescent images show that the nanoparticles indeed land in the unmasked area (in the holes).

typically greater than the target surface to force all the electric fields into the holes (unmask regions). The results are shown in Fig. 12.13 demonstrating the effect of adding dielectric mask on the cloud spray. This fluorescence image shows that the nanoparticles indeed only land in the unmasked area.

12.2.4 Magnetic Control

Silicon or gold nanoparticles do not respond to a magnet. Those when loaded with magnetic structures, such as Fe-based may. In the procedure, H-terminated silicon nanoparticles in isopropyl alcohol and iron (ferric and ferrous) chloride in water are mixed. In the process, the H–Si terminated clusters of silicon nanoparticles act as a reducing and growth element of the iron ions followed by oxidation and formation of magnetic cladding in water [10].

The reaction shows in room light the brown/orange color. We placed a stack of permanent magnets to the side of a glass container. With time on the order of a few minutes the material gets pulled by the magnet and gets pinned down at the wall as shown in Fig. 12.14A. Under irradiation by UV light the

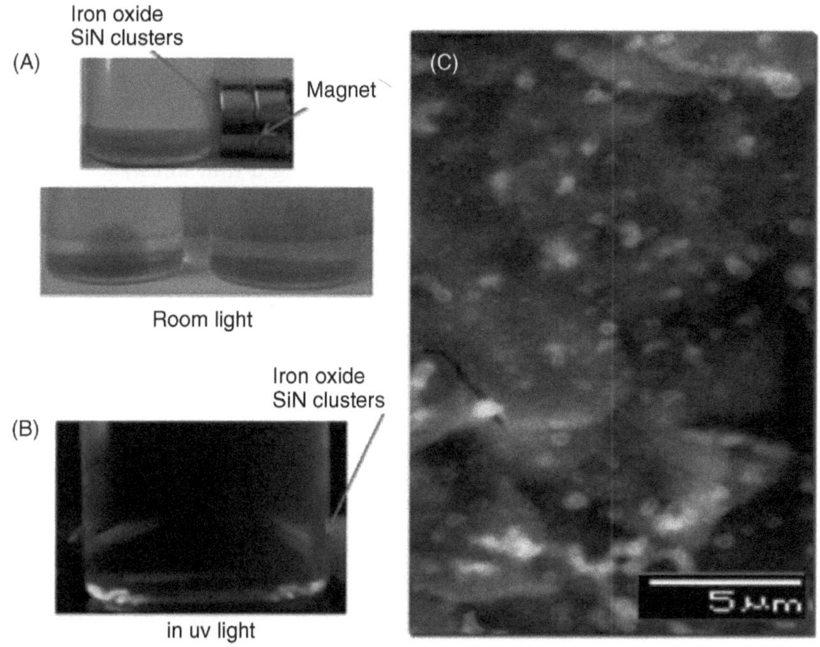

FIGURE 12.14 Magnetic control of silicon nanoparticles. (A) SEM image of core–shell structures with the core being a cluster of silicon nanoparticles and the shell coating of iron oxide. (B) The core–shell structures get pulled to the side wall under the effect of a magnet; (C) under irradiation by UV light the drawn material show red luminescence.

drawn material show red luminescence as shown in Fig. 12.14B. Under UV irradiation the colloid gives a luminescence band over 550–750 nm. A variety of cluster sizes ranging from 100 to 500 nm are observed by SEM as shown in Fig. 12.14C. Thus, silicon–iron oxide hybrid nanoparticles core are prepared with optical, luminescence, and magnetic functionality: brown/orange tint of iron oxide in room-light; characteristic luminescence of Si nanoparticles in UV; and magnetic control of suspension using permanent magnets. The results show that hybrid iron oxide-nanoparticles with luminescent silicon nanoparticle core are useful for providing optical, luminescence, and magnetic functionality.

12.3 PATTERNING: OPTICAL LITHOGRAPHY WITH NANOPARTICLES

The process starts with a p-type device- quality (100) Si wafer of 4–8 cm conductivity 10 dopant cm as a substrate. An oxide layer of 500-nm thickness is grown on the substrate. Imitating common lithography techniques, positive photoresist, masks, and a UV source are used to pattern holes and contacts. The substrate is placed in a buffered oxide etch (NHF: 1HF) for five minutes to etch through the oxide to the silicon substrate. Complete oxide removal is tested with a hot point probe [11].

The hole is filled with Si nanoparticles using the electrode position technique. The particles deposit in the hole, producing a film nm thick with an active region of 5 mm in diameter. The deposition time required depends on several factors, including the concentration of the nanoparticles and the current flow. A current flow in the range of 50–200 A is established. The process may require 20–60 min. The device is then finished by depositing on the particle film a semitransparent gold film of 4-nm thickness. Gold pads of 300-nm thickness are then deposited on the gold layer and on the back of the substrate, providing means to bias the device, as shown in Fig. 12.15A. Fig. 12.15B is a general schematic of the process showing multiple holes. The process was tested using a mask defined by letters "TED 60" as shown in Fig. 12.15C.

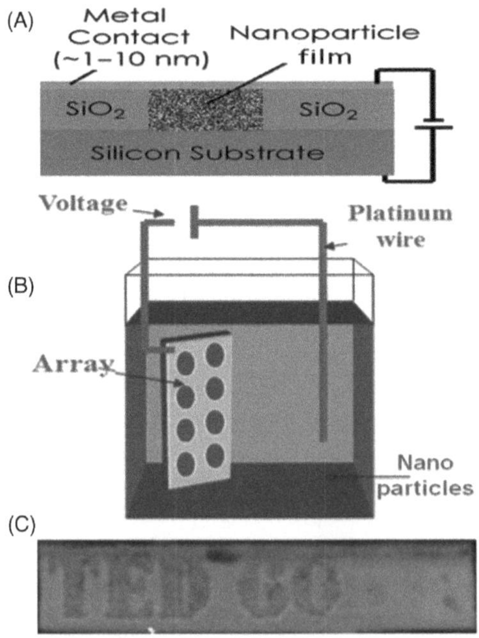

FIGURE 12.15 Patterning with optical lithography. (A) Schematic of a photodetector device built by drilling a hole in an oxide to expose the underlying silicon substrate, and filling the hole with silicon nanoparticles by electrodeposition, followed by metallization and electrical connections. (B) Schematic of the electrodeposition in an array of holes. (C) Enlarged luminescent photo of a system of irregular holes (TED 6O) filled with silicon nanoparticles.

12.4 DELIVERY BY DISPERSION IN POLYMERS AND ADHESIVES

Delivery of nanoparticles to surfaces may be realized by first dispersing the nanoparticles into a carrier polymer or adhesive [12]. The measurements were conducted in collaboration with Meta Bakir of the Jasiuk's group of the University of Illinois, the Kocyigit' group of Igedir University (Turkey), and the Nayfeh's group. Application areas are as diverse as cosmetics, coatings, polishing, and catalysis. Fig. 12.16A shows luminescence spectra of the carboxyl C1 and acetoxy a1 polymers and silicon nanoparticles, as well as their mixture, all in THF and taken under excitation with UV radiation at a wavelength of 365 nm. The polymers exhibit weak blue-green luminescence, whereas the nanoparticles exhibit red/orange luminescence while the mixture shows a pink color. Fig. 12.16B shows luminescent photos of the three solutions individually and their mixture under the same conditions. The figure shows that mixing the silicon nanoparticles with the polymer solution did not compromise; rather it preserved the strong PL of the Si nanoparticles. In room light, the solutions are transparent white. Fig. 12.16C shows the structures of the polymers and Fig. 12.16D presents a TEM image of the silicon nanoparticles.

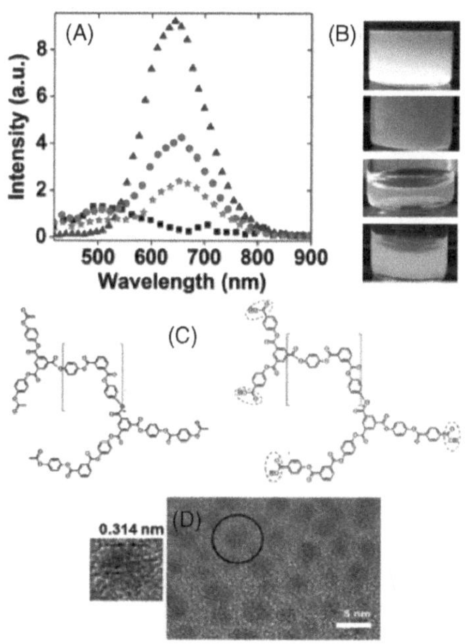

FIGURE 12.16 Delivery by dispersion in polymers and adhesives. (A) Luminescence spectra of the components of the adhesive (carboxyl-based and acetoxy-based polymers) and that of silicon nanoparticles as well as their mixture. (B) Luminescent photos of the three solutions individually and their mixture under the same conditions. (C) The molecular structures of the polymers. (D) TEM image of the silicon nanoparticles.

The three components: C1 carboxyl, a1 acetoxy polymer, and silicon nanoparticles are mixed. A certain volume of the mixture was spread on a preheated glass plate at 50°C, forming a thin film. The film was subjected to a thermal treatment. The temperature of the film was jacked up to 200°C over 30 min. The film was then optically tested. In room light the film is transparent as shown in Fig. 12.17A; but under UV it displays a "red luminescence" as shown in Fig. 12.17B. The thickness of the film is measured in two different ways. First, a cross section was made by curring the glass plate then it was imaged by SEM. Fig. 12.17C shows the image. It shows a thickness of 66.5 nm. A dektak-styler is also used to measure the tackiness.

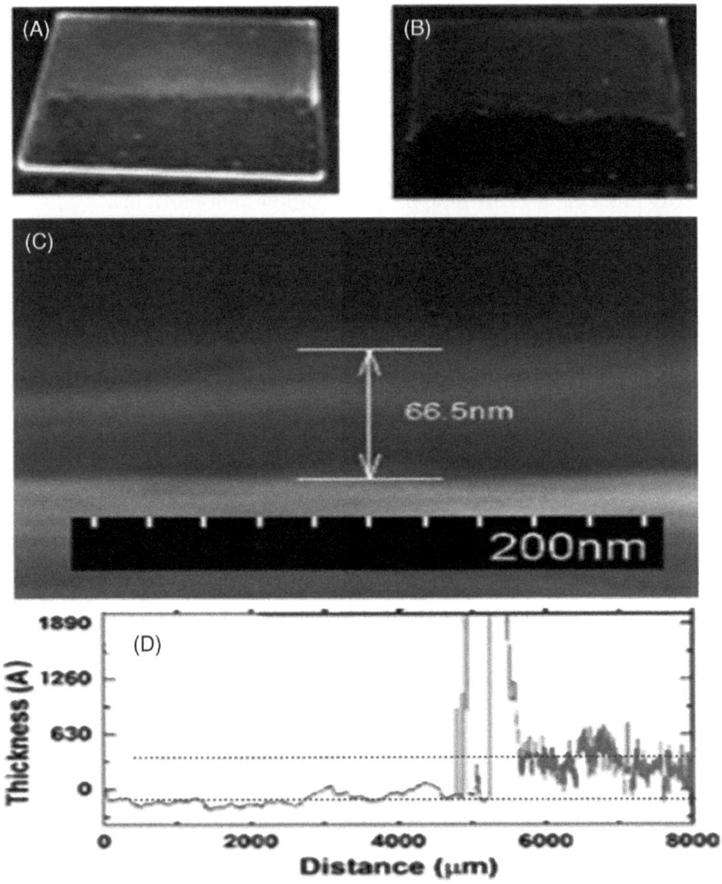

FIGURE 12.17 Characteristics of thin film of polymer impregnated with silicon nanoparticles (A and B) under UV light, displaying "red luminescence." (C) SEM image of a cross section made by cutting the glass plate, giving a film thickness of ~66 nm. (D) A dektakstyler profile giving a film thickness of 50–55 nm.

In addition to carrier of the nanoparticles, those are polymer/adhesive high- to low-k tunable material, and thermally stable from cryogenic to high temperatures, making them useful for Si microelectronics and spacecraft mission applications. The strong bonding, high k, the ability to tune k, as well as stability at high temperatures makes the material compatible with CMOS silicon microelectronics technology, allowing applications in integrated circuits and microelectromechanical systems (MEMS). The high resistance to micro cracking, at cryogenic temperatures is used in next generation spacecraft missions.

12.5 DIFFUSION-BASED DELIVERY IN BODY

We describe in this section a procedure for delivery in body, which was developed by the University of Illinois group in collaboration with the Gorin's group of the Saratov State University and the Ates group of Ankara University [13]. Bare calcium carbonate clusters are incubated in a Si nanoparticle solution in isopropanol (IPA) until IPA dried [13]. A schematic is shown in Fig. 12.18A. They are then recovered and imaged using fluorescence microscopy with excitation at a wavelength of 365 nm. The clusters are expected to exhibit the characteristic red luminescence of the Si nanoparticles, indicating the formation of calcium carbonate–SiNP composite clusters. The red luminescence emanates from the surface as well as the interior of the clusters, which implies that silicon nanoparticles have been drawn to the calcium carbonate, with some penetrating through the larger pores.

The calcium carbonate–SiNP composite clusters are then incubated in a PEI solution in water for 2 days. While in suspension in water, fluorescence

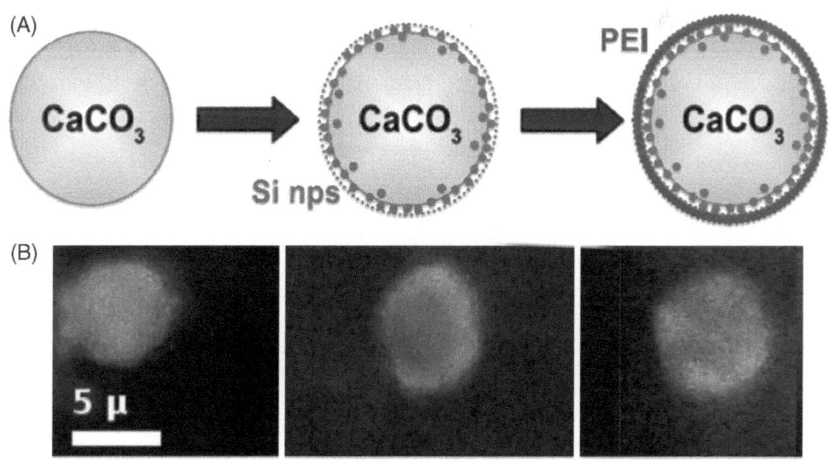

FIGURE 12.18 Polymer containers. (A) Schematic of a calcium carbonate cluster impregnated with silicon nanoparticles and wrapped with a polymer in the form of a container. (B) Synthesized containers.

microscopy images were taken using UV excitation at 365 nm. The fluorescence microscopy images show an inhomogeneous superposition of the red luminescence of the nanoparticles, the green fluorescence of the PEI polymer, and the blue emission of the $CaCO_3$, which points to growth of the PEI polymer.

A strong HCl acid is added to dispersion in IPA of $CaCO_3$–SiNP–PEI composite clusters. Calcium carbonate is practically insoluble in water, but when acid is added, it dissolves producing carbon dioxide, water, and calcium chloride. The $CaCO_3$ template core is removed (dissolved) using a strong hydrochloric acid (HCl) treatment. The carbonate is removed through the polymer, while blocking the passage of the silicon nanoparticles. Fig. 12.18B shows example clusters imaged by fluorescence microscopy (digitally enhanced contrast) while suspended in liquid after the acid treatment. The reduction in the blue tint indicates dissolution of the calcium carbonate. The images show that the microcontainers are mechanically intact and that Si nanoparticles are in the interior of the microcontainers. It is to be noted, however, that dissolution process reduced the strength of red luminescence. Since the red luminescence of the Si nanoparticles is stable in HCl environment due to their Si–H termination, the drop in brightness is probably due to the volume dilution and/or to the fact that a fraction of the nanoparticles may have diffused out through cracks in the polymer coating during the dissolution process. This is plausible since a single PEI layer is expected to be significantly

Fig. 12.19 shows luminescent images of microcontainers. Calcium carbonate microparticles impregnated with silicon nanoparticles followed by encapsulation in a PEI polymer layer, and then, the calcium carbonate was dissolved in HCl. The image was taken using excitation at 365 nm while being suspended in water.

Polyaniline–silicon nanoparticle capsules — a novel organic/inorganic material were also prepared. The capsules are produced by miniemulsion of aniline monomers in the presence of Si nanoparticles. Thin films of the capsules

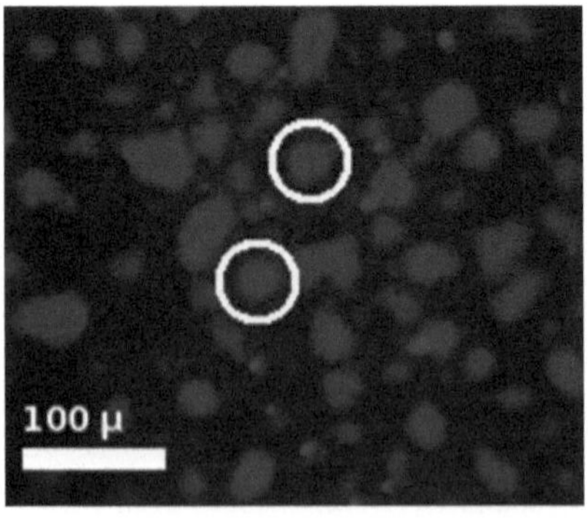

FIGURE 12.19 Polymer containers with the calcium carbonate core are dissolved by an HCL treatment while preserving the polymer wrapping and retaining much of the silicon nanoparticles inside of it.

were formed on a variety of substrates [14,15]. The above procedure covered in this chapters show that the nanoparticles may be delivered on any type of surface, including conducting, non-conducting, flexible, folded, three dimensional, as well as spatially selective deposition with programmed patterning.

REFERENCES

[1] Silicon nanoparticles: New photonic and electronic material at the transition between solid and molecule, in: M.H. Nayfeh, L. Mitas, V. Kumar (Eds.), Nanosilicon, Elsevier, 2007, p. 1; M.H. Nayfeh, Lasing effects in ultrasmall silicon nanoparticles, in: L. Pavesi, S. Gaponenke, L.D. Negro (Eds.), Towards the first silicon laser, Kluwer Academic Publishers, 2003, p. 165; M.H. Nayfeh, J. Therrien, G. Belomoin, O. Akcakir, N. Barry, E. Gratton, Stimulated blue emission and second harmonic generation from films of ultrasmall Si nanoparticles, Mater. Res. Soc. Symp. Proc. 638 (2000) F9.5; M. Nayfeh, L. Mitas, E. Gratton, Si-29 nanoparticles and reconstituted films: a highly fluorescent new form of silicon, Quantum Confinement VI: Nanostructured Materials and Devices, in: P. Braun, M. Cahay, J. Leburton, D. Lockwood, S. Bandyopadhay, J. Harris (Eds.), Electochemical Society Series (2001); M.H. Nayfeh, Laser oscillation in aggregates of ultrasmall Si nanoparticles, Mater. Res. Soc. Symp. Proc. 728 (2004) S6.6.

[2] Kolíbal M, Konečný M, Ligmajer F, Škoda D, et al. Guided assembly of gold colloidal nanoparticles on silicon substrates prepatterned by charged particle beams. ACS Nano 2012;6:10098–106.

[3] Osama MN, Dimitri AA, Mantey K, Munir HN. Uniform delivery of silicon nanoparticles on device quality substrates using spin coating from isopropyl alcohol colloids. Appl. Phys. Lett. 2009;94:043112.

[4] Mantey K, Munir HN, Al-Hreish B, Boparai J, Kumar A, Larry DS, Andrew JN, Salman AA, Khalid MA-S. Silicon nanoparticle-functionalized fiberglass pads for sampling. J. Appl. Phys. 2011;109:064321.

[5] Therrien JM, Belomoin GA, Nayfeh MH. Synthesis of ultra-small Si nano particle colloids and thin films-high temperature single electronics. Mater. Res. Soc. Symp. Proc. 1999;H11:4.

[6] Park J-U, Hardy M, Jun SK, Barton K, Kurt A, Kishore DM, Young CL, Michael SS, Andrew GA, John GG, Placid MF, John AR. High-resolution electrohydrodynamic jet printing. Nat. Mater. 2007;6:782.

[7] A. Smith, G. Belomoin, M. H. Nayfeh, T. Nayfeh, Spatially selective electrochemical deposition of composite films of metal and luminescent Si nanoparticles, Chem. Phys. Lett. 372 (2003) 415–418; S. Chaieb, M. Nayfeh, A. Smith, Assembly of silicon nanoparticles roll up into flexible nanotubes, Appl. Phys. Lett. 87 (2005) 062104-1.

[8] Smith A, Chaieb S, Alaql A, Alsalhi M, Nayfeh MH. Observation of assembly of fluorescent Si nanoparticles under the influence of electric current. Nanosci. J. Nanotechnol. 2002;2:471.

[9] Hoang T, Stupca M, Maximenko Y, Elhalawany N, Carr C, Yu H, Nayfeh MH. Complex of heavy magnetic ions and luminescent silicon nanoparticles. J. Appl. Phys. 2013;114:164319.

[10] Elhalawani N, Enders B, Hong T, Hassan LA, Nayfeh M. Synthesis of core-shell structures of silicon nanoparticles and erbium: 1.54 μm cathode luminescence. Appl. Phys. Lett. 2016;109:261103.

[11] O. Nayfeh, S. Rao, A. Smith, J. Therrien, M. Nayfeh, Thin film silicon nanoparticle photodetector, Photon. Technol. Lett. IEEE 16 (2004) 1927; M. Nayfeh, S. Rao, O. Nayfeh, A. Smith, J. Therrien, UV photodetectors with thin film Si nanoparticle active medium, IEEE Trans. Nanotechnol. 4 (2005) 660.

[12] A. Kocyigit, M. Bakir, O.S. Cifci, B. Enders, I. Jasiuk, M.H. Nayfeh, Imparting optical functionality to aromatic thermosetting copolyester by luminescent silicon nanoparticles crosslinked via *in situ* thermal polymerization reaction (to be published).

[13] Maximenko Y, Gorin D, Ates H, Svenskaya Y, Elhalawany N, Yamani Z, Yau S-T, Munir HN. Layer-by-layer growth of charged polymers and silicon nanoparticles. BioNanoScience 2016;6:142–7.

[14] Maximenko Y, Elhalawany N, Zain Y, Yau S-T, Munir HN. Polyaniline – Si nanoparticles nanocapsules as a dual photovoltaic sensitizer. Mater. Res. Symp. Proc. 2013;1500. https://doi.org/10.1557/opl.2013.182.

[15] Elhalawany Noha, Maximenko Yulia, Yau S-Tung, Nayfeh Munir H. Soluble silicon nanoparticles-polyaniline capsules for biosensing and imaging. Journal of Materials Research 2013;28:210–5.

Chapter 13

Advanced and Low Cost Energy and Lighting Devices

13.1 ENERGY HARVEST-PHOTOVOLTAICS

13.1.1 Standard Solar Cell

Solar PV is now the third most important renewable energy source, after hydro and wind power, in terms of global installed capacity [1]. The standard solar cell is a thin semiconductor device that is based on silicon material and metal as depicted in Figure 13.1. It is a sandwich of silicon material between two conducting metal electrodes. The front metal electrode is structured such that it allows entrance of light. The silicon material is made of two layers, one of which is silicon that is doped with phosphorous atoms (N-type silicon) and the other is doped with boron atoms (P-type silicon), which results in a PN junction. When light from the sun strikes the front side of the cell, part of it is reflected back, part of it is absorbed by the silicon material, and part of it passes through the cell if the cell is thin enough. There are two obvious design refinements. First the front surface of the solar cell is coated with an ultrathin coating to minimize or prevent reflection of light to allow better coupling of the light with the silicon material. The second refinement is the optimization of the amount of silicon material used such that it has the minimum thickness yet it is thick enough to stop light from passing through.

In the process, the absorbed light excites the bound electrons into a higher energy state, making them free electrons. These free electrons move about in all directions within the crystal, leaving holes where the electrons used to be, and the holes also shift around the crystal. The electrons (−) collect in the N-layer, the holes (+) in the P-layer. When the outside circuit is closed, electricity flows.

13.1.1.1 P-N Junctions

When an n-type and p-type silicon materials are joined as shown in Figure 13.2, several effects occur includingcharge diffusion and buildup of an electric field right at the interface. For instance, electrons diffuse from the n-type to the p-type

Fundamentals and Applications of Nano Silicon in Plasmonics and Fullerines.
http://dx.doi.org/10.1016/B978-0-323-48057-4.00013-X

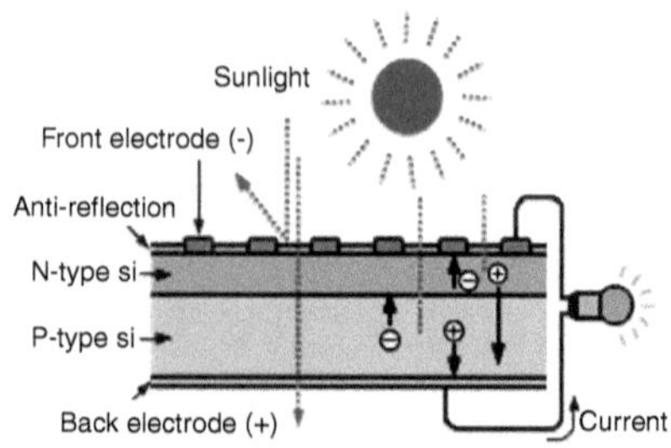

FIGURE 13.1 The standard solar cell, a thin semiconductor device that is in the form of a sandwich of silicon material between two conducting metal.

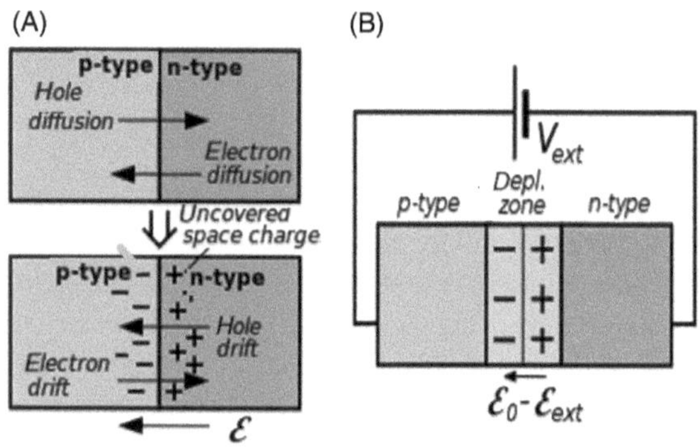

FIGURE 13.2 Schematic of P-N junctions (an *n*-type and *p*-type silicon materials are joined).

and holes diffuse by from the *p*-type to the *n*-type.The process cannot continue because when enough electrons migrate they stop the diffusion of more electrons by repulsion. The same thing happens to the holes. In other words, an electric field Ê forms, with a corresponding potential between the positive ion cores in the *n*-type and negative ion cores in the *p*-type. This region is called the "depletion region" since the electric field quickly pushes out free carriers out(depleted of free carriers).

13.1.1.2 Photovoltaic Effect- Versus Photoconduction

Generation of current in a device due to incident light, without the ability of storage of the charge is not considered power generation. This device is called current source (will be discussed below). In order to generate power, a voltage must be generated as well as a current. The photo generatednegative charge (electrons) are first separated from the photo-generated positive charge (holes) and then the electrons move to the *n*-type side and the holes to the *p*-type side where they are collected.If the external contacts are shorted(short circuit), there is no buildup of charge, as the carriers exit the device as current. If separation, motion and collection of charge are permanent when the circuit is open, then there will be extra electrons at the n type and extra holes on thetype creating, which creates voltage difference across the device is generated which eventually balances the voltage difference at the junction (interface).The voltage required to cause these two currents to balance is called the "open-circuit voltage". The process is known as the "photovoltaic effect"

13.1.1.3 Efficiency of Solar Cells

Because some degree of reflection and transmission are in general are present solar cells, three efficiencies are defined for any solar cell to quantify the amount of electricity produced for a given amount of light are characterized by an important parameter: the efficiency.

- "Quantum efficiency" (Q.E.) is the ratio of the number of carriers collected by the solar cell to the number of photons of a given energy incident on the solar cell.
- The "external" quantum efficiency of a silicon solar cell includes the effect of optical losses such as transmission and reflection. However, it is often useful to look at the quantum efficiency of the light left after the reflected and transmitted light has been lost.
- "Internal" quantum efficiency refers to the efficiency with which photons that are not reflected or transmitted out of the cell can generate collectable carriers. By measuring the reflection and transmission of a device, the external quantum efficiency curve can be corrected to obtain the internal quantum efficiency curve.

13.1.1.4 Types of Silicon

Silicon solar cells are classified according to the type of the silicon material used for solar cells. Those include the highest quality single crystalline, multicrystalline, polycrystalline or amorphous. The key difference between these materials is degree to which the semiconductor has a regular, perfectly ordered crystal structure, and therefore semiconductor material may be classified according to the size of the crystals making up the material. It is to be noted that the cost of the cell correlates with the quality of the silicon used.

Terminology for various types of crystalline silicon (c-Si).

Descriptor	Symbol	Grain Size	Common Growth Techniques
Single crystal	sc-Si	>10 cm	Czochralski (CZ) float zone (FZ)
Multicrystalline	mc-Si	1mm-10 cm	Cast, sheet, ribbon
Polycrystalline	pc-Si	1 μm-1 mm	Chemical-vapour deposition
Microcrystalline	μc-Si	<1 μm	Plasma deposition
Amorphous	a-Si		

Ribbon silicon is a type of polycrystalline silicon—it is formed by drawing flat thin films from molten silicon and results in a polycrystalline structure. These cells are cheaper to make than multi-Si, due to a great reduction in silicon waste, as this approach does not require sawing from ingots [48]. However, they are also less efficient.

Mono-like-multi silicon (MLM)
This form was developed in the 2000s and introduced commercially around 2009. Also called cast-mono, this design uses polycrystalline casting chambers with small "seeds" of mono material. The result is a bulk mono-like material that is polycrystalline around the outsides. When sliced for processing, the inner sections are high-efficiency mono-like cells (but square instead of "clipped"), while the outer edges are sold as conventional poly. This production method results in mono-like cells at poly-like prices.

Thin film
Thin-film technologies reduce the amount of active material in a cell. Most designs sandwich active material between two panes of glass. Since silicon solar panels only use one pane of glass, thin film panels are approximately twice as heavy as crystalline silicon panels, although they have a smaller ecological impact (determined from life cycle analysis).

Amorphous silicon solar cells: Amorphous silicon solar cells are cells containing non-crystalline silicon, which are produced using semiconductor techniques. Amorphous silicon solar cells are mostly used as power sources for devices requiring little electricity or as modulated light sensors. They are common in pocket calculators, watches, light detectors for cameras, and television and car navigator screens

13.1.1.5 Multiple-Junction Solar Cells

The solar spectrum is wide extending from the UV to deep infrared. There is no single material that can be used as the active material that could address or be sensitive to all portions of the spectrum. To alleviate this problem complicated designs which consists of stacks of cells sensitive to different portion of the solar spectrum combine to significantly increase the conversion efficiency. We present in Figure 13.3A a sketch of a solar module with three solar cell layers [2]. Each layer has a different band gap to capture sunlight with different levels of energy. As can be seen from the figure, the solar cell layer with the highest band gap need to be on top. This way, the first layer will capture the highest energy sunlight while letting the lower energy portion pass through to be captured by a lower bandgap material and so on. If the lowest band gap layer is placed on top, then a lot of heat is generated, which reduces the efficiency. The use of multiple materials allows absorbance of a broader range of wavelengths, which affords better conversion efficiency of sunlight to electrical energy.

We need a high band gap to get a high voltage, but a low band gap to get a high current. Since power is the product of voltage and current, one can optimize

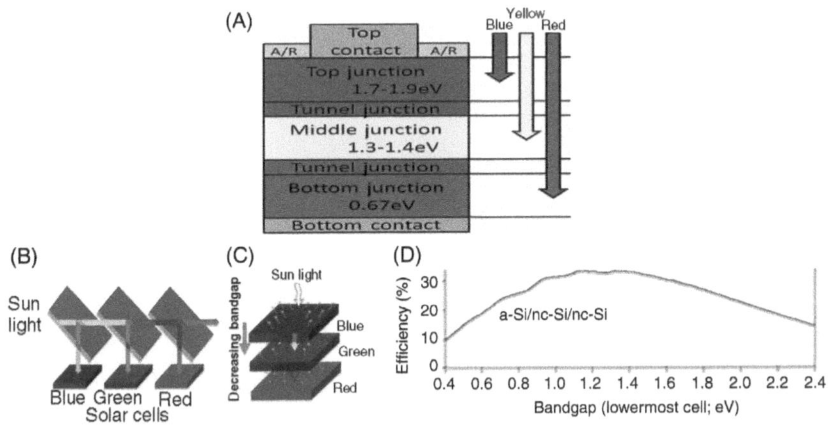

FIGURE 13.3 Multiple-junction solar cells (A) sketch of a solar module with three solar cell layers. (B) Spectral splitting approach with filters used to direct different wavelength bands in sunlight to appropriate cells. *(From Martin A. Green, Stephen P. Bremner, Energy conversion approaches and materials for high-efficiency photovoltaics, Nature Mater, Springer Nature)* (C) the analogue of splitting scheme in terms of a stacked tandem cells. (D) the limiting conversion efficiency of a stack of a-silicon and nc-silicon.

the band gap to generate maximum power. If we do the math, it turns out that an 1.4 eV band gap is ideal – but even this optimized band gap only allows us to capture 31% of the energy (Shockley-Queisser limit). Since the 31% efficiency limit is for a single solar cell, we can stack multiple solar cells on top of each other to capture more of the sun's energy. This is what we call a "multi-junction" solar cell. It should be noted that the traditional single-junction cells have a maximum theoretical efficiency of 33.16% [3]. Theoretically, an infinite number of junctions with graded bandgaps would have a limiting efficiency of 86.8% under highly concentrated sunlight [4].

Spectral splitting approach is also used. The working of the cell can be understood with the aid of Figure 13.3B. In this approach filters are used to direct different wavelength bands in sunlight to appropriate cells. Figure 13.3C gives the analogue of this splitting scheme in terms of a stacked tandem cells, showing light transmission and radiative coupling. Figure 13.3D gives the limiting conversion efficiency of a stack of a-silicon and nc-silicon.

13.1.1.6 Thin Film Architecture

A transition from bulk crystalline Si solar cells toward thin-film technologies aims at reducing usage of active material and may introduce new concepts based on nanotechnologies [5]. Consider an example of the multijunction solar cell shwn in Figure 13.4A [6]. It consists of a stack of two solar cells (so-called tandem cell). One of which is a thin amorphous silicon, which is placed on top. The second is a thicker cell (~ 2 μm) made from microcrystalline silicon and placed on the bottom. Since amorphous layer has a bandgap of 1.7 eV, then the

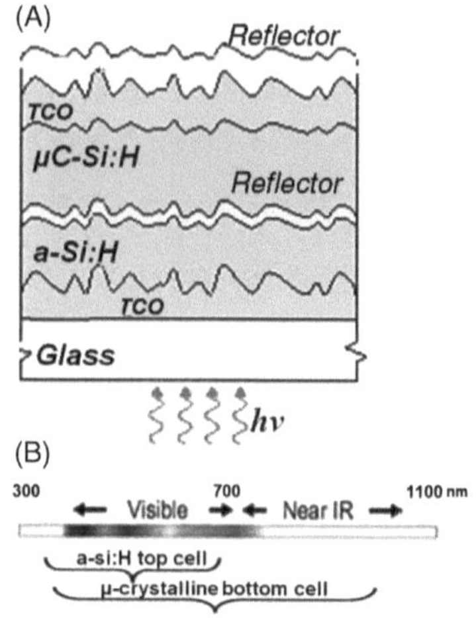

FIGURE 13.4 Thin film architecture (A) schematic of a thin cellconsisting of a stack of two silicon solar cells (thin amorphous silicon, and thicker microcrystalline silicon) (tandem cell). (B) the absorption band regions of the individual cells.

top cell absorbs the blue light. On the other hand, microcrystalline has a bangap of 1.1 eV, then it will absorb the red and near-infrared light (see Figure 13.4B). Typical stable efficiencies are about 9% (world record 11.7%). That is well over the stable efficiencies of single junction thin film silicon solar cells which are around 6%. One reason of the low costs of silicon thin film solar cells is its very low thickness (2 μm) compared to silicon wafer (200 μm). It is to be noted that the theoretical limit of this kind of cell (Shockley-Queisser limit) allows conversion efficiencies of over 30%, well above the measured values.It is belived that light does not couple well because it is too thin, with a significant fraction

Amorphous silicon: a-Si:H

Hydrogenated amorphous silicon (a-Si:H) differs from crystalline silicon by the lack of long-range order (disorder) and the high content of bonded hydrogen (typically around 10% in device-quality a-Si:H). The absence of long-range order in a-Si:H causes a relaxation in the transition rules, so that it behaves as a direct gap semiconductor with a band gap around 1.75eV. This behavior results in a much larger absorption in the visible part of solar spectrum, therefore much less material is needed to collect the same amount of incoming light. Active (intrinsic) layer thicknesses below 0.5 μm are commonly used in a-Si:H solar cells, whereas much thicker layers (around 200 μm) are required for crystalline silicon ones.

Comparing the crystalline silicon and amorphous silicon it must be emphasize the main difference in understanding of the band gap of these materials. While in case of silicon, almost no density of states is presented in the band states, a significant density of states can be observed in a-Si:H in the band gap. The presence of defects in a-Si:H has origin in two mechanisms: (i) the lack of order in the material cause a broadening of the valence and conduction bands leading to band tails, where localized energy states exist, (ii) unpassivated dangling results to defect states near the centre of the band gap.

Advantages of amorphous silicon PV technology are as follows the low deposition temperatures (typically 200–300 °C), which permit the use of low-cost substrates; the possibility to easily integrate such modules into facades, roofs, and other structures; the relatively low production energy and low material quantities that need to be invested in the fabrication of a module; the abundance of all raw materials involved; the potential for realizing a large-scale manufacturing operation with an excellent ecological balance sheet.

One of the main disadvantages of amorphous silicon is degradation of its performance during illumination, which is called Staebler-Wronski effect (SWE) or light-induced degradation. Due to the SWE effect, the electivity of the solar cell usually decreases by 15–35%. Subsequent annealing at 100–250 °C can restore the original efficiency. It has become clear that the SWE is due to the creation of new defects (dangling bonds) that act as additional recombination centers.

Ribbon silicon is a type of polycrystalline silicon—it is formed by drawing flat thin films from molten silicon and results in a polycrystalline structure. These cells are cheaper to make than multi-Si, due to a great reduction in silicon waste, as this approach does not require sawing from ingots. However, they are also less efficient.

Epitaxial silicon development: Epitaxial wafers of crystalline silicon can be grown on a monocrystalline silicon "seed" wafer by chemical vapor deposition (CVD), and then detached as self-supporting wafers of some standard thickness (e.g., 250 μm) that can be manipulated by hand, and directly substituted for wafer cells cut from monocrystalline silicon ingots. Solar cells made with this "kerfless" technique can have efficiencies approaching those of wafer-cut cells, but at appreciably lower cost if the CVD can be done at atmospheric pressure in a high-throughput inline process. The surface of epitaxial wafers may be textured to enhance light absorption.

transmitted through. In the red and infrared wavelength range 2 μm of silicon are not enough to absorb all light and therefore light trapping is needed.

13.1.1.7 Hetrostructures With Intrinsic Thin Film (Layer) (HIT)

The low efficiency as well as the low stability of amorphous silicon solar cells has led to the future study of silicon wafer-based technologies, which has the ability to solve problem with silicon shortage. One of promising technology is HIT (heterojunction with intrinsic thin layer) [7]. Current trend in wafer-based technology is decreasing of wafer thickness. Decreasing the thickness of the wafers by 70 μm brings 10–15% savings of silicon material and thus decreasing overall price. However, when the thickness of the wafer is decreased under 150 μm, the high temperature processes used during fabrication have to be omitted, because they cause the wafer bending. HIT solar cell is composed of a single thin crystalline silicon wafer (~120 μm) surrounded by ultra-thin intrinsic silicon layers and n-type and p-type doped amorphous silicon layers, which can be deposited using temperature below 300 °C [8] and so can be used in processing of thin wafers. The p/n junction of the p-type substrate HIT solar cell is formed by the p-type amorphous silicon a-Si:H layer and the n-type crystalline silicon substrate. The impurity-free intrinsic amorphous silicon thin layer is inserted between the highly doped a-Si:H layer and the substrate in order to reduce recombination of charge carriers at the interface. The intrinsic amorphous layer has ability to decrease the defect density at the interface due to three orders of magnitude lower defect density presented in this layer [9].

13.1.1.8 Crystalline Silicon Solar Cell (Heterojunction Back-Contact Type)

Shows the structure of the cell [10]. The top of the cell is a thin crystalline silicon (in black). The front of the cell is coated in amorphous silicon and an anti-reflective layer to improve efficiency to allow entrance of light in the cell. Below this layer, there is a layer (in yellow) of intrinsisc amorphous silicon (a-Si) that acts as a buffer. The next layer of the cell structure consists of an n-p amourphous structure: p-type a-Si/n-type alternating pattern layer (in pink/blue color). Finally both the negative and positive electrodes in an alertnateing pattern are placed at the botton (rear) end. This crystalline silicon solar cell that incorporates heterojunction technology and back-contact technology brings to the photovoltaic field a structure that has the following advantageous features:

- Heterojunction technology: which binds together plural kinds of semiconductors having different material properties such as combining crystalline silicon with amorphous silicon to reduce defects which lower conversion efficiency or combining materials which convert the light energy of different wavelength ranges into electricity to obtain a higher conversion efficiency.
- Back-contact technology, which forms electrodes only on the back face of solar cells, can make the light receiving area of the top face larger by gathering up all the electrodes on the back face and as a result enhance conversion efficiency.

With this design Kaneka Corporation [11] has surpassed the world record by 0.7 % to a new world record of world's highest conversion efficiency of 26.33% in a practical size (180 cm2) crystalline silicon solar cell.The theoretical efficiency limit of this type of cell as calculated is 29%.The difference of 2.7 % is attributed to a number of losses. First, a resistive loss is estimated at 0.5 %.The optical loss was put at 1%, and finally an extrinsic recombination or collection loss was estimated at 1.2% in which a free electron is prevented from going on for current collection.

13.1.1.9 Solar Concentrator

Concentration photovoltaics uses lenses and curved mirrors to focus sunlight onto small, but highly efficient, multi-junction solar cells. [12] A cartoon is shown in Figure 13.5 [13].

- Fresnel lenses: are shaped like a dart board, with concentric rings of prisms around a lens that's a magnifying glass. They focus scattered light to a tight beam.

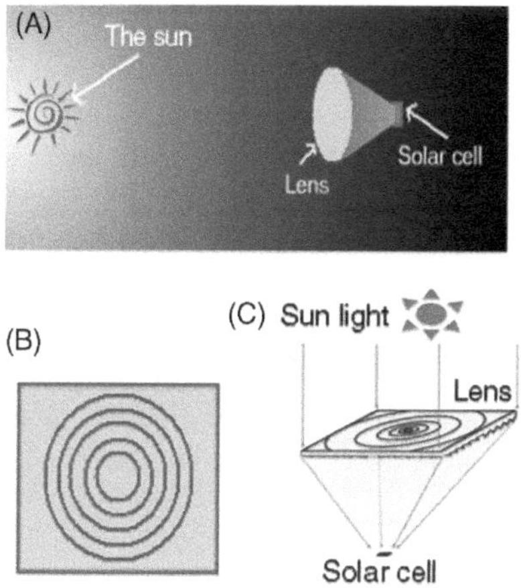

FIGURE 13.5 Solar concentrator: (A) cartoon of the solar concentrator system (B) diagram of a Fresnel lens (C) close-up of lens-solar cell head.

- A Fresnel lens is placed on top of every solar cell to focus sun light on the solar cell.
- Solar cells used are smaller, but highly efficient, multi-junction (MJ) solar cells.
- Solar cells can be spaced farther apart since light can be focused on each cell. This will decrease the number of solar cells needed. Moreover systems can be mechanized to track the sun and for cooling to further increase the efficiency.
- The high intensity allows the use of thicker glass or plastic over the solar panel, which protects it from damage by oncoming fast microparticles
- More suitable in sunny places as direct (not diffuse) sunlight is needed.

It is to be noted that Fresnel lenses have been around since Augustin Jean Fresnel invented them in 1822. Theaters use them for spotlights and lighthouses use them to make their lights visible at greater distances.

Non-silicon TF solar cells

The other TF solar cells, which will be not further discussed, are organic and the dye based solar cells. According to the National Renewable Energy Laboratory institute in the USA (NREL), the promising candidates for the best price/power ration are CdTe and CIGS solar cells.

The potential of Dye-sensitized solar cells depends on possibility to replace indium-tin-oxide (ITO) contact with another material and can by pushed into the terawatt range. However, the dye-sensitized solar cells suffer from degradation during operation, and further research needed to be carried out to make it effective for commercial usage.

13.1.1.10 Spectra Conversion and Down Shifting [14]

The problem with conventional single-junction silicon solar cells is that they effectively convert photons whose energy in near resonance with (close to) the bandgap (E_g) energy of silicon (1.1 ev). Therefore there is a mismatch between a significant portion of the incident solar spectrum and the spectral absorption properties of the material. For example, photons with energy (E_{ph}) smaller than the bandgap are not absorbed, hence they are not used for carrier generation. On the other hand, photons with energy (E_{ph}) larger than the bandgap are absorbed with an electron freed, but the excess energy $E_{ph} - E_g$ is lost due to relaxation and thermalization of the electron. These fundamental spectral losses are approximately 50%. Several approaches have been suggested to overcome these losses

- Multiple-junction cells
- intermediate bandgap material
- multiple exciton generation
- quantum dot concentrators: Luminescent solar concentrators doped with CdSe/CdS quantum dots provide a potentially low-cost and high-performance alternative to costly high-band-gap III–V semiconductor materials to serve as a top junction in multijunction photovoltaic devices for efficient utilization of blue photons. In this study, a photonic mirror was coupled with such a luminescent waveguide to form an optical cavity where emitted luminescence was trapped omni-directionally. By mitigating escape cone and scattering losses, 82% of luminesced photons travel the length of the waveguide, creating a concentration ratio of 30.3 for blue photons in a waveguide with a geometric gain of 61. Further, we study the photon transport inside the luminescent waveguide, showing unimpeded photon collection across the entire length of the waveguide.
- spectral converters (down- and up- converters) and (downshifters): Up-conversion sums two or more low energy (long wavelength) photons, each of energy $<E_g$ into one photon of energy $>E_g$. Down-conversion convers one photon of energy $>>E_g$, into two or more photons of energy $\sim > E_g$, i.e., with sufficient energy to be absorbed by the photovoltaic cell, with negligible thermalisation loss. Downshifters are materials that absob the high energy photons and convert it to lower energy photon $\sim >E_g$

In these so-called third- or next-generation PV concepts, nanotechnology is deemed essential in realizing most of these concepts.

Metal nanostructures have been studied rigorously for light scattering application in solar cells. Recently semiconductor nanostructures have gain attention for this purpose as they have the capability of downshifting high energy photons to lower energy photons and increase the intensity of incoming light at a wavelength range for which solar cell has its maximum response. These nanostructures absorb part of the incoming light which has higher energy than its band gap, making electrons to excite within from lower energy state to higher energy state. Couple of things can happen afterwards, i) these electrons can relax back to their original state by radiative recombination and in the process, emit light with energy lower than it band gap, or ii) once these electrons get excited to higher energy state, there is a charge separation and collection stage. Electrons within the nanostructure get collected directly by the electrodes.

13.1.1.11 Nano Si Solar Cells

A thin nanosolid of silicon nano particles placed on top of a solar cell can serve in three different ways as an enhancer of the efficiency:

Active effects:

- Top junction in multijunction photovoltaic devices for efficient utilization of blue / UV photons.
- Downshifter are materials that absorb the high energy photons and convert it to lower energy photon $\sim >$Eg.

Passive effects:

- UV filter.
- Forward scattering and collimation of incident light.
- Strong scattering and anti-reflection coating to increase the coupling of incident photons for use in thin film cells.

Nanostructuring can be achieved by delivering pre-prepared nanoparticles on the surface such as silicon or metal nanoparticles, or by nano texturing of the surface using acid treatment or laser ablation.

13.1.1.12 Nano Polycrystalline Cell

The use of prepared silicon nanoparticles to enhance commercial solar cells was conducted by the Nayfeh's group at the University of Illinois in collaboration with Mohammad Alsalhi and Turki Al Saud and Abdulrahman Almohanna of KSU and KACST in SaudiArabia [15]. A thin film of silicon nanoparticles or capsules of silicon nanoparticles, for example, are placed on top of a silicon-based p-n junction (amorphous, polycrystalline or mono crystalline silicon solar cell). Figure 13.6A shows BP Solarex Si cell. It is made of smaller and randomly oriented crystallites (polycrystalline). It is in contact with a heat sink. It is *p*-type Si with a thin *n*-type Si layer on top and coated with a very thin Si_3N_4 to filter out the heat generating infrared. The units are 12×12 cm^2 sheets, with 50 grid lines and two thicker buss lines (made of a silver/copper alloy). The cell's surface is textured with 2–4 μm across pyramidal islands, which reduces the reflectivity and enhances light coupling. The metal grids are bare and rough with a 1 μm typical substructure.

A nanocrystal film is now laid down on top of the cell. In these demonstrations, 2.9-nm and 1-nm silicon nanoparticles, prepared by electrochemical

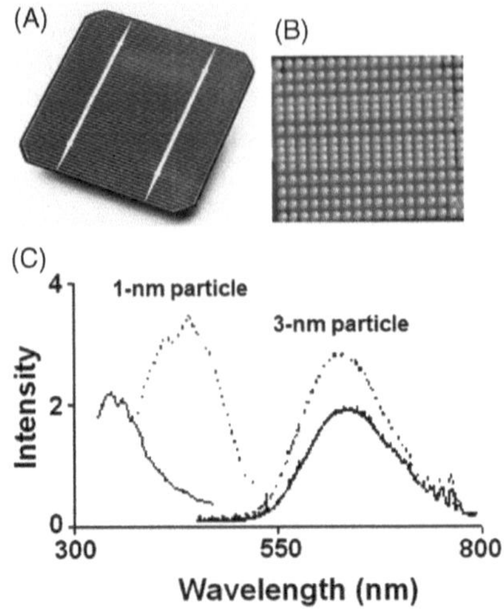

FIGURE 13.6 Nano polycrystaline cell (A) photo of a commercial British Petroleum (BP) polycrystalline solar cell. (B) a cartoon of the nanoparticles on the solar cell. (C) the photoluminescence of the precursor colloids for both 1- and 3-nm nanoparticles. Excitation at 365 nm (dash). Excitation at 254 nm (solid).

etching of silicon wafers, were used. Because the surface of the cell is rough due to texturing and presence of grid and bus lines, the nanoparticles were delivered from a volatile alcohol solution/colloid by dispensing rather than spin coating. We spread a certain volume of the colloid. Alcohol is allowed to dry.

Figure 13.6B presents a cartoon of the nanoparticles on the solar cell. Actual SEM images taken near an edge of a grid line for coated and un-coated cells show random clusters or islands of nanoparticles rather than superlattices of isolated nanoparticles (Figure 13.7). Drying from a liquid forms a film of closely packed nanoparticles. They show nanoparticle material overlaying and making contact with the metal lines. Away from grid line, the film is expected to be smooth and nearly transparent. Unlike on rough metal

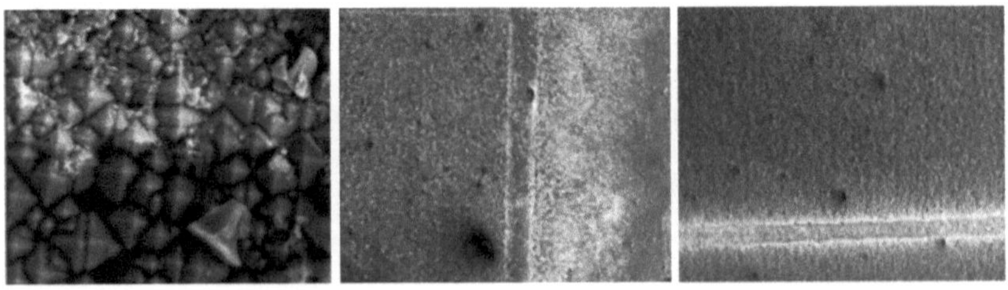

FIGURE 13.7 SEM images taken near an edge of a grid line for coated and un-coated cells show random clusters or islands of nanoparticles.

surfaces, particles are routinely observed to make smooth films on Si wafers. Under 365 nm irradiation, the luminescence is seen by the naked eye to cover the treated portion. A fluorescence microscope using excitation at 365 nm shows similar features. Figure 13.6C gives on the left and right the photoluminescence of the precursor colloids for both 1- and 2.85-nm nanoparticles respectively. Excitation at 365 nm (dash). Excitation at 254 nm (solid). The luminescence of the 2.85 nm particle is an orange/red band with a peak at 630 nm for both excitations. The 1 nm particle gives a blue band which red-shifts monotonically with the excitation. The solid or broken lines connecting the point measurements are just eye guides. The intensity of the spectra is in arbitrary units and is not to scale.

The performance was analyzed by monitoring the open circuit voltage *V*oc and the short circuit current *I*sc. Normally, it is determined from the current-voltage (*I*-*V*) relationships, with *I*sc and *V*oc as important characteristics. The delivered power is computed using $P = IV$, with the maximum typically approximated by $P \sim 0.75\ I\text{sc}V\text{oc}$. The film thickness is increased by repeating this procedure and is estimated from the volume of the colloid that was dispensed. The responses is measured before and after each incremental increase. Figure 13.8A and Figure 13.8B give the change in *V*oc and *I*sc with film thickness under

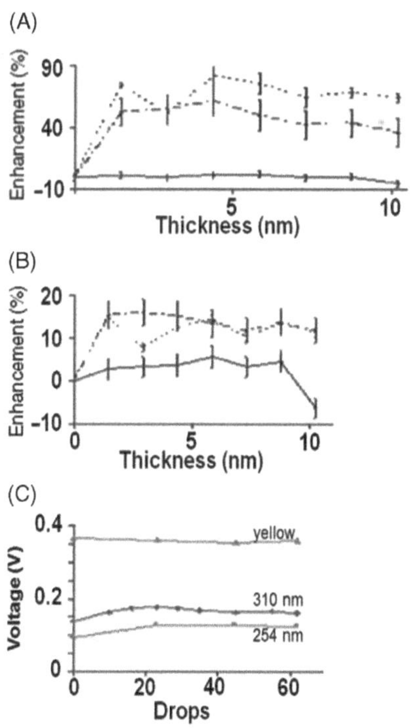

FIGURE 13.8 (A) and (B) the change in *V*oc and *I*sc with film thickness under ~50 W/m^2 UV irradiation. (C)*V*oc as a function of the thickness of the nanolayer.

~50 W/m^2 UV irradiation. Figure 13.8C gives *V*oc as a function of the thickness of the nanolayer. Increases in *V*oc by ~ 60% are observed at 254 and 365 nm. On the other hand, there is only ~14%–17% increases in *I*sc. This gives a power enhancement of ~ 62%. We also measured the response in the visible (~50 W/m^2 at 550–650 nm). The results showed a power enhancement of less than 3%.

The effect of films of 2.85 nm red particles was also tested. For those the absorption edge is 3.2–3.4 eV, while the confinement band gap is 2.15 eV (in the visible at ~570 nm excitation). Those are expected to capture a different slice of the solar spectrum because of the different bandgap. The luminescence spectra the alcohol colloid precursors were shown above in Figure 13.6C (on the right); it shows a red band peaking at 630 nm. The results show that, in the UV, *V*oc is enhanced by ~80% with a slight increase of 5% in *I*sc, giving a power enhancement of ~67%. We measured the response in the visible range. We find an effect only under blue-green part of the spectrum, where *V*oc is enhanced by 10%–15%, while *I*sc increase by less than 5%, giving a power enhancement of 7.5%–13.5%.

The enhancement for both size particles in the UV is due to strong direct band edge absorption of Si (at 3.2–3.4 eV). The lack of enhancement for 1 nm in the visible, on the other hand, is due to the fact that the film is transparent (wide band gap material),with a large absorption edge of 3.2–3.4 eV and a large confinement band gap of ~3.6 eV.The larger particles respond more strongly in the blue/green due to the fact that they have a confinement band gap of 2.15 eV (in the visible at ~570 nm excitation).

The analysis shows that the improvement is dominated by voltage rather than current. For 1 nm particles, the increase in *I*sc is 17%. We determined the corresponding voltage using a standard solar cell relation, which relates *V*oc to *I*sc logarithmically. The increase in voltage was only 30%, short of the measured enhancement (60%–70%). For the 2.85 nm particles, the 5% increase in the current is much too small to account for the 80% voltage enhancement. Thus wavelength down conversion may not entirely account for device enhancement. It is likely that direct charge separation contributes significantly to the voltage enhancement. Due to strong confinement, the rate of radiative recombination of excitons in ultrasmall Si particles rises with size,while nonradiative recombination dramatically drops, allowing fast luminescenceand allowing charge separation to compete favorably as well.

When light strikes the nanoparticles some light is absorbed. As a result, electron hole (e-h) pairs (excitons) are produced in the nanoparticles. If the electron and hole are separated from each other completely before they could recombine to produce down shifted light (luminescence) or recombine non-radiatively to produce vibrations and heat then the electrons and holes may be transported, collected and stored on two external electrodes appropriately constructed and positioned. The voltage difference between the electrodes can be harnessed at a later time as a battery. The electric field in a p-n junction plays a pivotal role in charge separation and collection. This architecture using 2.9-nm and 1-nm silicon nanoparticles showed relative power enhancements of the efficiency of the under lying solar cell in the UV and in the visible. It is to be noted that there is a contributing component due to down conversion effect resulting from the radiative recombination of the e-h pairs before they are separated.

13.1.1.13 Nano (Silicon) a-Si:H [16]

The effect of Si NPs on thin-film n + -i-p+ a- Si:H solar cells was carried outby Ammar Nayfeh group at Masdar Institute of S and T in collaboration with the Nayfeh's group at the University of Illinois. Figure 13.9A shows the top view of fabricated solar cells (the layer stack of the solar cell). For this cell, 20 nm heavily doped p-type a-Si, 100 nm undoped a-Si, 20 nm heavily doped n-type a-Si and 80 nm ITO were grown on a p+ type Si substrate using plasma enhanced chemical vapor deposition (PECVD). The p+ type Si substrate serves as back contact and does not contribute significantly to the electron–hole-pair (EHP) generation. Silver (Ag) was then deposited using electron-beam evaporator tool and lift off technique was used to form the finger contacts. Three

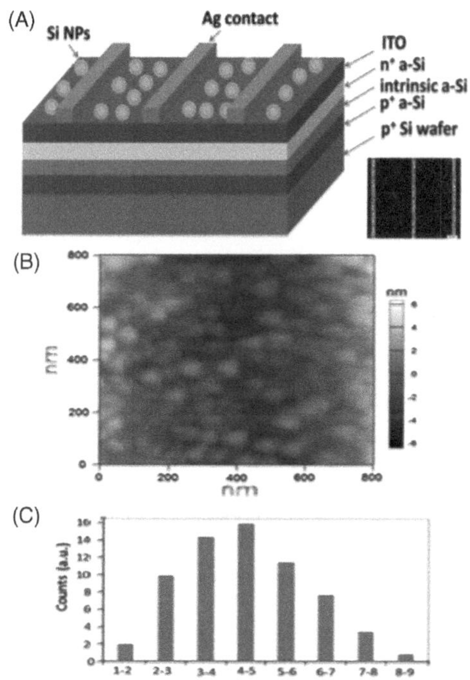

FIGURE 13.9 Nano (silicon) a- Si:H: The effect of Si NPs on thin-film n + -i-p+ a- Si:H solar cells was carried out at Masdar Institute of S and T in collaboration with the University of Illinois. (A) topview of fabricated layer stack of the solar cell. (B) AFM Topographic image of Si NPs coating cell showing Si NPs forming clusters during the multiple spin coating steps. (C) Histogram of height of Si NP clusters.

different cell area-sizes were fabricated i.e. 1 cm × 1 cm, 0.5 cm × 0.5 cm and 0.25 cm × 0.25 cm. Solar cells of size 1 cm × 1 cm are presented in this work. After the solar cell fabrication, Si NPs were spin coated on top of the solar cell. In order to increase the surface coverage, the spin coating step is repeated 25 times using a total of about 4 mL of the Si nanoparticle solution. The solution has ~ 10 μM concentration hence giving us a total of ~ 2.49×10^{16} deposited Si NPs. Topographic image of Si NPs coated n + -i-p+ a-Si:H solar cell taken by VHX-5000 digital microscope is shown in Fig. 13.9B. The figure shows an Atomic Force Microscopy (AFM) image of the Si NPs coated on top of the solar cell. This figure shows that Si NPs have formed clusters and some of the clusters have width over 100 nm. The clusters are formed during the multiple spin coating steps. Histogram of this image is shown in Fig. 13.9C, which shows the variation of height of these Si NP clusters. While majority of the clusters have a height between 2 and 7 nm, some of the clusters are over 8 nm, which means these clusters have 3 layers of Si NPs placed on top of each other.

Current-voltage characteristics (J–V) of n^+-i-p^+a-Si:H reference cell and cell with Si NPs under one sun i.e. AM1.5 is shown in Figure 13.10A. Moreover, Fig. 13.10 gives the percentage increase of both both the internal and external quantum efficiency due to the addition of Si nanoparticles.

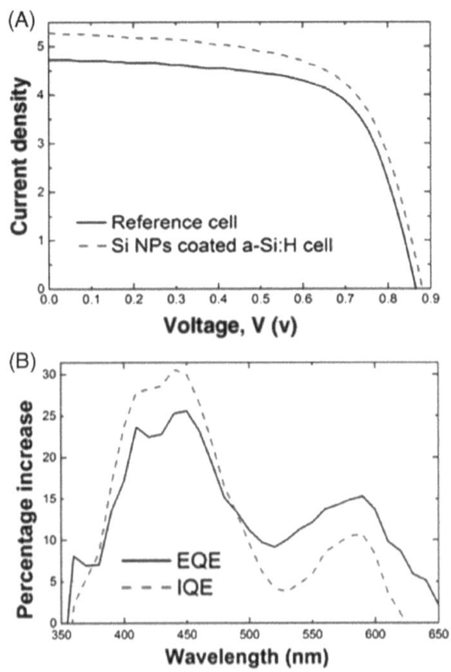

FIGURE 13.10 Characteristics of the solar cell in Figure (A) Current-voltage characteristics (J–V) of n^+-i-p^+a-Si:H reference cell and cell with Si NPs under one sun i.e. AM1.5. (B) percentage increase of both the internal and external quantum efficiency due to the addition of Si nanoparticles.

This comparison shows that there are two distinct areas of enhancement. Between 360–530 nm range Si nanoparticles have low photoluminescence but shows greater boost when spin coated on solar cells, suggesting charge separation in Si nanoparticles and subsequent collection by electrodes. On the other hand, enhancement in performance between 530–610 nm is considered due to the photoluminescence of Si nanoparticles.

13.1.1.14 Nano (Gold) a-Si:H [17]

Light interaction with nano metal is also emerging as a useful intermediate agent for improving light coupling to thin film solar cellsdemonstrated at Masdar Institute of S and T; hence improving their efficiency. Thin film solar cells utilize 1-2 μm thick semiconductor material placed over substrates of cheap material compared to silicon, such as glass, plastic or steel. Metal nanoparticles are deposited on the top of the semiconductor as shown schematically in Figure 13.11 (inset). When light hits these metal nanoparticles at their surface plasmon resonance, light is scattered in many different directions. This allows light to travel along the active semiconductor and bounce between the substrate and the nanoparticles enabling the semiconductor to absorb more light even though it is a thin layer.

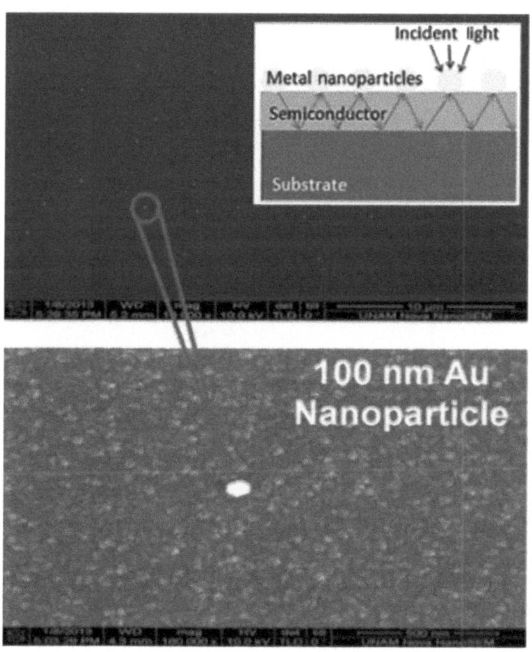

FIGURE 13.11 (Top) Nano (gold) a-Si:Hplasmonic gold nanoparticles placed on the top surface of thin film a-Si:H solar cells using spin coating. (inset) schematic of metal nanoparticles deposited on the top of the semiconductor. (Bottom) zoom on the gold nanoparticles. *(From Kazi Islam, Aaesha Alnuaimi, Enes Battal, Ali Kemal Okyay, Ammar Nayfeh, Effect of gold nanoparticles size on light scattering for thin film amorphous-silicon solar cells, Solar Energy 103 (2014) 263–268)*

As an example,plasmonic gold nanoparticles size are placed on the top surface of thin film a-Si:H solar cells using spin coating. Colloidal gold nanoparticles dispersed in Phosphate Buffered Saline (PBS) solution were spin-coated on top of the ITO as shown in Figure 13.11. Seven different diameter-size i.e., 10 nm, 20 nm, 50 nm, 80 nm, 100 nm, 200 nm and 400 nm particle sizes were investigated. The spin coating recipe was optimized to obtain uniform distribution as well as optimum concentration of particles on the surface.The solar cell stack, shown in Figure 13.12A is grown on a p+ type Si wafer and consists of 20 nm heavily doped p-type a-Si, 500 nm

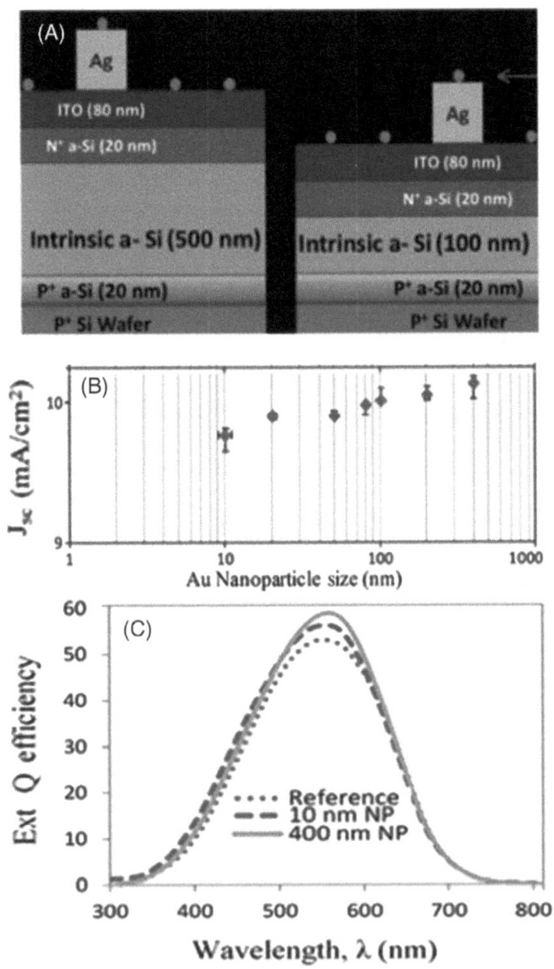

FIGURE 13.12 Characteristic of the nano gold cell in Figure 13.11. (A) The cross section of the solar cell stack. (B) short circuit current density Jsc vs. Au nanoparticle size (plotted in log scale). (C) The external quantum efficiency (EQE) of the fabricated devices. *(From Kazi Islam, Aaesha Alnuaimi, Enes Battal, Ali Kemal Okyay, Ammar Nayfeh, Effect of gold nanoparticles size on light scattering for thin film amorphous-silicon solar cells, Solar Energy 103 (2014) 263–268)*

undoped a-Si, 20 nm heavily doped n-type a-Si and finally 80 nm Indium Tin Oxide (ITO) on the top.

Figure 13.12B gives the short current densityJsc vs. Au nanoparticle size (plotted in log scale). The Jsc increases from 9.34 to 10.1 mA/cm2. Measurements show that the median reflectance of the device drops from 23.19% with no nanoparticles to less than 20% for all nanoparticles sizes. The drop shows the improved surface scattering due to the gold nanoparticles. The efficiency increases with increasing nanoparticles size due to the increase in the Jsc, from 4.28% to 5.1%. Analysis shows that the fill factor (FF) also increases from 51% for the reference cell up to 56% for the gold nanoparticles based cells of all sizes.

The external quantum efficiency (EQE) of the fabricated devices given in Figure 13.12C shows an improved spectral response with increasing nanoparticles size. The peak EQE is 59% for the cell with 400 nm nanoparticles while the reference cell has 53% peak EQE as shown in the figure. Two prominent mechanisms can explain the contribution of metallic nanoparticles based on application: scattering mechanism and the near-field localization effect. Photocurrent enhancement by metal nanoparticles on the top surface of solar cells can be explained by the light scattering mechanism. Metal nanoparticles are strong scatterers of light at wavelengths near their resonant frequency. The Au nanoparticles allows for additional scattering at the surface thus reducing the overall reflectivity and increasing the overall efficiency.

13.1.1.15 Nano (Indium Nitride) c-Si HIT [18]

The nanoparticles are first prepared using high-pressure chemical vapor deposition (HPCVD). It was used to grow InN nanoparticles, in which laser ablation was carried out by focusing a nanosecond pulsed ND:YLF laser beam at a wavelemgth of 527 nm on an InN target. Figure 13.13A shows an SEM image of the InN nanoparticles, showing nanoparticles of ~ 15 nm across on the average. Figure 13.13B shows the layer stack of the fabricated solar cell. It is a-Si:H (n +)/a-Si:H (i)/c-Si (p) HIT solar cell.

The solar cell is fabricated as follows. A Si wafer with doping concentration (p +) of approximately1 $\times$ 10^{19} cm^{-3} was used as substrate. 2 μm epitaxial Si layer was grown on the p+ substrate using low-pressure chemical vapor deposition (LPCVD) with nominal boron concentration of 1 $\times$ 10^{16} cm^{-3}. An undoped a-Si layer of 5 nm and n+ doped a-Si layer of 15 nmwere grown afterwards. The p+ type Si substrate serves as a back contact and does not contribute significantly to electron–hole pair generation. In addition, 80 nm ITO was deposited using plasma-enhanced chemical vapor deposition (PECVD). InN NPs were then spin-coated on top of ITO. After the coating of NPs, silver (Ag) was deposited using an electron-beam evaporator tool and a lift off technique was used to form the contacts. Three different cell area-sizes were fabricated, i.e. 1 cm $\times$ 1 cm, 0.5 cm $\times$ 0.5 cm and 0.25 cm $\times$ 0.25 cm. Solar cells of size 0.25 cm $\times$ 0.25 cm are presented in this work unless stated otherwise.

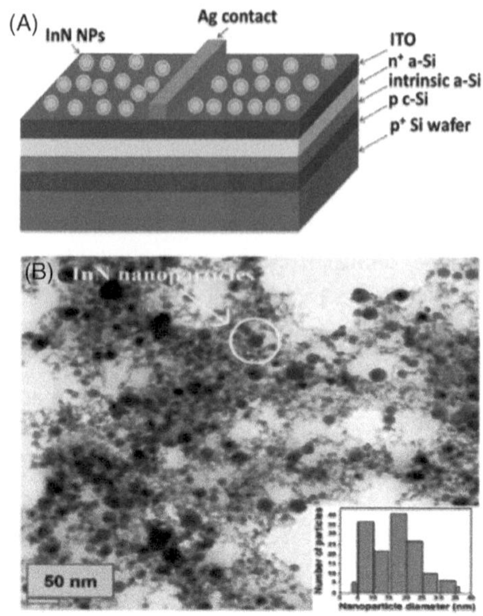

FIGURE 13.13 Nano (Indium nitride) c-Si HIT. (A) layer stack of the fabricated a-Si:H (n +)/a-Si:H (i)/c-Si (p) HIT solar cell. (B) SEM image of the InN nanoparticles, showing nanoparticles of ~ 15 nm across on the average. Inset is a histogram of the particle size.

J–V characteristics of the c-Si HIT reference cell with ITO and the cell with ITO + InN NPs under one sun, i.e. AM1.5G are given in Figure 13.14A. Enhancement factor of external and internal quantum efficiency of c-Si HIT cells due to the addition of InN NPs are given in Figure 13.14B. The external and internal enhancement factors of the quantum efficiencies due to the addition of InN nanoparticles are compared in Fig. 13.14B. According to a team **at Masdar Institute,** with this nanoparticles, there is only one distinct region of enhancement and it is outside the range over which InN nanoparticles have higher photoluminescence. Authors claimed, though photoluminescence of InN nanoparticles contributes to this enhancement, suggested by the increase in internal quantum efficiency, major boost occurs due to the light scattering ability of these nanoparticles. They suggested that agglomerated InN nanoparticles have the necessary size required to efficiently scatter light and a significant reduction in reflection from the solar cells after the addition of InN nanoparticles validates their theory.

13.1.1.16 Nano (CuO) HIT Cell [19]

Copper oxides can be used in thin film heterojunction solar cells. All three forms of copper oxides are semiconductors. CuO has a monoclinic crystalstructure. It has a direct energy band gap of ~1.5 eV. Since for a single p-n junction solar cell, the optimum band gap is 1.4 eV, the optical properties of CuO make

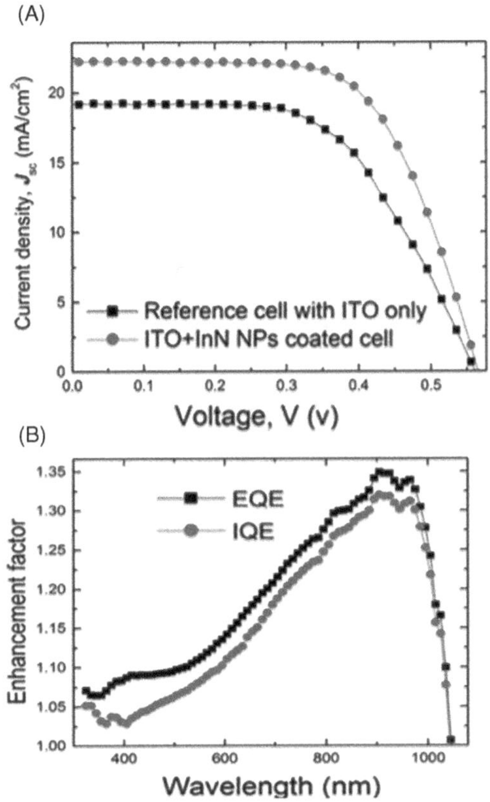

FIGURE 13.14 Characteristic of the nanoInN cell in Figure 13.13. (A) J–V characteristics of the c-Si HIT reference cell with ITO and the cell with ITO + InN NPs under one sun, i.e. AM1.5G. (B) Enhancement factor of external and internal quantum efficiency of c-Si HIT cells due to the addition of InN NPs.

it a very suitable semiconductor absorber material for solar cell applications [11]. In CuO, the absorption coefficient is expected to be high. Moreover, electically, the majority carrier mobility and minority carrierdiffusion length are also expected to be adequate.

But building the devices proved to be problematic. The interface turned out to be of low quality. For example, reactive magnetron sputtering was used to depositthe CuO layer from a Cu target and a sputtering plasma consisting of Ar and O_2. The resultis a poor quality interface between copper oxide and silicon (Figure 13.15A) as the result of silicon oxide on the silicon surface. In fact, the deposited CuO has a nanocrystalline structure with a crystallite size of ~8 nm. The optical band gap of the CuO layer was measured at 1.07 eV. Figure 13.15B depicts a solar cell stack based on CuO. The photovoltaic performance has been of low quality because of the tendency of holes and electrons to recombine

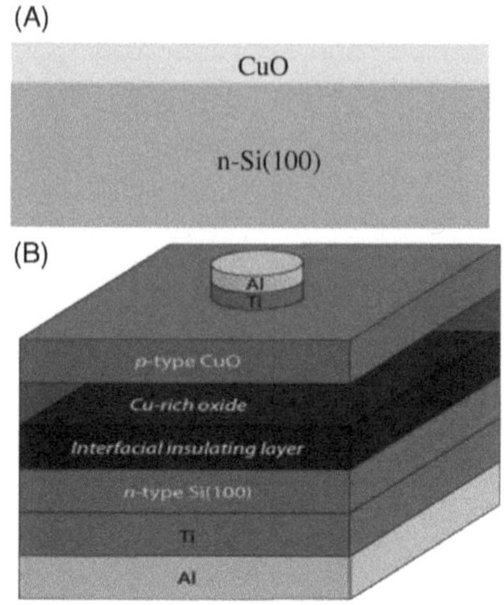

FIGURE 13.15 Nano (CuO) HIT cell. (A) interface between copper oxide and silicon. (B) depicts a solar cell stack based on CuO.

before they are separated and collected, severely limiting the production of electricity.

Recent reports however claimed improvements in the deposition conventional procedures. They realized this by increasing the pressure during the deposition stage of fabrication, which they found enhances both the crystal and interface quality, thereby reducing the charge recombination rate which allowed production of high-performance solar cells that employ cupric oxide as the p-type material and silicon as the n-type material. Despite the significant improvements showed that the interface quality was still limited by the unwanted formation of a copper-rich oxide layer as well as by the production of a silicon oxide layer on silicon surface. Further improvements of the photovoltaic performance is expected when those are reduced, or eliminated.

13.1.1.17 Solar Spectrum versus Particle Size

As discussed above, nanostructured thin films on top of solar cells are used to enhance the efficiency of solar cells in a variety of ways, including passive processes, such as antireflection coating and forward scattering, scattering/trapping to enhance coupling/absorption in thin solar cells, and UV-filtering (ultra small luminescent particles) and active processes such as down-conversion, top photovoltaic cell.Those processes are governed by three parameters: Wavelength λ of incident light, size of the particle, the relative size parameter $x = 2\pi a/\lambda = ka$, the particle optical property relative to the surrounding medium(refractive index $m = n\text{-}ik$), the bandgap of the particle and the underlying solar cells.

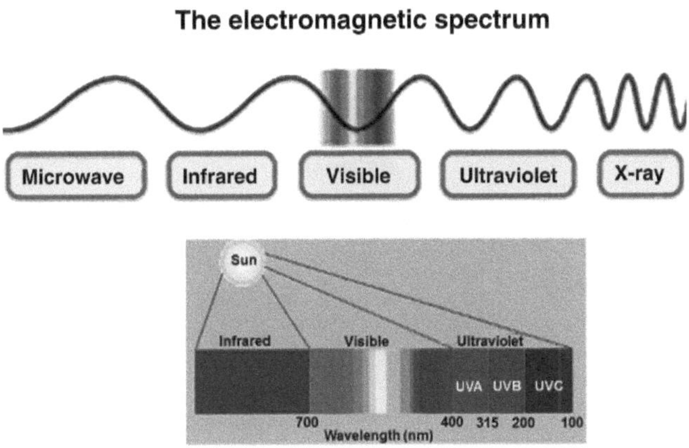

FIGURE 13.16 The relevant portion of sunlight for harvesting by solar cells lies in three ranges: UV, visible and infrared.

The relevant portion of sunlight for harvesting by solar cells lies in three ranges: UV, visible and infrared as shown in Figure 13.16. Sun light is a section of the wider spectrum. This continuous range of frequencies is known as the electromagnetic spectrum. The entire range of the spectrum is often broken into specific regions. The subdividing of the entire spectrum into smaller spectra is done mostly on the basis of how each region of electromagnetic waves interacts with matter. The diagram below depicts the electromagnetic spectrum and its various regions. In addition to the X-ray region, the visible portion is one of the narrowest regions within the spectrum.

The visible spectrum is called as such because it is the portion of the electromagnetic spectrum that is visible to the human eye. The visible light is also called simply light. A typical human eye will respond to wavelengths from about 390 to 700 nm (430–770 THz).

Maxwell's equationsare the basis of theoretical and computational methods describing light scattering. Exact analytical solutions to Maxwell's equations (Mie theory) are only known for selected geometries (such as spheres, cluster of spheres, infinite cylinders). In the case of more complex geometries and for inhomogeneous particles the original Maxwell's equations are discretized and solved. Multiple-scattering effects of light scattering by particles are treated by radiative transfer techniques (see, e.g. atmospheric radiative transfer codes).

The type of particle's material that have been of interest for light scattering/absorption/emission are ice crystals, dust, planetary dust, blood cells, metal/oxide and silicon and silicate. Those cause observable phenomena such as rainbows, the color of the sky, halos, galaxy's tint, and electricity.

In general scattering is characterized by a non-dimensional size parameter, $x = 2\pi a/\lambda$, where a is the radius of the spherical particle, and λ is the

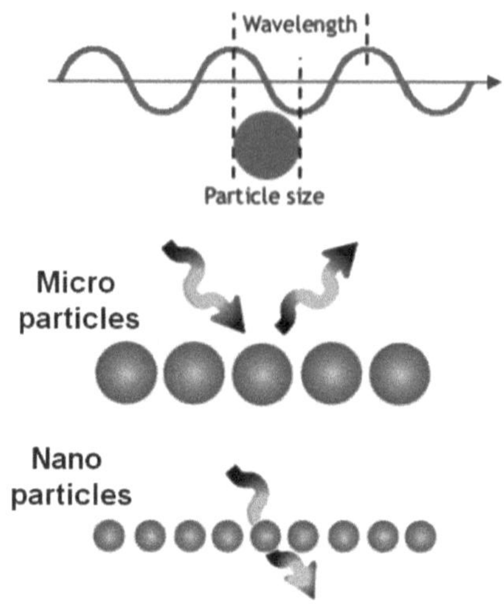

FIGURE 13.17 Cartoon of light wavelength and scattering strength VS particle size.

wavelength of incoming light. Relative size of a scattering particle is defined by size parameter which is the ratio of its characteristic dimension and wavelength. Large particles in general and silicon particles in particular are mot luminescent but they exhibit very interesting light scattering phenomenon. Light scattering from molecules to nanoparticles and microparticles (**see** chapter 9, Figures 9.11 and 9.12). Qualitatively, the maximum scattering occurs when the particle diameter is equal to half a wavelength, while large particles become reflective while ultrasmall ones become transparent as depicted in Figure 13.17.

There are a number of interesting and technologically useful features for solar cell that arise. First the angular distribution (phase function) derived from the Mie theory can be sketched for different sizes of particles or specifically in terms of the parameter x as sketched in Figure 9.12. Second,directional light scattering by spherical silicon nanoparticles in the visible spectral range. These unique scattering properties arise due to simultaneous excitation and mutual interference of magnetic and electric dipole resonances inside a single nanosphere. Directivity of the far-field radiation pattern can be controlled by changing light wavelength and the nanoparticle size. Forward-to-backward scattering ratio above 6 can be experimentally obtained at visible wavelengths as shown in Figure 9.13 as explained in Example 9.9. These unique properties of silicon nanoparticles make them promising for design of novel low-loss visible- and telecom-range nanoantenna devices. Third, a core shell configuration of metal and silicon or silica has been used to fine tune the process. Tuning the size of the core as well as the shell thickness and the material composition of the nanoparticles along with the incident wavelength can be used to bring simultaneous resonance of magnetic and electric response as well as

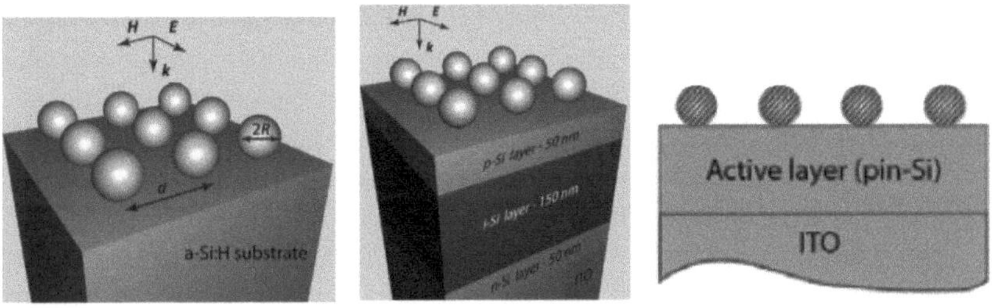

FIGURE 13.18 Layout of the solar cell architecture periodic arrays of spherical nanoparticles on top of the active layer.

match their strength. Detailed examples are given in Figure 9.17-Figure 9.18 in Example 9.11. This is found to enhance he directionality in the forward direction. The effect of the conductivity of the material on light scattering and plasmonresonances are covered in Example 9.10 by comparing such responses between Si and gold.

13.1.1.18 Nano Structured-Anti-Reflective Coatings [20]

Thin film a-Si-based solar cells were built and theoretically analyzed to optimize antireflection coatings in the visible range, hence optimize light coupling (**see Section 11.4**). Two interesting coatings were examined. A dielectric coating based on nano crystalline silicon and plasmonic coatings based on silver nanoparticles. Figure 13.18 shows the layout of the solar cell architecture. It used periodic arrays of spherical nanoparticles on top of amorphous silicon. The studies showed that the size of the nanoparticles must be big enough such that the electric- and magnetic-dipole responses of nanoparticles are significant. This usually happens when the diameter is a significant fraction of the wavelength of the incoming light. In this case interference phenomenoa take place between three waves:

- wave reflected from the substrate
- electric-dipole response of nanoparticles
- magnetic-dipole response of the

For appropriate sizes the interference among the three may result in zero reflection (blooming effect) from the structure. This is called substrate-mediated Kerker effect. Silver nanoparticle coatings also demonstrate strong antireflective properties in the visible spectral range. In this case, the effect is due to substrate-induced bi-anisotropy and blooming, caused by substrate-induced magnetic response.

Overall, we observe that both narrow- and broad-band transmittance increases can be obtained with nanoparticle coatings. For the increase in the entire visible range, Ag-NPC with radiuses 70–80 nm give the best transmittance enhancement up to 15%, and Si-NPC with radiuses 30–40 nm causes up to 8% of enhancement. At the same time, at the narrow spectral range, Si-NPC enables up to 40% of transmittance increase.Thus, one can see that placing NPC on top of active layer of silicon photovoltaic element can efficiently increase absorbance inside it: up to 8–10% with Ag- or Si-NPC of small radius (30–40 nm) and up to 30% with Ag-NPC of radius 80 nm having advantages of multipole resonances of plasmonic localized modes.

From practical point of view, silicon nanoparticles provide more advantages in comparison with silver ones becauseof severe degradation of silver in ambient conditions. Silicon nanoparticles, on the other hand has high chemical stability and lower optical losses. Moreover, recent advances in methods of their cost-effective fabrication by laser printing open up a possibility to vary nanoparticle size and crystallographic phase, making silicon nanostructures even more prospective for photovoltaic applications.

PV Si market

The current world population of 7.2 billion is predicted to increase to 9.6 billion in 2050,[1] and world energy consumption is projected to grow by 56% between 2010 and 2040 to 820 quadrillion British thermal units (Btu).[2] Although fossil fuels constitute almost 80% of energy use,[2] they are depleting, and it is becoming more urgent to ensure that energy supply is sustainable. Renewable energies already belong to the world's fastest-growing energy sources and for electricity generation are increasing by 2.8% per year. The most important renewable energy sources are hydro, wind, and photovoltaic (PV).

Silicon wafer-based modules dominate the PV market with an 86% market share. The other 14% belongs to thin-film PV, some of which is based on hydrogenated amorphous silicon (a-Si:H) and hydrogenated microcrystalline silicon (μc-Si:H), which share the advantage of silicon's abundance as a raw material. An additional advantage is that less silicon is required than for conventional wafer silicon solar cells. Thus, silicon-based thin-film technology has the potential for low-cost, rapid, large-area industrial production and low energy-payback time.

Current solar cell efficiencies are 10.1% for single-junction a-Si:H in the p-i-n configuration (meaning a combination of positively doped, intrinsic, and negatively doped silicon), 10.8% for single-junction μc-Si:H p-i-n,[7] and 12.3% for tandem a-Si:H/μc-Si:H (micromorph).

13.1.1.19 Nano Texturing/Trapping Tradeoff for Silicon Material: [21]

Light scattering can be utilized to trap incident light which is a useful tool for increasing efficiency, especially in thin film solar cells. The concept, also termed light in-coupling, Incorporating a textured surface increases the effective light path and hence absorption in the silicon absorber material. Consequently, even thinner absorber layers can be used, an advantage for amorphous silicon. This is quite important for silicon since silicon has a low absorption coefficient in the near-IR region, which makes light trapping especially relevant. The effect is enhanced further by a surface capable of scattering several wavelengths (broadband) simultaneously. This can be achieved by incorporating a textured surface with features of different sizes (**see** Figure 13.17). The most commonly applied method, of randomly texturing transparent conductive oxides (TCOs). For this approach, the TCO has to be sufficiently thick for adequate conductivity and degree of texturing. Most recently, it has been noted that textured glass surfaces might be used in the case of thin-film solar cells because they are too thin, and carbon nanotubes may be used for front contacts. Texturing glass surfaces can be achieved by acid, such as hydrofluoric acid, as well as by laser ablation texturing.

13.1.1.20 Solar Thermos Photovoltaics [22]

In thermos photovoltaic, an absorber's / emitter system is placed in front of the solar cell as shown in Figure 13.19A [23]. The absorber is designed such that (i) its absorption range covers the entire solar radiation spectrum. (ii) Effectively heating the absorber/emitter system to the required temperature.

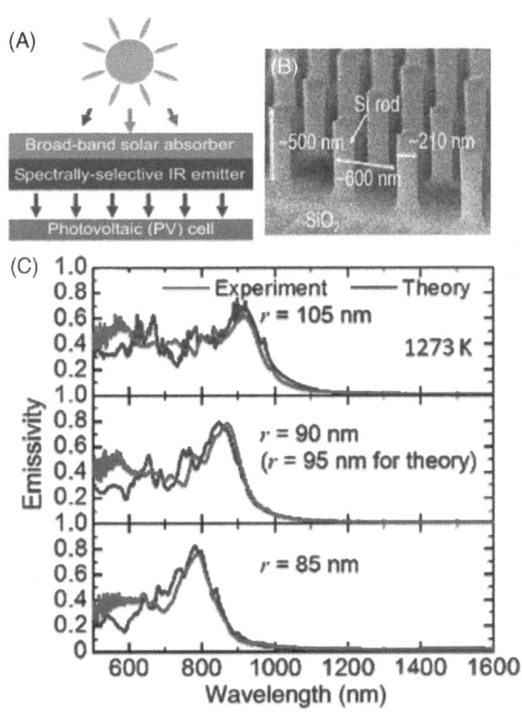

FIGURE 13.19 Solar thermos photovoltaics. (A) an absorber's / emitter system is placed in front of the solar cell. (B) fabricated array of silicon rods using silicon-on-insulator (SOI) substrate. (C) Measured thermal emissivity of the Si rod arrays (solid red lines) along with theoretically calculated emissivity spectra (solid blue lines).

(iii)The emitter is configured to produce IR radiation with the energy of individual photons slightly exceeding the bandgap of the chosen photovoltaic cell. Such IR radiation can be converted into electric power with high efficiency, theoretically over 80%. This effectively narrows the wavelength bandwidth and tune it to a particular wavelength or energy. For example if the solar cell is made of crystalline silicon, the device is designed such that the entire spectrum of solar power is concentrated into a narrow spectrum with photon nergy of ~ 1.1 eV, the bandgap of silicon. For this to occure, the A/E device must be heated by sun light to high temperatures. For hot objects to emit visible wavelengths, a temperature of 1000°C is needed; the material must stand the temperature without melting or evaporating. One promising example is a nanostructured So-based device.

Manufacturing of such device starts with silicon-on-insulator (SOI) substrates with a 500-nm-thick top Si layer and a 1-mm-thick SiO2 layer on a 700-mm-thick Si substrate. Standard silicon electron lithography and plasma etching techniques is used to create rod-array patterns. The area of the pattern is 1-mm × 1-mm. Figure 13.19B shows a fabricated array. The height h = 500 nm, radius r = 210mn, and interspacing 600 nm. The height of the rods is short of the silicon thickness to keep the system robust. Thinner rods were also fabricated including the radii r of 85, 90, and 105 nm, while keeping the height and intersapacing the same (a = 600 nm, h = 500 nm, and t Si = 0 nm).

The thermal emissivity of the Si rod arrays were measured as givem in Figure 13.19C (solid red lines) withrvalues of 105, 90, and 85 nm at 1273 K in surface normaldirection in the near-infrared range. The figure also shows theoretically calculated emissivityspectra (solid blue lines). These temperatures are well below silicon's melting temperature, which is over 1400°C. Using this material, their nanoscale semiconductor raises the energy conversion rate of solar cells to at least 40%. A peak emissivity of 0.77 at a wavelength of 790 nm (see Figure 13.20) and a low background emissivity of <0.02 to 0.05 for the 1100- to 7000-nm wavelength range, under operation at 1273 K has been achieved.

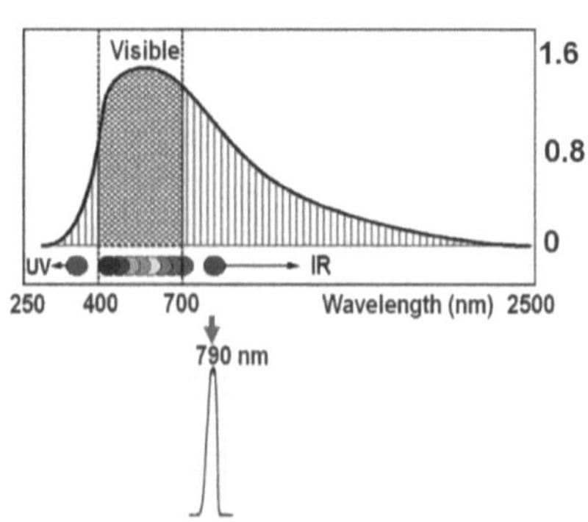

FIGURE 13.20 Measured peak emissivity of the array of silicon rods.

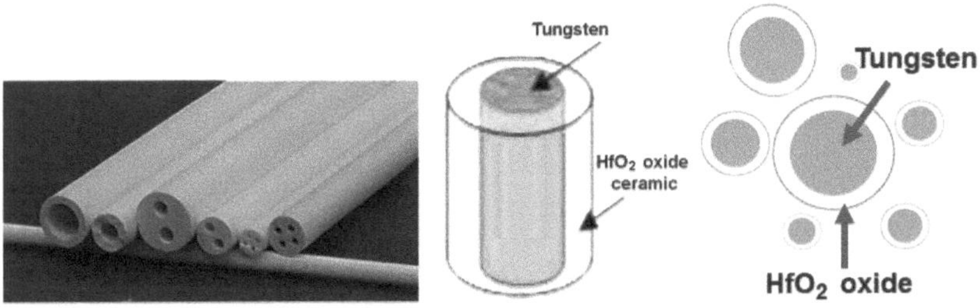

FIGURE 13.21 Hafnium dioxide coated tungsten-based absorber as an absorber/emitter for thermal photovoltaics applications. *(From http://tatehoozark.com/wp-content/uploads/sites/55/2015/11/HUH-2-e1446906950302.jpg)*

Theoretical calculations of intrinsic Si rod-array emitters with a rod radius of 105 nm (190 nm) can convert 59% (84%) of the input power into emission shorter than 1100 nm (1800 nm) at 1400 K.

Tungsten-based absorber has been also proposed as an absorber/emitter for thermal photovoltaics applications. Theoretical treatments show that this material can provide an 80% efficiency rating. This compares very favorably to athe maximum (~ 35%) which is provided by present-day silicon-based solar panels. However, in practice, the tungsten system yielded a mere 8% efficiency. The drastic drop is believed to be due to failure of the material at the extreme temperatures involved(>1000 °C). The stability of the tungsten wire is improved by coating it by a nanolayer of hafnium dioxide (Figure 13.21). Hafnium oxide is a very refractory high temperature material which is stable in air and inert environments. The ceramic added significant structural durability at high temperatures. Coating tungsten nanoparticles with hannifium oxide improves the heat tolerance of tngston asan emitter. With particle sizes ranging from 10 nanometers to 10 micrometers, the multiscale structure traps and absorbs light more efficiently and at temperatures greater than 700 degrees Celsius. (Image: Renkun Chen, mechanical engineering professor, UC San Diego Jacobs School of Engineering). Engineers at UC San Diego have developed a nanoparticle-based material for concentrating solar power plants that converts 90% of captured sunlight to heat.

13.1.1.21 Light Channeling in Si Nanorod (Elimination of Front Electrode Reflectivity) [24]

A 16-nanometre-thick film of gold is placed on a flat sheet of silicon. The surface is then riddled with an array of nano-sized holes, the silicon and the perforated gold film together are immersed in a solution of hydrofluoric acid and hydrogen peroxide. The gold film immediately began sinking into the silicon substrate, and silicon nanopillars began popping up through the holes in the film.The nanopillars grow to a height of 330 nanometres above the front surface as shown in Fig. 13.22.

FIGURE 13.22 Si nano pillars/rods as light channeling to eliminate front electrode reflectivity.

When the solar cell is exposed to sun light, the nanorods intercept the light and channel it inside the solar cell without going through the metal layer. In other words they effectively funnel light down into the silicon substrate underneath.

13.1.1.22 Electrode Reflection Effects

Conventional solar cells are panels with a square grid of wires interlaced over the top of them. These wires are necessary to conduct the electrical charge absorbed by the cell, but while they might not look like they're really blocking the cell's capacity to absorb sunlight, in reality they are – obstructing 5–10 percent of the light that would otherwise hit the cell. If only you could make those wires… disappear? Well, it seems that we can.

- Back electrode: Back-contact technology, which forms electrodes only on the back face of solar cells, can make the light receiving area of the top face larger by gathering up all the electrodes on the back face and as a result enhance conversion efficiency.
- Blooming effect: In this case interference phenomenoa take place between three waves: wave reflected from the substrate; electric-dipole response of nanoparticles; magnetic-dipole response (**see** Figure 13.18 **and** section 11.4). For appropriate sizes the interference among the three may result in zero reflection (blooming effect) from the structure. This is called substrate-mediated Kerker effect. Silver nanoparticle coatings also demonstrate strong anti-reflective properties in the visible spectral range. In this case, the effect is due to substrate-induced bi-anisotropy and blooming, caused by substrate-induced magnetic response.
- By-passing: By creating nanowires /rod that extend well above the front surface of the solar cells where the electrodes are located shields them from the impinging light. When the solar cell is exposed to sun light, the nanorods intercept the light and channel it inside the solar cell without going through the metal layer. In other words they effectively funnel light down into the silicon substrate underneath.

Metal nanostructures have been studied rigorously for light scattering application in solar cells. Recently semiconductor nanostructures have gain attention for this purpose as they have the capability of downshifting high energy photons to lower energy photons and increase the intensity of incoming light at a wavelength range for which solar cell has its maximum response. These nanostructures absorb part of the incoming light which has higher energy than its band gap, making electrons to excite within from lower energy state to higher energy state. Couple of things can happen afterwards, i) these electrons can relax back to their original state by radiative recombination and in the process, emit light with energy lower than it band gap, or ii) once these electrons get excited to higher energy state, there is a charge separation and collection stage. Electrons within the nanostructure get collected directly by the electrodes.

13.2 CURRENT SOURCE (UV AND INFRARED PHOTOCURRENT SENSORS) [25]

In the above section nanoparticles are placed on a p-n junction, which resulted in charge storage, hence photovoltaic effects. In this section the nanoparticles are simply placed on a conducting substrate. A thin oxide layer is grown on a silicon wafer, followed by etching out a hole or multiple holes using a mask. The holes are filled with silicon nanoparticles and an aluminum thin layer is grown as a cap/ electrode as shown in Figure 13.23(A). Continuous flow of electrons may proceed which may act as an instantaneous current source.

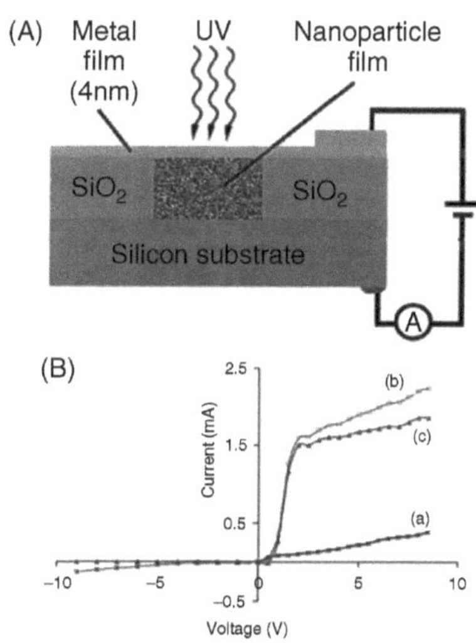

FIGURE 13.23 Nano photoconductor (UV and infrared sensors). (A) schematic of Si nanoparticles-based device for photodetection. (B) characterization of the device using current–voltage (I-V). **(a)** under dark conditions. **(b)** taken with 2 mW at 365 nm. **(c)** the difference of the responses taken with and without 365-nm light.

Architectures of this type have been used as highly sensitive photo-detectors for various applications.

The device is placed under UV light from the Hg lamp of a fluorescence microscope at powers of1–3mW, focused to a spot 2 mm in diameter. Figure 13.23(B) **(a)** shows the current–voltage (I-V) curves under dark conditions. They show somewhat rectifying behavior. Similar behaviorwas observed in silicon nanoparticle-based devices of equivalent configuration constructed to conduct electroluminescence studies [19]. The forward current, beginning to rise at 0.75 V, increases nonlinearly and saturates locally to 0.075 mA at 2 V. Beyond 2 V, the current rises linearly to 0.4 mA at 9 V, with an impedance of 21 k. At 2 V, the ratio of forward to reverse current is seven. Fig. 13.23(B) **(b)** was taken with 2 mW at 365 nm. It shows very strongrectification. The forward current is enhanced by a factor of 20, while the reverse current is essentially unaffected. The ratio of forward to reverse current at 2 V is now 160. Figure 13.24(B) **(c)** gives the difference of the responses taken with and without 365-nm light. Other measurements showed that the device has good visible

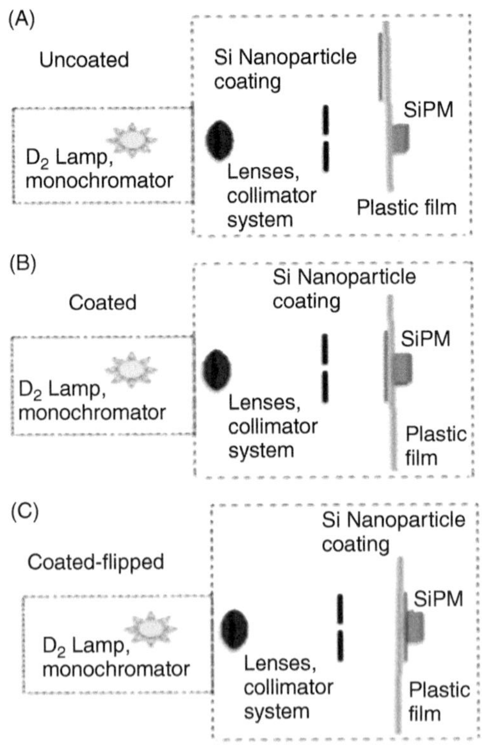

FIGURE 13.24 Layout of Cerenkov light detector. (A) uncoated configuration in which light passes through plastic film only to SiPM. (B) is the coated configuration case in which light passes through Si nanoparticle coating then plastic film to SiPM. (C) is the coated-flipped configuration in which light passes through plastic film then Si nanoparticle coating to SiPM.

blindness. The responsitivity, measured using Hg or Xe lamps drops to 0.036 at 560 nm compared to 22 at 250 nm. Due to propagation problems of deep UV, the gain at 200 is underestimated.

Cerenkov one application of this sensitivity to UV radiation is the detection of Cerenkov light [26]. The measurements were carried out by Magill andXieat Argonne National Laboratory in collaboration with the Nayfeh's group. Cerenkov light is emitted in crystals and glasses, scintillation light in neutrino, dark matter, and rare decay.In those high energy physics (HEP) applications, clear, dense crystals are used. When charged particlepass through the crystal, both visible scintillation light and Cherenkov UV light are produced by the charged particles. The Cherenkov intensity, being weak is dominated by the much stronger scintillation radiation. Simulations show that a homogeneous calorimetermade entirely of crystal could contain high-energy jets in a detector volume similar to those in existing experiments. In fact, it has been demonstrated in Using Dual Readout capability (simultaneous detection of both Cerenkovand scintillator light), unprecedented energy resolution for hadrons produced in e + e interactions can be achieved in these simulations. It is anticipated that the amount of (visible wavelength range) scintillation light produced by charged particles in the crystal will be more than is needed, so a few small photosensors (e.g., SiPMs) located on a crystal face or at corners could be used to collect that component while maintaining a high degree of homogeneity. For the Cerenkov light, however, it is foreseen that light collection will be required on several sides of the crystals with a collection area comparable to that of the crystal face. A photosensor is needed which 1) must be able to detect photons in the UV wavelength range ($<$ 350 nm), 2) must be thin so as to preserve the homogeneity of the calorimeter, 3) must be of large area ($\sim$few cm2), and 4) should be blind to visible light so as not to detect light from scintillation.

The photosensor used was a Hamamatsu MPPC, which is a Silicon PhotoMultiplier (SiPM) of 3mm $\times$ 3mm square area with 50 micron pixel pitch mounted in a ceramic enclosure [26]. The photon detection efficiency is $\sim$ 10% for 320 nm light, rising to a maximum of $\sim$37% at 450 nm, then falling off to $\sim$ 5% near 900 nm in wavelength. The same SiPM was used in all of the tests reported here. The match of 3 nm diameter Si nanoparticles with this particular SiPM is not ideal, since the peak emission of the nanoparticles is at a wavelength ($\sim$650 nm) not optimal for detection by the SiPM.

The layout of the test is shown in Figure 13.24. The tests were performed on a fully automated optical-electrical measurement system consisting of a Newport 30W continuum deuterium (D2) lamp, a Newport Apex monochromator, a low noise Femto DLPCA-200 current amplifier and a Keithley 2701 multimeter. A detection range of $\sim$ 160 nm to 900 nm can be achieved with the above system. The D2 lamp intensity is a smoothlyfalling function of increasing wavelength with a few narrow spikes in intensity at wavelengths of 490 nm, 580 nm, and 660 nm. Figure 13.24 **(top)** is the case of uncoated configuration in which light passes through plastic film only to SiPM. Figure 13.24 **(Center)** is the coatedconfiguration case in which light passes through Si nanoparticle coating then plastic film to SiPM, and Figure 13.24 **(bottom)** is the coated-flipped configuration in which light passes through plastic film then Si nanoparticle coating to SiPM.

The first run was taken in 5 nm wavelength steps from 250 nm–400 nm. In this test, response fromplastic-only propagation (Uncoated) was compared to the response with the coating preceding theplastic film (Coated) and is shown in Figure 13.25. Figure 13.25A is the response of SiPM through uncoated plastic film (squares), and through plastic film with Si nanoparticle coating (dots), and Figure 13.25B gives the ratio of coated to uncoated signals. In both cases, the response as shownwas normalized by the SiPM response to the D2 lamp. In this case, the relative response between the coated and uncoated configurations is seen as evidence for the wavelength-shifting properties of the Si nanoparticles. An increase in the response of the SiPM for decreasing wavelengths less than 320 nm is seen in the coated response in contrast to the decreasing response of the SiPM only and the flat response of the Uncoated signal. At the lowest wavelength tested here (250 nm), the Coated response is

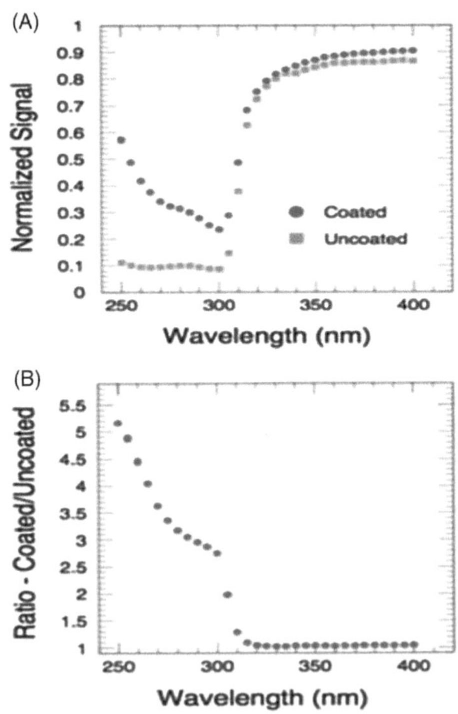

FIGURE 13.25 Characterization of the Cerenkov sensor. (A) response of SiPM through uncoated plastic film (squares), and through plastic film with Si nanoparticle coating (dots). (B) ratio of coated to uncoated signals.

a factor of ~5 greater than the Uncoated response. Recall that the wavelength has been shifted to ~ 650 nm — not optimal for this SiPM. Note that a significant rise in response is seen at the lower limit of the test (250 nm)—this is where the maximum in response of the first Si nanoparticle photodiode was seen.

Enhancement of the wavelength response of a Hamamatsu MPPC was observed when Si nanoparticles were deposited on a plastic film immediately preceding the photosensor. It appears that the Si nanoparticles absorbed the UV light with wavelengths as low as the D2 lamp and monochromatorsystem could deliver (>160 nm) and then emitted light at a wavelength detectable by the SiPM, turning the nanoparticle coating/photosensor combination into a UV photon detector. Further tests are now planned to determine the optimal size of the nanoparticles (for tuned wavelength response), the optimal thickness of the nanoparticle coating, and measurements of the efficiency of the response as a function of wavelength and for the SiPM operated with its normal bias voltage in Geiger mode.

This demonstration showed that silicon nanoparticles nanosolid films provide a sensitive detector for UV radiation while being blind to visible radiation, property which is useful for elementary particle collisions in neutrino, dark matter, and rare decay experiments. In those weak UV Cherenkov radiation is produced that gives vital information about momentum and geometry of collision [26]. Silicon-based sensitive UV photodetectors have military as well as commercial applications. Military applications include missile warning systems, biological attack warning systems and jet engine sensors.

Antenna In another development, silicon nanowire arrays were used [27]. Efficient procedures for fabrication of nanowires were recently developed [28,29]. Hemispherical gold deposits are made on the end of the wires. When a near-infrared optical field falls onto the gold deposits, it excites plasmon resonances, which remarkably amplify the intensity, effectively making them like antennas. The same gold deposit can form a p-n like Schottky junction with the nanowire, which enhances charge collection. Thus the system acts as an effective near-infrared photodetector [27].

13.3 BIOFUEL CELLS [32]

A block diagram of the concept of a fuel cell is shown in Figure 13.26. It shows two electrodes (anode and cathode) and a liquid that conducts positive charge but not electrons (electrolyte). Two chemicals are injected continuously: a fuel (hydrogen gas for example) and oxygen or air. The output is water and a current of electrons in the wires externally connecting the electrode and (or carbon dioxide depending on the type of fuel). Thus hydrogen and oxygen are consumed. The hydrogen gas may be substituted by another chemical rich in hydrogen, such as ethanol, methanol, sugars, etc.

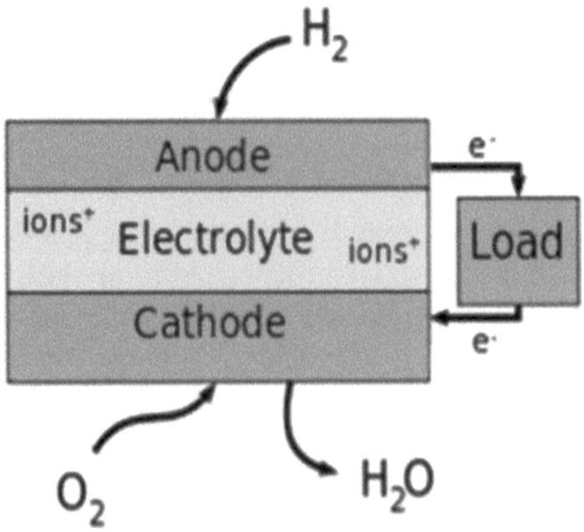

FIGURE 13.26 A block diagram of the concept of a standard fuel cell.

There are more than 20 varieties of fuel cells, but they all work in the same general manner. Two chemical reactions occur at the interfaces of the three different regions. At the anode a catalyst oxidizes the fuel turning the fuel into a positively charged ion and a negatively charged electron. The electrolyte used in the fuel cell is a one-way electrolyte: the substance is specifically designed such that ions can pass through it while electrons cannot. But the freed electrons would travel through the external wire connecting the electrodes, which creates the electric current in the circuit. The ions travel through the electrolyte to the cathode. Once the ions reach the cathode, another chemical reaction takes place between the ions, electrons, and the injected oxygen. The result of the reaction at the cathode is the creation of water or carbon dioxide depending on the fuel injected.

Thus one can see that fuel cells are clearly different from batteries. A battery is a closed device: the chemicals needed for its entire lifetime are already packed in it. They react with each other to generate an electromotive force (emf or voltage). When the chemicals are consumed then the battery stops. On the other hand, fuel cells continue to work as long as the input chemicals are being fed in (injected). An example of a popular fuel is ethanol. Ethanol is a hydrogen-rich liquid and it has a higher specific energy (8.0 kWh/kg) compared to methanol (6.1 kWh/kg).

	Equation
Anode	$C_2H_5OH+3H_2O \rightarrow 12H^+ + 12e^- + 2CO_2$ oxidation
Cathode	$3O_2 + 12H^+ + 12e^- \rightarrow 6H_2O$ reduction
Overall reaction	$C_2H_5OH + 3O_2 \rightarrow 3H_2O + 2CO_2$ redoxreaction

Glucose cells are devices that convert chemical energy from glucose fuel to electricity. They work by oxidizing glucose at one electrode (anode) and reducing oxidant at another (cathode).

$$C_6H_{12}O_6 + H_2O \rightarrow C_6H_{12}O_7 + 2H^+ + 2e^- \text{(Anode)}$$

$$O_2 + 2H^+ + 2e^- \rightarrow H_2O \text{(Cathode)}$$

$$C_2H_{12}O_6 + O_2 \rightarrow C_6H_{12}O_7 + H_2O \text{(Overall reaction)}$$

This reaction provides 2 electrons per molecule of glucose and a maximum reversible cell voltage of 1.3 V. In principle, the complete oxidation of glucose to carbon dioxide and water could provide 24 electrons per molecule, although in practice this reaction is difficult to achieve.

13.3.1 Metal Catalyst

Fuel cells use precious (noble) metal catalysts, such as Pt, Pd and Au; with the most popular being platinum. The support materials for Pt catalysts areprimarily carbon-based. Theyare normally manufactured by the pyrolysis of hydrocarbons, such as natural gas or oil fractions. One popular commercial carbon support material is Valcan XC-72. There are several procedures for synthesis of carbon support material for use in Pt and/or Pt-based alloy catalysts:

- Chemical synthesis methods, such as impregnation-reduction, colloidal methods, etc.improvements in dispersion and size distribution are still required for achieving high mass activity.
- Physical syntheses, such as the spattering method, are also a promising way to fabricate small and active Pt catalysts.
- Plasma, which is generated in liquid, is also applied for forming catalysts, such as Pt and PtAu.

13.3.1.1 Bimetallic Catalysts

There are other interesting support materials for platinu. For example, nanoporous gold has been used as support for binary PtBi. Studies showed that porous gold decorated by the Pt-Bi alloy constitutes a high performance anode catalyst for direct glucose fuel cell. It is prepared by electroless and electrochemical plating methods. First a dense Pt layer is deposited. This is followed by depositing a small amount of Bi, which results in scattered Bi particles. As to the cathode, one can use a standard commercial platinum on carbon substrate (Pt/C) catalyst electrode. It is interesting to compare the performance of the thePtBi electrode with the standard Pt electrode. With the PtBi anode, measurements give 18 mW mg^{-1} for the maximum gravimetric power density. In comparison, for noble metal loading of only 0.45 mg cm^{-2} (Au 0.3 mg and Pt 0.15 mg), an open circuit voltage of 0.9 V is obtained with a maximum power density of 8 mW cm^{-2}. This is about 4.5 times lower than that of commercial Pt/C electrode

Although Pt has been used extensively it is susceptible to poisoning. In fact, its active sites are so easily poisoned by unexpected intermediates such as CO causing performance deterioration. Therefore, bimetallic electrocatalysts, such as PtRu, PtAu, PtPd, PtBi are developed to reduce the poisoning by suppressing adsorption of poisoning intermediates.PtBi/C, PtAu/C, PtRu/C, PdBi/C, all on carbon support have been used; but the bismuth system has been more popular because of its low cost.New nanostructured electrocatalysts have been developed, which are based on non-noble metals, preferentially mixtures of Fe, Co, Ni at the anode, and Ni, Fe or Co alone at the cathode.

13.3.1.2 Nano Silicon Catalyst [30–34]

The use of silicon nanoparticle-based electrode was developed at by the Yau's group at Cleveland State University in collaboration with the University of Illinois. Silicon nanoparticle catalysts are prepared as follows. An*n*-type heavily doped (~ 0.005 Ω cm) silicon wafers with its surface containing the native oxide was cleaned with ethanol, isopropanol, and deionized water. A 0.1ml drop of a 1-nm silicon nanoparticle (Si1) water colloid was spread on the wafer surface. The sample was then incubated in a container under moisture for 10 hours and then rinsed with deionized water. The surface of the particle-immobilized wafer was covered with a mask to form the Si1–Si electrode with a working area of about 1mm × 1mm. The cathode electrode is prepared using glucose oxidase enzyme (GOx). It is an oxido-reductase that catalyzes the oxidation of glucose to hydrogen peroxide and D-glucono-δ-lactone. This enzyme is produced by certain species of fungi and insects and displays antibacterial activity when oxygen and glucose are present.

Fig. 13.27 is anatomic force microscopy (AFM) image of the Si29–Si electrode, showing a typical particle distribution on the silicon surface, which is only partly covered with the particles. Fig. 13.27 is an AFM image showing the GOx electrode, where the entire electrode surface is covered with a monolayer of the enzyme. Estimationshows that typically there are 4000 GOx molecules or 8000 flavinadenine dinucleotide (FAD) centers within an electrode surface area of 1 × 1 μm^2, while there are only 1000 Si29 particles within the same area.

The schematic of the fuel cell is depicted in Figure 13.28A. It shows the electrodes and the reactions taking place. Glucose is dissolved in a phosphate buffer solution (PBS), and then introduced into the cell for measurements. The cyclic voltammograms (CVs) of the electrode are taken at *p*H 7 with the results given in Figure 13.28B. The increase of the anodic current due to the presence of glucose (curve *b*) above the background (curve *a*) is clearly shown.

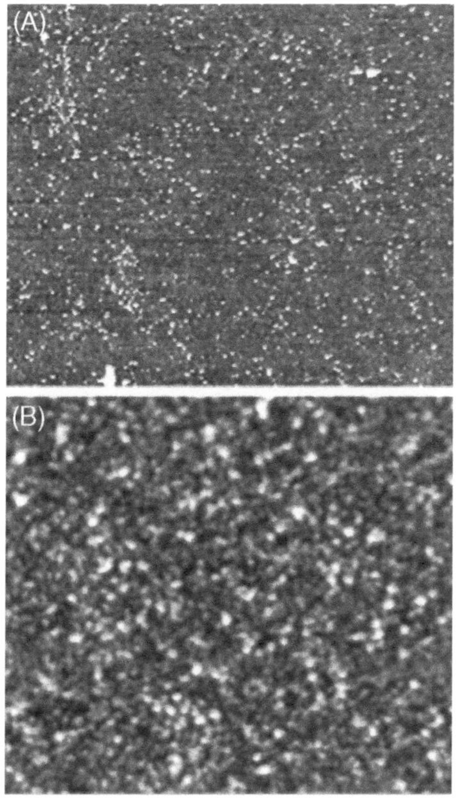

FIGURE 13.27 (A) Nano silicon catalyst:atomic force microscopy (AFM) image of the Si29–Si electrode, showing a typical particle distribution on the silicon surface. (B) AFM image of a GOx electrode, where the entire electrode surface is covered with a monolayer of the enzyme.

The response of the cell may be calibrated against the concentration of glucose, relevant to the human body. The inset shows the results over a range of 0–50 mM, which covers the physiological level of 3–8 mM. The calibration curve of the electrode at *p*H 7 as shown in the **inset** is linear. Secondly, the results are useful for drawing some information about electrode poisoning. The absence of electrode poisoning is reflected in the electrode's reversible response to varying the glucose concentration as indicated by the arrows. Curve *b* in Figure 13.28B indicates that glucose is oxidized at the anode electrode. Thus, the Si29 particle exhibits a catalytic character in the detection of glucose.The current response with concentration of the Si29–Si electrode and the GOxelectrode to glucose are compared in Figure 13.28C. The current values were measured at a potential of 0.7 V for both electrodes, showing a good enhancement die to the nanoparticles. Note that the bare silicon wafer showed no electrochemical response to the substances used in this work up to a potential as high as 1.6 V.

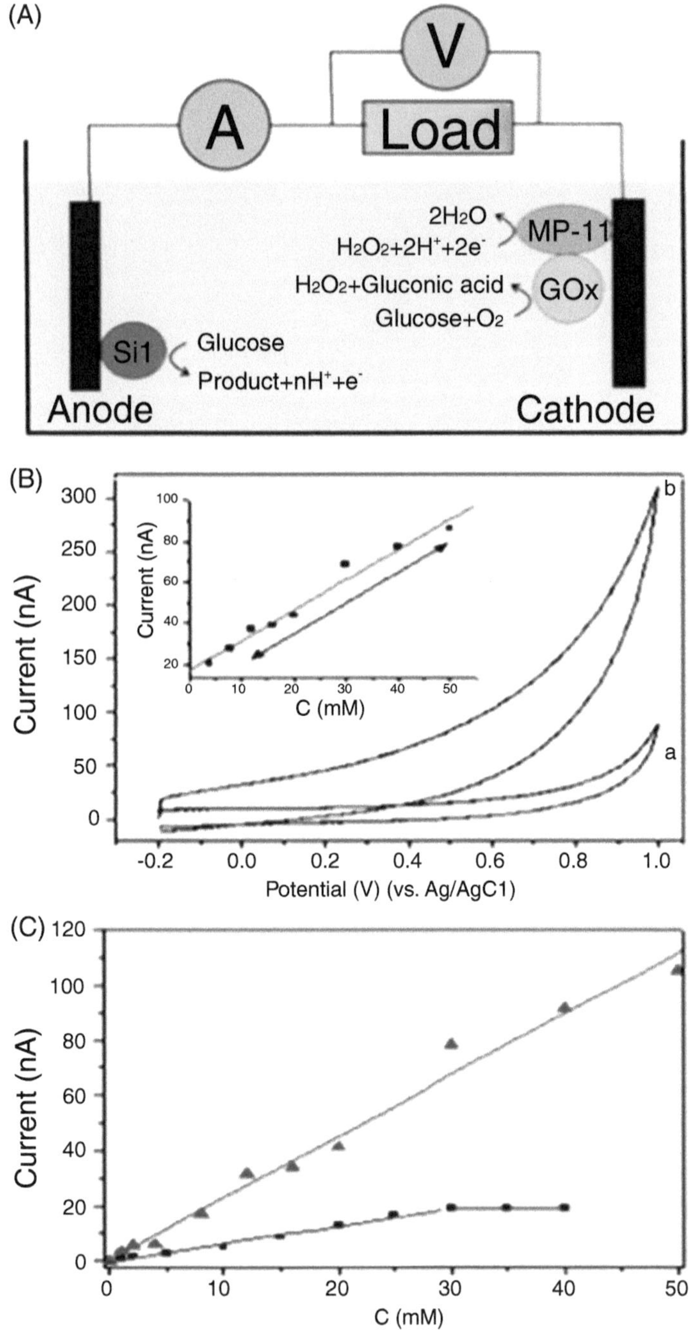

FIGURE 13.28 Nano fuel cell. (A) schematic of the nano fuel cell, showing the electrodes and the reactions taking place. (B) (top) The cyclic voltammograms (CVs) of the electrode. (inset) The calibration curve of the electrode. (C) The current response with concentration of the Si29–Si electrode and the GOx electrode to glucose are compared.

This type of fuel cell works well with several other fuels including glucose, ethanol, and methanol as shown in Figure 13.29A. The figure gives the current density as a function of the concentration. In addition, the figure shows that the fuel cell works with anodes using nanofilm catalysts made from other sizes of silicon nanoparticle. Two types of silicon electrodes were made: silicon wafers immobilized with the Si1 particle (the Si1–Si electrode)and those immobilized with the Si2.8 particle (the Si2.8–Si electrode). Carbon paper (Toray, TGPH-120) was also used as electrode for its porosity, which leads to improved particle loading. Atomic force microscopy of the electrode surface showed the presence of the particles at a sub-monolayer level.The particle-immobilized carbon paper electrodes (the Si1-CP electrode and the Si2.8-CP electrode) have a particle surface density of 28 g cm^2. Measurments using nanoparticles of 1 nm and 2.8 nm particles as well as several fuels are shown in Figure 13.29A.

The fuel cell is also sensitive to external light. In the previous sections, the silicon nanoparticle films were shown to be responsive to light. For example, they act as a photoconductor material as well as a photovoltaic active material. The effect of external light on its catalyst activity is therefore expected.

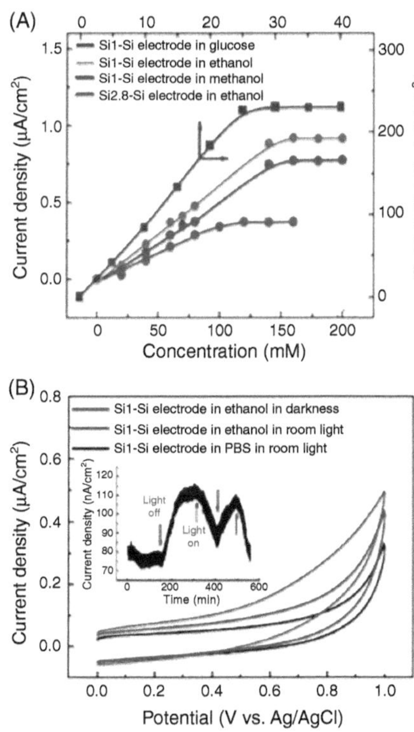

FIGURE 13.29 Fuel cell with several other fuels including glucose, ethanol, and methanol. (A) the current density as a function of the concentration. (B) CVs in an ethanol with a Si1–Si electrode in both room light (white fluorescent light with an intensity of 0.5 mW/cm2) and darkness.

Figure 13.29B shows the CVs obtained in an ethanol experiment with a Si1–Si electrode in both room light (white fluorescent light with an intensity of 0.5 mW/cm^2) and darkness at pH 7. Room light inhibits the oxidation current from darkness conditions by nearly a factor of two as can be seen the figure. The effect is also fast and reversible under weak light irradiation that can follow switching on and off of the light. This light-dependent effect is observed with a Si1-CP electrode in ethanol at pH 13. However, when the electrode is illuminated with strong white light (with intensity of 200 mW/cm^2 for example), the oxidation current shows a 14-fold increase in the current as compared to the room-light current. The increased catalytic current caused by high intensity light points to development of conditions, such as heating that overcome the activation barrier.

Other electrodes have akso been tested. In one test, an electrode using carbon support was similariy prepared. A basal plane of HOPG was covered with 1-nm silicon nanoparticle (Si1) colloid. This is followed by incubation. The particle-immobilized HOPG(Si1-HOPG) electrode had a surface area of 2mm × 5mm.

13.3.1.3 Hybrid Fuel Cells

Hybrid biofuel cell can be used in which one electrode is based on nano silicon while the other is based on standard enzyme. In fact, Si nanoparticle were used as the anode catalyst. The anode was an epHOPG immobilized with Si1 and the cathode was an epHOPG immobilized with horseradish peroxidase (HRP). HRP is an enzyme.The anode and cathode were immersed in a PBS contained in a beaker and electrically connected through a load resistance RL. This prototype single-compartment fuel cell operated on glucose and H2O2, which were dissolved in the PBS. Glucose was oxidized by Si1 and H2O2 was reduced by HRP. It is known that HRP in the presence of an electron donor catalyzes the reduction of H2O2 to water. The output power of the hybrid cell, was found to improve when HRP was replaced by the heme-containing polypeptide microperoxidase-11 (MP-11); it reduces H2O2 at the cathode.

13.4 SUPERCAPACITOR STORAGE

13.4.1 Standard Capacitor

A capacitor, in its simplest form, consists of metallic plates or surfaces separated by a dielectric medium as shown in Figure 13.30. In terms of the surface area S, electrode spacing d and the permittivity of the fillining material ε, the capacitance is device is defined as $C = \varepsilon S/d$ There are many candidates for a conductor. This includes a foil, thin film, sintered bead of metal, or an electrolyte. The spacer non-conducting dielectric can be vacuum, air, glass, ceramic, plastic film, paper, mica, and oxide layers. Capacitors store charge of opposite

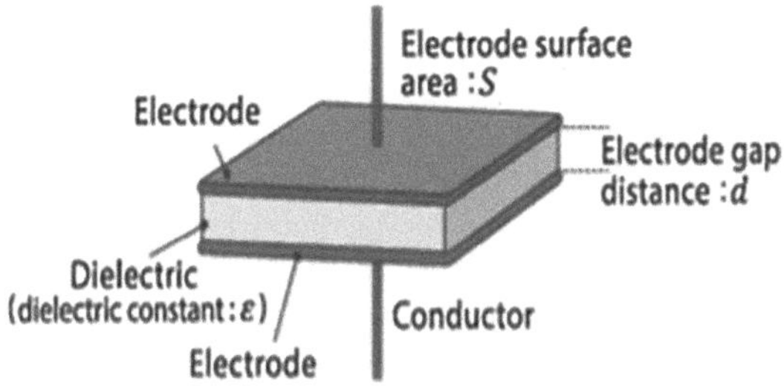

FIGURE 13.30 Schematic of a Standard Capacitor

sign on the plates, hence store electric energy, which can be recovered as a useful electric current. An ideal capacitor has a perfect dielectric spacer, which does not allow charge to dissipate between the plates through it. The capacitor's charge capacity depends on the material spacer in terms of direct dependence on its refractive index, being higher for higher index.

13.4.1.1 Electrolyte Capacitor

An electrolyte is a substance that produces an electrically conducting solution when dissolved in a polar solvent, such as water [35]. The dissolved electrolyte separates into cations and anions, which disperse uniformly through the solvent. Electrolytes effectively include most soluble salts, acids, and bases; as well as DNA, polypeptides and synthetic polymers. Polyelectrolytes which contain charged functional groups function also as electrolytes. Electrically, such a solution is neutral. Figure 13.31A is a sckematic diagram of an electrolyte capacitor. The cathode electrode (positive) has an insulating oxide layer (shown in yellow). The space beteween the plates is filled with an electrolyte.

Because of the dielectric layer that always exis in the cathode, the electrolyte does not form an ionic conductive connection between the two electrodes.

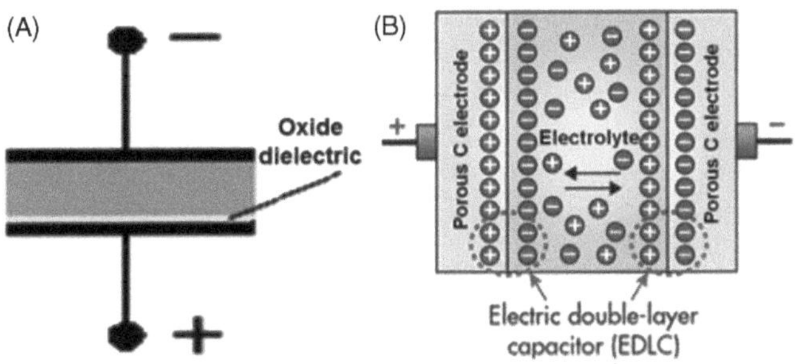

FIGURE 13.31 Electrolyte-based capacitors (A) schematic diagram of a standard electrolyte capacitor. (B) A schematic of standard supercapacitor (SC).

13.4.1.2 Standard Supercapacitor (SC)

A schematic is shown in Figure 13.31B. Unlike electrolyte capacitors, in supercapcitors the oxide layer over the cathode is removed. This allows the electrolyte to make a conducting path between the two eelctrodes [36]. This creates a layer of +, – charge layer on one electrode and a layer of –, + charge layer on the other other as can be seen in the figure. This distinguishes supercapacitors from conventional electrolyte capacitors. This is why they are also known by electric double-layer capacitor, or ultracapacitor. The electrodes are generally made of carbon or its derivetives. We numerate some features of supercapacitors.

- Very high capacitance (rated in farads (F)) but low voltage limits and store more energy than electrolytic capacitors
- The metal electrodes consist of two metal plates, which only are coated with a porous material known as activated carbon. As a result, they have a bigger area for storing much more charge.
- Because the separation takes place at the interface, the separation of charge is therefore of the order of a few ångströms (0.3–0.8 nm), much smaller than in a conventional capacitor.
- It is a high-capacity capacitor with capacitance values much higher than other capacitors (but lower voltage limits). They typically store 10 to 100 times more energy per unit volume or mass than electrolytic capacitors,
- accept and deliver charge much faster than batteries. can be cycled hundreds of thousands times with minimal change in performance, A supercapacitor's lifetime spans 10 to 20 years, and the capacity might reduce from 100% to 80% after 10 or so years.
- tolerate many more charge and discharge cycles than rechargeable batteries.
- Used where rapid charge/discharge cycles, such as in regenerative braking; memory backup for static random-access memory.
- low energy density. Thus, they can't be used as a continuous power source. One cell has a typical voltage of 2.7 V; if higher voltage is needed, the cells must be connected in series.
- Unlike capacitors and supercapacitors, batteries store energy in a chemical reaction. This way, ions are inserted into the atomic structure of an electrode, instead of just clinging to it like in supercapacitors. This makes supercapacitors (and storing energy without chemical reactions in general) able to charge and discharge much faster than batteries. Due to the fact that a supercapacitor does not suffer the same wear and tear as a chemical reaction based battery, it can survive hundreds of thousands more charge and discharge cycles.

13.4.1.3 Supercapacitors for Flexible and Portable Electronics

Energy storage for portable electronic devices, which arebecoming increasingly important to the present society, forms by far the largest mobile energy storage market today and is experiencing innovation for future applications such as flexible/printed electronics and display [37]. To operate flexible devices, flexible energy storage needs to be integrated with the active devices.

Electrochemical supercapacitors with high energy and power capabilities are potentially capable of powering flexible devices. In order to be integrated with flexible electronics, supercapacitors should be made flexible with sheet-like structures, which are lightweight. Also, the fabrication method of the supercapacitors should be inexpensive and high-throughput.

13.4.1.4 CNT-Paper-Like

Paper-like nanocomposite material can be used as the basic unit for assembling flexible capacitors. The first step is to grow films of vertically aligned carbon nanotubes (CNT) on silicon substrates using thermal–chemical vapor-deposition. The electrolyte is now prepared by dissolving cellulose in an ionic liquid. The electrolyte is now infiltrated into the CNT to form a film of cellulose and electrolyte, embedding the CNT. After solidification on dry ice, the resulting nanocomposite paper is peeled from the substrate. A supercapacitorwas made by bonding two pieces of the nanocompositepapers back-to-back. Another method may be used to for. CNT is first dissolved in water then sprayed on plastic substrates to form flexible electrodes. Supercapacitors werethen formed by sandwiching flexible gel/solid electrolyte with two flexible electrodes.The use of silicon nanoparticle-based electrode was developed at Cleveland State University in collaboration with the University of Illinois.

13.4.1.5 Brushed on Sheet (Nanococktail)

A simple approach for constructing flexiblesheets of supercapacitor was developed by the Yau's group at Cleveland State University in collaboration with the University of Illinois [38]. The construction is based on painting a sheet of flexible plastic electrolyte with a composite material made of a conducting polymer and carbon nanotubes.

The two-step construction of the capacitor sheet is schematicallydescribed in Figure 13.32A. In the first step a thin film solid electrolyte is made. Polyvinyl alcohol (PVA) powder (Sigma–Aldrich) is mixed with water with a ratio of 1:9 by weight. The PVA mixture is then mixed with phosphoric acid(H3PO4, 85%) with a ratio of 1:1 by volume. Films of PVA are now formed by first casting the final mixture on a glass surface. Once water had evaporated, the the cast is peeled off from the surface. This procedure gives a free-standing films of a thickness of about 0.3 mm.

In the second step, a nanocmposite of CNT in PANI is made. A volume of 0.1 ml of polyaniline (PANI) dispersed in xylene (Sigma–Aldrich)and 0.003 g of single-walled carbon nanotubes (CNT) (Carbon Solutions) are mixed in N-dimethylformamide. Both materials can be used as purchased.

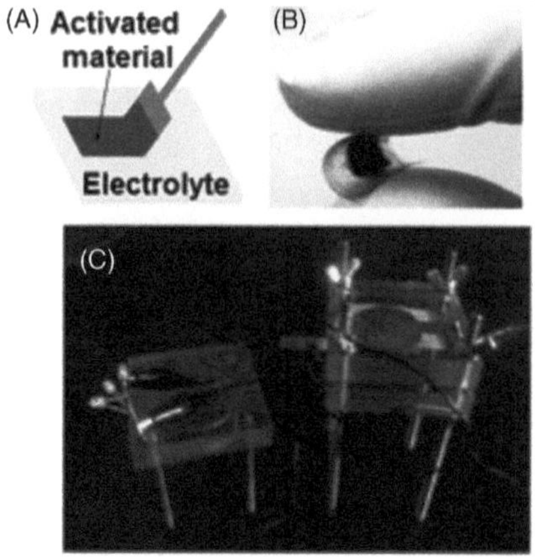

FIGURE 13.32 Brushed on sheet super capacitor (A) schematic of the construction of the super-capcitr capacitor sheet with a paint brush. (B) a capacitor sheet. (C) a stack of three capacitor sheets drive the three red LEDs.

A painting brush is then used to transfer thePANI–CNT aqueous mixture on both sides of a PVA film to form anddefine the shape of a capacitor. Water is evaporated by putting the painted sheets in an oven at 40 C. Figure 13.32B shows a capacitor sheet.The two-dimensional size of the capacitor is about 5mm × 5mm. Electrical contacts can be made by either using metallic clips or coppersheets.

One potential application of the capacitor sheet is now demonstratedby using the capacitor to power light-emitting diodes (LED). For this demonsta-ration, however a single supercapacitance sheet as prepared is not sufficient. The problem is the low operationvoltage (~1V) provided by a single sheet. To satisfy the operatingrequirement of LEDs, a stack of several capacitor sheets connected in series can be used to increase the voltage. Measurments of dis-chargingcharacteristics of stacks of one, two, and three capacitor sheets showed that the initial voltages of the stacksare proportionally higher, reaching 4V.In fact, a stack of three capacitor sheets was needed to drive the three red LEDs as shown in Figure 13.32C.

It is believed that this kind of device is a good step in the diection of ful-filling the requirements of flexible electronics applications. Solid electrolytes are preferred over liquid electrolytes, which evaporate and therefore require enclosures. Carbon nanotubes were used to reduce the internal resistance of the conducting polymer. The fabrication approach is inexpensive and potentially high-throughput.

To increase the specific capacitance of the sheets so they may handle higher loads, flexible supercapacitor sheets, the active electrode material is enhanced bysupplementing the CNT/polymer with additional nano structures. For example, a nanocomposite material consistingof the conducting polymer,

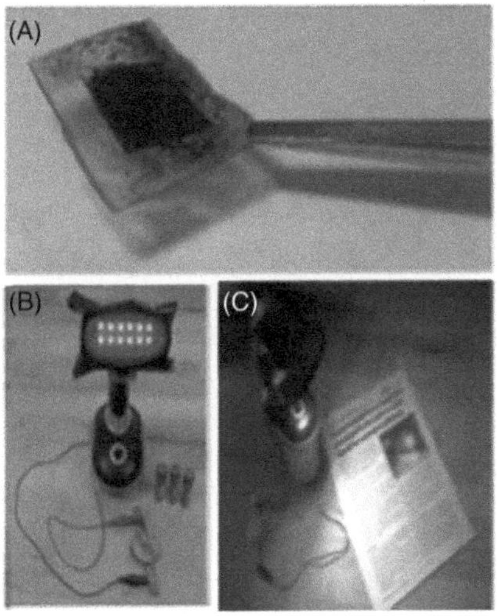

FIGURE 13.33 Nano cocktail-based supercapacitor (A) supercapacitor sheet using nanocomposite material consisting of conducting polymer, polyaniline, and nanomaterials with quasi-zero, one and two dimensionalities. (B) Stacks of capacitor sheets are now able for the first time to power a LED lamp consisting of 12 LEDs. (C) demonstrates the potential of the nanocomposite-based capacitor sheet in illumination applications.

polyaniline, and nanomaterialswith quasi-zero, one and two dimensionalities, includingcarbon nanotubes, silicon nanoparticles, titanium oxideparticles and grapheme flakes was prepared. The nanocomposite materials is prepared as follows. First, 0.4 ml of PANI and 0.003 g of single-walled CNT are mixed in 99% N-dimethylformamide (final volumewas 1 ml) to form the base mixture. Then three individualnanomaterials (0.6 ml of Si2.8, 0.02 g of GR or 0.02 g ofTiO2) are mixed with 1 ml of the base mixture to formcomposite mixture, i.e. PANI/CNT/Si2.8, PANI/CNT/GP andPANI/CNT/TiO2. The PVA sheets are made as described above. A painting brush was used to transfer thecomposite material to both sides of a PVA sheet and dried inan oven.The two-dimensional size of the capacitor was about5 mm x 5 mm. Electrical contacts are made by metallic clips or coppersheets. Figure 13.33A shows a supercapacitor sheet.The produced capacitor sheets show a high specific capacitance of 477.1 F/g. Stacks of capacitorsheets are now able for the first time to power a LEDlamp consisting of 12 LEDs as shown in Figure 13.33B. Figure 13.32C demonstrates the potential of the nanocomposite-based capacitor sheet in illumination applications.

13.4.1.6 *Graphene Supercapacitor [39]*

Graphene is often suggested as a replacement for activated carbon in supercapacitors, in part due to its high relative surface area (which is even more substantial than that of activated carbon). The surface area is one of the limitations of capacitance and a higher surface area means a better electrostatic charge storage. In addition, graphene based supercapacitors will utilize its lightweight nature, elastic properties and mechanical strength.

Graphene-based supercapacitors are said to store almost as much energy as lithium-ion batteries, charge and discharge in seconds and maintain all this over tens of thousands of charging cycles. One of the ways to achieve this is by using a highly porous form of graphene with a large internal surface area (made by packing graphene powder into a coin-shaped cell and then dry and press it).

13.4.1.7 Supercapacitor Thread [40]

Super capacitors in the form of flexible wires can be constructed. The wires or "threads" can be hundreds of meters long and can be weaved, knitted to form stand-alone patterns or patterns attached to rigid or flexible substrates according to preselected programming. The backbone (electrode) of the capacitor is low cost ultrathin industrial bare metal wires, while the active material is composites of ultrasmall silicon nanoparticle in polymers. The patterns can be integrated in standard fabric manufacturing technologies. The process enables wearable recharging energy storage devices.Stainless Steels have excellent ductility and can be supplied by California Fine Wire in ribbon and round forms. Round diameters can be drawn as small as .0003" and ribbons from as small as .001" x .005" to as large as 0.010" x 0.050".

In the process, industrial commercially available ultrathin metal wires are used. The metal wire is first coated with silicon nanoparticle-polyaniline composite. Polyaniline is a conducting polymer. This is accomplished by immersing the wire in the silicon nanoparticle-polyaniline solution for a certain period of time, or letting the wire pass through the silicon nanoparticle-polyaniline solution. The coated wire is then put in air for the water content in the composite to evaporate. This is followed by immersing coating of the processed wire with a solution of a solid electrolyte, such as polyvinyl alcohol (PVA). The twice-coated wire is then put in air for the evaporation of water. The flexible one-dimensional thread is now ready to be used to construct fabrics by weaving or knitting using conventional standard technologies, or for any type of patterning.

For mass production, a spool of the long wire is transferred to another spool by driven motors while the segment of the wire between them is exposed to the solution. The inter distances of the configuration as well as the speed of the thread are chosen to allow enough (prescribed) thicknesses of the composite and the PVA to build while at the same time allow drying of the layers.

A prototype using a double wire was constructed for demonstrationby the Yau's group at Cleveland State University in collaboration with the University of Illinois [40]. A plastic thin rod of 0.5 mm diameter was used as a support. Two segments of the super capacitor wires are twisted on the plastic rod. The entire structure of the twisted pair was coated again with PVA. After coating, the plastic rod can be removed. On the other hand, the super capacitor thread can be twisted about each other without the plastic supporting rod. Figure 13.34A

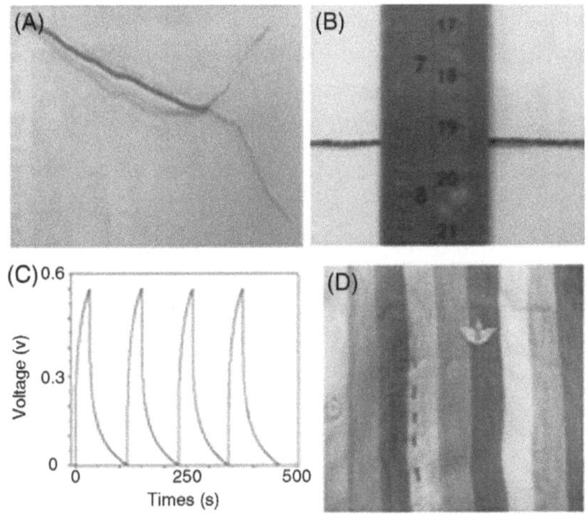

FIGURE 13.34 Supercapacitor thread (A) photo of thread super capacitor segment. (B) a scale with the supercapacitor. (C) charging-discharging curves of the capacitor. (D) photo of thread supercapcitor sewed on the shirt.

shows the finished super capacitor segment, while Figure 13.30B shows a scale. Figure 13.34C shows the charging-discharging curves of the capacitor, confirming the operation of the capacitor.

As a demonstration of wearable devices, we used a simple low-tech stainless needle to knit the super capacitor thread into the pocket of a household shirt. Two clip contacts were bonded to the capacitor for connection to external devices. The photo in Figure 13.34D shows the shirt, and capacitor thread device.

It is interesting to mention other non-silicon based wire supercapaciors. Wire-based electrochemical micro supercapacitor was developed using a variety of materials. For example, in one development, a flexible plastic wire and a Kevlar fiber were both covered with arrays of high-quality ZnO nanowires. The ZnO nanowires were grown by hydrothermal method. The ZnO nanowires on the Kevlar fiber was coated with a thin gold film. Capacitors were assembled from the two. In a second development, textile fibers, such as cellulose or polyester woven in the form of ordinary textile are coated with single-walled carbon nanotube (SWNT). Each cotton fiber is comprised of multiple individual cotton fibrils, which are in turn composed of multiple microfibrils bundled together. This is done by "dipping and drying" of everyday textile into a solution of the nanotubes. The resulting material is a conductive textile. Super capacitors are made from these conductive textiles. In a third development, Au nanowires are first grown inside the channels of commercially available AAO templates (nanopore) using electrodeposition. After the electrodeposition of the Au nanowires, CVD was carried out to grow multi-walled carbon nanotubes (MWNTs) inside the template, by the pyrolysis of acetylene. The template is removed. The presence of an evaporated metal film prevents the CNT/AuNW hybrid structures from collapsing after the removal of templates.

13.4.1.8 Supercapacitor Applications

Applications of supercapacitors stem from the nature of the devices. Unlike fuel cells and batteries, which deliver low power but over long times (continuous), supercapacitors deliver large peak powers but over short times (non-contineous in burst manner). Thus the domain of action of supercapacitors lies in systems requiring many rapid charge/discharge cycles within transportation systems such as cars, trains, aircraft, cranes and elevators. Moreover it can be used in backup power systems, and many more

13.5 SOLID-STATE-LED WHITE LIGHTING

In this section we present the use of nanoparticles to enhance the characteristic of white bulbs produced in solid state white lighting [41,42]. The LED [43] consists of a chip of semiconducting material of a certain bandgap doped with impurities such that it creates a p-n junction as show in Figure 13.35A. The p-type side is doped such that it has excess positive charge (holes), while the n-type side is doped such that it has excess electrons. When it is forward biased with the p side connected to the positive side of the battery while the n side connected to the negative end, a current flows across the junction from the p-side (anode) to the n-side (cathode). If the biasing is reversed, no current flows. Charge-carriers, electrons and holes, flow into the junction from the external electrodes with different voltages. Electrons enter in the conduction band, the upper edge of the bandgap. Holes on the other hand enter at the bottom edge of the band. Thus when they meet they combine and relates the energy equal to the bandgap energy in the form of an electromagnetic wave or light (photon). When an electron meets a hole, it falls into a lower energy level and releases energy in the form of a photon.

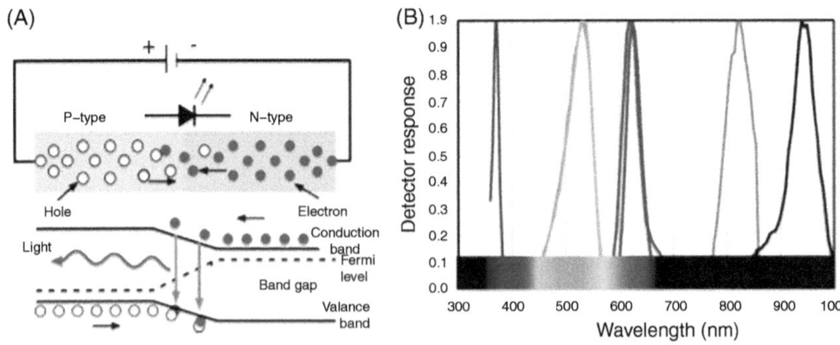

FIGURE 13.35 Solid state LED white lighting (A) Schematic of an LEDp-n junction. (B) a variety of available LED in near-infrared, visible, or near-ultraviolet regime. *(From https://en.wikipedia.org/wiki/Light-emitting_diode#/media/File:PnJunction-LED-E.svg)*

The wavelength of the light emitted depends on the band gap energy of the materials forming the p-n junction. The materials used for the LED have a direct band gap with energies corresponding to near-infrared, visible, or near-ultraviolet light (**see** Figure 13.35B). Blue and most green LEDs are made from gallium nitride (GaN). The brightest red LEDs are made from aluminum gallium arsenide (AlGaAs). The LEDs used in near-infrared remote controllers are also AlGaAs devices; their peak emission is about 880nm and peak detection around 820nm.

The figure shows that the energy level of the holes is less than the energy levels of the electrons. This phenomenon is generally called electroluminescence, the electrons dissipate energy by emitting photons in gallium arsenide phosphide (GaAsP) and gallium phosphide (GaP) semiconductors. If the semiconductor is translucent, the junction becomes the source of light as it is emitted, thus becoming a light-emitting diode. However, when the junction is reverse biased, the LED produces no light and—if the potential is great enough, the device is damaged.

It is to be noted because silicon and germanium are indirect band gap materials, parallel silicon or germanium diodes do not produce optical emission, the electrons and holes usually recombine by a non-radiative transition. In other words, in electroluminescence, the electrons dissipate energy in the form of heat for silicon and germanium diodes.

13.5.1 LED Configurations and Phosphors for Producing White Light

The objective of using electronic chips for lighting is to manufacture light bulbs that are superior to conventional bulbs yet with characteristics that match the sun's white light spectrum,given in Figure 13.36A. Light emitting diodes (LED) are chips powered by electricity. They produce light of specific color in the UV, blue, green or red light range. Figure 13.36B shows merging of red, green and blue light beams via a prism to produce white light. The goal of solid state white lightings is to mix Led-based light sources or derivatives to generate white light that approaches the quality of sun light spectrum. There are several schemes available.

The first scheme involves mixing the light from three independent LEDs: blue, green and red is one scheme for generating while light as shown in Figure 13.36C,D. But the sum has several short commings. First, because the emission of each LED is not wide enough, the mix produces "finger shaped" spectrum, not smooth as normal sun light, with much of the constituents in the sunlight spectrum is missing. This spectrum is far from the spectrum of sun light given in Figure 13.36A. Moreover because independent diodes degrade at different rates, with time the balance of colors will change with usage time causing change in the characteristic of the white light.

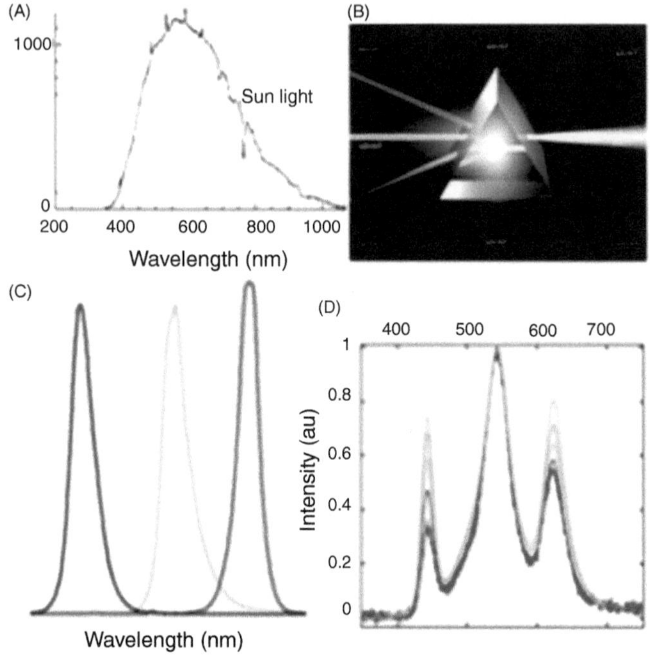

FIGURE 13.36 LED configurations and phosphors for producing white light: (A) sun's white light spectrum. (B) merging of red, green and blue light beams via a prism to produce white light. (C) three independent LEDs: blue, green and red. (D) mixing the three LEDs produces "finger shaped" spectrum, not smooth as normal sun light, with much of the sunlight spectrum is missing.

The second scheme involves the use of a single blue Led covered with a yellow or amber phosphor. Phosphors are usually made from a suitable host material with an added activator. The best known type is a copper-activated zinc sulfide and the silver-activated zinc sulfide (*zinc sulfide silver*). The phosphor coating intercepts/absorbs part of the blue light and reemit it in the form of a wideband yellow/amber light. The emerging light is actually a mixture of the yellow/amber light with the leftover of the blue light as schematically shown in Figure 13.37A. The spectrum of the emerging light is shown is shown in Figure 13.37B. The phosphor provides very little red.This posed a fundamental limit to reach high correlated color temperature (CCT) and color rendering index (CRI) at the same time. It is to be noted that there is no red phosphor that can be pumped by the blue Led.

Although this solution produces a "hand-shaped" spectrum and short of full match with sun light as it is missing a red component, the human eye senses the spectrum as white, and as such developers of this process Isamu Akasaki, Hiroshi Amano and Shuji Nakamura were awarded the 2014 Nobel Prize in Physics. The first LED (red) was invented by Nick Holonyak at the University of Illinois several decades earlier [44], but it was deemed that the more recent blue LED (invented by the awardees) and its use in this configuration afforded a very important service, namely filling the world with new "white" light.

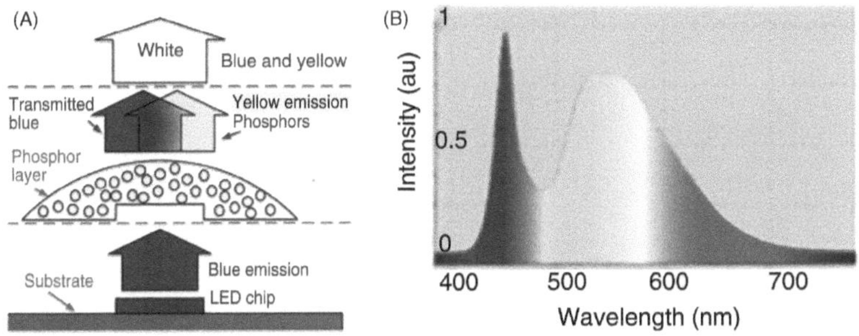

FIGURE 13.37 (A) schematic of a white LED system utilizing a yellow phosphor pumped by a blue LED (mixture of the yellow/amber light with the leftover of the blue light) (B) The spectrum of the emerging light. *(From https://electronics.stackexchange.com/questions/124095/how-to-generate-red-light-from-a-blue-dominated-spectrum)*

Isamu Akasaki, Hiroshi Amano and Shuji Nakamura
2014 Nobel Prize in Physics

Bright lights
The 2014 Nobel Prize for Physics has been awarded to Isamu Akasaki, Hiroshi Amano and Shuji Nakamura for their development of blue LEDs. The prize is worth SEK 8m (£690,000) and will be shared by the three winners who will receive their medals at a ceremony in Stockholm on 10 December.

Akasaki is a Japanese citizen and works at Meijo University and Nagoya University. Amano is a Japanese citizen and works at Nagoya University. Nakamura is a US citizen and works at University of California, Santa Barbara.

The third scheme involves adding a red LED to the blue LED-pumped phosphor to provide a red component. This solution is presently being used in filamentary LEDs. This system suffers from the fact that the blue and red LEDs are independent and degrade at different rates. In addition the spectrum has two sharp sources and one broad band, which will not be able to mimic the sunlight spectrum. This posed a fundamental limit to reach high correlated color temperature (CCT) and color rendering index (CRI) at the same time.

The fourth emerging procedure is to use a single master UV LED [42], which pumps and control three phosphors as shown in Figure 13.38A. In this development, solidstate white lighting is achieved by mixing the output of three

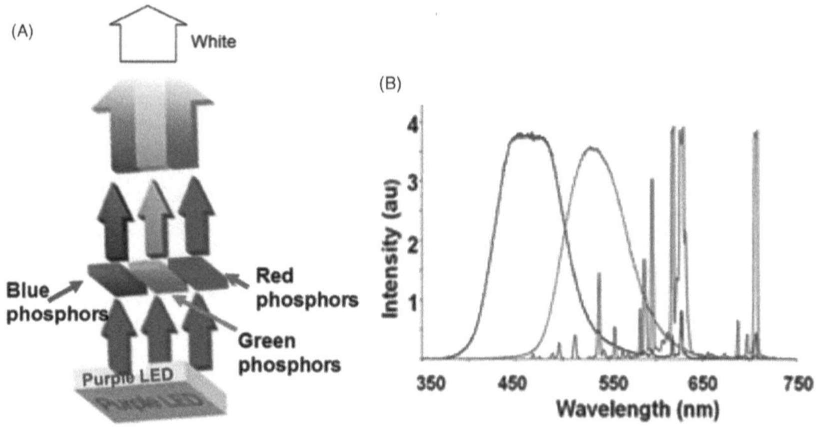

FIGURE 13.38 UV LED-pumped white light (A) schematic of bulbs that use three phosphors pumped by a single master purple UV/blue LED. (B) sum of the individual spectra. *(From LUGER RESEARCH e.U., LED professional Review 57, Technologies for Engineering an LED Light Closest to Sunlight, p 68 & www.led-professional.com/LED Light Closest to Sunlight © 2016 Toshiba Materials Co., Ltd)*

wide band phosphor converters in the blue, green, and red, which are pumped by single UV LED. UV-solid state lighting enable better color mixing and control, as well as production of stronger white sources that handle bigger areas. In addition, UV LED suffers less blue shift at high driving current condition, hence reducing the color shift between bare LED die and the encapsulated die.

13.5.2 Red Phosphor Challenge [45]

There are several challenges facing UV-based lighting. Those include the development of higher power UV LEDs,safe configurations, as well as efficient wide band red converters. As to the red phosphor, few good efficient red phosphors, whether sulfide-, nitride-, or oxide based have been known. Unlike blue and green phosphors which produce wide-band emission, red phosphors spectra are dominated by sharp line spectra with branching ratios that depend on the UV wavelength, not ideal for color mixing as shown in Figure 13.38B. The red phosphor yttrium oxide-sulfide activated with europium (Y2O2S:Eu), for example, has been investigated in UV-based lighting. Co-doped phosphate materials were recently synthesized for near UV pumping, which provided a peak wavelength of 610 nm.It has been conjectured that the phosphate material might contribute to a white-light source with a luminous efficiency of 45 lm/W and color rendering index greater than 90 at a color temperature of 5600 K. Also, multi-layered red, green, and blue phosphors showed high CRI of over 90.

In addition to providing spectra dominated by sharp red linesnot smooth enough as normal white sun light, there are problems of availability and stability (Figure 13.38B). In addition, phosphor films are found to have appreciable reflectivity causing a non-negligible fraction of the pumping LED light to be reflected backwards. This in turn causes heating and requires more power input. More energy efficient and rigorous designs must be used to minimize heating.

13.5.3 Nanosilicon-based Solution [46]

Silicon nanoparticles have also been demonstrated at **the University of Illinois** as in collaboration with Abdullah Aldwayyan and Mohammad AlSalhi of KSU, Saudi Arabia as a substitute for or additive to Eu-based "red phosphor" as sketched in Figure 13.39A,B.

The phosphors we used are commercial grade. Colloids of ZnS:Ag and ZnS:Cu,Au,Alas well as of silicon nanoparticles are prepared. Figure 13.40A gives SEM images of the blue (top) and green phosphors (middle) placedon single crystal silicon substrate. The TEM image (bottom) of the silicon nanoparticles is taken with the particles placed on a graphite grid.As was shown in Chapter 9, the solution of silicon nanoparticles is brightly red-orange luminescent under 254-422 nm irradiation with band head at ~ 630 nm. The phosphors ZnS:Ag andZnS:Cu,Au,Al are luminescent under excitation with 365 nm with band heads at 440 and 525 nm, respectively. Figure 13.40B gives the excitation spectra of thin films of arbitrary thicknessand area of the three components blue and green phosphorsand red Si nanoparticles when each is monitored at its maximumresponse at 440, 525, and 610 nm, respectively. The relativesignal size is not indicative of the relative efficiency. The figure shows that when the wavelength of excitation at 365 nm is changed to 254 nm, the efficiency of the blue phosphorand the nanoparticles each drop to the 50% level,whereas the green phosphor efficiency drops more strongly tothe 30% level.

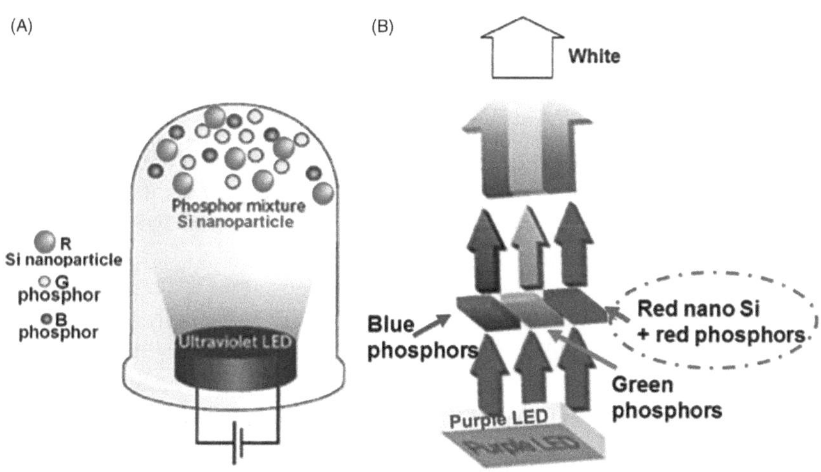

FIGURE 13.39 Nano silicon-based solution: (A) Schematic of bulbs and (B) phosphors utilizing blue and green phosphots and silicon nanoparticles (as red phosphor) pumped by a single master purple UV/blue LED (b).

(A) (B)

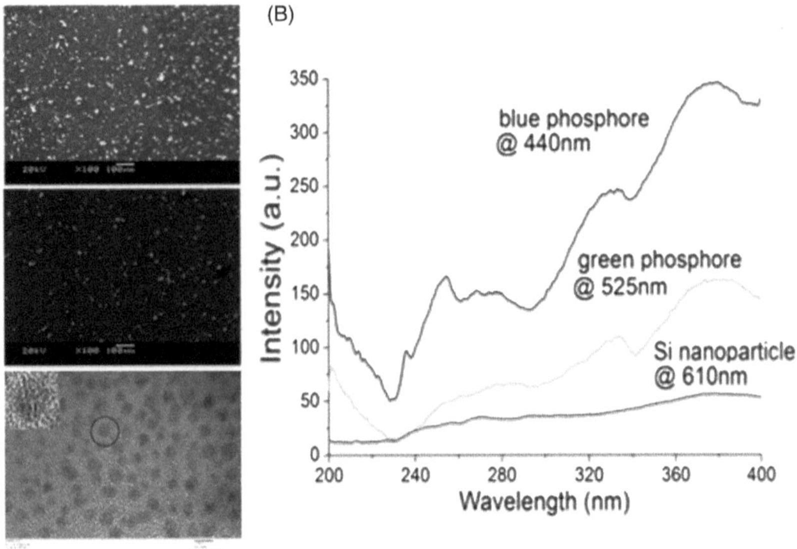

FIGURE 13.40 (A) SEM images of the blue (top) and green phosphors (middle) and TEM of silicon nanoparticles (bottom) placed on single crystal silicon substrate. (B) excitation spectra of thin films of arbitrary thicknessand area of the three components blue and green phosphorsand red Si nanoparticles when each is monitored at its maximumresponse at 440, 525, and 610 nm, respectively.

The green, blue and nanoparticles are now mixed in RTV individually. The response of the mixtures under excitation with the 365 nm wavelength as well as the spectra are recorded. Then the three phosphors are mixed together. Figure 13.41A gives the normalized emission spectra of ZnS:Ag (blue), ZnS:Cu,Au,Al(green) and Si nanoparticles (red) individually dispersed in RTV under the excitation of 365nm radiation. The gray spectrum is due to a mixture of the three components dispersed in RTV under excitation of 365 nm. The figure also shows the normalized spectrum of pure RTV (light blue) for reference. Figure 13.41B gives (left to right) the corresponding luminescent images of the individual components and of the mixture under excitation of 365nm radiation.

Figure 13.41C gives luminescence spectra of three shades of white light using differentbranching ratios of the blue and green phosphors and red Si nanoparticles. The spectra are analyzed for CCT (top) 7981 K; (middle) 6722 K; and (bottom) 5560 K. In (middle), the spectrum of the CIE standardilluminant D65 standard of solar radiation is also presented.The CCT temperature of the white samples is obtained from the SPD and utilizing the standard observer's CMFs.

13.5.3.1 Local/Remote Phosphor

There are two configurations for applying the phosphorconverter [47]. As shown in Figure 13.42A,B, phosphors can be deposited onto an LED die surface

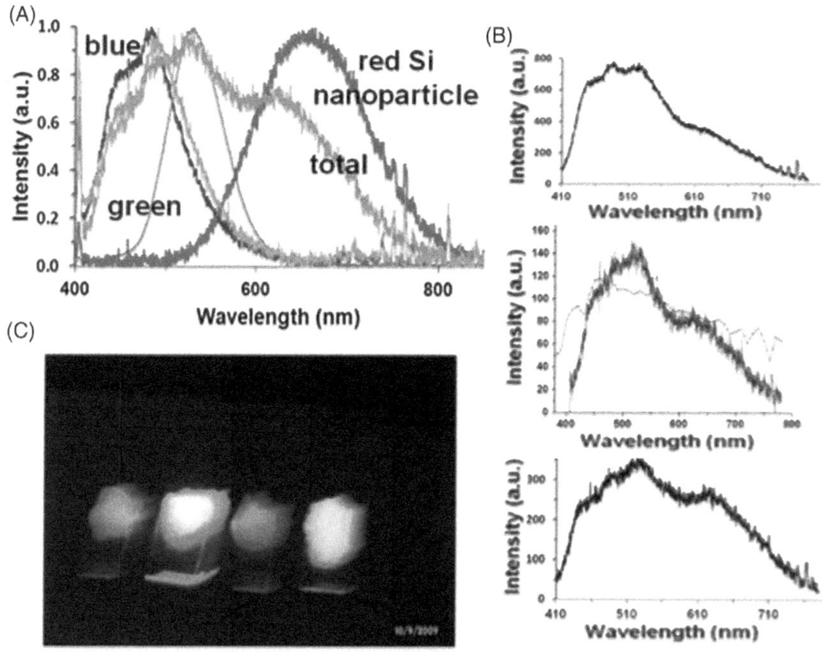

FIGURE 13.41 (A) the normalized emission spectra of ZnS:Ag (blue), ZnS:Cu,Au,Al (green) and Si nanoparticles (red) individually dispersed in RTV under the excitation of 365nm radiation. The gray spectrum is due to a mixture of the three components dispersed in RTV under excitation of 365 nm. The figure also shows the normalized spectrum of pure RTV (light blue) for reference. (B) (left to right) the corresponding luminescent images of the individual components and of the mixture under excitation of 365nm radiation. (C) luminescence spectra of three shades of white light using different branching ratios of the blue and green phosphors and red Si nanoparticles.

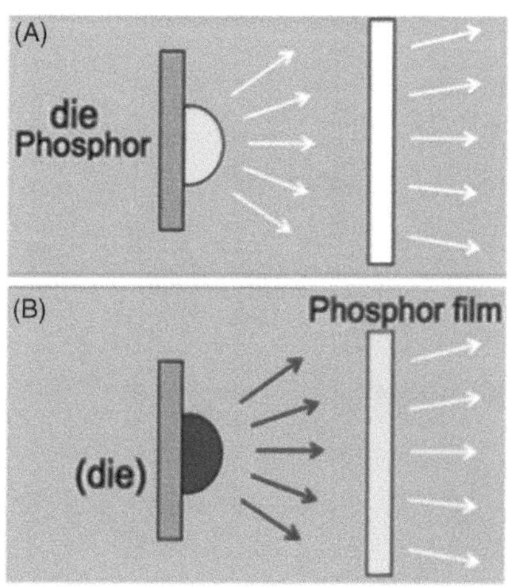

FIGURE 13.42 (A) Phosphors deposited onto an LED die surface itself ('local phosphor'), (B) a free-standing phosphor structure positioned over the LED ('remote phosphor'). *(From https://www.ledinside.com/press/2015/8/light_polymers_introduces_liquid_crystal_based_remote_phosphor_flexible_film_for_led_lighting. Used with permission from LEDinside. © LEDinside)*

itself ('local phosphor'), or made into a free-standing structure and positioned over the LED ('remote phosphor') respectively. The latter configuration is more expensive, but provides better efficiency. The reason for the improved efficiency is that in the local configuration half the light goes backwards, which most of it gets absorbedon die surface. Not only the light is lost, but it also heats the Led, which may cause heat damage. On the other hand, in the remote phosphor configuration very little of the phosphor emission hits the Led.

13.5.4 LED Filament

Manufacturing of an LED "Filament" [48] starts from a transparent substrate made of glass or sapphire materials as shown in Figure 13.43A. Unlike regular LED sources, which are mounted metal substrate, the filament substrate is insulating. Many (sometimes hundreds) of tiny unpackaged LED chips are mounted on the substrate. This configuration is called Chip-On-Glass (COG) [49]. Unlike metal substrates, the transparency of glass is pivotal in this application as it allows the emitted light to disperse evenly and uniformly in all directions. This innovative design use both blue LEDs as well as red LEDs in an alternating pattern. The LED filament is then encapsulated in a resin made up of a silicone and phosphor mixture that performs the usual transformation of the LED chips' blue light into white light. Pumping of the phosphor by the blue led does not provide sufficient amount of red light. The red LED is used to directly compensate for the lack of red color modulate the color temperature. Most manufacturers rely solely on the phosphor to set the color temperature. This added degree of control allows the lamp maker to provide a more accurate level of color temperature, and flexibility during the manufacturing process. The downside is that the CRI performance will not be as consistent when using the combination of both blue LED and red LED chips, as they degrade independently from each other. Presently

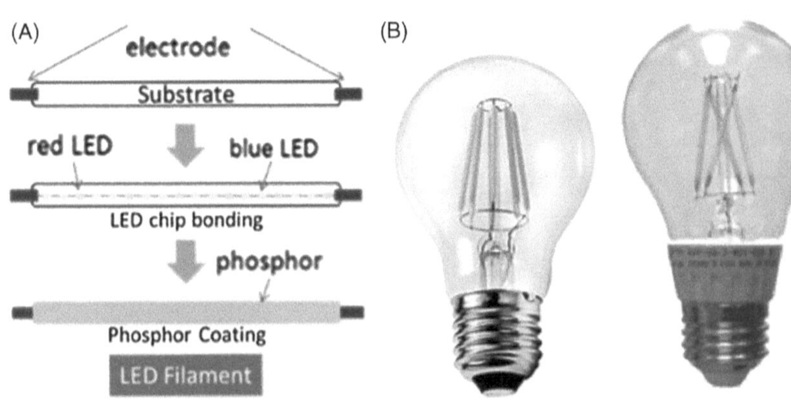

FIGURE 13.43 LED filament (A) Schematic of the use of red and blue LEDs and a yellow phosphor on a transparent glass or sapphire filament. (B) photos of example bulbs that utilize filaments. *(From The Next Generation of LED Filament Bulbs http://www.ledinside.com/knowledge/2015/2/the_next_generation_of_led_filament_bulbs. Used with permission from LEDinside © LEDinside)*

ongoing work at LED Systems (Kazakhstan), NanoSi Advanced Technologies, and Reach-NanoSi is focusing on the incorporation of red luminescent silicon nanoparticles as a substitute to alleviate those degradation effects.

Typically each filament has 28 LEDs. The voltage per blue LED is $2.48 < \Delta V < 3.7$ and per red LED is $1.63 < \Delta V < 2.03$. Two filaments with a mix of red and blue LEDs would need ~ 110 V, while four filament would need ~ 220 V to 240 V. A traditional LED lamp would require 3 V to 12 V. Four filaments are typically used as shown in Figure 13.43B. To give better color rendition, a mix of phosphors are used, which improves on the early blue LEDs with yellow only phosphor. First introduced in 2008 by Ushio Lighting

A number of positive issues arise.

- Potential higher efficiencies by the use of more LED emitters from lower driving currents.
- Near full "global" illumination can be obtained from arrays of filaments. [2]
- Standard socket ((as incandescent lamp sockets)) which provide mechanical support and electrical connections, as well as compatibility with present technology.
- Transparent glass bulbs glass filled with gas for thermal dissipation
- No heat sink
- Lighter weight lamp that costs less, and has the look of a classic incandescent lamp

13.5.5 Heat Management and its Implication [50]

Poor thermal dissipation and incomplete flux geometry resulting from the elimination of heat sink results in higher temperature. The increased temperature accelerates the degradation of silicone binder, and leakage of blue light. Moreover, heat reduces emission efficiency, limiting the performance of LEDs itself. The high operating temperatures reduce the lifespan of the LED. The absence of a heat sink therefore requires the use of a high thermal conductivity gas inside the bulb to aid heat dissipation.

Nano-structuring or use of nanoparticle layers can provide heat transfer interfaces. Conventional methods to increase heat transfer in cooling systems employ features such as fins and grooves on the surfaces, increasing their surface area. Also, fabrication of these features can be expensive. Because the surface to volume in nanoparticles is extremely high, their ability to transfer heat is very high. Mixing phosphors with nanoparticles is one configuration that might be useful. In a recent development, MIT researchers Jacopo Buongiorno and Lin-Wen Hu used magnetic nanoparticles to provide improvement in heat transfer, much more than using conventional procedures [51]. The nanomaterial consists of tiny particles of magnetite, a form of iron oxide. A magnetic fluid was prepared by dissolving the nanoparticles in water. Cooling systems generally rely on water pumped through pipes to remove unwanted heat. The magnetite nanofluid flowed through tubes and was manipulated by magnets placed on the outside of the tubes. The magnets attract the particles closer to the heated surface of the tube, greatly enhancing the transfer of heat from the fluid, through the walls of the tube, and into the outside air. Without the magnets in place, the fluid behaves just like water, with no change in its cooling properties. But with the magnets, the heat transfer coefficient is higher. Improvements as high as 300 percent has been achieved. This method could prevent hotspots that can lead to system failures. The system could also be applied to cooling everything from electronic devices to advanced fusion reactors.

In addition, the magnets may be used to target certain areas or certain times (switch on-off), making the procedure useful for systems where hotspots appear by placing magnets next to the hotspot. This could be useful for microchips and other electronic systems, such as "lab on a chip" microsystems.

13.5.6 Plasmon-Enhanced Emission From Quantum Wells (LED) [52–54]

Metallic thin films were applied directly to LEDs. In this arrangement, the metal is found to enhance quantum yield (QY) by enhancing the spontaneous emission rate of excitons in the active layer of the LED (quantum wells). The metal layer must be grown at a distance smaller than the evanescent decay length of the SPPs, namely about 10 nm (see chapter 8). This allows the combining electron–hole pair to give their energy to the SPPs of the metal, which enhances the emission. Figure 13.44A shows the photoluminescence intensity spectrum for different metal layers deposited over the blue LED at a distance of 10 nm. Photoluminescence spectra of blue LEDs coated with Ag, Al and Au. The enhanced emission of the coated LEDs is due to the coupling of SPPs with the roughness of the metal to far-field radiation.The metallic surface must also be made rough

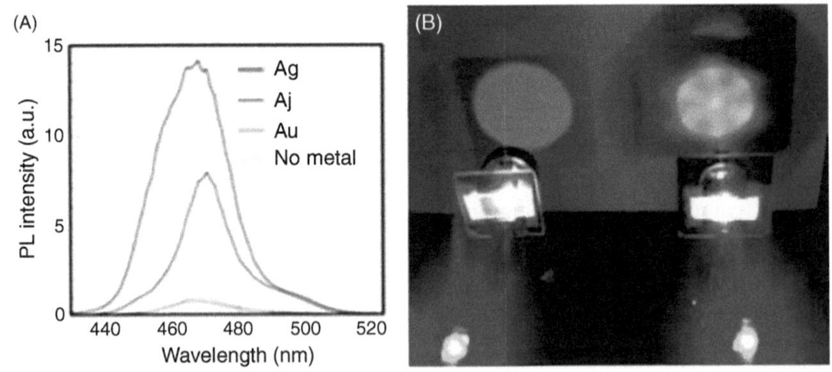

FIGURE 13.44 (A) photoluminescence intensity spectrum for different metal layers deposited over the blue LED at a distance of 10 nm. (B) Photograph of the emission of a standard phosphor-converted LEDs (pcLED) (left) and a pcLED that exhibits enhanced emission due to the integration of a hexagonal array of metal nanoparticles (right) Reprinted from Ref. 52 Reprinted with permission from Ref. 54. Copyright 2013, AIP Publishing LLC.

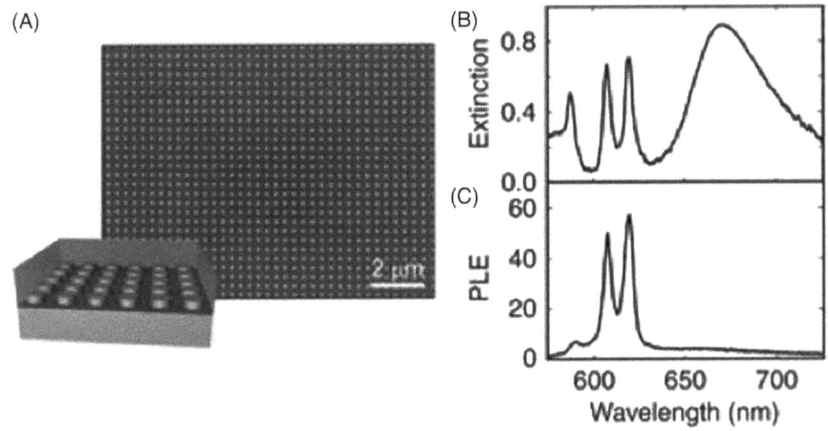

FIGURE 13.45 (A) scanning electron microscope image and a sketch of a metal nanoparticle array pcLED. The phosphor layer is represented by the transparent red layer. (B) the extinction. (C) photoluminescence.

to allow the emission to propagate efficiently to free space and enhance the far field extraction emission intensity. Figure 13.44B presents Photograph of the emission of a standard phosphor-converted LEDs(pcLED) (left) and a pcLED that exhibits enhanced emission due to the integration of a hexagonal array of metal nanoparticles (right)

13.5.7 Plasmon-Enhanced Emission From Phosphor Layers [55]

An array of aluminum or gold nanoparticles are placed over the phosphorous light converters in the solid state lighting system. The collective plasmonic resonances modify the emission of the luminescence layers. Figure 13.45A gives a scanning electron microscope image and a scketch of a metal nanoparticle array pcLED, where the phosphor layer is represented by the transparent red layer. Figure 13.45B,C give results, which illustrates the enhancement for 700-nm-thick phosphor layer consisting of a high-QY and a photo-stable dye. The ratio of the emission from the dye layer with and without the nanoantenna array is measured. The spectra are taken for normal incidence. Both the extinction and photoluminescence enhancement (PLE) spectra are displayed. The aluminum layer enhances the emission by more than a factor of 60 at certain wavelengths and in defined directions.

13.6 RENEWABLE ENERGY AND NANOTECHNOLOGY [56]

Renewable energy sources consist of: solar, hydrogen, geothermal, wind, hydro, and bioenergy. In the renewable sector, nanotechnology research is primarily driven by solar photovoltaic needs, withthe solar is largerthan all of the rest combined.

13.6.1 Solar

Efficiency is one of the biggest drawbacks with photovoltaic solar power devices. Studies conducted to enhance the efficiency try to optimize any of the parameters: absorption, reflection, conversion, and heat generation. These can absorb a maximum of about 48 percent of incoming solar radiation, about half of which can be converted into electricity. to convert sunlight into an electric potential.

Most solar cells in use use wide thin crystals of silicon. Nanostructured silicon solar cells however have showed that nanotechnology could be able to dramatically increase the absorptivity of solar cells. For example.

- Nanowires and nanocones-based solar cell: can absorb ~ 90 percent of incoming solar radiation (at certain angles of incidence), almost twice the amount that can be absorbed by traditional thin-film solar cells. Silicon nanomaterials used ~ one percent the amount of needed to build conventional solar cell.
- Nano-engineered nanowires solar cells: absorb up to 96 percent of incoming solar radiation.
- Carbon nanotubes, when used as a coating on silicon, can absorb an astonishing 99 percent of the ultraviolet, visible, infrared and far-infrared light that strikes it.
- Nano colloids: Replacing fluids used in solar collectors (typically water) with a nanofluid (colloid), thermal efficiencies is enhanced by up to 88 percent (or by over 200 percent at higher temperatures).

13.6.1.1 Geothermal Power

One futuristic concept of extracting geothermal heat is by heat exchange. In the process, a nanofluid is injected into naturally heated rocks. The nanofluid gets heated by the hot rocks. The hot nanofluid may now be used to generate electricity. The use of nanofluid is expected to increase the efficiency over a simple liquid because the presence of nanocomponents boost the heat-retaining properties of fluids.

13.6.1.2 Hydrogen Energy [58]

In this energy source hydrogen is used as a fuel. When it is reacted in a fuel cell with oxygen electricity is generated, with water byproduct.

- Clean burning
- Higher efficiency: Hydrogen fuel cells are 2 to 3-fold more efficient than conventional gasoline
- Storage problem: It is difficult to store large volumes of hydrogen in bulk material in such a way to quickly release it to generate power in an economically on-line fashion.
- Nanostructured storage and recovery: Instead of bulk storage, storage in core-shell nanoparticles could enable quick release to generate power in an

economically on-line fashion. This is made possible because, due to the extremely small size of the nanoparticles (large surface-area-to-volume ratio), the stored atoms always remain relatively close to their surface. Examples of this include nanoblades carbon nanotubes and carbon nanofibers

13.6.1.3 Magnesium Engine [58–61]

The combustion of Mg powder in water has been used as the basis of the "magnesium engine", which has been demonstrated to afford auto- as well as space propulsion-applications. The process produces high-energy steam and hydrogen, which power the engine.Also Mg powder has been burned directly with carbon dioxide in new rocket jet engines, developed for in situ Mars propulsion. while simultaneously suffering combustion with water to produce $Mg(OH)_2$ and hydrogen:

- $Mg(s)+2H_2O(g)\rightarrow Mg(OH)_2(aq)+H_2(g)$.
- Unlike bulk, Mg in powder or sub-micron form reacts vigorously with water exothermally

13.6.1.4 Wind Energy

Nanotechnology has a role in enhancing the efficiency of harvesting wind energy through upgrading a number of components of a wind turbine, including the blades and the electric system.

- Lighter blades: Exploit the strength of carbon nanotubes to produce stronger, longer and light weight blades. Nanotube-filled epoxy can be used.
- Surface mechanical attrition treatment (SMAT) by nano-structuring of surfaces. Polymer nanocomposites include fibreglass or carbon fibre reinforced epoxy resin and polyester resin
- Lower mechanical resestence: use nanomaterial for tribological coatings, and wear protection layers of bearings and gear boxes.
- Lower power loss: Incorporate carbon nanotube in electric transmission lines to lower the electric resistance
- Enhanced coupling to the grid: optimization of power storing devices using nano devices to enhance feeding efficiency of wind power in the grid.
- Weather Protection: Use conductive nanomaterial for enhancing lightening protection.

REFERENCES

[1] Reinders Angele, Verlinden Pierre, Freundlich Alexandre. Photovoltaic Solar Energy: From Fundamentals to Applications. John Wiley & Sons; 2017. Feb 6.

[2] King RR. 40% Efficient Metamorphic GaInP/GaInAs/Ge Multijunction Solar Cells. Appl. Phys. Lett 2007;90:183516.

[3] Rühle Sven. Tabulated Values of the Shockley-Queisser Limit for Single Junction Solar Cells. Solar Energy 2016;130:139–47. doi: 10.1016/j.solener.2016.02.015. 02-08, Bibcode:2016SoEn.130.139R.

[4] Rühle Sven, Green Martin A. Third Generation Photovoltaics: Advanced Solar Energy Conversion. Springer; 2003. 65.

[5] Murri Roberto. SiliconBased Thin Film Solar Cells. Bentham Science Publishers; 2013. Mar 20.

[6] Thin Film Silicon Solar Cells on Glass, PHOTOVOLTAICS AND THIN FILM ELECTRONICS LABORATORY. http://pvlab.epfl.ch/thin_film_on_glass PV-LAB.

[7] van Sark Wilfried GJHM, Korte Lars, Roca Francesco. Physics and Technology of Amorphous-Crystalline Heterostructure Silicon Solar Cells. Springer Science & Business Media; 2011. Nov 16.

[8] Miroslav Mikolasek. Current status and progress in the new generation's silicon based solar cells 2009; 06. Júl, Autorčlánku: Elektrotechnika, Študentské práce Ročník 2 číslo 7: http:// www. posterus. sk/?p=1247.

[9] Jensen N, Hausner R, Bergmann R, Werner J, Rau U. Optimalization and Charakterization of Amorphous/Crystalline Silicon Heterojunction Solar Cells. In: Progress in Photovoltaics: Research and Applications 2002;10:1–13.

[10] http://www.nedo.go.jp/english/news/AA5en_100109.html.

[11] Yoshikawa Kunta, Kawasaki Hayato, Yoshida Wataru, Irie Toru, Konishi Katsunori, Nakano Kunihiro, Uto Toshihiko, Adachi Daisuke, Kanematsu Masanori, Uzu Hisashi, Yamamoto Kenji. Silicon heterojunction solar cell with interdigitated back contacts for a photoconversion efficiency over 26%. Nature Energy 2017;2:32. doi: 10.1038/nenergy. 2017. Article number: 17032.

[12] López Antonio Luque, Andreev Viacheslav M. Concentrator Photovoltaics. Berlin Heidelberg: Springer; 2009. Sep 2.

[13] http://www.qrg.northwestern.edu/projects/vss/docs/power/1-how-do-solar-concentrators-work.html.

[14] Wehrspohn Ralf B, Rau Uwe, Gombert Andreas. Photon Management in Solar Cells. John Wiley & Sons; 2015. Jun 8.

[15] Stupca M, Alsalhi M, Al Saud T, Almuhanna A, Nayfeh MH. Enhancement of polychrystalline silicon solar cells using ultra thin films of silicon nanoparticle. Appl. Phys. Lett 2007;91:063107.

[16] Chowdhury FarsadImtiaz, Nayfeh Munir, Nayfeh Ammar. Enhancened performance of thin film silicon solar cells with a top film of silicon nanoparticlesdue to down-conversion and near resonance charge transport. The Journal of Solar Energy 2016;125:332–8. doi:10.1109/PVSC.2014.6925364, Efficiency enhancement in thin-film c-Si HIT solar cells using luminescent 2.85 nm silicon nanoparticlesAuthor(s), Farsad I. Chowdhury; AaeshaAlnuaimi; Kazi Islam; Ammar Nayfeh, Photovoltaic Specialist Conference (PVSC), 2014 IEEE 40th.

[17] Chowdhury Farsad Imtiaz, Alnuaimi Aaesha, Alkis Sabri, Ortaç Bülend, Aktürk Selçuk, Alevli Mustafa, Dietz Nikolaus, Okyay Ali Kemal, Nayfeh Ammar. Enhancement in c-Si solar cells using 16 nm InN nanoparticles. Materials Research Express 2016;3. Number 5, Published 18 May.

[18] Islam Kazi, Alnuaimi Aaesha, Battal Enes, Okyay Ali Kemal, Nayfeh Ammar. Effect of gold nanoparticles size on light scattering for thin film amorphous-silicon solar cells. Solar Energy 2014;103:263–72. May.

[19] Masudy-Panah S, Dalapati GK, Radhakrishnan K, Kumar A, Tan HR, et al. p-CuO/n-Si heterojunction solar cells with high open circuit voltage and photocurrent through interfacial engineering. Progress in Photovoltaics: Research and Applications advance online publication 2014; doi: 10.1002/pip.2483. 23 February.

[20] Baryshnikova KV, Petrov MI, Babicheva VE, Belov PA. Plasmonic and silicon spherical nanoparticle antireflective coatings. Scientific Reports 2016;6 doi: 10.1038/srep22136. Article number: 22136.
[21] Solanki Chetan Singh, Singh Hemant Kumar. Anti-reflection and Light Trapping in c-Si Solar Cells. Springer; 2017. Jul 30.
[22] Solar Thermophotovoltaic System Using Nanostructures, United, States., Department of, Energy., Office of Science, 2015
[23] Asano' Takashi, Suemitsu' Masahiro, Hashimoto Kohei, De Zoysa Menaka, Shibahara Tatsuya, Tsutsumi Tatsunori, Noda' Susumu. Near-infrared–to–visible highly selective thermal emitters based on an intrinsic semiconductor. Science Advances 2016;2 doi: 10.1126/sciadv.1600499. 23 Dec, no. 12, e1600499.
[24] Narasimhan Vijay K, et al. Hybrid Metal–Semiconductor Nanostructure for Ultrahigh Optical Absorption and Low Electrical Resistance at Optoelectronic Interfaces. ACS Nano 2015; doi: 10.1021/acsnano.5b04034.
[25] Nayfeh M, Rao S, Nayfeh O, Smith A, Therrien J. UV photodetectors with thin film Si nanoparticle active mediaum. IEEE Transcations on Nanotechnology 2005;4:660.
[26] Magill S, Xie J, Nayfeh M, Yu H, Fizari M, Malloy J, Maximenko Y. Enhanced UV light detection using wavelength-shifting properties of Silicon nanoparticles. J of Instrumentation 2015;10:1. P05008.
[27] Jee S-W, Zhou K, Kim D-W, Lee J-H. A silicon nanowire photodetector using Au plasmonic-nanoantennas. Nano Convergence 2014;1:29.
[28] Qiu T, Wu XL, Mei YF, Wan GJ, Chu PK, Siu GS. From Si nanotubes to nanowires: Synthesis, characterization, and self-assembly. J Cryst Growth 2005;277:143.
[29] Mantey K, Shams S, Nayfeh MH, Nayfeh O, Alhoshan M, Alrokayan S. Synthesis of wire-like silicon nanostructures by dispersion of SOI using electroless etching. J Appl Phys 2010;108:124321.
[30] Choi Y, Wang G, Nayfeh M, Yau S-T. Electro-oxidation of organic fuels catalyzed by ultrasmall silicon nanoparticles. Appl. Phys. Lett 2008;93:164103.
[31] Choi Yongki, Wang Gang, Nayfeh Munir H, Yau Siu-Tung. A hybrid biofuel cell based on electrooxidation of glucose using ultra-small silicon nanoparticles. Biosensors and Bioelectronics 2009;24:3103–7.
[32] Wang G, Mantey K, Nayfeh MH, Yau S. Enhanced aperometric detection of glucose using Si-29 particles. Appl. Phys. Lett 2006;89. 243901.
[33] Liu Qiang, Nayfeh Munir H, Yau Siu-Tung. A silicon nanoparticle-based polymeric nano-composite material for glucose sensing. Journal of Electroanalytical Chemistry 2011;657:172–5.
[34] Yau S, Song.china Yang, Nayfeh M. Electrostatically enhanced performance of yeast-based microbial fuel cell. RSC Advances 2015;5:19203–5. doi: 10.1039/C4RA01545B.
[35] Zhong Cheng, Deng Yida, Hu Wenbin, Qiao Jinli, Zhang Lei, Zhang Jiujun. A review of electrolyte materials and compositions for electrochemical supercapacitors. Chemical Society Reviews 2015;(21).
[36] Frackowiak Elzbieta. Supercapacitors: Materials Systems and Applications Francois Beguin. John Wiley & Sons; 2013. Apr 2.
[37] Caironi Mario. Large Area and Flexible Electronics. John Wiley & Sons; 2015. May 4.
[38] Liu Qiang, Nayfeh Munir H, Yau Siu-Tung. Supercapacitor electrodes based on polyaniline–silicon nanoparticle composite. Journal of Power Sources 2010;195:3956–9.
[39] Liu Qiang, Nayfeh Munir H, Yau Siu-Tung. Brushed-on flexible supercapacitor sheets using a nanocomposite of polyaniline and carbon nanotubes. Journal of Power Sources 2010;195:7480–3.

[40] Liu Qiang, Osama Nayfeh, Nayfeh Munir H, Yau Siu-Tung. Flexible supercapacitor sheets based on hybrid nanocomposite materials. Nano Energy 2013;2:133–7.

[41] Žukauskas Artûras, Shur Michael S, Gaska Remis. Introduction to Solid-State Lighting. Wiley; 2002. Apr 18.

[42] Shur Michael S, Zukauskas Arturas. UV Solid-State Light Emitters and Detectors. Springer Science & Business Media; 2004. May 31.

[43] Cangeloso Sal. LED Lighting: A Primer to Lighting the Future,. O'Reilly Media Inc.; 2012. Jul 10.

[44] Makittrick J, Shea-Rohwer L. Review: Down Conversion Materials for Solid-State Lighting 2014; doi: 10.1111/jace.12943. 16 April.

[45] Schmitt Laura. The Bright Stuff: The LED And Nick Holonyak's Fantastic Trail Of Innovation. BookBaby; 2012. Aug 24.

[46] Stupca Mathew, Nayfeh Osama M, Hoang Tuan, Nayfeh Munir H, Alhreish Bahjat, Boparai Jack, AlDwayyan Abdullah, AlSalhi Mohamad. Silicon nanoparticle-ZnS nanophosphors for UV- based white LED. J. Appl. Phys 2012;112:074313.

[47] Local/remote phosphor Light Polymers Develops First Liquid-Crystal-Based Remote Phosphor Flexible Film for LED Lighting, Sergey Fedotov, Light Polymers, Inc., http://www.lightpolymers.com/light-polymers-develops-first-liquid-crystal-based-remote-phosphor-flexible-film-for-led-light.

[48] LED "Filament" The Next Generation of, LED., Filament Bulbs, AXP., Lighting, February 5, 2015, http://www.ledinside.com/knowledge/2015/2/the_next_generation_of_led_filament_bulbs.

[49] Jin Yufeng, Wang Zhiping, Chen Jing. Introduction to Microsystem Packaging Technology. CRC Press; 2016. Apr 19.

[50] Weng Chun-Jen. Advanced thermal enhancement and management of LED packages. International Communications in Heat and Mass Transfer 2009;36(3):245–8. https://doi.org/10.1016/j.icheatmasstransfer.2008.11.015.

[51] http://news.mit.edu/2013/magnetic-nanoparticles-could-aid-heat-dissipation-1119.

[52] Lozano Gabriel, Rodriguez Said RK, Verschuuren Marc A, Rivas Jaime G6mez. Metallic nanostructures for efficient LED lighting. Science & Applications 2016;5. doi: 10.1038/lsa.2016.80. e16080; http://www.nature.com/lsa/journal/v5/n6/full/lsa201680a.html?foxtrotcallback=true.

[53] Okamoto K, Niki I, Shvartser A, Narukawa Y, Mukai T, et al. Surface-plasmon-enhanced light emitters based on InGaN quantum wells. Nat Mater 2004;3:601–5.

[54] DiMaria J, Dimakis E, Moustakas TD, Paiella R. Plasmonic off-axis unidirectional beaming of quantum-well luminescence. Appl Phys Lett 2013;103:251108.

[55] Lozano G, Louwers DJ, Rodriguez SRK, Murai S, Jansen OTA, et al. Plasmonics for solid-state lighting: enhanced excitation and directional emission of highly efficient light sources. Light Sci Appl 2013;2. e66.

[56] Zang Ling. Energy Efficiency and Renewable Energy Through Nanotechnology. Springer Science & Business Media; 2011. Sep 6.

[57] Gandia Luis M, Arzamedi Gurutze, Dieguez Pedro M. Renewable Hydrogen Technologies: Production, Purification, Storage, Applications and Safety 2013. May 3, Newnes.

[58] TIT and Mitsubishi Prototypes Pollution-free Engine Excluding Fossil Fuel "by Motohiko Hamada and Nikkei Monozukuri, techon.nikkeibp.co.jp; CO2-Free Engine Powers Up" https://www.mitsubishi.com/mpac/e/monitor/back/0608/green.html.

[59] Levenstein Steve. Clean Magnesium Energy Cycle Hints at Fossil Fuel Freedom 2007. by, July 27, InventorSpot. com.

[60] Miller TF, Herr JD. Green rocket propulsion by reaction of Al and Mg powders and water. AIAA paper 2004. arc.aiaa.org, 40th AIAA/ASME/SAE/ASEE Joint Propulsion Conference and Exhibit 11–14 July 2004, Fort Lauderdale, Florida; Characteristics of reaction between magnesium and water at high temperature, X. Zhou, W. Zhang, S. Li, Journal of Solid Rocket Technology, 2009 - en.cnki.com.cn.

[61] Guan-peng Liu, Xiao-de Guo, Hong-zhen Duan, Wei Jiang, Feng-sheng Li. Experimental investigation and mechanism exploration on reaction between Mg and water. Journal of Solid Rocket Technology 2008-05 - en. cnki. com. cn.

Chapter 14

Electronics and Communication

14.1 SCALING DOWN INTEGRATED-CIRCUIT (IC)

The microelectronics industry has great interest in nanofabrication because its own mission requires making devices smaller and smaller. Specifically, the industry has the desire to continuously shrink both the transistor and the memory cells. Key transistor dimensions employed in current technologies are well into the nanoscale regime. This scalability has been enabled through the introduction of the 3-D FINFET architecture which results in improved electrostatic channel control. For example, 14 nm CMOS technology which features finFET architecture on an SOI substrate for a diverse set of SoC applications including HP server microprocessors for IBM server machines have FIN pitch of 42 nm and reasonable short-channel effects for gate length down to gate length down to ~20 nm [1], which is well below the definition of the nano regime of <100 nm. The reason behind this concerted effort is the desire to build denser and denser super-high-density microprocessors and memory chips. The increased density is needed to handle the ever increasing amount of information being generated by modern societies.

Since the 1970s, the microelectronics industry has been adopting the top-down method discussed above to scale down the integrated-circuit (IC). It effectively removes "one atom at a time" until the desired structure emerges. This effort has been so aggressive to the degree that the industry is rapidly approaching inherent physical limits of miniaturization.

14.1.1 Shrinking and Making Transistors Faster

For the past 50 years, electronic computers have grown more and more powerful. The cause of this is the fact that the transistor, the basic subunit, has become smaller and smaller hence faster and faster. Fig. 14.1A shows the linear dimension of the transistor for different years since 1970 [2]. Fig. 14.1B shows three parameters from bottom to top: the cost of the area of the processed chip, the area of a transistor, and the cost of a transistor. All of the parameters are in normalized US$. Thus it shows that despite the fact the processing of a chip per unit area keeps going up, the price per transistor continues to drop. The increase in processing cost is offset by the larger number of transistors that can be accommodated. This is in line with Moor's law. Beyond the 7 nm node, this will no longer be the case, and the cost of a transistor begins to rise, making the process not cost effective.

Fundamentals and Applications of Nano Silicon in Plasmonics and Fullerines.
http://dx.doi.org/10.1016/B978-0-323-48057-4.00014-1

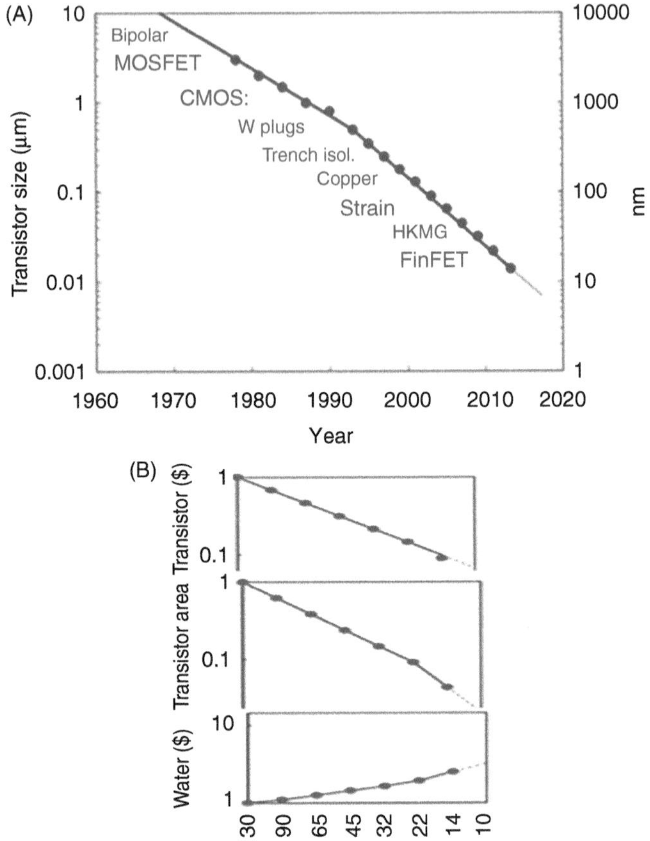

FIGURE 14.1 (A) Linear dimension of the transistor over the years, since 1970. (B) Three parameters from bottom to top: the cost of the area of the processed chip, the area of a transistor, and the cost of a transistor.

The industry utilizes a variety of advanced procedures to shrink transistors, as well as uses advanced material composition to achieve faster electron movements. We briefly discuss some of these procedures.

14.1.2 Strained Faster Nanotransistors

There is room for improving the speed of the transistor without making it smaller. This can be done by stretching them such that such that the interspacing between silicon atoms become larger than the normal spacing. Stretching creates strain in the silicon material and as such the transistor becomes faster [3–6]. The transistor is called strained. Plastic is stretched by mechanical force. Silicon, the second hardest material, is stretched by growing germanium over it. Ge is a material whose atoms are spaced out more than those of silicon. It should be noted

Moore's Law

Moore's Law OR

how overall processing power for computers will double every two years

The first transistor was a *point-contact transistor* that measured half an inch (1.27 cm) in height. The transistor wasn't very powerful, but physicists recognized the potential of the device. Before long, physicists and engineers began to incorporate transistors into various electronic devices. And as time passed, they also learned how to make transistors smaller and more efficient.

By the 1960s, computer scientist (and Intel co-founder) Gordon Moore made an interesting observation. He noticed that every 12 months, engineers were able to double the number of transistors on a square inch piece of silicon. Like clockwork, engineers were finding ways to reduce the size of transistors. It's because of these small transistors that we have electronic devices like personal computers, smartphones and mp3 players. Without transistors, we would still be using vacuum tubes and mechanical switches to make calculations.

Since Moore's observation, the shrinking trend has continued. But it hasn't kept up with the pace Moore observed. These days, the number of transistors doubles every 24 months. But that raises an interesting question: How small can transistors – and by extension, CPUs – get? In 1947, a single transistor measured a little over one-hundredth of a meter high. Today, Intel produces microprocessors with transistors measuring only 45 nm wide. A nanometer is one-billionth of a meter!

Intel and other microprocessor manufacturers are already working on the next generation of chips. These will use transistors measuring a mere 32 nm in width. But some physicists and engineers think we might be bumping up against some fundamental physical limits when it comes to transistor size.

Moore's Law Is Dead. Now What?

Shrinking transistors have powered 50 years of advances in computing—but now other ways must be found to make computers more capable.

that for the effect to take place it is sufficient to move silicon atoms away from each other only by a tiny amount; Ge is therefore ideal for this task because its electronic structure is similar to silicon, while its atoms are further apart by a small percentage of 4% than Si.

Fig. 14.2A outlines how straining takes place [7]. First, a silicon germanium (SiGe) alloy is grown by chemical vapor deposition (CVD) (Fig. 14.2A; right). The interatomic spacing in the alloy is slightly larger than that of silicon. A thin layer of silicon is now grown over the alloy. Depositing the layer of silicon over a substrate of silicon germanium (SiGe), in the plane parallel to the surface of

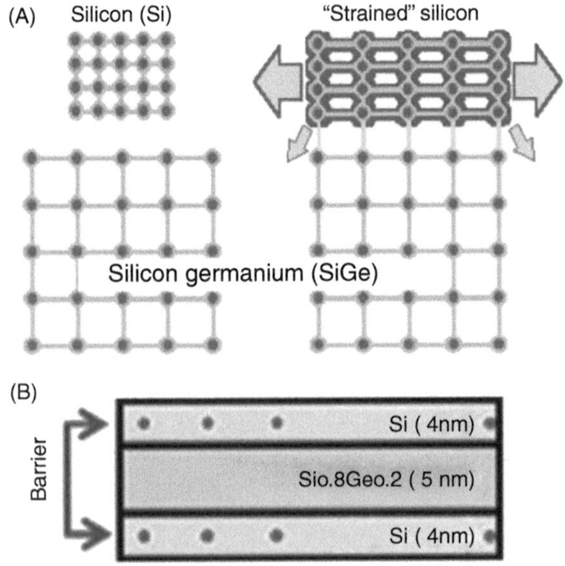

FIGURE 14.2 (A) Straining silicon by depositing a layer of silicon over a substrate of silicon germanium (SiGe) alloy. (B) The basic component of strained diode devices.

the alloy, forces the silicon atoms to move away from each other to conform to the underlying alloy. This forced motion strains the silicon film. Since atomic forces between atoms drop when they move away from each other, electrons or current flow in the chip, begin to move faster, which result in better chip performance and lower energy consumption. In terms of numbers, studies show that good successful strained alloy (Si1 $-$ x Get) where x is the Ge fraction may have for example a Sage ratio of 43%:57%. The speed of electrons in the strained material increases by 70%, which allows strained transistors to switch 35% faster. The indirect bandgap at room temperature $E_{g,indirect}$ of the alloy $Si_{1-x}Ge_x$ changes according to the phenomenological expression as follows: $E_{g,indirect} = 1.12{-}0.41x + 0.008x^2$ eV with $x < 0.85$, enabling prediction of the threshold voltage (turn-on voltage) of strained-Si devices and hence facilitating circuit design. These effects were demonstrated by a group at MIT (Hasan Nayfeh and Dimetri Antoniadis). A sizable transistor speed improvement was achieved experimentally. The concept of strained-Si channel transistor devices were was implemented in practice into at IBM and other chip manufacturers starting at the 65-nm technology node [4].

14.1.2.1 Strained Diodes

Construction of oscillators and switching devices are among applications of strained silicon. Those devices can operate at terahertz frequencies. This kind of technology enables novel applications in the areas of wireless and optical communication.

Fig. 14.2B depicts the basic component of diode devices. It consists of a thin SiGe region (5 nm). This layer plays the role of a quantum well. The well is then sandwiched between two layers of Si (each 4 nm).Those act as potential barriers. This configuration is called a double barrier. The well allows electrons or holes to be trapped inside, while the potential barriers prevent them from escaping. This sandwich is sandwiched again between two highly doped contacts to provide a Fermi sea of electrons, one is doped negatively to supply electrons (emitter) and the other is doped positively to accept electrons (collector). As to the electronic structure, inside the well there is a number of resonant energy levels each lies at a different energy as well as with a small width (like levels encountered when we deal with the elementary concept of particle in a box). The system is called resonant-tunneling diode (RTD), which is effectively a diode in which electrons can tunnel through some resonant states at certain energy levels.

In operation, electrons and holes have discrete energy values inside the quantum well. When a voltage, such that the energy value inside the quantum well is equal to that of the emitter side (resonance case), is applied across an RTD, a terahertz wave is emitted. As voltage is increased away from the resonance condition, the terahertz wave dies out because the energy value in the quantum well is outside the emitter side energy. This strained IV–IV heterostructures (SiGe/Si) system offers sufficient high frequency and opto-electronic properties far better than conventional silicon technology and can be integrated much better than III–V materials systems (InP, GaAs, etc.) heterostructures.

14.1.3 Germanium Nanotransistors on Silicon

Electrons move faster in pure germanium than in silicon. This is why germanium (Ge) has been emerging as a viable candidate to augment silicon for electron device applications. However, there are still major obstacles to its full scale adoption. Those include device requirement, cost, and environmental instability due to the fact that germanium oxide dissolves in water create. Moreover, growth over silicon and integration of germanium in silicon is hampered by the large mismatch between their structures since the interspacing of atoms in Ge is larger than that of silicon by 4%. The mismatch results in growth that is dominated by "islanding" and misfit dislocations. These dislocations are normally formed at the Si substrate/Ge film interface. They terminate the film surface as threading dislocations. Those degrade device performance. Recently a group at Stanford University (Ammar Nayfeh and Krishna Saraswat) developed new methods for solving this problem [8]. In these methods, Ge growth on Si, which is usually hampered by the large lattice mismatch (4%), has been accomplished by the incorporation of hydrogen annealing in the growth recipe. This reduces the Ge diffusion barrier, dramatically improving the defects and surface roughness. The Stanford group successfully used this novel heteroepitaxial growth technique to demonstrate high-performance transistors in germanium grown directly on Si. This is a good step in the direction of integration of Ge with the Si electronic standard technology for high-performance applications.

As a follow up to these development, recent work at Masdar Institute developed a low temperature processing of Ge/Si interfaces that demonstrates a great potential for p-channel transistor applications in monolithically integrated CMOS platform. The process was demonstrated by building high quality capacitors [8]. In the procedure, low-temperature radio-frequency plasma-enhanced chemical vapor deposition (RF-PECVD) was used. Intrinsic germanium was deposited on a silicon wafer, followed by a metal oxide dielectric (Al2O3). Then a metal gate electrode was deposited (Ti/Al). The total thickness of the device is only 9 nm. The electrical characteristics of the (Ti/Al)/Al_2O_3/i-Ge/Si MOSCAP device exhibits n-type (p-channel). Post annealing of the Ge layers at different temperatures in H_2 and N_2 gas ambient were applied. The trap density at Ge/Al_2O_3 interface is measured; the measurements show very low threading dislocation density (TDD) of $\sim 1 \times 10^6$ cm^{-2}, affording high-quality devices.

- The Transistor Developed in 1947 by engineers working for Bell Laboratories to amplify sound over phone lines
- The transistor replaced vacuum tubes (non-reliable, bulky, generated a lot of heat).
- The first transistor was a point-contact transistor half an inch (1.27 cm)
- The world's first IC came along (in 1958): two transistors were attached to a silicon crystal
- Gordon Moore (by the 1960s) stated that every 12 months, number of transistors on a square inch piece of silicon would double
- Transistors double every 24 months.
- Metal-insulator-semiconductor (MIS) quantum tunneling becomes significant delocalizing agent at 5 nm. Electrons cannot be controlled or confined. This begins to answer the question of "How small can transistors – and by extension, CPUs – get?"

When one reaches the limit of size and quantum dynamics of motion of charge kicks in strongly and begin to dominate, we can take advantage of the various quantum effects to build nanometer-scale electronic switching devices for use in building ultra-densely integrated electronic computers. On another front, many groups at academic institution and industry have been experimenting with silicon nanoparticles. They embedded them in the bulk of the transistors to upgrade it to quantum-effects and single-electron solid-state devices. The transistor becomes digital and operates at extremely low current of a single electron, becomes faster, and needs much less energy to run.

Fig. 14.3 shows a cross section schematic of a single transistor on the wafer. It is made up of a p-n-p structure, with a thin insulator above the n-type silicon

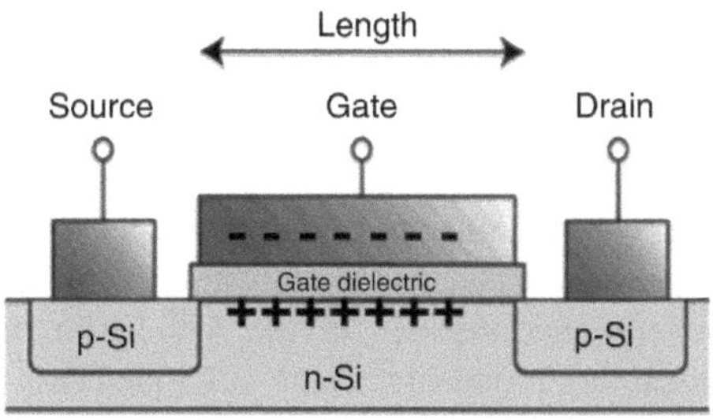

FIGURE 14.3 A cross section schematic of a single transistor shows source, gate, and drain regions. The p-n-p structure, a thin insulator above the n-type (gate dielectric) and heavily-doped poly-crystalline silicon or metals form the conducting contacts regions.

region, known as the *gate dielectric*. Heavily-doped poly-crystalline silicon or metals form the conducting contacts regions: source, gate, and drain regions.

When a negative potential is placed on the conductor above the gate dielectric, the n-region in the silicon inverts to p-type and current can flow through the transistor. With no potential above the gate dielectric, the p-n-p structure blocks current flow. In modern devices, the gate region itself is often times less than 100 nm in length. These devices are embedded in the silicon and then connected together in a variety of configurations with alternating layers of insulators and conductors to form a wiring scheme on top of the transistors themselves.

In a modern IC processing plant, a state-of-the-art device may require more than 400 processing steps. The fabrication process utilizes the same equipment for different steps. Even so, due to the high tech nature of the equipment and extremely clean environment in which devices are manufactured, the factory may cost upwards of $5 billion to build, with approximately $1 billion for the building, and $4 billion in equipment!

14.1.3.1 On Chip and Chip to Chip Communication

As devices become smaller and smaller metal interconnect wires will become the limiting factor and the bottle necks. Alternative optical interconnects architecture for on-chip data communication is of substantial interest. Use of light as interconnections overcome speed and signal integrity concerns in future electrical interconnects. In fact to achieve chip-to-chip and board-to-board optical interconnects development of highly-efficient nanomaterial-based devices as nano sensors and photodetectors are required. It is estimated that by 2010, when processors exceed clock frequencies of 11.5 GHz, optical interconnects may be required to avoid chip-to-chip and on-chip bottlenecks. Moreover, one of the most significant factors limiting future microprocessor performance and one of the great consumers of microprocessor power is the clock distribution network. The use of optical distribution networks at the global level is emerging as an alternative to balanced electrical clock distribution networks. Light can be distributed to multiple receivers across the chip with low skew. If this light can be efficiently converted to an electrical signal, this idea becomes a viable alternative.

Current photodetector and photo transmitter technology utilizes III–V systems, such as GaAs. Recently, researchers have improved the silicon-based detector technology. They monolithically integrated a photodetector that achieves a much higher data detection rate (10 Gbps) than previously achieved [9]. Such advancements are very encouraging for the realization of chip-to-chip and board-to-board optical interconnects. Nano photodetector systems based on films of Si, Ge, and GeSi nanoparticles or quantum dots have been recently demonstrated, but with moderate efficiency. Nayfeh's group at Illinois constructed nano photodetectors by deposition of 1 nm Si nanoparticle films on Si p-type substrates, with good conversion efficiency under UV irradiation [10].

14.1.3.2 Laser on a Chip

Now that light can come out of silicon nanoparticles and efficiently, Lorenzo Pavesi in Italy was working to test for laser action from silicon. He implanted silicon ions or atoms using an accelerator in a glass slab and heated it to high temperatures till the silicon ions started to move more freely in the substrate. When the substrate was subsequently cooled, the Si atoms nucleated to form silicon nanoparticles of ~3 nm diameter (like rain droplet formation). He tested the glass slab and reported signs of gain, which implied that laser action is feasible if other conditions are provided for (see Section 9.7). Meanwhile the Nayfeh's group a series of papers in Applied Physics Letters, using the already made silicon nanoparticles, reported observation of stimulated emission, a sign of feasibility of laser action and also observation of very small "micro" shooting beams from collection of particles (see Section 9.7). Thus, in addition to being ultra bright, reconstituted films or aggregates of the particle family appear to exhibit the phenomenon of stimulated emission, which is a necessary condition for the production of laser beams. Light-emitting Si devices, if proven, could eventually result in a laser on a chip, new generation of Si chips, and extend the functionality of Si technology from microelectronics into optoelectronics.

14.1.4 Molecular Organic Electronics [11]

As the transistor size approaches 5 nm, they become so physically close together each in dense chips. In this limit, electrons experience quantum tunneling and therefore cannot be localized in the transistor. In other words the transistor ceases to operate and as such one need to come up with a completely different idea. An alternative to the field-effect transistor (FET)–based on solid silicon is what has been coined a molecular electronic device. The new paradigm uses molecules. It is currently one area being studied as a potential successor to conventional silicon-based electronics technology. One aspect of molecular electronics is the fabrication of devices whose function is governed by single molecules.

We mention here some example of electronics components. A chain of aromatic rings connected together by acetylene linkages may act as a wire that conducts electricity and can be used to connect devices. Thiol (–SH) functional groups attached both ends of the molecular wire act as "alligator clips" for attaching molecular electronic units to metal substrates. Carbon nanotubes might be used as electric wires or as support for molecular circuit elements – for example, filled with conducting metal atoms to create among the structurally strongest nanowires chemically possible. Other formations of molecules may act as a diode rectifier or amplifier as in a transistor. On the other side of the field, DNA molecules and nanoparticles can be used as molecular electrical characterization.

Unlike nanostructures built from bulk solids, molecules can be made identically, cheaply, and easily, by the billions of trillions that will be needed for industrial-scale production of ultra-dense nano electronic computers. Two significant challenges are to devise molecular structures which act as electrical switches, and to assemble these molecules into the precise extended structures needed for reliable computation. Exciting theoretical and experimental progress toward these goals is just beginning.

Not surprisingly, molecular electronics was named "the breakthrough of the year" by Science Magazine in 2001. The unique potential of molecular electronics is the "bottom-up" self-assembly of inherently small objects such as molecules into devices and circuits. This represents a paradigm shift compared to the semiconductor "top-down" approach. Furthermore, the simplicity of this approach also offers significant economic edge over the conventional approach in which the escalating costs of wafer processing facilities are approaching tens of billions of dollars.

Although quite promising, fundamental new challenges are in the way of the realization of single molecule devices. These challenges place estimates of their mainstream applications way in the future, nearly around the middle of the 21st century.

14.1.4.1 Moore and Quantum Mechanics at 5 nm: Metal-Insulator-Metal (MIM) [12]

Indeed, the electronics industry is approaching transistors that are smaller than 7 nm. The subcomponents in the device will be even smaller than that. In this size regime problems arise, where silicon transistors become so physically close together that electrons experience quantum tunneling. So instead of staying in the intended logic gate, the electrons can continuously flow from one gate to the next, essentially making it impossible for the transistors to have an off state. Due to the costs involved in development, 5 nm is predicted to take longer to reach market than the 2 years estimated by Moore's law. Fig. 14.4A shows two metal (M) separated by a thin insulating material of thickness db. The insulator creates a barrier against electron transfer between the two metals. This is called a tunnel junction. It is a barrier, such as a thin insulating layer or electric

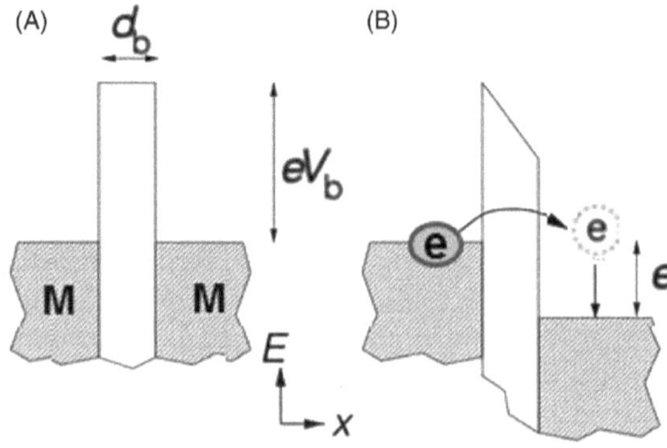

FIGURE 14.4 Moore and Quantum mechanics at 5 nm: metal-insulator-metal (MIM). (A) Two metal (M) separated by a thin insulating material, creating a barrier against electron transfer between the two metals (tunnel junction). (B) Voltage-induced electron quantum tunneling.

1 nm Transistor [13,14]

Companies, including Intel had originally announced that they would be *exploring other materials for producing 7 nm semiconductors* and beyond.

A-team at Lawrence Berkeley National Laboratory has successfully built a functional 1 nanometer long transistor gate (see Fig. 14.5). According to the team, it is believed to be the smallest working transistor ever made. The device uses carbon nanotubes and molybdenum disulfide (MoS_2) to create a sub-7 nm transistor. The MoS_2 functions as the semiconductor, with the hollow carbon nanotube functioning as the gate to control the flow of electrons. Yet it is not clear if it would be possible to develop a viable method to mass produce the new 1 nm transistors or a chip using them.

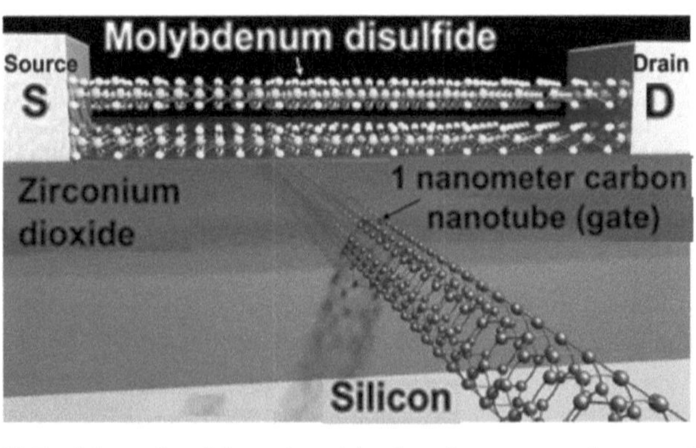

FIGURE 14.5 Schematic of 1 nm transistor based on uses carbon nanotubes and molybdenum disulfide (MoS_2).

potential, between two electrically conducting materials (M). Fig. 14.4B shows that when a voltage difference is applied between the two metals, the energy inside is elevated relative to the other metals. Electrons pass through the barrier by the process of quantum tunneling. Classically, the electron has zero probability of passing through the barrier. However, according to quantum mechanics, the electron has a non-zero wave amplitude in the barrier, and hence it has some probability of passing through the barrier. Tunnel junctions serve a variety of different purposes. Quantum mechanical analysis shows that the process becomes significant for a db < 5 nm. Quantum mechanically electrons might become transferable between the two. However, according to pure classical EM theory such transfer is not possible unless electrons are given enough energy to overcome the barrier energy. It is to be noted that in the absence of the external potential, there is no resultant electron flow either way because the energy of the two metals are at this level Because of the energy of the two metals are at this level, a balance takes place.

7 nm Strained SiGe Transistor Chip:

7 nm strained SiGe transistor chip: It's the generation after the next generation.

Presently, the most advanced commercial CPUs are built using a 14 nm process and there are plans afoot for 10 nm chips in 2016. Shrinking manufacturing any further for example to reach 7 nm node has proven challenging and not at all straightforward has remained out of reach due to a number of fundamental technology barriers with the most notable among them being the material properties of silicon itself. IBM, Samsung, and the SUNY Polytechnic Institute team used a

- Silicon–germanium (SiGe) alloy for the channel to improve electron mobility
- Extreme ultraviolet (EUV) lithography was employed to etch the microscopic patterns into each chip.
- Offers ~50% area scaling improvements over today's most advanced 10 nm technology (twice the power density of one built at 10 nm)
- At least a few years from commercial production

5 nm (nanometer) node:

In semiconductor manufacturing, the International Technology Roadmap for Semiconductors defines the 5 nm node as the technology node following the 7 nm node. Transistors at the 7 nm scale were first produced by researchers in the first decade of the 21st century – the process scale may represent the end of Moore's Law scaling for electronic devices. As of 2017, no 5 nm scale devices have been commercially produced. The 5 nm node was once assumed by some experts to be the end of Moore's law. Transistors smaller than 7 nm will experience quantum tunneling through their logic gates. Due to the costs involved in development, 5 nm is predicted to take longer to reach market than the 2 years estimated by Moore's law.

So What's After FinFETs?

Chipmakers exploring nanosheet, nanoslab, nano-ring, and hexagonal FETs.

Chipmakers face a multitude of uncertainties and challenges at 5 nm, 3 nm and beyond. Even today, traditional chip scaling continues to slow as process complexities and costs escalate at each node. In theory, *finFETs* are expected to scale to 5 nm as defined by Intel. (A fully-scaled 5 nm process is roughly equivalent to 3 nm from the foundries). The finFET likely will run out of steam when the fin width reaches 5 nm. So at 5 nm or beyond, chipmakers will need a new solution. Otherwise, traditional chip scaling will slow down or stop completely. Chipmakers have been exploring various transistor options for 5 nm and beyond. Samsung has opted for aFET for 4nm by 2020.

- nanosheet
- nanowire (*gate-all-around* category)
- hexagonal
- nano-ring
- nanoslab

For now, gate-all-around technology appears to be the most practical technology after finFETs. It's an evolutionary step from finFETs and shares many of the same process steps and tools. A lateral gate-all-around technology is basically a finFET on its side with a gate wrapped around it. Tiny wires or sheets serve as the channels.

There are other transistor options, as well. Gate-all-around, III-V materials, or tunnel FETs?

14.2 NONVOLATILE FLASH MEMORY CELLS: SILICON- AND METAL-BASED NANOPARTICLES

CMOS memories can be categorized into two major categories: Volatile memories in which information is lost when the power is turned off, and non-volatile memories in which information is retained even when the power is switched off [15,16]. We begin the discussion with non-volatile memory. Flash memory, which is a nonvolatile form of memory (it requires no power to store information) is increasingly common in consumer devices. They find applications in many portable electronic devices such as mobile phones, cameras, laptops etc. Today's flash devices store information by applying an electric field to a "floating gate" – basically a chunk of polycrystalline silicon – at the center of a transistor. This gate is surrounded by an insulating material, which needs to be relatively thick so that small defects do not allow the charge to leak out. As a result, a device like the flash-based iPod Nano, which packs four gigabytes of memory into its small frame is still carrying around a lot of inactive material (baggage).

Floating gate memory devices consist of FET with dual gate architecture, namely, floating gate and control gate for its memory characteristics. These Current traditional memory solutions are reaching their limits in terms of both scaling and the speed, and capacity requirements of next-generation mobile communications. Scaling has now become a serious issue for the memory

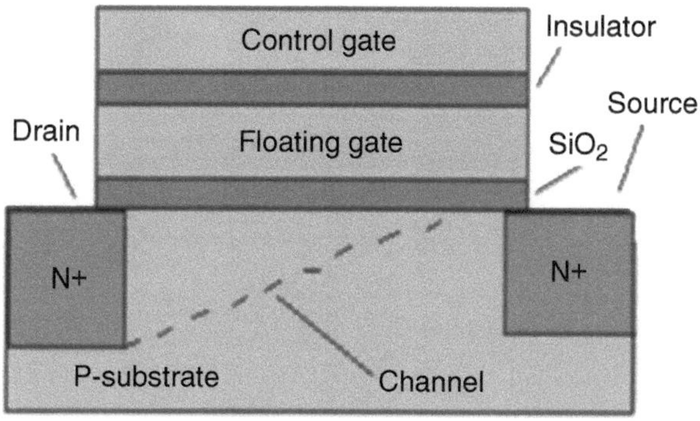

FIGURE 14.6 Schematic of conventional flash memory, consisting of field effect transistor with dual gate architecture, namely, floating gate and control gate.

industry. 65-nm fabs will be up and running in just a year or so. At 65 nm leakage is a major hurdle. 3D structures offer one solution, but there's a limit on how far you can go with this, and this remains to be only a holding action until something better comes along. In addition, Flash has a serious architectural scaling problem that seems likely to become critical well below 90 nm.

14.2.1 Conventional Flash Memory

Flash memory in use today consist of FET with dual gate architecture, namely, floating gate and control gate for its memory characteristics as shown in Fig. 14.6. In other words, each cell flash memory resembles a standard MOSFET, except the memory has two gates instead of one. As in ordinary MOS transistors the control gate lies on top. The new gate, called floating gate, lies below this control gate. In the design, this gate is insulated all around by the oxide layer (SiO_2). The floating gate is interposed between the control gate and the MOSFET channel. The basic idea here is that any electrons placed on the floating gate are trapped there and, under normal conditions, will not discharge for many years. This is made possible because the floating gate is electrically isolated by the oxide layer.

The device works as follows. Depending on the presence or absence of electrons in floating gate, memory state can be read as 0 or 1, respectively. Programming or erasing (P/E) operation is performed by controlling the flow of electrons from or to the substrate through the tunneling oxide. An injection of an electron occurs from the inversion layer via direct tunneling when the control gate is forward biased with respect to the source and drain. The resulting stored charge screens the gate charge and reduces the conduction in the inversion layer, that is, it effectively shifts the threshold voltage of the device to be more positive. See example, which estimates the magnitude for a single electron per nanocrystal.

It is to be noted that the basic characteristics can be studied using a capacitor configuration instead of a transistor. A capacitor is chosen instead of a transistor

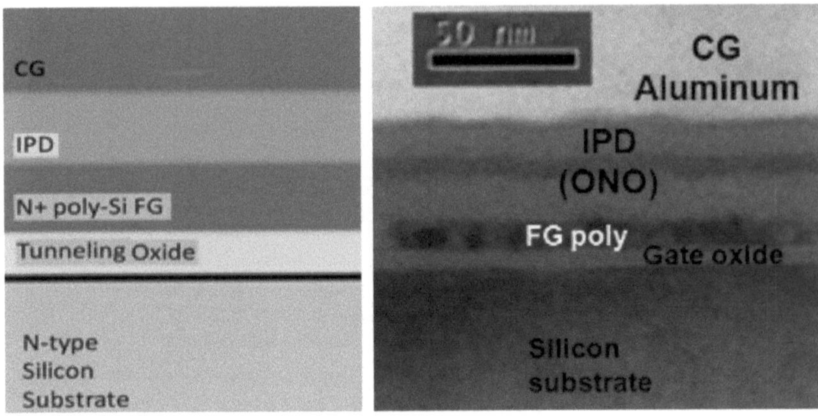

FIGURE 14.7 Floating-gate capacitor memory as the test structure.

because the capacitor enables the study of program/erase characteristics, retention and endurance with reduced complexity of fabrication compared to the transistor. The state of the device is sensed by measuring the flat-band voltage from a capacitance–voltage sweep. The schematic of the floating-gate capacitor shown on the left in Fig. 14.7 is chosen as the test structure [17]. The starting substrate is an (100) 1–5 Ω cm n-type silicon substrate. The device isolation is performed by growing a 0.5 micron thick thermal oxide and opening up active areas with a wet buffered oxide etch. Following that, the tunnel oxide is grown in dry oxygen at 800°C to obtain ~7 nm oxide. This thickness is limited by reliability concerns. Following this, the critical step of poly-silicon floating-gate deposition is performed. An oxide layer is deposited at low temperature followed by annealing, then a silicon nitride layer is deposited and annealed then a silicon oxide layer is deposited and annealing. The 950°C nitrogen anneal also in addition to improving the quality of the IPD stack. All these higher temperature steps serve to activate the dopants in the floating-gate and also help crystallize the amorphous silicon into poly-silicon (Fig. 14.8).

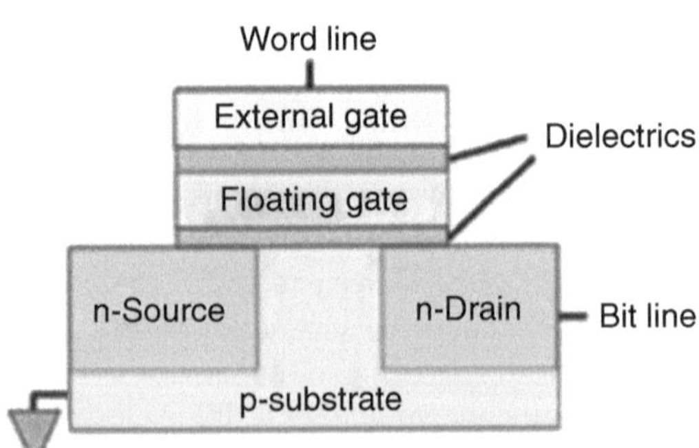

FIGURE 14.8 Schematic of conventional non-volatile memories.

Following the 950°C nitrogen anneal, the control gate is deposited. In the case of these capacitors, it was usually aluminum. Then, the aluminum is patterned into the control gate. Fig. 14.7 shows on the right high-resolution TEM images of the gate stack of an ultra-thin 7 nm poly-silicon floating-gate device. As the image reveals, the polysilicon floating-gate is obtained to be very smooth at such a low thickness. This was the thinnest reported poly-silicon floating-gate device at the time.

Conventional Non-Volatile Memory:

CMOS memories can be categorized into two major categories: Volatile memories in which information is lost when the power is turned off, and non-volatile memories in which information is retained even when the power is switched off. The non-volatile flash memory contains the NAND and NOR technologies. The NAND flash memory utilizes the Fowler–Nordheim tunneling for programming operation while the NOR flash memory uses the channel-hot-electron injection process. Both erase by the Fowler–Nordheim tunneling process.

There are many types of the non-volatile memories. One of which, the flash memory, is quite popular. Flash memory devices are the future for any type of portable application as they are non-volatile. However, they are slow and usually a combination of more than one type of memory is required in order to create a mobile device. Among the non-volatile memories, flash has been proved to be the most suitable one for portable applications, the main reasons are:

Advantages:

- Flash memory has a very simple structure requiring only one transistor, thus making it easier to achieve the highest chip density in comparison to other non-volatile memories.
- Compatibility with the current CMOS technologies. The materials and the fabrication process of the flash memory are compatible with conventional CMOS processes
- They can be programmed and erased many times (more than 100,000 times) while at the same time they offer the smallest cell size.
- Flash memory devices are the future for any type of portable application since they are non-volatile. However, they are slow and usually a combination of more than one type of memory is required in order to create a mobile device. Among the non-volatile memories, flash has been proved to be the most suitable one for portable applications, the main reasons are:

Drawbacks:

- Require high programming and erasing voltages and their program/erase speed is slow in contrast to the rest of the non-volatile memories.
- Further scaling looks impossible including such as the scaling of the tunneling dielectric layer thickness.
- Slow and usually a combination of more than one type of memory is required in order to create a mobile device.
- High process temperatures and the high thermal energy generated.

- Due to the reduction of the dimensions, higher operating voltages are required that can cause undesirable device behavior.
- A minimum channel length has to be maintained, which impedes the further scaling of the cell. A technological challenge is additionally
- The tunneling dielectric layer (usually an oxide) is problematic. It should be thin to allow the charges to tunnel through (for the programming and erasing processes) but, at the same time, must be thick enough to ensure data retention by preventing charge tunneling. Defects in the oxide may cause failure of the device and are among the most important factor that cause reliability concerns, which can lead to high operating voltages and high leakage currents. Another issue is the limit of the channel length.

In order to overcome these difficulties, new materials and device architectures are being explored. For example,

- Charge-trapping devices, such as silicon-oxide nitride-oxide-silicon (SONOS)
- SONOS devices exploiting the intrinsic defects of the silicon nitride that can trap charges.
- Metal or silicon nanocrystals for the floating gate
- Multilevel cell where the memory arrays are vertically stacked.
- Ferroelectric and phase change memories.
- High-k materials are mainly considered for the dielectric layer because a material with higher dielectric constant (k) can be physically thicker (reduces the leakage current) while having the potential to increase the gate capacitance and therefore improve the device performance. Three-dimensional (3D) approach hopes to solve the obstacle of the minimum value of the channel length.

14.2.2 Silicon Nanoparticle Memory

Until now, miniaturization was accomplished by the reduction of feature sizes with concomitant reduction of tunneling oxide thickness. Devices face the limitation that all stored charge will leak back to the channel through a path formed by repeated operation. Moreover, with the feature size reaching sub-25 nm, further device scaling is hampered by: (1) significant cell-to-cell interference as the spacing between the floating gates has now become comparable to memory cell size, and (2) non-scalability of tunneling oxide thickness due to unreliable charge retention after several program/erase cycles. Use of nanoparticles as charge-trapping nodes will aid in both minimizing interference and enhancing reliability through compartmentalization of charges.

Fig. 14.9 shows the transformation from a standard memory (A) to memory-cell structures employing discrete traps (B) as charge storage media [18].The nano-crystals floating gate is considered to be an alternative approach to replace the traditional planar floating gate for information storage. Nanocrystal memory possesses many advantages. Incorporating Si nanoparticles is an example of charge storage elements in non-volatile memory devices. Nanoparticle-based memories benefit from reduced lateral discharge paths, and promise larger retention time, lower power consumption, and faster operation. Moreover, it alleviates

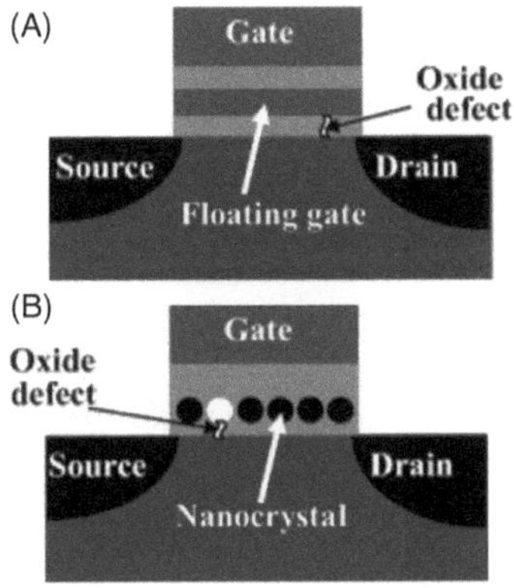

FIGURE 14.9 The transformation to memory-cell structures employing discrete traps as charge storage media. (A) Traditional memory-cell using oxide defects. (B) Nanomemory-cell using nanocrystals.

scaling problem in conventional floating gate memory. However, there are still some challenges that must be overcome; they will be discussed later.

Nanocrystal nonvolatile memory can be classified into three major categories according to the materials employed:

- Semiconductor: silicon (Si), germanium (Ge), Ge–Si core–shell, Ge/Si hetero-nanocrystal

Nanocrystal memory based on Si and Ge has been extensively investigated because silicon (Si) and germanium (Ge) are fully compatible with the semiconductor industry. The semiconductor (Si-base) NC-NVM is compatible with the IC industry; the source of the earliest research and the most mature fabrication technology.

- Metal, metal oxide, and metal silicide: The memory effects of Ag, Au, Pt, W, Co, Ni, $NiSi_2$, $Ni_{1-x}Fe_x$, Hf, TiN, and Al metal nanocrystals have all been fabricated and studied

The major advantages of metal nanocrystals over semiconductor nanocrystals include: (i) higher density of states around the Fermi level; (ii) scalability of the nanocrystal size; (iii) a wide range of available work functions and; (iv) smaller energy perturbations due to carrier confinement. In addition, electrostatic modeling (from both analytical formulation and numerical simulation) has demonstrated that metal nanocrystals significantly enhance the electric field between the nanocrystal and the sensing channel set up by the control gate bias, and hence such a system can achieve a much higher efficiency in low-voltage program/erase operations. In metal fabricated technology, the program efficiency and data retention can be improved by engineering an asymmetrical barrier structure. Also, the NC formation oxidation temperature can be reduced by the metal/silicon mix.

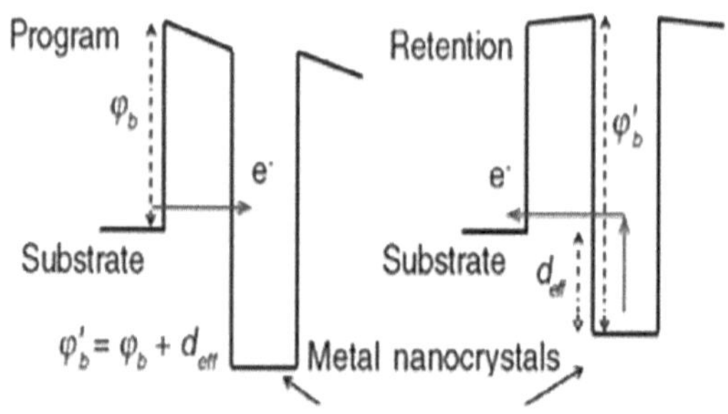

FIGURE 14.10 **The energy band diagram for metal nanocrystals embedded between a control oxide and tunnel oxide during (*left*) programming and (*right*) retention.**

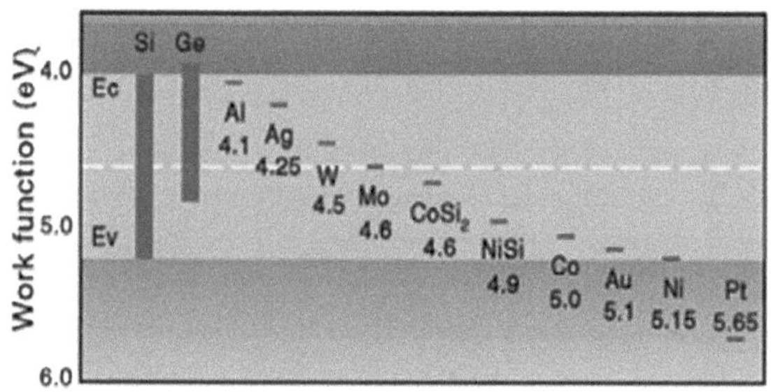

FIGURE 14.11 **Workfunction of a variety of metals.**

The depth of the potential well (d_{eff}) at the storage nodes, is dependent on the metal material, correlating with the metal work function. An asymmetrical barrier is created between the substrate and the storage nodes, that is, a small barrier for programming and a large barrier for retention. Fig. 14.10 shows the energy band diagram for metal nanocrystals embedded between a control oxide and tunnel oxide during (left) programming and (right) retention. Fig. 14.11 shows work functions of various nanocrystal materials.

- High-*k* dielectric nanocrystals: HfO_2 and CeO_2 nanocrystals embedded in SiO_2. High k dielectric devices can restrain the stored charge migration after an endurance test. Further, such cells could be useful for high-density two-bit nonvolatile memory applications.

14.2.2.1 Transistor (FET) Nanosilicon Memory

The first silicon nanocrystals based memory was reported by IBM in 1996 [19]. The schematic cross section of the concept is described in Fig. 14.12 using *n*-channel silicon FET. The figure shows the source, drain, channel and gate of the transistor. On top of the inversion surface of the *n*-channel a thin silicon oxide ~1.1–1.8 nm thick is laid by thermal growth (tunneling oxide). A distributed film of nanocrystals of silicon is now laid over the tunneling oxide. The nano crystals cover the entire surface channel region. A thicker tunneling oxide ~4.5 nm or

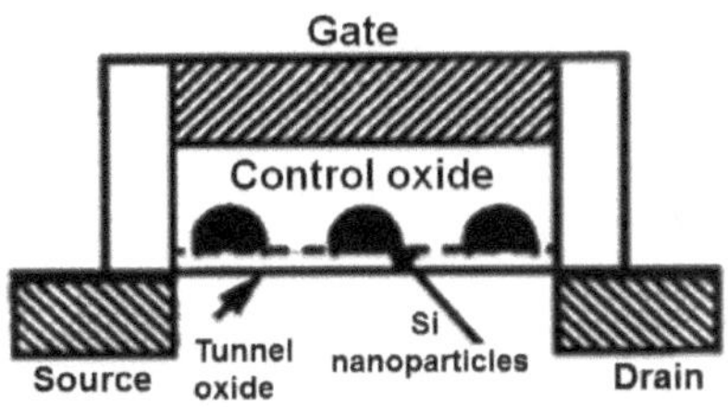

FIGURE 14.12 **Schematic cross section of Si Transistor (FET) nano silicon memory first reported by IBM in 1996.**

higher directly deposited is laid over the nano crystals (control oxide). Thus the nanocrystals are insulated from the channel and from the gate using the oxide. The nanocrystals are formed or deposited through in situ spontaneous decomposition during CVD. The nanocrystals, resulting from this procedure are half spheres. This increases the contact area with the tunneling oxide, hence to the tunneling source (the channel) leading to the most efficient injection condition. The nanocrystals are approximately 5 nm in size, with a density of slightly less than 1012 cm^2, which gives an approximate inter crystal spacing of 5 nm. Thus the crystals are effectively isolated from each other with little leaking.

Each floating gate cell (flash memory) resembles a standard MOSFET transistor, except it has two gates instead of one. On top is the control gate, as in ordinary MOS transistors. Below this control gate, situates the new gate called floating gate, which is insulated all around by the oxide layer (SiO_2). The floating gate is interposed between the control gate and the MOSFET channel. Because the floating gate is electrically isolated by the oxide layer, any electrons placed on it are trapped there and, under normal conditions, will not discharge for many years.

The device works as follows. An injection of an electron occurs from the inversion layer via direct tunneling when the control gate is forward biased with respect to the source and drain. The resulting stored charge screens the gate charge and reduces the conduction in the inversion layer, that is, it effectively shifts the threshold voltage of the device to be more positive. See example, which estimates the magnitude for a single electron per nanocrystal.

Memory Nano Cell (Voltage Shift)

$$\Delta V_T = \frac{q n_{\text{well}}}{\epsilon_{\text{ox}}}\left(t_{\text{cntl}} + \frac{1}{2}\frac{\epsilon_{\text{ox}}}{\epsilon_{\text{iS}}} t_{\text{well}}\right)$$

ΔVT is the threshold voltage shift, t_{cntl} is the thickness of the control oxide under the gate, t_{well} is the linear dimension of the nanocrystal well, εs are the permittivities, q is the magnitude of electronic charge, and n_{well} is the density of nanocrystals.

For nanocrystals that are 5 nm in dimension, 5 nm apart, that is, a nanocrystal density of 131012 cm^2, with a control oxide thickness of 7 nm, the threshold shift is nearly 0.36 V for one electron per nanocrystal, a large value that in a good transistor changes the subthreshold current by five orders of magnitude.

Number of nanocrystals

In a nano memory of 0.18 mm effective channel length, with a 0.25 mm channel width, the total number of nanocrystals in our model example are 800,

Coulomb blockade

Coulomb blockade effect can be very significant at these dimensions. The Coulomb charging energy (*q*2/2*C*tt, where *C*tt is the nanocrystal capacitance! for a 5 nm diameter nanocrystal of silicon in a matrix of SiO_2 is 74 meV. Not only is it larger than thermal energy, but it also limits additional injection of carriers into the nanocrystal. Thus, for a given write pulse, applied as a voltage signal between gate and source-drain, there is a saturation of a number of electrons that are accommodated in the nanocrystals, with each single electron causing a threshold shift of 0.36 V for this example.

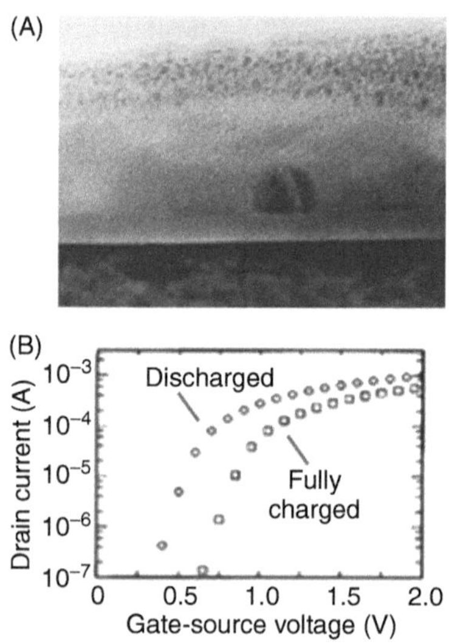

FIGURE 14.13 (A) FET nano memory device constructed and measured. (B) Characteristics of the device drain current versus grate-source voltage.

A device (Fig. 14.13A) was constructed and measured. The gate length is 0.4 mm and gate width is 20 mm The 1.6 nm tunneling oxide, nanocrystals approximately 5 nm in size, a density of slightly less than 10^{12} cm^{-2}, and a control oxide of 7.3 nm. A voltage of 1.25 V is applied to the gate while the channel is grounded. Electrons are drawn from the channel injected across the tunnel oxide and trapped in the nanocrystals (charging or writing step). The drain current–gate voltage characteristics are shown in Fig. 13.13B.The characteristics are obtained using 20 ns pulses, applied between gate and grounded source, with the drain maintained at 100 mV. For the fully charged condition at 1.25 V, the gate is pulsed from the 1.25 V condition.

The shift in threshold of 0.25 V under 1.25 V static gate bias shows the effect of the storage of electrons which we estimate to be one electron per nanocrystal. Larger gate voltages lead to an increase in the number of stored electrons, that is, in this same device structure, at 3 V static gate bias, the threshold shift is nearly 1.2 V, corresponding to larger number of electrons per nanocrystal.

14.2.2.2 Capacitive Nanosilicon Memory

The above results use a complex FET architecture as well as in situ synthesized nanoparticles which do not allow full control on the size shape or density, as well as are limited to 5 nm average size. In this section we present the capacitive architecture for charge-based nano memory cell, which was developed by a collaboration between MIT and University of Illinois.

There are several procedures that may be used for synthesis of silicon nanoparticles, each with advantages and limitations.

- CVD, aerosol, plasma assisted, ion implantation, and laser ablation methods do not produce nanoparticles or dots of constant size and shape.
- The in situ, *methods,* such as CVD, plasma assisted or ion-implantation is not fully CMOS compatible. Since synthesis of the particles involve high temperature anneals, such steps may interfere with the standard CMOS fabrication process flow, which may result in non-uniform thickness or defects. In fact ion implantation may result in formation of particles in direct contact with the substrate.
- CVD or plasma assisted processes produce non-spherical structures (hemispherical) because they involve growth on a surface, which provides poorer charge retention. It is to be noted that in situ procedures, such as CVD, plasma assisted or ion-implantation require higher temperature conditions to achieve narrower size
- Distributions, which are often diabolic with conditions for uniform density distributions, which hampers independent control of either.
- Ex-situ prepared

A second generation has been achieved using which utilizes ex situ synthesis and more simple capacitive as well as ultra small silicon nanoparticles. The work was conducted by Osama Nayfeh in the Demetri Antoniadis group at the Massachusetts Institute of Technology (MIT), presently at the SPAWAR Systems Center Pacific (San Diego), in collaboration with the Nayfeh's group at the University of Illinois. A capacitive architecture may also be used as a charge-based nano memory cell [20]. In this case nanoparticles are embedded in a metal-oxide-semiconductor (MOS) capacitor system as shown in Fig. 14.14A. We start the device fabrication with RCA cleaned 150 mm Si wafers with 10^{15} cm^{-3} boron levels, followed by 800°C dry oxidation to grow an 4.2 nm SiO_2 tunnel oxide

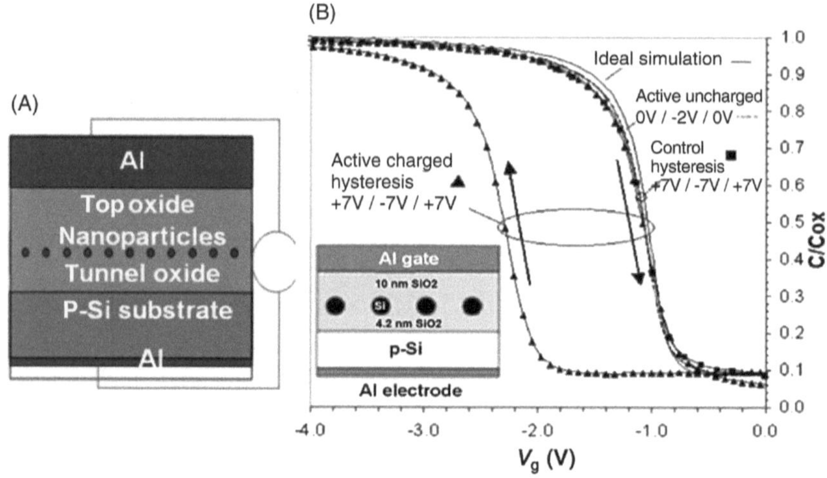

FIGURE 14.14 Capacitive nano silicon memory. (A) Schematic of Si nanoparticles-based metal-oxide-semiconductor (MOS) capacitor memory. (B) The charge storage in the MOS capacitors using *C-V hyesteresi loop* measurements. (inset) actual device.

layer. We then dispense via spin coating the nanoparticle/isopropanol (IPA) colloid directly over the tunnel oxide. An ~10 nm SiO_2 capping layer was then deposited by low-pressure CVD at 400°C. 500 nm gate and back side aluminum films were then deposited and a forming gas (N_2/H_2) annealed at 450°C for 10 min completed the process. Control devices were also fabricated by dispensing IPA containing zero nanoparticles and continuing the identical process.

Atomic force microscopy (AFM) and ellipsometry are used to test the particle distributions and to calibrate the colloid concentration. After calibration, we used colloids with decreasing molar concentration to approach sub monolayers.

Realization of complementary metal-oxide-semiconductor (CMOS) compatible devices puts stringent conditions on the quality, uniformity, and density distribution of the embedded nanoparticles. Recently, an ex situ electrochemical technique has been developed to produce colloidal, highly monodisperse, spherical, 1 nm in diameter hydrogen-terminated Si nanoparticles ($Si_{29}H_{24}$) that may alleviate these problems. Size reduction may also offer scaling advantages: however, recent calculations predict that electron storage may be inhibited when particle size is reduced below 3 nm.The majority of previous experimental studies utilized 5–10 nm Si nanocrystals fabricated by CVD or ion-implantation techniques. In these devices, electrons and holes are injected and stored in the nanoparticle as a signature of writing and erasing. More recently, memory effects in 2–3 nm Si nanocrystals have been studied, and mixed electron and hole storage was also observed.

The charge storage in the MOS capacitors can be seen using *C-V* measurements. Fig. 14.14B presents *C–V* hysteresis loops at a frequency of 1 MHz for an active device (incorporating nanoparticles) and a zero-nanoparticle (control) device. Both have a negligible hysteresis for uncharged conditions (forward and backward between 0 and −2 V). Under program/erase conditions (forward and backward between +7 and −7 V), the control device has only a slight counterclockwise hysteresis of ~0.01 V; however, the active device has a clockwise hysteresis of ~1.2 V. The positive half cycle (+7 to − 7 V) of the active device hysteresis loop involves hole charging while the negative half (−7 to + 7 V) involves complete erasing and returning to the uncharged state. The negligible hysteresis in the control device indicates that the effect is nanoparticle related. We estimate by integration, a stored charge density of ~5×10^{11} cm_2 for the active device. Fig. 14.14B also shows an ideal simulated *C–V*. The persistent large hysteresis, and the similarity in the control, active, and simulated curves, in regard to the absence of kinks, humps, or differences in slope point to the fact that the charging in the active device is due to the charging of the nanoparticles and not the interface states.

The left shift in the *C–V* for the programed (+7 V) active device indicates hole charging of the particles. With erase bias (−7 V), stored holes are removed and the curve shifts back to the right to the uncharged state. We performed more measurements of the charging state after programming for the range of 0–12 V, and observed only hole and no electron charging. Also, with negative-bias erase conditions up to −12 V, we observed only removal of stored holes and no electron storage, that is, no right shift. The effectively pure hole-charging characteristic may be understood in terms of the properties of the highly monodisperse 1 nm Si particles.

We tested the charge retention and durability characteristics of several devices. We measured at room temperature the stored hole charge loss rate (measured as voltage shift)after programming with +3 V. Fig. 14.15 shows 25% charge loss after ~100 s, and 50% in ~1 h. Extrapolation indicates that full charge loss should occur in ~ 1 yr. We also measured the endurance characteristics in Fig. 14.15 with 35 ms +7 V/ −7 V program/erase pulses. We observed only slight decrease in the hysteresis after 10^5 cycles.

Quantum Monte Carlo and time dependent local density theory calculations predict that for 1–3 nm Si particles, the energy gap rises and the electron affinity decreases. Belomoin et al. measured 3.5 eV for the energy gap (E_g) of 1 nm $Si_{29}H_{24}$ by excitation/emission, and Therrien measured ~1.8 eV for the charging energy (E_{sc})by ultrahigh vacuum scanning tunneling microscope tunneling spectroscopy.

Ex-Situ Si Nanoparticles in Memory

Smaller / Higher Capacity/Faster. When using nanoparticles, much less insulation is needed, so that the memory occupies half or third as much space while running faster. Or, in other words, a flash-based device can carry twice or three times as many information. This development opens up the possibility to put larger memories onto an economically sized chip.

Lower Cost: Cutting down on insulation also decreases the voltage needed to store information. This makes it much easier to integrate flash memory with information processing on the same chip, which reduces costs since the number of steps needed to make the device drops by more than half.

Higher Density: Larger density is enables, so overall it's a much better way of making the embedded nonvolatile memory technology.

Nanoparticle memories can bring non-volatile memory right onto the CPU chip, increasing data access times and reducing power and chip count

The new technology is inherently more scalable, and well suited to the next generation of mobile computing and communications, which are hungry for high capacity memories capable of storing and rapidly accessing video and large databases without overburdening battery power sources.

These contemplated features explain the large demand for such inexpensive embedded memory. Several groups and companies are racing to implement the silicon nanoparticle technology. Important firms such as Intel, Freescale, Micron, Samsung, and STMicroelectronics are beginning to settle on new technology platforms for the post-Flash era and are finding ovonic and nanocrystal line / nanoparticle memories increasingly to their liking.

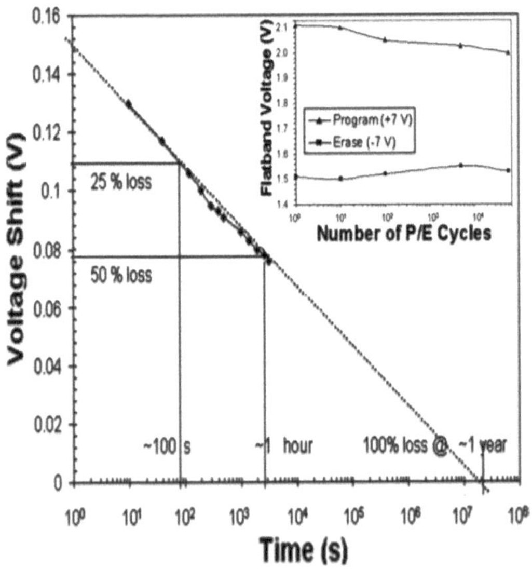

FIGURE 14.15 The charge retention and durability (inset) characteristics of the device (voltage shift as a function of time).

We have seen the use of silicon as floating gates for non-volatile memory devices to achieve high data retention and faster program/erase speeds. Ultra small silicon nanoparticles in the sub 3-nm regime are optically active. Therefore, group-IV semiconductors (Si and Ge) are useful for silicon planar technology compatible light-emitting and floating gate memory devices. The size-dependent electrical and optical properties of Si and Ge nanoparticles were introduced above in Chapters. We outline some of the schemes to achieve light emission from indirect band gap Si and Ge nanoparticles embedded in different high band gap oxide matrices.

14.2.3 Metal-Based Nanoparticles Nanomemory [21]

Metal-based nanoparticles are used as discrete nodes for trapping charge for floating gate flash memory.

The major advantages of metal nanocrystals over semiconductor nanocrystals include:

- higher density of states around the Fermi level
- scalability of the nanocrystal size
- wider range of available work functions
- smaller energy perturbations due to carrier confinement.
- Plasmonic enhanced fields: significantly enhance electric fields between the nanocrystal and the sensing channel set up by the control gate bias, resulting in much higher efficiency in low-voltage program/erase operations.
- Asymmetrical barrier structure for improved program efficiency and data retention. Also, the NC formation oxidation temperature can be reduced by the metal/silicon mix.
- Deeper potential well (d_{eff}) at the storage nodes due to larger metal work function (see Fig. 14.11).
- Charge leakage can be suppressed by improving the NPs/dielectric interface
- Metal, metal oxide, and metal silicide have been used. The memory effects of Ag, Au, Pt, W, Co, Ni, $NiSi_2$, $Ni_{1-x}Fe_x$, Hf, TiN, and Al metal nanocrystals have all been fabricated and studied. A schematic of an all metal-based floating gate nano memory is shown in Fig. 14.16. It uses both Al_2O_3 and HfO_2, as well as SiO_2 oxides. Moreover the trapping nodes are RuOx nanocrystal oxides.

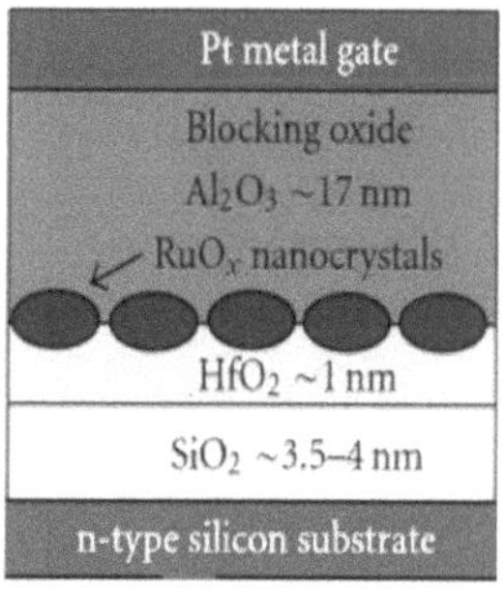

FIGURE 14.16 A schematic of an all metal-based floating gate nano memory utilizing Al_2O_3 and HfO_2 as well as SiO_2 oxides with RuOx nanocrystal oxides as trapping nodes.

The disadvantages of metal-based memory (especially with regard to flexible electronic applications) include

- High charge leakage from metal nano particles
- High leakage current from metal oxide
- Encapsulation of metal/metal oxide with SiO_2 (core–shell nanoparticles) require high temperature processing >400°C
- Composites of metal oxide (high-k inorganic nanoparticles) in polymer matrix have a rough surface and high gate leakage current to have a low on/off current ratio.
- Fixed-charge trapping defects generated by high-k nanoparticles in dielectric layer. These fixed-charge associated trapping defects in dielectrics will become deadly shielding of preventing the mobile charging/discharging progress in memory transistors.

We now present some issues of metal-based nanomemory by discussing two examples, one of which is gold-based and the other is ZnO-based memories.

14.2.3.1 ZnO-Based Memory With Embedded Si-Nanoparticles [22]

A fabricated ZnO-based memory device with Si nanoparticles embedded within a ZnO charge trapping layer is demonstrated by the Ammar Nayfeh's group at Masdar Institute of S and T in collaboration with Ali Okyay group at Bilkent University.

The structure and fabrication is briefly presented. Fig. 14.17A shows a cross-section of the final device structure with the Si nanoparticles. The channel-last memory cells were fabricated on highly doped p-type (111) Si wafer which is used as a back-gate electrode. First a 15-nm-thick Al_2O_3 blocking oxide is first deposited by ALD followed by a 2-nm-thick ZnO charge trapping layer. Then, Si-NPs were spun on the ZnO. Again, a 2-nm-thick ZnO charge trapping layer was ALD deposited so that the Si-NPs are embedded within the charge trapping ZnO. This was followed by ALD deposition of a 3.6-nm-thick Al_2O_3 tunneling oxide and an 11-nm-thick ZnO channel at 250°C. The channel was made using optical lithography. The source and drain contacts were created by depositing 100 nm Al by thermal evaporation followed by lift off. Using plasma-enhanced chemical vapor deposition (PECVD), a 360-nm-thick SiO_2 layer is deposited for device isolation. Finally, rapid thermal annealing (RTA) was performed.

In total darkness, gate voltage is swept in a desired range in order to *I–V* characteristics of a memory cell. Memory cells are programmed by applying a negative gate voltage and erased by applying a positive gate voltage, which suggests that holes are being trapped. *I–V* measurements were taken at $V_d = 10$ V. The memory is programmed by applying $V_g = -10$V for 5 s with source and drain being grounded, and erased by applying $V_g = 10$V for 5 s. Fig. 14.17B presents the I_{drain}–V_{gate} curves of the programmed and erased states of memory devices with and without Si NPs. The ΔV_t is extracted at a drain current of

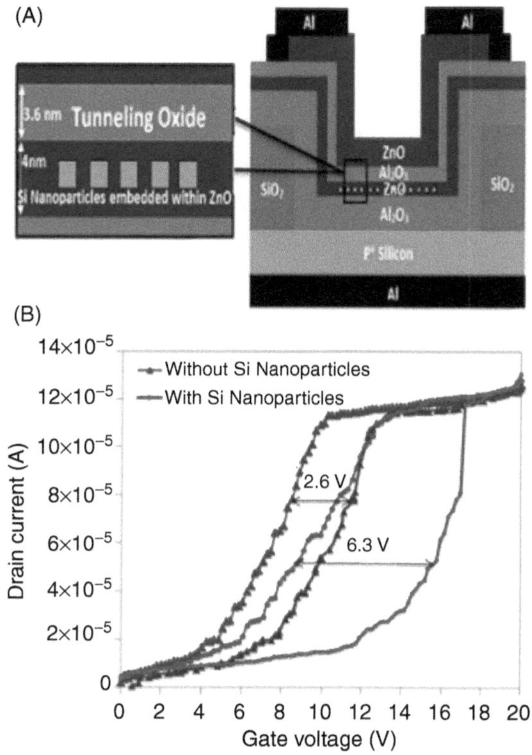

FIGURE 14.17 ZnO-based memory with embedded Si-nanoparticles. (A) A cross-section of the final device structure with the Si nanoparticles. (B) The I_{drain}–V_{gate} curves of the programmed and erased states of memory devices with and without Si NPs.

4×10^{-5}A which is near the extrapolated turn on of the device. The figure shows that the presence of the silicon nanoparticles increases the shift from 2.6 eV to 6.3 eV. Increasing V_t by an amount of ~3.7 V (from 2.6 V) means that the Si nanoparticles behave as charge trapping centers with a high trapping density within the bandgap of ZnO.

The retention time characteristic with and without nanoparticles is studied. Fig. 14.18 shows the V_t shift versus time after a single programming event at −10 V. The plot shows that the memory with Si NPs loses 36% of its initial charge in one year while that takes only 70 min in devices without NPs. Also 41% of the charge of the memory device with NPs is lost in 10 years while that only takes 100 min for devices without NPs. The plot indicates that the slope of the retention time curve is improved with NPs which means that the rate of charge loss is reduced due to Si-NPs better confinement.

The retention time characteristic with and without nanoparticles is studied. Fig. 14.18 shows the V_t shift versus time after a single programming event at −10 V. The plot shows that the memory with Si NPs loses 36% of its initial

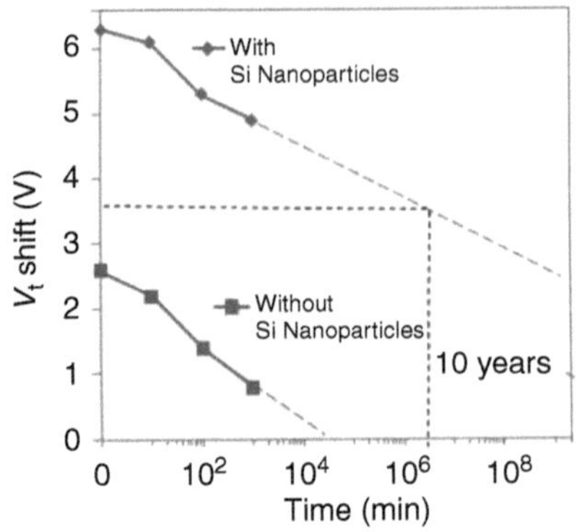

FIGURE 14.18 The retention time characteristic with and without nanoparticles of the device in Fig. 14.17.

charge in one year while that takes only 70 min in devices without NPs. Also 41% of the charge of the memory device with NPs is lost in 10 years while that only takes 100 min for devices without NPs. The plot indicates that the slope of the retention time curve is improved with NPs which means that the rate of charge loss is reduced due to Si-NPs better confinement.

Memory Charge Density

Assuming the threshold voltage shift is mainly due to the stored charge in the trapping layer, the charge trap states density can be calculated using equation (2.3). At a programming voltage of −10 V and with a capacitance of the charge trapping layer per unit area: $C_t = 560$ nF/cm^2, the ΔV_t is 6.3 V which corresponds to a charge trap states density of 1.1×10^{13} cm^{-2} or equivalently 1.67×10^{-6} C/cm^{-2}, and at a programming voltage of −1 V, the ΔV_t is 2.6 V, which corresponds to a charge trap states density of 5.95×10^{12} cm^{-2} or 9.52×10^{-7} C/cm^{-2}.

A low power ZnO-based charge trapping memory with Si nanoparticles is fabricated and studied. With 2 nm Si-NPs, the memory cells show a much higher V_t shift and a longer retention time (>10 years). The results show that Poole–Frenkel effect is the dominant mechanism for hole emission at low electric fields allowing for low voltage programming. The large V_t shifts obtained with Si nanoparticles at low voltages and the excellent retention highlight a promising technology for future ultra-low power memory devices.

14.2.3.2 Nanogold Flash Memory [23]

Nanoparticle MIS capacitors were fabricated via the following procedure: clean p-type silicon with a thin layer (approx. 3 nm) of SiO_2 and an Ohmic aluminum back contact was salinized, followed by deposition of a layer of gold nanoparticles (dodecanethiol capped, purchased from Sigma Aldrich) via Langmuir–Blodgettde position. A PVP insulating layer approximately 50 nm thick was the evaporated on top of the nanoparticle layer, followed by evaporation of an aluminum gate electrode. All evaporations were carried out in an Edwards Auto 306 evaporator at a pressure below 5×10^{-6} Torr. All *C–V* measurements were performed on an HP4192A impedance analyzer under computer control. All *C–V* scans were performed by starting the scan with a positive gate voltage (inversion region for a p-type semiconductor). In order to ensure accurate *C–V* measurements, the leakage current through the capacitors was measured in the range of gate voltages used. Leakage currents of MIS capacitors in the voltage ranges used were measured on an HP4140B picometer and were found to be below 1 nA in all cases.

The nanoparticles used in the fabrication of the MIS capacitors and the structure of the capacitors are shown in Fig. 14.19. In order to assess the quality and coverage of the gold nanoparticle layer, atomic force microscopy (AFM) images were taken of the Langmuir–Blodgett deposited layer before the final insulator and gate contacts were fabricated. In Fig. 14.19, the gold nanoparticle layer is clearly visible as the light areas on the darker substrate. From the analysis of the image, the surface coverage of nanoparticles is approximately 80%.

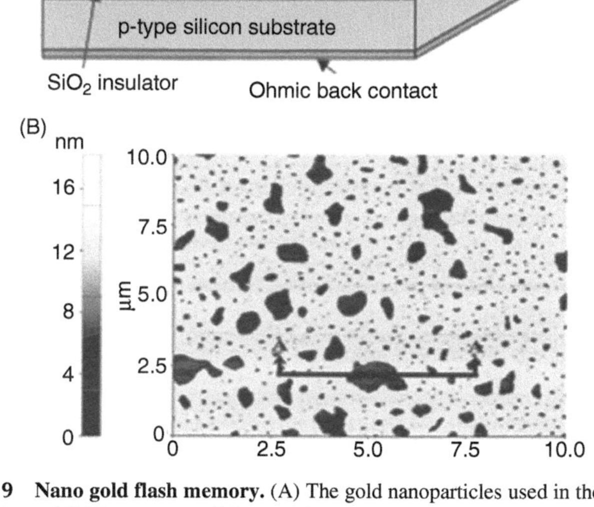

FIGURE 14.19 Nano gold flash memory. (A) The gold nanoparticles used in the fabrication of the MIS capacitors. (B) The structure of the capacitor memory.

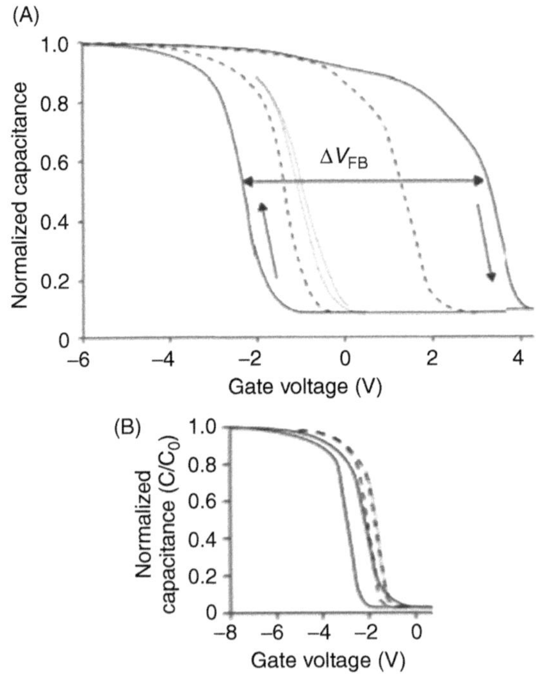

FIGURE 14.20 (A) *C–V* curves for the gold nanoparticles in the MIS capacitors in Fig. 14.19B. (B) The corresponding *C–V* without the nanoparticles.

Scans of other areas of the substrate also confirm that this level of coverage is consistent across all the devices.

As can be seen from Fig. 14.20A, which shows typical *C–V* curves for the gold nanoparticles in the MIS capacitors, as the voltage scanning window is increased, the amount of hysteresis also increases. This large amount of hysteresis can be attributed to charge trapping and de-trapping of the gold nanoparticles; when nanoparticles are not present in the capacitors, very little hysteresis can be seen (Fig. 14.20B).From the amount of hysteresis present in the C–V curves during the maximum gate voltage sweep, the density of trapped charges attributed to the nanoparticles is approximately $3.3 \times 10^{12} cm^{-2}$.

14.2.3.3 A Multilayer Nanocrystal

It is an approach in which the floating gate is made from several nanoparticle layers. In this case Coulomb blockade repulsion by a lower-layer of nanocrystals can repel charge from the higher layers and hence suppress the charge leakage from upper-layer nanocrystals into the substrate. Therefore, multi-layer nanocrystal memory can achieve a better charge storage ability than typical single-layer nanocrystal memory.

14.2.3.4 Challenges of Nanoparticles Floating

Nanocrystal memory properties are influenced by nanocrystal size, shape, and configuration. Nanocrystals are not formed by patterning, hence fluctuations between individual devices becomes important. According to the International Technology Roadmap for Semiconductors (ITRS 2009), memory cell size is likely to reduce to 1000 nm^2 by 2020. This means that only about ten nanocrystals can be contained in every memory cell. On such a small scale, it becomes increasingly challenging to synthesize suitable materials with uniform size and shape, and assemble them into a well-ordered nanocrystal matrix.

High-density nanocrystals can store more charge in a memory device and mitigate the influence of fluctuations between individual devices because there are more nanocrystals in each memory cell. However, the charges stored in nanocrystals leak though the surrounding oxide more easily in high density nanocrystal systems. If the surrounding oxide cannot effectively prohibit the charge transport between nanocrystals, nanocrystal memory will lose the advantage of having a discrete charge storage node; that is, preventing the loss of total stored charge through a leakage path in the tunnel oxide. Even though surrounding oxide passivation can alleviate the charge transport between nanocrystals, eventually the charges in nanocrystals will directly tunnel through the thin surrounding oxide if the nanocrystal density continues to increase.

Metallic nanocrystals are considered the most promising for commercial production. In addition, in order to simultaneously achieve high program/erase speeds and long retention times without sacrificing other memory properties, tunneling barrier engineering, multi-layer nanocrystals, work function engineering, and surrounding oxide passivation have been proposed. However, nanocrystal memory still faces some challenges upon scaling down, such as fluctuations from device to device and charge leakage at high density. It is, therefore, important to continue to find methods that strategically match the scaling limits with appropriate requirements in the NVM industry.

14.2.4 Flexible Flash Memories: Metal Versus Silicon-Based [24]

Electronic devices are traditionally fabricated using rigid semiconductors and high-temperature manufacturing method. Data storage is one key requirement in many electronic devices. After the establishment of flexible thin film transistors (TFT), some of them were studied in order to develop into flexible flash memory devices. Plenty of efforts have been spent on printed memory devices. The requirements and challenges for processing and operating printed memories include the following

- Low temperatures $<400°$ C
- Polymeric substrates
- Large-area
- Flexible

- Low-cost electronic devices
- Large enough memory window with program and erase voltages below 20 V (<10 V)
- Low-power power and low voltage operation
- Thin thickness high-dielectric constant

There have been several architectures using a variety of silicon and metal-based materials. The significance of dielectric layer is attracting more and more interests in order to develop high-performance printable floating gate memory transistors. Many studies have tried to find better gate dielectrics with polymeric materials, inorganic/organic hybrids, inorganic/organic bilayers, and organic/inorganic hybrids via sol–gel reaction. We mention here some of those:

A polymer dielectric film, one of the drawbacks is the high operating voltage in devices due to its thickness and low dielectric constant.

High-k inorganic materials, such as Al_2O_3 and Ta_2O_5 had been used as gate insulators due to their dielectric constants which are much higher than those of polymers. But metal oxide has relatively high leakage current and these films can be produced with the vacuum techniques, such as CVD, sputtering or flame hydrolysis. Vacuum equipment is not suitable for a low-temperature and low-cost process required by printing techniques. Alternatively, high-k inorganic nanoparticles can be dispersed in polymer matrix. However, a nanocomposite dielectric layer with high-k NPs embedded in polymer matrix has a rough surface and high gate leakage current to have a low on/off current ratio. In addition, it is difficult to eliminate fixed-charge trapping defects generated by high-k NPs in dielectric layer. These fixed-charge associated trapping defects in dielectrics will become deadly shielding of preventing the mobile charging/discharging progress in memory transistors.

An Au—SiO_2 core–shell NC capacitor but temperature for this memory device is higher than 400° C. and then this device is not able to be fabricated by printing techniques.

14.2.4.1 Si/SiO_2 Hybrid Core/Shell Trapping/Tunneling Element [25]

As discussed above metal nanoparticles suffer from charge leakage in their use in memory devices. In order to avoid the metal contamination in dielectrics, Si nanocrystals (see above) have been used in floating gate. As a variation to the pure H-terminated Si nanoparticles, Si-SiO2 core–shell nanostructures are proposed to work as trapping sites in floating gate memory, as shown in Fig. 14.21.

Uniform SiO_2 shell of 3–5 nm in thickness surrounding each Si NPs is formed by silicon cores simultaneously oxidized in solutions. Measurements showed that this action reduces lateral discharging with retention properties improved. More importantly, with monolayer of Si NPs trapping layers, this ultra-thin/uniform SiO_2 shell functions as tunneling layer of flash memory devices and leads to quite low-operation voltage for printable NPs floating-gate memory

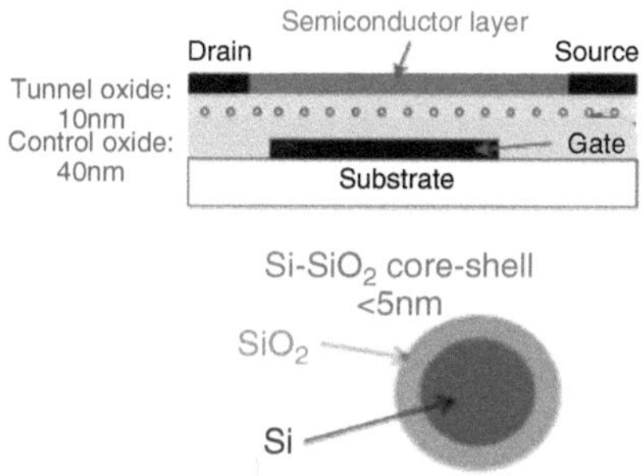

FIGURE 14.21 **Si-SiO_2 core–shell nanostructures as trapping sites in floating gate capacitor memory.**

devices. Compared to traditional memory stacks (substrate/tunnel layer/charge trapping layer/control layer/gate electrode), the present proposal does not involve separate tunnel layer, and the high-quality SiO_2 shell acts as tunnel layer. Hybrid dielectric non-volatile memory with nano particles (Si/SiO_2 core/shell) as charge trapping layer US 9406765 B1.

Accordingly, a first aspect of the presently claimed invention is to provide a printable floating gate transistor for non-volatile memory. A second aspect of the presently claimed invention is to provide a method for fabricating a floating gate transistor. The memory transistor of the present invention exhibits a memory window of 4.2 V, long retention time (100 ks) with low operating voltage (<7 V).

14.2.4.2 Poly-Si Memory

An interesting study by Dr. Young [26] on the fabrication of poly-Si TFTs and memory devices on glass using a silicon nitride layer with a high trapping state density, reported a high field-effect mobility (170 cm^2/Vs), as well as a threshold voltage of 4.2 V. The process temperature did not exceed 300°C, but the poly-Si was produced by the crystallization of amorphous silicon films using a scanned excimer laser annealing process.

14.2.4.3 Silica-Coated Copper Nanowires (Conductivity-Based) [27]

In this information is encoded in states of conductivity or resistance. The active material is silica-coated copper nanowires, which is encased in a polymer matrix. The resistance of this wire can be changed by applying a small voltage. In other words it can be switched between a state of high resistance, and a state of low resistance. If the wire is connected to a current source, then none can stop an electric current or allow it to flow, based on the external biasing voltage. Another interesting feature is the fact that the polymer can be dissolved in methanol, creating a colloid of the wires. This enables spraying the wires through the nozzle of a printer. A memory device is constructed as follows. A commercially available gold nanoparticle ink is used to print a series of gold electrodes onto a glass slide.

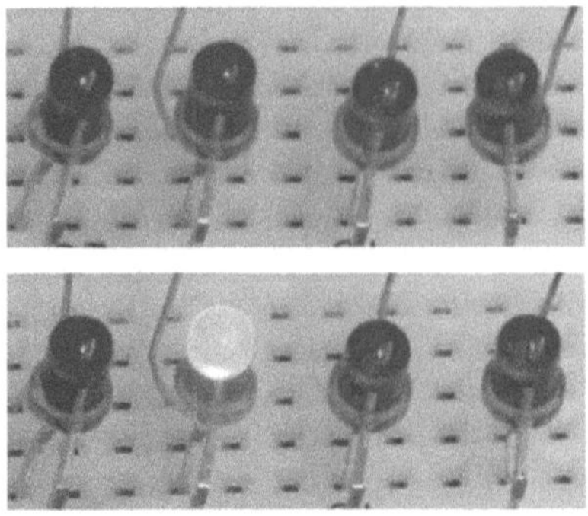

FIGURE 14.22 **Silica-coated copper nanowires (conductivity-based memory).** *(Used with permission from Matthew Catenacci, Duke University)*

The copper-nanowire memory material is then printed over the gold electrodes. A second set of copper electrodes are then printed over the active medium.

The device was developed and demonstrated at Duke University by turning on four LED lights according to a programmed schedule (Fig. 14.22). Application may include apply it to fabrics, RFID tags, curved and flexible substrates, or substrates that can't sustain high heat.

The time it takes to switch back and forth between states is measured to be around three microseconds. This rivals the speed of the charge-based flash drives. The retention time reaches up to ten years, and the material can be rewritten many times without degrading. The memory capacity is too small for most application, but they may be useful in low-memory demanding applications, where low cost and flexibility are the key.

14.3 VOLATILE MEMORY DYNAMIC RANDOM-ACCESS MEMORY (DRAM)

Volatile memories are storage devices in which information is lost when the power is turned off. A schematic drawing depicting the cross-section of the original device is shown in Fig. 14.23 [28]. It goes back 50 years ago to 1968.

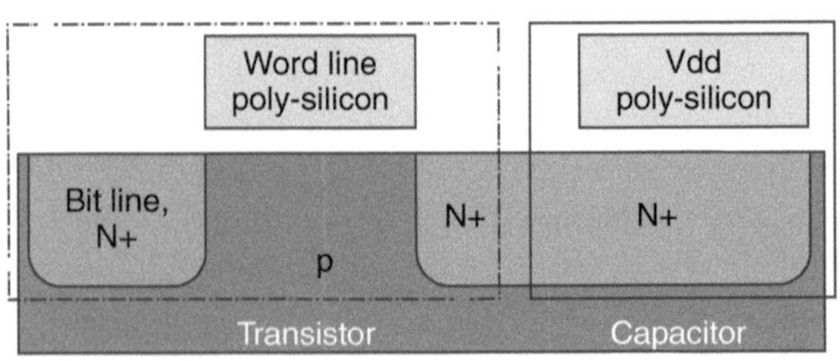

FIGURE 14.23 **A schematic drawing depicting the cross-section of the original volatile memory dynamic random-access memory (DRAM).** (See: Robert Dennard, IBM, United States Patent 3,387,286)

As the skeptic shows, the unit cell consists of one-transistor and one-capacitor (1T1C) cell. As one can see it does not involve an electrically isolated layer, such as a floating gate. The two components are connected such that the transistor is used to admit current into and discharge the capacitor during writes and reads respectively. The second terminal is grounded or as most commonly used at present time is connected to the battery $V_{CC}/2$. The filling of the capacitor is made from an oxide-nitride-oxide (ONO) dielectric. The capacitor plates are made of polysilicon.

In this scheme the capacitor is constructed above the surface of the substrate. This scheme is limited with regard to size reduction. Another geometrical configuration has been developed, which is more amenable to scaling. This is called the trench capacitor depicted in [29]. The construction begins by etching a deep hole into the silicon substrate. One of the plates of the capacitor is produced by heavily doping the substrate volume surrounding the hole. This produces a buried n^+ plate, which reduces resistance. The filling of the capacitor is now made by growing or depositing a layer of oxide-nitride-oxide dielectric. The top plate of the capacitor is now made by filling/depositing the hole with doped polysilicon. Finally, a polysilicon strap is grown to connect the top of the capacitor to the drain terminal of access transistor. The depth-to-width ratio of trench capacitors can be varied to need. A ratio of 50:1 has been achieved in DRAMs in recent years. Despite the fact that it is more difficult to construct trench capacitors within deep holes and to connect to the access transistor's drain terminal trench capacitors have numerous overriding advantages.

- Economical surface area: The occupied area of buried capacitors, for a given capacitor's size and capacitance, can be minimized better compared to that of surface capacitors, while meeting the requirements to connect it to the drain terminal without decreasing the capacitor's size, and thus capacitance.
- Enhanced capacitance: The capacitance can be increased by etching a deeper hole without any increase to surface area.
- Better integration: Being under the layers of metal interconnect allows them to be more easily made planar, which enables it to be integrated in a logic-optimized process technology, which have many levels of interconnect above the substrate.
- Less degradation of logic transistors: High- temperature demanding embedded capacitors are constructed before the transistors.

In a recent development in 2014, IBM researchers exploited these advances and more to optimize the development of volatile memory. They used a 4th generation deep trench embedded DRAM to a memory solution featuring finFET architecture on an SOI substrate. It provided the following memory features:

- Ultra-dense (0.0174 μm^2) memory

- Allows gate length scaling to the sub 20 nm regime and superior low V_{dd} operation
- Substantial (>35%) performance gain compared to predecessor technology
- Solution for industry leading "scale-out" processor design
- Optimized HP and LP devices simultaneously on chip

14.4 COMMERCIALIZATION ISSUES STAGES OF DEVELOPMENT OF SI-BASED NANOMEMORY [29]

We first briefly present some numbers on the market size and share of nano-memory. According to a 2010 report from NanoMarkets, an industry analyst firm the market for nano-enabled memory will grow to $7 billion ($US) in 2010, up from $1.4 billion on 2008. NanoMarkets also projected that each of the competing technologies will have a share in market revenues in 2010 as follows:

- MRAM $1.5 billion
- nanocrystalline/nanoparticle $980 million
- Holographic $980 million
- Ovonic memory $877 million.

NanoMarkets saw 2010 as a breakout year for nanomemory because, by that time, conventional solutions would have become simply be unable to scale further or provide the memory requirements needed for ubiquitous computing. In fact technical improvements, such as lower power requirements for ovonic memory, and real commitments from larger firms to specific nanomemory platforms (Freescale to nano crystalline/particle memories) was the reason nanomemory was brought much closer to reality. In fact both semiconductor firms and OEMs took nanomemories much more seriously than they did a couple of years ago.

Flash memory, which is a nonvolatile form of memory (it requires no power to store information), is increasingly common in consumer devices. Today's flash devices store information by applying an electric field to a "floating gate" – basically a chunk of polycrystalline silicon – at the center of a transistor. This gate is surrounded by an insulating material, which needs to be relatively thick so that small defects in it don't allow the charge to leak out. As a result, a device like the flash-based iPod Nano, which packs four gigabytes of memory into its small frame, is still carrying around a lot of inactive material.

These current traditional memory solutions are reaching their limits in terms of both scaling and the speed, and capacity requirements of next-generation mobile communications. Scaling has now become a serious issue for the memory industry.

By 2010 NanoMarkets predicts that four key segments will have emerged in the nano-enabled memory markets: MRAM, Ovonic, Holographic, and Nanocrystalline.

One of the most important nanotechnology solutions is silicon nanoparticle technology. This nanotechnology replaces the solid silicon gate with a large number of tiny silicon crystals separated by minute amounts of insulation. The result: much less insulation is needed, so that the memory occupies half as much space. Or, in other words, a flash-based gadget can carry twice as many songs. Cutting down on insulation also decreases the voltage needed to store information. This makes it much easier to integrate flash memory with information processing on the same chip, which reduces costs nanoparticle memories can bring non-volatile memory right onto the CPU chip, increasing data access times and reducing power and chip count.

Several groups and companies are racing to implement the silicon nanoparticle technology. Important firms such as Intel, Freescale, Micron, Samsung, and STMicroelectronics are beginning to settle on new technology platforms for the post-Flash era and are finding ovonic and nanocrystalline memories increasingly to their liking. As a prototype to memory cells, MOS capacitor devices containing the ex-situ produced, identical (constant size), spherical, 1 nm Si nanoparticles were fabricated at MIT (by a joint team from MIT (Osama Nayfeh and Demetri Antoniadis) and from the University of Illinois at Urbana-Champaign (Kevin Mantey and Munir Nayfeh) for use in future flash EEPROM devices, without changes to the standard CMOS fabrication process. With constant size 1nm particles, we demonstrated the characteristics of ideal nanomemory devices, opening the way for integration of Si nanoparticles in standard Si CMOS platform, with potential scaling down to 1 nm.

Freescale demonstrated a 4-megabit flash memory device as early as 2003 and has most recently (2006) demonstrated a 24-megabit flash memory device using silicon nanoparticle materials. Moreover, at least 5 of the top 10 flash memory suppliers as well as several embedded memory suppliers and significant number of universities and labs have ongoing projects to demonstrate nanocrystal memory. The impetus is flash memory that is smaller, faster, and lower in cost. Despite these strides, successful commercial products have not been realized so far. For instance, experts forecast that devices may only become available as a technology for the future 45 nm technology node or beyond, while Freescale projected that production of smaller in size and lower in cost commercial products will be available by the end of 2008.

Speedy commercialization has been hampered by the unavailability of commercially viable, high quality nanoparticles. Creating the missing link to make the technology commercially viable is made possible by the discovery at the University of Illinois of uniform (i.e., high-quality) silicon nanoparticles. It is the key to the technology, and the reason the technology will seize the memory market. Uniformity is Holy Grail- the long-sought breakthrough. But not far behind are the requirements of low cost, no toxicity, and mass production/scalability. These pivotal conditions, which our technology meets could potentially make nanoparticle memory, finally, invade our homes.

The use of highly improved nanoscale silicon materials can realize a new generation of commercial flash memory that will be third the size of conventional flash devices and could cost much less. In addition, the technology would make it possible to affordably embed flash and logic devices on the same chip, a move that would save space and energy and improve chip speed.

There are several challenges. Advanced nanomemory device concepts put stringent conditions on the quality of the particles and the matrix in which they are embedded. The Holy Grail is manufacturing particles that are:

- highly monodispersed/identical or reproducible in size and shape
- uniformly distributed in the device without touching each other
- ultrasmall in size to allow extreme down scaling
- commercially viable amounts
- low cost
- silicon technology (CMOS) compatibility

Industrial prototypes have been developed using silicon nanoparticle technology by several groups and companies. Important firms such as Intel, Freescale, Micron, Samsung, and STMicroelectronics are beginning to settle on new technology platforms for the post-Flash era and are finding ovonic and nanocrystalline/nanoparticle memories increasingly to their liking.

There have been several strides in the manufacturing of prototypes but to date, there is no significant advance in the commercialization of nanosilicon memory structure devices. We mention here the highlights:

- Motorola demonstrated a 4-megabit flash device using the silicon nanoparticles (2003)
- Freescale demonstrated a 24-megabit flash device using silicon nanoparticles (2006)
- IBM demonstrated a flash memory using an array of silicon nanocrystals (2003)
- 5 of the top 10 flash memory suppliers: At least 5 of the top 10 flash memory suppliers as well as several embedded memory suppliers and significant number of universities and labs are seeking demonstrate nanocrystal memory, each with an aggressive ongoing project

There have been limitations of realization of commercial Si-based memory. Nanomemory has made a slow start with several promised product launches that have failed to materialize. Freescale announced delay of plans to produce commercial products to the end of 2008. In general, nanocrystal/nanoparticle storage is being forecast by some as a technology for the 45 nm technology node or beyond. One major reason for the delay of commercial products is the unavailability to manufacturers of commercially viable, reproducible quantities of nanoparticles of the highest quality. For instance, current device commercialization is hampered by particle size nonuniformity, low yield, distribution nonuniformity, impact on the matrix quality, or on CMOS compatibility.

At the same time there are other competing technologies. Flash nanoparticle could have some stiff competition from other new nonvolatile memory technologies. By 2010 NanoMarkets predicts that four key segments will have emerged in the nano-enabled memory markets: MRAM, Ovonic, Holographic and Nanocrystalline.

MRAM (magnetoresistive random access memory) chip, based on a type of memory that could eventually replace both the high-speed memory in a computer and flash memory in cell phones and digital cameras. MRAM stores information not by storing electrons but by changing the magnetic state of a material. MRAM promises a high-capacity next-generation memory that can replace SRAM/Flash combos and battery-backed up RAM, as well as supplying improved non-volatile memory solutions for high-end mobile products. MRAM is already sampling and there are as many as 20 firms actively pursuing this opportunity.

MRAM could even prove a threat to the disk-drive industry. A large bank of low-power, fast, non-volatile MRAM memories could hold the computer's system and applications software in addition to all data, enabling an "instant on" notebook that could last all day on a single battery.

Holographic or phase-change memory uses lasers to switch a material between crystalline and amorphous states. This technology could scale down to the "5-to-10-nanometer range," a fraction the size of today's memory elements. NanoMarkets also believes that holographic memory is likely to be a prime contender for both high-end data storage and consumer video media markets in the not-too-distant-future.

Metal particles could also replace the polysilicon gates in flash memory. Nanotube technology may replace on-chip SRAM in L2 cache and would considerably reduce the power consumption of today's CPUs.

It is interesting to present here some prototype results generated by a collaboration of MIT and UIUC. The development is based on a breakthrough technology, which was developed at the University of Illinois and NanoSi Advanced Technologies that provides silicon nanoparticles of the highest technical and commercial quality available today. The procedure produces *ex-situ silicon nanoparticle flash technology*. This nanotechnology replaces the solid silicon gate with a large number of tiny silicon crystals/particles separated by minute amounts of insulation. In this configuration, a defect in the insulation would let charge escape from only a couple of neighboring nanocrystals/particles, leaving most of the stored charge intact. We believe it meets all the conditions necessary for a successful commercial product. As a prototype to memory cells, MOS capacitor devices containing the ex-situ produced, identical (constant size), spherical, 1 nm Si nanoparticles were fabricated at MIT (by a joint team from MIT and UIUC) for use in future flash EEPROM devices, without changes to the standard CMOS fabrication process. Sub monolayers of particles were dispersed on a tunnel oxide by spin-coating followed by capping with a low-temperature deposited oxide.

High-frequency CV measurements were performed on devices fabricated with varying Si nanoparticle density. The measured hysteresis loops and programming characteristics, exhibit sharp single threshold, substantial voltage shift, and large charge retention. The rise and shape, and absence of substructure in the CV agree with similarly constructed control samples and with an ideal simulated device that assumes no interface states. This result suggests the fact that charging is due to charging of the nanoparticle bulk and not to charging of interface states between the nanoparticles and adjacent SiO_2, or at the other oxide interfaces.

The charge retention in several devices was tested. The study showed that full charge loss should occur in ~1 year. But this can be easily increased to many years by increasing the thickness of the bottom oxide to 7 nm instead of 4 nm. The endurance of the MOS memory device was also measured, which showed only slight decrease after 10^5 cycles.

With constant size 1 nm particles, we demonstrated the characteristics of ideal nanomemory devices, opening the way for integration of Si nanoparticles in standard Si CMOS platform, with potential scaling down to 1 nm. See chapter for the improved features.

14.5 SILICON QUANTUM COMPUTING [30]

The concept of traditional computing depends on the motion of electrons in circuits, which necessarily produce heat and loss of energy. This is unlike the concept of quantum computing which does not involve motion of electrons, rather involves properties of stationary electrons. The relevant property is intrinsic quantum mechanical properties, namely its spin. Thus quantum-based computing is expected to be faster and with less Joule energy (heat) losses. The classical and quantum computers may also be contrasted as follows. In classical computers, data are rendered as binary bits, which are always in one of two states: 0 or 1. In quantum computers, however, a quantum bit can exist in both of these states at once, a condition known as a superposition. This allows many computations to be performed in parallel, making the operation much faster.

Quantum computers were practically introduced by Daniel Loss and David P. DiVincenzo in 1997 [31]. The concept of the Loss–DiVincenzo spin quantum computer (or spin-qubit quantum computer) may be introduced via Fig. 14.24. It consists of a double quantum dot (nanoparticle) separated by ~100 nm. An electron spin is assigned to each quantum dot. Each electron spin S_L or S_R define one quantum two-level system, or *qubit*. Moreover, the two electrons form a collective overall quantum spin state. A narrow gate (electric

FIGURE 14.24 The concept of the Loss–DiVincenzo spin quantum computer (or spin-qubit quantum computer).

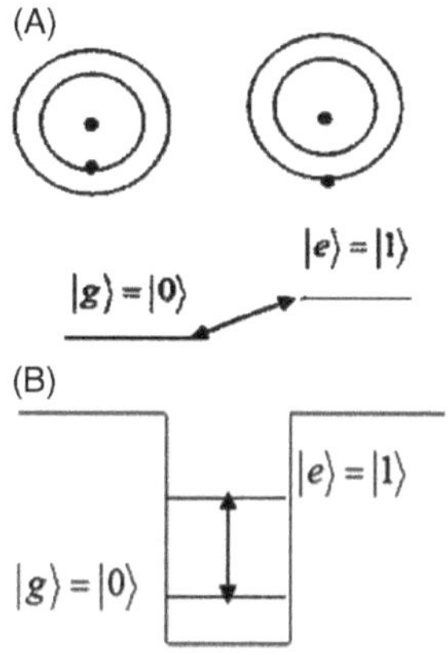

FIGURE 14.25 (A) Internal states of an atom i.e. ground stat |g> and excited state |e> as the qubits |o> and |1>. (B) Energy levels in artificial atoms (superconducting circuits or spins in solids (quantum dots or nanoparticles)) as qubits for quantum computers.

contact point) is placed between the two dots to apply an electric field to affect the two electrons. Because electrons are charged, the two electrons can interact with each other through Coulomb forces, exchange forces, or dipolar forces, while the collective overall spin state of both electrons is manipulated by an external magnetic or electric field. The gate field can for example modulate the coupling, allowing swap operations.

There are several schemes for quantum computing that have been employed or proposed in quantum computing. They utilize different elements from which electron spin is derived (qubits). The elements include:

- natural atoms (neutral atoms or ions): internal states of an atom, that is, ground state |g> and excited state |e> for showing the qubits |o> and |1> show in Fig. 14.25A
- artificial atoms (superconducting circuits or spins in solids (quantum dots or nanoparticles)): energy levels quantum dots (Fig. 14.25B) are mimics of those in atoms and can be used in the main structure of quantum computers
- nuclear spin [32]

This is displayed by observing the example depicted in Fig. 14.26. It shows a diamond lattice in which there is a pair of adjacent defects, namely, a nitrogen atom substituting for a carbon, together with a vacancy (a missing atom). It is a structurally NV center. Each NV center is composed of at least one nuclear spin, and an electron complex with total spin 1. Under a suitable driving field, the electron spins will constitute a channel by which the nuclei in neighboring NV centers can interact, thus achieving entanglement.

- Phosphorus impurities in silicon (both electron and nuclear spins) (see Fig. 14.27A).

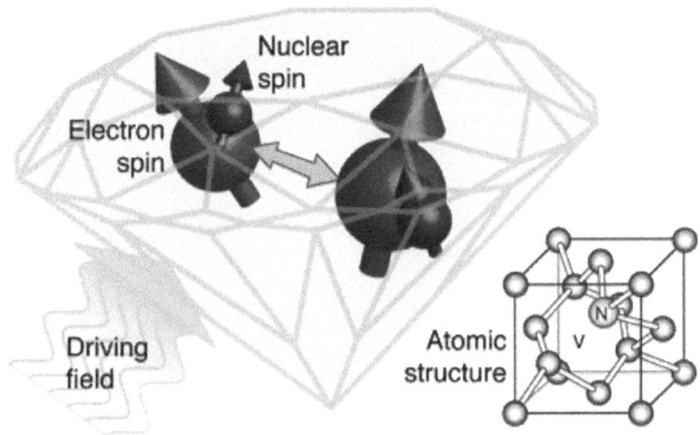

FIGURE 14.26 **NV center: a diamond lattice with a pair of adjacent defects, namely, a nitrogen atom substituting for a carbon, together with a vacancy (a missing atom).**

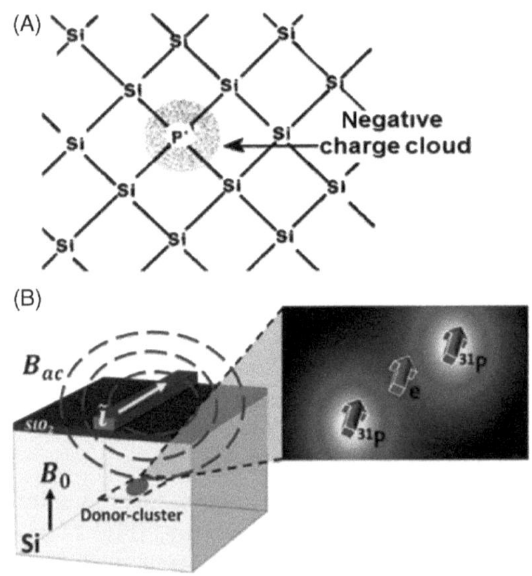

FIGURE 14.27 (A) A schematic of a donor dot qubit device in silicon [33] (collaboration of Purdue University and UNSW). (B) A simplified schematic of a donor dot qubit.

A schematic of a donor dot qubit device in silicon with ESR measurements using an ac-magnetic field (B_{ac}) is shown in Fig. 14.27B [33] (collaboration of Purdue University and UNSW). The figure shows a simplified schematic of a donor dot qubit with a microwave ESR line. The donor dot (shown as the red disk in the dashed rectangular region) is patterned into the silicon substrate (the pink box) by STM lithography. A static magnetic field (B_0) is applied to the system to polarize the electron spins bound to the dot. The metal wire (the blue bar) deposited on the SiO_2 layer carries an alternating current that can generate an electromagnetic field (B_{ac} represented by the blue dashed ellipses) to rotate the qubit electron spin with microwave frequency. Spin readout can be performed by spin dependent tunneling to a single-electron-transistor. The potential landscape of a 2P donor-dot with one bound electron (2P:1e), where light color represents high magnitude and dark color represents low magnitude. The arrows indicate the nuclear and electronic spins. The enlargement shows an electron spin bound by the quantum confinement of a 2P donor dot containing two nuclear spins. The background depicts its electrostatic potential landscape. The spins are coupled by the electron-nucleus hyperfine interaction which depends on both the nuclear spin orientation and the electronic probability density at the site of the nuclear spins.

- Photons have also been considered [34].

A photon qubit of information may be constructed using the horizontal (|H>) and vertical (|V>) polarization of a single photon (shown in Fig. 14.28A). An arbitrary state of a qubit can be represented on the Bloch sphere (known as the Poincaré sphere in optics). Examples of diagonal (D), anti-diagonal (A), right circular (R), and left circular (L) are shown in Fig. 14.28B. http://www.nature.com/nphoton/journal/v3/n12/fig_tab/nphoton.2009.229_F1.html

14.5.1 Spin Coherence

Spin coherence is characterized by two time scales T1 and T2. T1 is the amplitude damping. It measures the loss of energy. T2 is the phase coherence time, or the phase damping, which is more relevant for quantum computing. T1 constitutes a limit to T2. Under interacting or non-isolated spin conditions T2 < T1; otherwise, T2 approach T1. One pivotal problem of quantum

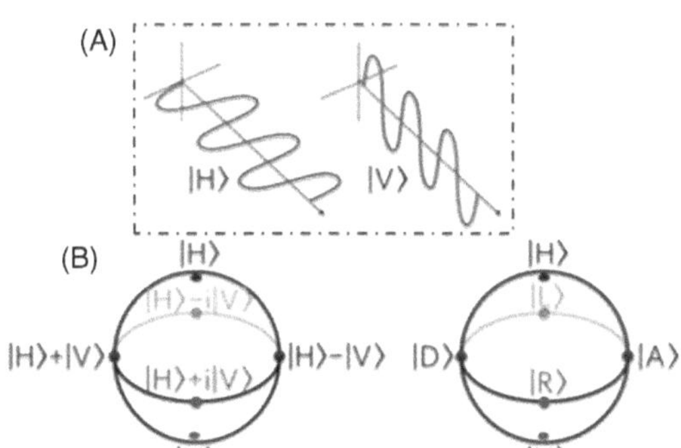

FIGURE 14.28 A photon qubit (A) information may be constructed using the horizontal (|H>) and vertical (|V>) polarization of a single photon. (B) An arbitrary state of a qubit can be represented on the Bloch sphere (known as the Poincaré sphere in optics). Examples of diagonal (D), anti-diagonal (A), right circular (R) ,and left circular (L).

computation is the degree of stability (coherence) of the spin of electrons under standard conditions. It is highly delicate and sensitive to thermal agitation (temperature), as well as to the forces that may be exerted on it from the environment especially if it is located in an atom or solid for example. This causes the spin configuration or direction to become strongly randomized and hard to precisely control it.

14.5.2 Spin in Silicon or Phosphorus Impurity Donors in Silicon

Spin in silicon or phosphorus impurity donors in silicon (P:Si) have the longest coherence time. Therefore, they are considered the leading candidates for quantum computers, as well as other spin transport-based spintronics applications. A candidate for such a quantum computer is a lateral quantum dot system. Therefore a successful candidate must afford identification of well-defined spins (qubits); reliable state preparation; low de-coherence; accurate quantum gate operations and strong quantum measurements.

The coherence time may be increased by external agents, such as strain. In the case of phosphorous, some believe that applying a local strain opens the way for control of spins on individual phosphorous donors or in quantum dots (nanoparticles), as shown in Fig. 14.29 [35]. The figure is a schematic diagram showing the effect of strain on a crystal donor. On the left there is no voltage applied to the actuator (yellow), and as such the crystal is unstrained (the electron donor wave function is depicted in red). On the right we show what happens when a potential is applied to the actuator. It strains the crystal and alters the shape and energy of the wave function (blue). The strain brings the electron spin (long arrow) into resonance with an appropriately tuned microwave field.

Electronic and nuclear spin of dopants in macroscopic samples of bulk silicon have recently been shown to have exceptional coherence or stability

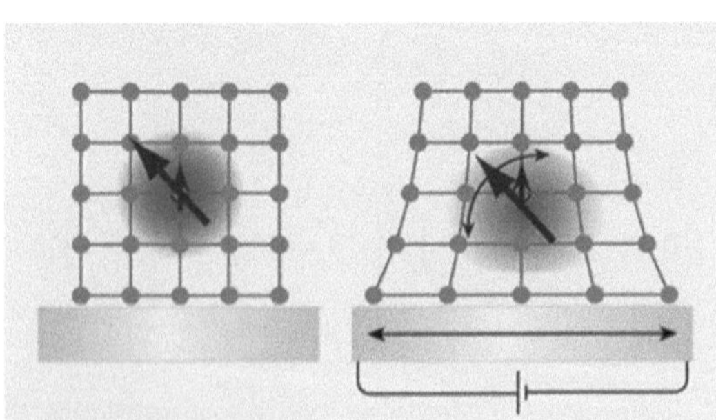

FIGURE 14.29 Local strain enabled-control of spins on individual phosphorous donors or in quantum dots (nanoparticles).

properties. It is therefore necessary to understand the physical behavior of dopants as functional elements in a nanostructured environment envisioned for nanoelectronics and quantum computation. is a schematic of FinFET device [36–38]. A phosphorous impurity has been implanted in the n+ silicon layer. The device is used to study the properties of the spin (qubit) by studying control of the wavefunction of a single donor bound electron. Such studies aim at enabling, for example, tunable hyperfine coupling between the donor electron and nuclear spin as envisioned in the Kane proposal [39]. Such problems are of central importance to both for pair-wise (mutual) entanglement, as well as coherent transport of spin in current proposals for quantum computation. is a schematic of two-qubit system of proximate donor spins, which under control of external gates A and J can be used to entangle the qubits. (B) Measured real-space coupled-hole transition density using cryogenic scanning tunneling spectroscopy. Hole spin qubits are of interest because of the possibility to perform qubit operations using dynamical electric fields and because of the relative insensitivity of their coherence to the spin of the host silicon nuclei. Rich pseudospin physics is imparted by the complex valence band structure.

14.5.2.1 Spin Coherence Time is Limited by Several Basic Factors

- For an atomic electron, the interaction of the electron spin with the nucleus's magnetic field is one decoherence factor. The magnetic field produced by the nucleus can be produced by its own spin (hyperfine effect) or the relativistic magnetic field brought about by the motion of the electron in the Coulomb field of the nuclear charge (spin orbit coupling – SOC) [40]. Silicon-based systems are the leading candidates for quantum computation because Si has the weakest SOC and its hyperfine effect is eliminated by using the isotope ^{28}Si, which has no nuclear spin.
- For silicon another decoherence challenge remains, namely the presence of conditions that relax the spin blockade (Pauli Exclusion Principle), such as spatial electronic energy degeneracies [41]. Because of its cubic symmetry, Si has sixfold orbital degeneracy at the first excited state at the bottom of the conduction band (valley degeneracy), which lifts the Pauli spin blockade. Lifting the cubic symmetry, hence valley degeneracy is achieved by two processes: mechanical straining and nanoscale spatial quantum confinement.

In pure silicon, lifting the cubic symmetry, hence the valley degeneracy is achieved using three processes: mechanical straining, nanoscale spatial quantum confinement, and internal E-fields due to interfaces [42]. Straining is achieved by growing the Si structure on a material having a different lattice constant. This is similar to the procedure used to create a strained silicon channel in strained transistor [4–6]. Typical designs are limited to 10–20-nm thick films or 6–10-nm quantum dots grown on SiGe substrates, providing moderate

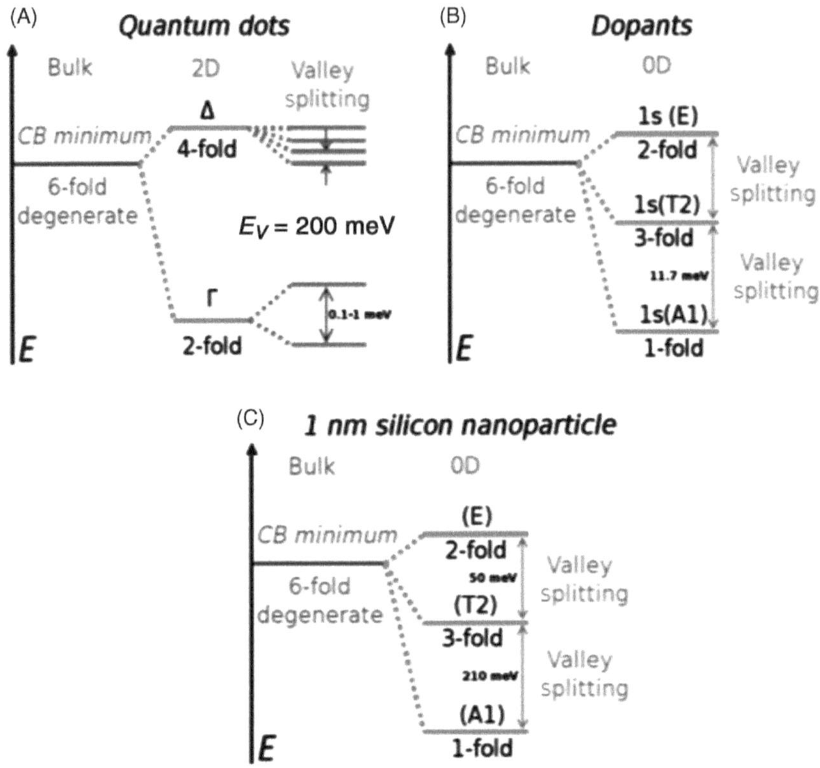

FIGURE 14.30 The energy diagram of the Si valley splitting with lifting the cubic symmetry, hence the valley degeneracy (A) Si quantum dots, (B) phosphor donors, (C) 1 nm silicon nanoparticle.

strain and confinement. Thus a small valley splitting <1 meV is achieved, which limits coherence to milliseconds and operation to cryogenic temperatures. Nevertheless, an electron's spin in silicon has the longest coherence time among semiconductors. In the (P:Si) donor system, the impurity breaks the cubic symmetry by locally subjecting Si to spherical symmetry, providing larger valley splitting (11.7 meV), which allows operation at near room temperature. So far the donor system in pure silicon isotopes at 2 K provides coherence time T2 of 0.6 s having a theoretical limit of T1 = 0.6 hours24, dropping to milliseconds at as low as 15 K [43]. The energy diagram of the valley splitting is depicted in Fig. 14.30A–B.

First, principle atomistic calculations were performed. The 1-nm particle is small enough to carry out such simulations. In the simulations, all configurations from $Si_{29}H_{24}$ (fully re-constructed) to $Si_{29}H_{36}$ (bulk-like non-constructed) were probed. The excited state potential energy surface was calculated at the time-dependent density functional theory (TD-DFT) level using the B3LYP functional with the TURBO-MOLE quantum computational package. The TZVP basis was used which is a triple split valence basis with polarization functions added for each atom. In the simulation, a dimer bond is stretched; then the nuclei are relaxed to minimize energy.

Second, principle atomistic calculations of the 1-nm particle were performed. The energy of the lowest unoccupied molecular orbital at the minimum of the conduction band (LUMO) splits into three valley states; those are non-degenerate *A*1-state, a double-degenerate *E*-state and a triple-degenerate T_2-state, with energies of ~3.29, 3.5, and 3.55 eV respectively relative to the top of the valence band (ground state) (Fig. 14.30C) [44], pointing to a large 210-meV splitting. It is known that in ultrasmall nanoparticles, where translational symmetry is significantly broken, such multiplicity of state manifold arises. In previous calculations, valley splitting was found to correlate with size reduction.

Silicon has the potential of providing very long spin coherence time because of weak spin-orbit interaction. Silicon, however, suffers from decoherence caused by the six-fold degeneracies resulting from the cubic symmetry that relaxes the Pauli Exclusion Principle. Present designs in quantum computing utilize straining, quantum confinement, or impurity centers to break the symmetry, providing moderate degeneracy lifting and singlet-triplet (S-T) splitting.

14.5.2.2 Silicon Thin Film Transistor mQubit [45]

In this application, instead of the use of quantum dots or nanoparticles, one uses the body of a silicon TFT. In other words the bits in existing silicon chips are reconfigured, and turned into qubits. Each TFT is used as a source for the spin.

Shows an array of thin film silicon transistors (TFT) as can be found in any display. In fact silicon processor chips in classical computers and smart-phones house millions of transistors and logic gates. A pair of those TFTs is shown in. Each of the TFT in the pair has a "spin," or magnetic field, indicated by the arrows and are labeled red and blue. The pair of spins constitute an electron qubit. It constitutes a two-qubit logic gate device. The device is the central building block of a quantum computer. The device is manipulated by metal electrodes on the surface of each TFT. This is essentially the same device technology as existing computer chips. As such it is expected to be much easier to manufacture a full-scale processor chip than to manufacture from scratch any other quantum computation technology. The ability to conduct one- and two-qubit calculations are required to realize quantum computers. Shows how an alternating electromagnetic *B* field rotates the qubit electron spin. A team from the University of New South Wales, Australia (UNSW) demonstrated the first ever silicon qubit of this type. Read more at: https://phys.org/news/2015-10-crucial-hurdle-quantum.html#jCp

Capacitance and Coulomb Blockade (Single-Electron Transistor)

The capacitance for a spherical nanoparticle of radius r = 10 nm is given by $C = 4\pi\varepsilon r$. For GaAs, $\varepsilon = 13.2\varepsilon_0$ which gives $C = 1.47 \times 10^{-17}$ F. When the nanoparticle receives a single electron, its potential changes by the amount V = e/C or = 0.011 = 11 mV. The energy needed for the electron to enter into the nanoparticle is $E = e^2/2C$. The effect of quantized current is called Coulomb staircase and the energy barrier preventing insertion of electrons is called Coulomb blockade. There are two criteria which need to be fulfilled for the staircase to appear: (1) stability against thermal fluctuations: the single-electron energy needs to be larger than the thermal fluctuations, otherwise thermal excitations can put in electrons washing away the effect, that is, $e^2/2C >> k_BT$. (2) Stability against quantum tunneling (with tunneling resistance R_T) to surrounding environment. The product of the particle energy and the capacitor charging time needs to be larger than the Planck constant h, otherwise the uncertainty principle will make atoms jump in and out of the particle at will. The capacitor charging time t is R_TC where R_T is the resistance of the tunnel junction between the particle and the environment, that is, Et = $(e^2/2C)\ R_TC >> h$ or $R_T >> h/e^2$. If the particle is placed between source and drain are directly connected to each other, this would constitute a single-electron transistor (SET) form of a field effect transistor (FET), the basic building block of modern electronics.

14.6 INTEGRATION OF SILICON WITH METAL

14.6.1 Core-Shell Silicon-Er (1.54 μm Emission)

Light emission from bulk Si is very inefficient due to its indirect electronic bandgap; hence one of the major challenges is the realization of efficient light sources using materials compatible with the Si technology. Silicon nanostructures are considered because they exhibit higher light emission efficiencies in the UV/visible regime due to quantum confinement. Another approach involves doping silicon with other materials, such as erbium. Er embedded in Si is considered for applications in the infrared regime as it emits at 1.54 μm, the absorption minimum wavelength in commercial optical fibers; hence this system has become a candidate for both optoelectronics and telecommunications. The Er system however suffers from several problems.

- Er (in isolation) has a relatively low excitation cross section ($\sim 10^{-21}$ cm^2) and narrow excitation bands because relevant electronic transitions are forbidden due to symmetry selection rules
- Er (in Er_2O_3 oxide environment) has enhanced absorption and emission because the crystal field of the oxide breaks the symmetry, making the otherwise forbidden infrared transitions allowed. Enhancement may also be achieved by incorporating C or N, which have higher electronegativity than Si, and thus change the charge distribution.

- Er near Si nanostructures has higher efficiency of emission due to resonant transfer between the nanocrystalline Si structures (strong wideband absorption) and the erbium center.

The integration of nano Si, Er, and O is therefore a means of enhancing the optical activity of erbium. Several morphologies have been developed to integrate the three components.

- Er was implanted in porous Si, a pore rich, undulated-wire-like nano-structured skeletal of Si. The structure prevents aggregation, and allows incorporation of a high degree of O or H on the surface due to large surface/volume ratio, as well as provides wide absorption bands.
- O and Er ions were co-implanted in nano grains of amorphous Si followed by annealing to create Er_2O_3 luminescent centers [46].
- Si and Er ions both were co-implanted in oxygen-rich host solids, such as SiO_2 followed by annealing to allow Si nanoparticle and Er_2O_3 centers to form.
- Creation of core–shell architecture with nano silicon core and Er_2O_3 shell architecture

The material can be excited by photons or electrons. Photoexcitation has been traditionally conducted with light sources extending from the UV to the visible, such the bands of an Hg mercury lamp. Excitation can also be performed using for example the Ar–Kr ion laser linesat 488 and 514.5 nm and those of HeCd laser lines at 325 or 480 nm in near resonance with Er-ion lines. On the other hand, excitation using low energy electron (electroluminescence EL) and using high energy electrons (cathodoluminescence CL) are required for integration on a Si chip.

After the implantation of Er and O in a silicon chip, the chip is first annealed at high temperature. The infrared spectrum is then recorded at lower cryogenic temperatures (liquid nitrogen temperature or lower) to reduce detection thermal noise effects. The CL shows much weaker spectrum [47].The emission was also recorded using laser excitation; the sample was prepared by annealing. The spectrum and emission intensity and resolution is much improved over what is obtained by electron excitation (CL-cathodoluminescence).

Core–shell architectures with nano silicon core and Er_2O_3 shell architecture were synthesized in the Nayfeh's group at the University of Illinois in collaboration with the Noha Elhalawany's group at the National Research Center in Cairo and the Ersin Bahceci group at Iskenderun University (Turkey). Nanoparticle suspensions are prepared in isopropyl alcohol (IPA) as was described in Section 9.9.2. In aqueous environment the particles form larger clusters due to hydrophobic forces. We use a solution of erbium salt ($ErCl_3$) in water as a source of metal Er^{3+} ions. In the procedure$ErCl_3$ solution is added to nanoparticle solution. The reaction results in core shell structures (see 11.16) in which the core is a silicon nanoparticle core and the shell architecture is Er_2O_3. Drop drying is used to form a thin film of the core–shell structures. The CL of films

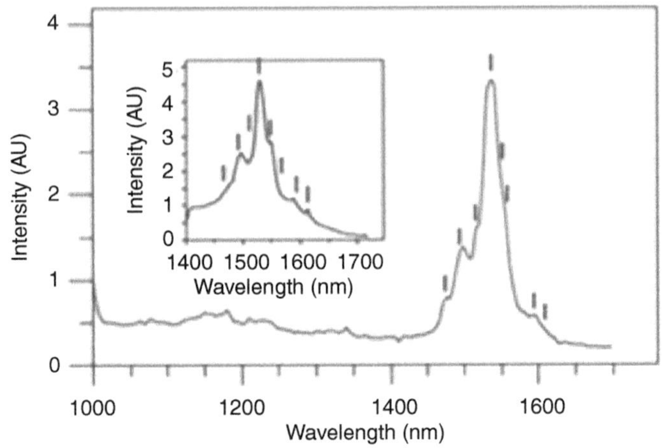

FIGURE 14.31 Core–shell silicon-Er (1.54 μm emission): strong emission in the 1543 nm region when excited by fast electrons (cathode luminescence).

was recorded at accelerating voltage of 25 kV and 200 nA. The temperature of the stage is set at 6 K and that of the sample at 10–30 K, while the infrared single element Ge detector is cooled to 77 K (liquid nitrogen) to reduce thermal noise effects, and has a spectral response covering up to 1800 nm with a maximum efficiency at 1550 nm and resolution of 4 nm.

Fig. 14.31 shows strong emission in the 1543 nm region. The figure shows that the 1540 band is split into several subcomponents at 1475, 1500, 1530, 1545, 1590 nm. The inset in the figure shows another spectrum taken from another region on the sample. This is possible since CL affords higher spatial resolution over EL because the diameter of an electron beam in CL is in nanometer scale ($\sim$100 nm). The spectrum in the inset resolves peaks at 1515 and 1610 nm.

Fig. 14.32A depicts a schematic of the energy scale of the nanoparticle and electron excitation, relaxation and energy transfer to the erbium ion. Fig. 14.32B gives a partial energy diagram of a single isolated Er^{3+} ion, showing the degenerate states: ground 4f $I^{15/2}$ and first and second excited states 4f $I^{13/2}$ and 4f $I^{11/2}$ respectively. Isolated atomic Er has the electronic configuration [Xe] $4f^{12}$ $6s^2$; thus Er^{3+} has 11f electrons with a configuration [Xe] $4f^{11}$. 4f to 4f transitions are dipole forbidden in isolated ions. Emission from those transitions, however are allowed due to a symmetry braking perturbation by the electric field of nearby oxygen atoms. The 1540 band represent transitions from the first excited state $4I^{13/2}$ to the ground state $4I^{15/2}$ of the Er^{3+} ion.

Detailed ab initio calculations were previously executed. It calculated the crystal electric caused by the presence of oxygen in the configuration shown in Fig. 14.32C. The results are shown in Fig. 14.32D. It shows how the transition is split by the crystal electric field produced by the presence of oxygen and it gives the wavelengths of the transition the transition component. The experimental results have very good agreement with the theoretical calculation.

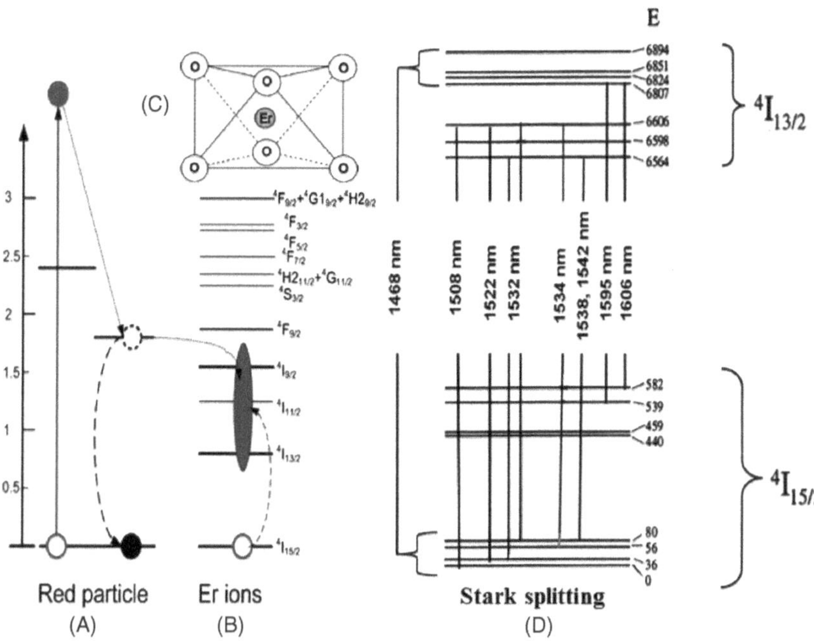

FIGURE 14.32 (A) Schematic of the energy scale of the nanoparticle and electron excitation, relaxation and energy transfer to the erbium ion. (B) A partial energy diagram of a single isolated Er^{3+} ion, (C) presence of oxygen in the configuration. (D) Transitions available due to splitting by the crystal electric field produced by the presence of oxygen

These results show that these core–shell architectures are amenable to electron excitation, and that they are very efficient emitters of infrared radiation. The two features are required for integration on a Si chip. Moreover, the emission is at 1.54 μm, which is the wavelength of absorption minimum in commercial optical fibers; hence, this system has become a candidate for both optoelectronics and telecommunications.

14.6.2 Noble Metal (Plasmonic) – Silicon Core Shell

There is current interest in integration of electronics and optics by integrating silicon with metals especially noble metals, such as gold and silver. Plasmonic fields within a few nanometers from the metal can approach atomic field strengths, which when applied to Si induces significant distortion in its crystal structure that breaks symmetry and modifies the electronic and optical properties, affording Si for example efficient visible photoluminescence, infrared antenna reception, or enhanced coupling to light.

Bulk semiconductors, especially silicon, form the backbone of modern electronics and computing. Bulk silicon however is a very dull material, being an especially poor emitter of light, turning added energy into heat. This makes integrating electronic and photonic circuits a challenge. There have been several proposals to alleviate this problem.

(i) One approach involves doping silicon with other materials. However, the emitted light is not in the visible rather in deep infrared. Moreover, the emission is not very efficient and can degrade the electronic properties of silicon.
(ii) Another approach is to use nanotechnology by making silicon devices that are very small, such as using luminescent nanoparticles, five nanometers in diameter or less as introduced above. As we have seen earlier, at that size quantum confinement effects allows the device to emit light. But making electrical connections at that scale is not currently feasible and may compromise the optical activity of the nanomaterial [48], as well as afford very low electrical conductivity.
(iii) Another scheme involves subjecting bulk silicon to concentrated fields of plasmonics. The high fields afforded by plasmonic nano metal material can cause distortion and modification of the crystal structure, hence the interaction with light.

We describe briefly an architecture that demonstrates some of these effects. First a pure silicon nanowire is wrapped with a coating of glass as shown in Fig. 14.33A. Then it is mounted on a glass substrate. Then it is coated with silver. Because of the glass substrate, silver does not wrap completely around the wire, effectively making the silver coating take the shape Ω. This leaves a narrow transparent glass window through which a laser beam can be sent in and emitted light can be extracted [49]. Practically, this narrow window did

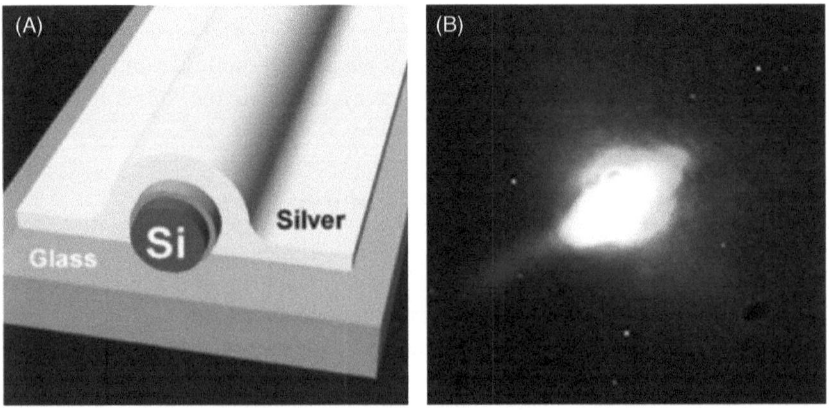

FIGURE 14.33 Schematic of a hybrid silicon–silver nanowire system. (A) Silicon nanowire wrapped with a glass coating followed by silver coating in the shape of Ω, on a glass substrate. (B) White light emission when a laser beam irradiates the system through the Ω window *(From C.-H. Cho, C.O. Aspetti, J. Park, R. Agarwal, Silicon coupled with plasma nanocavities generates bright visible hot luminescence, Nat. Photon. 7 (2013) 285–289. [49]).*

not compromise the silver coating from acting as a plasmonic cavity. When a narrow band blue laser beam is sent through the window, the silicon wire produces white light that spans the visible spectrum as shown in Fig. 14.33B. It is to be noted that bare silicon wires do not produce such white light when excited by the blue laser, indicating that plasmonic effects. The broad bandwidth light provides good potential for operation in photonic or optoelectronic devices.

REFERENCES

[1] C.-H Lin et al., "High performance 14 nm SOI FinFET CMOS technology with \$0.0174∼mu \$ m 2 embedded DRAM and 15 levels of Cu metallization", IEDM Tech. Dig., pp. 3.8.1-3.8.3, 2014.

[2] S. Anthony, Intel forges ahead to 10nm, will move away from silicon at 7nm, 2/23/2015, 9:31 AM https://arstechnica.com/gadgets/2015/02/intel-forges-ahead-to-10nm-will-move-away-from-silicon-at-7nm/; Moore's Law,http://www.mooreslaw.org/; https://www.technologyreview.com/s/601441/moores-law-is-dead-now-what/.

[3] Nayfeh OM, Chleirigh CN, Hennessy J, Gomez L, Hoyt JL, Antoniadis DA. Design of tunneling field-effect transistors using strained-silicon/strained-germanium type-II. Staggered heterojunctions. IEEE Electron Device Lett. 2008;29(9):1074–7.

[4] Strained silicon MOSFET technology, JL Hoyt, HM Nayfeh, S Eguchi, I Aberg, G Xia, T Drake, EA Fitzgerald, ..., Electron Devices Meeting, 2002. IEDM'02. International, 23-26; High performance 45-nm SOI technology with enhanced strain, porous low-k BEOL, and immersion lithography, S Narasimha, K Onishi, HM Nayfeh, A Waite, M Weybright, J Johnson, ...Electron Devices Meeting, 2006. IEDM'06. International, 1-4; W,-H. Lee, A. Waite, H. Nii, H. M. Nayfeh et. Al. "High performance 65 nm SOI technology with enhanced transistor strain and advanced-low- \$K\$ BEOL", IEDM Tech. Dig., pp. 56-59, 2005.

[5] Sun Y, Thompson SE, Nishida T. Physics of strain effects in semiconductors and metal-oxide-semiconductor field-effect transistors. J. Appl. Phys. 2007;101(10):104503. ISSN 0021-8979.

[6] Maiti CK, Chattopadhyay S, Bera LK. Strained-Si Heterostructure Field Effect Devices. NY: CRC Press; 2007.

[7] http://encyclopedia2.thefreedictionary.com/strained+silicon

[8] A. Nayfeh, C.O. Chui, K.C. Saraswat, T. Yonehara, Effects of hydrogen annealing on heteroepitaxial-Ge layers on Si: Surface roughness and electrical quality, Appl. Phys. Lett. 85 (14), 2815–2817; G.H. Dushaq, M. Rasras, A. Nayfeh, Germanium MOS capacitors grown on silicon using low temperature RF-PECVD, J. Phys. D.

[9] Wang J, Lee S. IBM Ge-Photodetectors for Si-Based Optoelectronic Integration. Sensors 2011;11(1):696–718.

[10] Nayfeh O, Rao S, Smith A, Therrien J, Nayfeh M. Thin Film silicon nanoparticle photodetector. IEEE Photon. Technol. Lett. 2004;16:1927.

[11] Does molecular electronics compute? Nat. Nanotechnol. 2013;8:377.

[12] Ohdomari I, Oshima M, Hiraki A. Control of Semiconductor Interfaces: Proceedings of the First International Symposium, on Control of Semiconductor Interfaces. Karuizawa, Japan: Elsevier; 1993.

[13] Desai SB, Madhvapathy SR, Sachid AB, Pablo JB, Wang Q, Ho GA, Pitner G, Kim MJ, Bokor J, Hu C, Philip HSW, Javey A. MoS2 transistors with 1-nanometer gate lengths. Science 2016;354(6308):99–102.

[14] S. Anthony, F. Sebastian, 7nm, 5nm, 3nm: The new materials and transistors that will take us to the limits of Moore's law. https://www.extremetech.com/computing/162376-7nm-5nm-3nm-the-new-materials-and-transistors-that-will-take-us-to-the-limits-of-moores-law
[15] Hong S, Auciello O, Wouters D. Emerging Non-Volatile Memories. NY: Springer; 2014.
[16] Nishi Y. Advances in Non-volatile Memory and Storage Technology. Amsterdam: Elsevier; 2014.
[17] Raghunathan S. Scaled Planar Floating-Gate N and Flash Memory Technology: Challenges and Novel Solutions. Stanford University; 2010. http://purl.stanford.edu/yy374yj9591.
[18] Dimitrakis Panagiotis. Charge-Trapping Non-Volatile Memories: Volume 1 – Basic and Advanced Devices. NY: Springer; 2015.
[19] Tiwari, Rana F, Hanafi H, Hartstein A, Crabbe EF, Chan K. A silicon nanocrystals based memory. Appl. Phys. Lett. 1996;68:1377.
[20] Nayfeh O, Antoniadis D, Mantey K, Nayfeh MH. Memory effects in MOS capacitors incorporating dispensed highly mono-disperse 1 nm Si nanoparticles. Appl. Phys. Lett. 2007;90:153105.
[21] Zhao C, Zhao CZ, Taylor S, Chalker PR. Review on non-volatile memory with high-k dielectrics: flash for generation beyond 32 nm. Materials 2014;7(7):5117–45.
[22] El-Atab N, Ozcan A, Alkis S, Okyay AK, Nayfeh A. ZnO low power zinc-oxide based charge trapping memory with embedded silicon nanoparticles via poole-frenkel hole emission. Appl. Phys. Lett. 2014;104:013112.
[23] Prime D, Paul S, Josephs-Franks PW. Gold nanoparticle charge trapping and relation to organic polymer memory devices. Phil. Trans. R. Soc. A 2009;367:4215–25.
[24] Dimitrakis Panagiotis. Charge-Trapping Non-Volatile Memories, vol. 2. NY: Springer; 2017.
[25] Hybrid dielectric non-volatile memory with nano particles (Si/SiO+ core/shell) as charge trapping layer US 9406765 B1.
[26] Panagiotis Dimitrakis Charge-Trapping Non-Volatile Memories Volume 2--Emerging Materials and Structures. Springer, NY; 2017.
[27] C. Scott, Duke University Develops 3D Printing Technique That Could Enable Flexible Digital Storage by, https://3dprint.com/170069/duke-university-3d-printing
[28] R. Dennard, IBM, United States Patent 3,387,286.
[29] What are Retention Time and Tests for SDRAM memory? SDRAM Technology. http://www.sdram-technology.info/retention-tests.html
[30] McMahon D. Quantum Computing Explained. NY: John Wiley & Sons; 2007.
[31] Loss D, DiVincenzo DP. Quantum computation with quantum dots. Phys. Rev. A 1997;57:120. (1998); on arXiv.org in Jan.
[32] Graphical representation of the qubit creation process using diamonds. Image courtesy of the American Physical Society.
[33] Wang Y, Chen C-Y, Klimeck G, Simmons MY, Rahman R. Characterizing Si:P quantum dot qubits with spin resonance techniques. Scientific Rep. 2016;6. 31830.
[34] Jeremy L. O'Brien, Akira Furusawa & Jelena Vučković, Nature Photonics 3, 687 - 695 (2009). doi:10.1038/nphoton.2009.229; N. Gisin and R. Thew, Nature Photon 1, 165-171 (2007); P. Kok, et al., Rev. Mod. Phys. 79, 135 (2007); O. Hosten, M.T. Rakher, J.T. Barreiro, N.A. Peters and P.G. Kwiat, Nature 439, 939 (2006).
[35] Dreher L, Hilker TA, Brandlmaier A, Goennenwein STB, Huebl H, Stutzmann M, Brandt MS. Electroelastic Hyperfine Tuning of Phosphorus Donors in Silicon. Phys. Rev. Lett. 2011;106:037601. https://physics.aps.org/articles/v4/6.
[36] Verduijn J, Tettamanzi GC, Rogge S. Wave Function Control over a Single Donor Atom. Nano Lett. 2013;13(4):1476.

[37] Verduijn J, Agundez RR, Blaauboer M, Rogge S. Non-local coupling of two donor-bound electrons. New J. Phys. 2013;15(3):033020.

[38] Mol JA, Salfi J, Miwa JA, Simmons MY, Rogge S. Interplay between quantum confinement and dielectric mismatch for ultra-shallow dopants. Accepted for publication in Phys. Rev. 2013;B7:245417.

[39] Kane BE. A silicon-based nuclear spin quantum computer. Nature 1998;393:133.

[40] Wilson DK, Feher G. Phys. Rev. 1961;124:1068. SOC.

[41] Pauli M, Li Q, Cywinski L, Culcer D, Hu X, Das SS. Phys. Rev. B 2010;81:085313.

[42] Zwanenburget FA, et al. Silicon quantum electronics. Rev. Mod. Phys. 2013;85.

[43] K. Saeedi, S. Simmons, J. Z. Salvail, P. Dluhy, H. Riemann, N.V. Abrosimov, P. Becker, H.-J. Pohl, J. J. L. Morton, M. L.W. Thewalt, Science 342, 830 (2013); A. Morello et al., Nature 467, 687 (2010); M. Raith, P. Stano and J. Fabian, Phys. Rev. B 83, 195318 (2011)

[44] Mantey K, Zhu A, Boparai J, Nayfeh M, Marsh C, Chaar Al G. Observation of linear solid-solid phase transformation in silicon nanoparticles. Phys. Rev. B 2012;85:085417.

[45] Veldhorst M, Yang CH, Hwang JCC, Huang W, Dehollain JP, Muhonen JT, Simmons S, Laucht A, Hudson FE, Itoh KM, Morello A, Dzurak AS. A two-qubit logic gate in silicon. Nature 2015;526:410–4.

[46] Kik PG, Brongersma ML, Polman A. Strong exciton-erbium coupling in Si nanocrystal-doped SiO_2. Appl. Phys. Lett. 2000;76:2325.

[47] Hoang T, Elhalawany N, Enders B, Bahceci E, Abuhassan L. Strong 1.54 μm cathodoluminescence from core-shell structures of silicon nanoparticles and erbium. Appl. Phys. Lett. 2016;109:261103.

[48] Hilliard JE, Nayfeh HM, Nayfeh MH. Re-establishment of photoluminescence in Cu quenched porous silicon by acid treatment. J. Appl. Phys. 1995;77:4130.

[49] Cho C-H, Aspetti CO, Park J, Agarwal R. Silicon coupled with plasma nanocavities generates bright visible hot luminescence. Nat. Photon. 2013;7:285–9.

Chapter 15

Biomedicine and Chemical Sensing

15.1 CONFINED LIGHT IN SERVICE OF SUBSTANCE DETECTION

Understanding interactions between strong light and matter is central to many fields. Deeper understanding of strong light coupling with matter is expected to afford the creation of advanced tailored applications [1]. The designation "strong field" applies to an external electromagnetic field that is sufficiently strong to cause significant alterations in atomic or molecular structure and dynamics of the material. Present-day laser technology as well as light enhancement due to plasmonic effects has made it possible to study the behavior of atoms and molecules in fields that have peak electric field strength of the order of atomic fields inside atoms or molecules. Under these conditions, even tightly-bound ground states must be greatly altered by the presence of the field [2].

Understanding interactions between light and nano matter is central to many fields, providing invaluable insights into the nature of matter as well as the nature of light. Indeed, greater understanding of light–matter coupling has enabled creation of tailored applications, resulting in a variety of devices such as microscopic lasers, switches, sensors, modulators, fuel and solar cells, and detectors. As discussed above, hybrid-plasmonic monolithic nanowire optical cavities highlight recent progress made in tailoring light–matter coupling strengths.

Detailed Maxwell's theory was used to calculate the enhancement in intensity of light and the corresponding electric fields of light in the proximity of gold nanoparticles [3]. The results for 10-nm nanoparticles are given in Fig. 15.1A using a gray/white color code for incident light of 1V/m electric field. The enhancement based on these theoretical estimations can be as large as a factor of 10, which provides enhancements of two orders of magnitude in intensity. Enhancements can be even more dramatic when there is more than one metal particle. Fig. 15.1B–C displays a schematic of the field lines between two particles as well as actual results using the gray/white color code respectively. It shows, in space between two closely placed plasmonic nanoparticles the electric field may be enhanced by nearly a factor of 30, providing nearly three orders of magnitude enhancement in intensity. Fig. 15.1D

Fundamentals and Applications of Nano Silicon in Plasmonics and Fullerines.
http://dx.doi.org/10.1016/B978-0-323-48057-4.00015-3

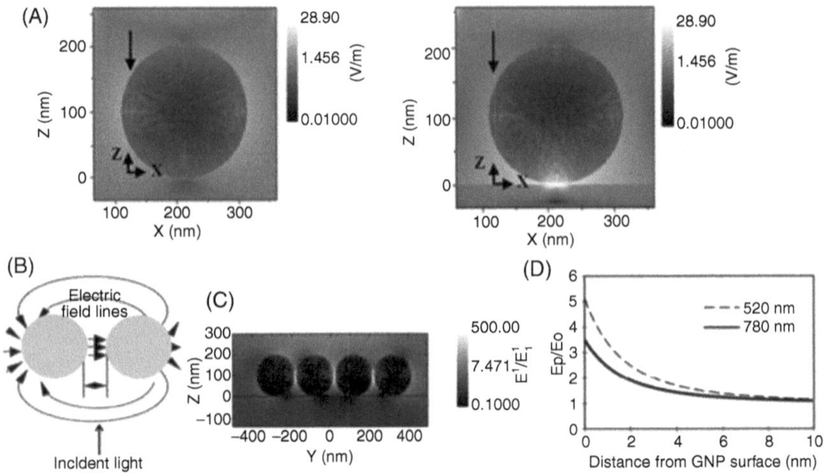

FIGURE 15.1 Detailed Maxwell's theory was used to calculate the intensity of light and the corresponding electric fields of light in the proximity of 10-nm gold nanoparticles. (A) near a single nanoparticles (B) schematic of the field lines between two particles as well as (C) actual results. Gray/white color code is used for incident light of 1V/m electric field (D) detailed dependence, as an example, of the electric field near a gold nanoparticle illuminated with light of two different wavelengths. *(From N. Nedyalkov, S. Imamova, P. Atanasov, M. Obara, Gold nanoparticles as nanoheaters and nanolenses in the processing of different substrate surfaces', J. Phys. Conf. Ser. 223 (2010) 012035. [3])*

gives detailed dependence on the wavelength of the light. As an example, the figure gives the electric field near a gold nanoparticle illuminated with light of two different wavelengths [3].

Thus the confined light to nano metal spaces is short range (dies out exponentially) and does not propagate to large distances. Because of the short range of the field, it interacts only with nanostructures next to it, within ~ 10-30 nm. Thus it provides spatial nano resolution. This opens up many applications, especially those optical processes or phenomena that depend very sensitively on the magnitude of the electric field, such as fluorescence, Raman scattering and infrared absorption, resulting in plasmon-enhanced fluorescence (PEF), surface-enhanced Raman scattering (SERS), and surface-enhanced infrared absorption spectroscopy (SEIAS).

The Raman scattering effect, for example depends on the fourth power of the field; thus they are enhanced if scattering is performed with the confined light near a metal nanoparticle rather than with ordinary propagating light. The Raman scattering effect is used for sensing and identification of substances. When a substance is irradiated with light, it scatters light called Raman light at a wavelength or frequency slightly shifted from the wavelength of the original irradiated light by an amount equal to the natural frequency of molecules that make up the substance. So, detecting the Raman light and analyzing its spectrum allows the identification of the substance. Since Raman light is usually

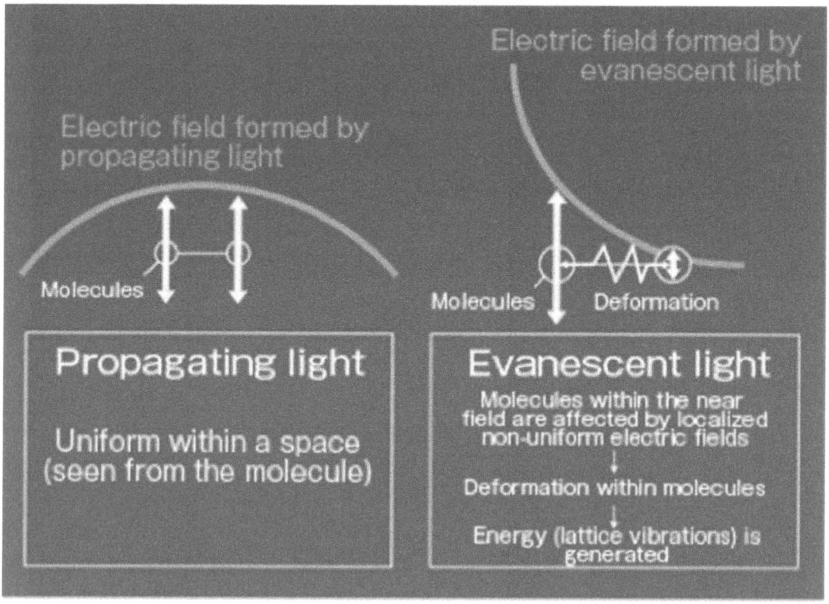

FIGURE 15.2 Schematic of a large molecule in the proximity of metal nanoparticle. It shows the profile of the plasmonic electric field. Different parts of the molecule experience different electric field strengths, introducing strong distortion within the molecules. *(From Nanophotonics, http://www.hamamatsu.com/eu/en/technology/innovation/nanophotonics/index.html. [4], Courtesy of Hamamatsu Photonics K.K)*

very weak, detecting its intensity directly is quite difficult. If the electrical field strength is increased 10 times by surface plasmon resonance, then the Raman light is intensified 104 times so the intensity becomes 10000 times higher.

Another interesting phenomenon of strong fields is the ability to optically manipulate the structure of molecules. Since the electric field in the vicinity of the plasmonic nanoparticles drops very rapidly over 10 nm, different parts of large molecules will experience different electric field strengths hence different electric stress. This causes distortion within the molecules and consequently causes new ways to deliver energy to the molecule including vibrations, non-linearities and electronic structure as shown in Fig. 15.2. The additional energy supplements the energy of the incident photon [4].

15.2 PLASMONICHYPERTHERMIC-BASED TREATMENT AND MONITORING OF ACUTE DISEASE

One serious problem inherit in plasmonics is energy loss due to heat generation resulting from strong absorption especially in the infrared – visible and UV spectral ranges [5–6]. So far, the losses are effectively too high to make practical plasmon-based electronic or photonic devices. However, these losses turned out to be a blessing for other applications, especially for cancer therapy (hypothermic-based treatment).

In hypothermic-based treatment, gold nano structures are implanted in a certain part of the body. The structures are then irradiated by infrared radiation to excite usurface electron in the gold. The excited surface plasmon electrons relax in two different ways:

- Surface electons can either emit the light absorbed. This happens at the same energy as the incident light. The process is known as surface plasmon resonance (SPR) scattering
- Surface electrons transfer the absorbed energy to the bulk of the metallic lattice in the form of thermal energy. This followed by cooling the hot lattice via phonon–phonon interactions. The heat is transferred to the surrounding medium. This forms the basis of all plasmonic nanoparticle-based photothermal therapy applications.

The process is strongly dependent on the shape of the nano structures. Two shapes are popular, one of which is spheres while the other is in the form of nanorods. The dependence on shape stems from the fact that the SPR band of nanoparticles contains both scattering and absorption components. Thus tuning the shapes and sizes of the nanoparticles can change their scattering and absorption properties dramatically. For example, when the aspect ratio of the AuNRs increases, the longitudinal band undergoes a red shift and whereas the scattering component increases, the absorption component of the longitudinal band decreases. Therefore, for NIR-imaging applications, larger AuNRs that have high scattering efficiency are preferred, and for photothermal applications, smaller AuNRs that have high absorption efficiency are preferred.

In this treatment, a near-infrared laser is used, which penetrates deep into the tissue, heating implanted nanoparticles to about 49 degrees Celsius as shown in Fig. 15.3A [7]. This is the temperature level which is needed to kill many targeted cancer cells. This results in a threefold increase in killing cancer cells and a substantial tumor reduction within 30 days. Current hypothermic techniques are not selective, i.e., they involve applying heat to the whole body, which heats up cancer cells and healthy tissue, alike. Thus, healthy tissue tends to get damaged. By using gold nanoparticles, which amplify the low energy heat source efficiently, cancer cells can be targeted better and heat damage to healthy tissues can be minimized.

Nanorodsare also used in hyperthermia treatment in cancer therapy. In that case, the nanorods are conjugated with some tumor targeting substance and ingested by the patient. Then, the patient is exposed to infrared light, which locally heats the tumor cells containing the nanorods. The goal is to destroy the cancerous tissue only. To highlight the heat effect, we present in Fig. 15.3B the 2D model of a laser activated gold nanorod. The temperature map after 2s irradiation with a 0.5 mW/cm2 diode laser.

It is interesting to compare the two shapes in a little bit more depth. We showed in Fig. 8.12A–B the surface plasmon oscillations in spherical gold nanoparticles (AuNPs) and gold nanorods (AuNRs). SPR frequency highly depends on the size and shape of the nanoparticles, as it was indicated before

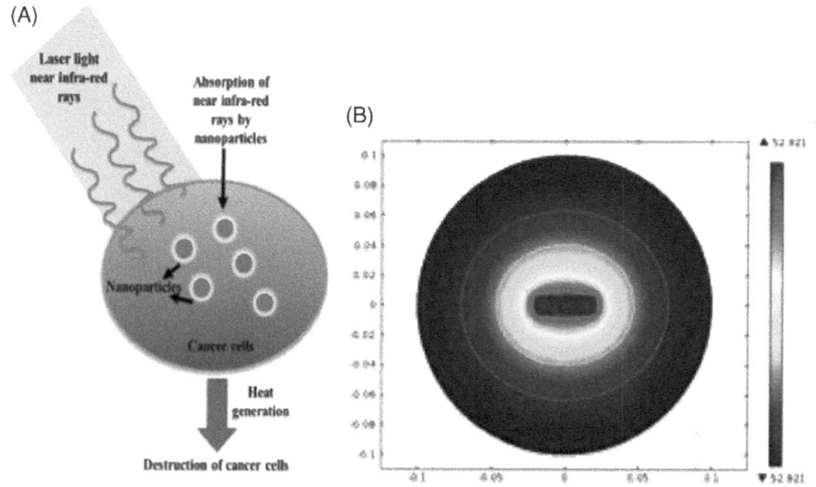

FIGURE 15.3 (A) Schematic of plasmonic nanoparticle-based photothermal therapy for cancer treatment. A non-consequential weak infrared light gets concentrated in gold nanoparticles, heating them enough to kill cells. *(From Javed Masood Khan, Md.Javed Equbal, Aqeel Ahmad, Nanotechnology as a Next Generation Therapeutics: Hope for Cancer Treatment In Modern Technology: Present and Future of Cancer, Dr. Muzafar Ahmad Macha (Eds.) OMICS International)* (B) 2D model of the temperature map of a gold nanorod heated by laser light (see Fig. 8.14). *(From F. Rossi, F. Ratto, R. Pini, Proceedings of the Comsol Conference 2012: "Laser Activated Gold Nanorods for the Photothermal Treatment of Cancer,"ISBN: 978-0-9839688-7-0 COMSOL Conference 2012 Milan Proceedings (2012))*

for the absorption maxima. When AuNPs absorb light, the oscillating electromagnetic field of the light triggers polarization of the conduction band electrons on the surface of the nanoparticles, and the polarized electrons go through collective coherent oscillations with respect to the positive ions in the metallic lattice; these oscillations are called as surface plasmon oscillations. Because surface plasmon oscillation has the same frequency as the incident light, it is also known as surface plasmon resonance (SPR).

For spherical AuNPs, there is only one SPR frequency, around 520 nm in the visible region, which is responsible for the intense red color of the spherical AuNPs. For gold nanorods (AuNRs), there are two SPR frequencies, known as the longitudinal and transverse bands. The transverse band is for electron oscillations that take place along the transverse direction and is a weak absorption band in the visible region similar to the SPR frequency of the spherical AuNPs. The longitudinal band is related to the electron oscillations along the longitudinal direction and is a strong absorption band in the vis-NIR region.

15.3 DRUG DELIVERY AND IMAGES

Gold nanoparticles can be ferromagnetic (like in a permanent magnet), paramagnetic or diamagnetic. The magnetism of the particles depends on exactly how they are prepared. But gold is only weakly magnetic under normal

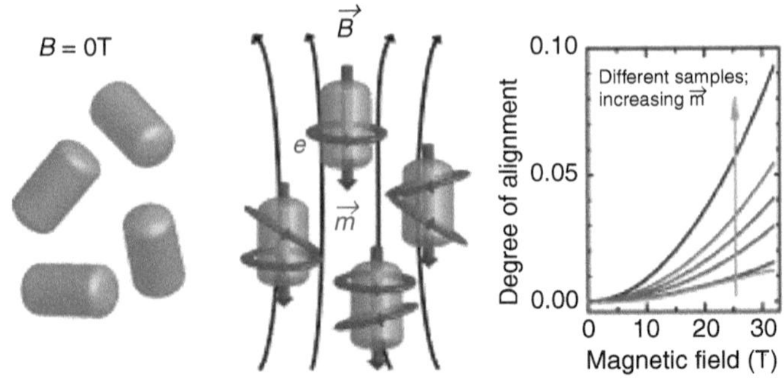

FIGURE 15.4 **Enhanced diamagnetism activity of gold in the nano regime.**

conditions (diamagnetic material). Gold and iron can be integrated in the configuration of core-shell structures to produced double functionality of plasmonic and magnetic functionlities. In the structures an iron core is rapped by a gold shell (Gold shell-iron core). The nanoparticles (Fe@Au) were prepared by reverse micelle method with aqueous reactants, surfactant, co-surfactant and oil phase. Magnetic measurements show that the particles are superparamagnetic at room temperature. In some tests, an anti-cancer drug (doxorubicin) was loaded onto the nanoparticles. The loading is achieved by connecting the amine (-NH2) group of DOX binds to the gold shell. The nanoparticles may now be manipulated magnetically, promising a convenient targeted drug delivery system to fight acute diseases including cancer [8] in a liquid. In an More recently, the weak diamagnetism known in bulk gold, was found to be stronger in nanogold. Fig. 15.4 shows that in the absence of an external magnetic field (B = 0) the nanorods are randomly oriented external magnetic field B as well as using optical excitation to affect the plasmon effects, gold mamorods can be made to align. The degree of alignment is recoreded using optical means as a function of the external field. The resulting magnetic moment m was found to be 14 times larger than what is known [9].

15.3.1 Molecular Signature of Death of Cancer Cells

Improvement of the sensitivity of Raman analysis due to the enhanced electric field in the proximity of gold nanoparticles (as discussed above in section 4.2) has recently been used to monitor and confirm death of cancer cells. The extra sensitivity allows following changes in the molecular content inside cells, including destruction or formation of molecules in cancerous cells during their death [6].

15.3.2 Containers for Drug Delivery

It has been shown that drug capsules or microcontaines can be produced using porous calcium carbonateparticles as templates. The particles are first incubated in a drug colloid, which results in impregnatingthe carbonates with the drug.

This is followed by encapsulation in a PEI polymer layer, and then, the calcium carbonate is dissolved in HCl. This results in a drug capsule. In section 12.5 it was shown that the same procedure can be used to encapsulate luminescent silicon nanoparticles [10]. Now using the same procedure to simultaneously impregnate a drug and Si nanoparticles in porous calcium carbonate templates would result in a capsule useful fortargeted drug delivery and controlled release. Fig. 15.5A shows schematics of two possible conficuration for incorporation of the silicon nanoparticles (marked red) in the capsule. The capsules in Fig. 15.5B show red luminescent particles, which are wrapped by the polymer. Microcontainers display red luminescence, characteristic of the nanoparticles, indicating the retaining of the Si nanoparticles in the interior of the polymer containers.The figure also shows that multiple layersof shell formation, following the encapsulation with PEI, produces luminescent microcapsules with enhanced luminescence that allows more sensitive detection and tracking. Optical tracking of microcontainersis easily achieved. This is achieved because the nanoparticles are luminescent.

Capsules may thenbe cracked open by ultrasound or other procedures upon delivery of the capsules to the point of interest as shown in Fig. 15.5C. Tracking of liquid flow upon breaking the microcontainer open is possible with sensitive detection. Electrical interaction is possible due to polarizability of Si nanoparticles.

As to safety issues, due to their ultrasmall size and spherical shape, Si nanoparticles are expected to flush more easily than larger, heavier, or needle-like

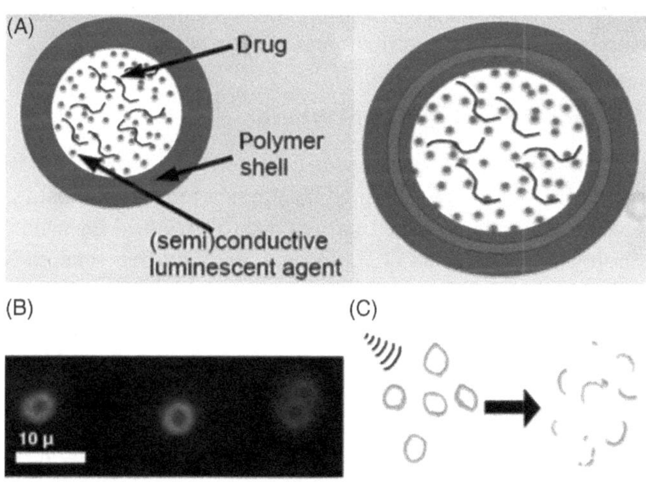

FIGURE 15.5 (A) schematic of drug capsule in the form of drug-polymer core-shell in which silicon nanoparticle markers may be incorporated in the core or the h shell or both. It shows the drug molecules, red luminescent particles, which are wrapped by the polymer. (B) Core shells synthesized with nanoparticles wrapped by a polymer layer. (C) Schematic of using ultrasound to break the shell for controlled delivery of the load

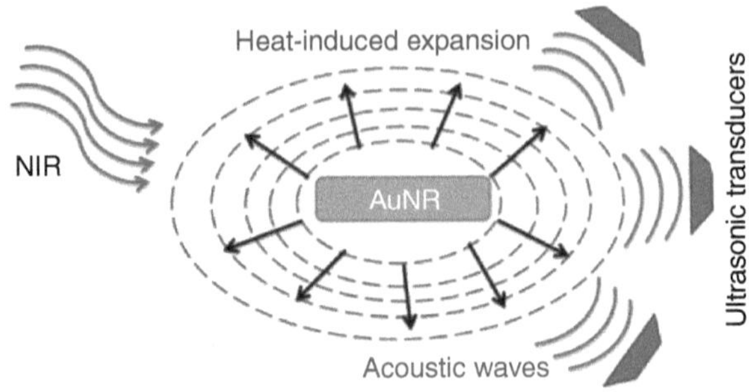

FIGURE 15.6 **The use of gild nanorods in hyperthermic application and imaging.**

nanomaterial. Also, Si nanoparticles form silicic acid (SiH4O4) when it interacts with body fluid in the body.

15.3.3 Photoacoustic Imaging of Cancer Cells

The use of AuNRsin hyperthermic application can also be used for imaging as explained in Fig. 15.6. When the rod is irradiated by a laser beam, sonic wavesare generated by the thermo-elastic expansion of the heated tissue. Typically nanordos are chosen with dimentions such that they have an absorption band in the range of 700–840 nm. The sonic waves are detected by ultrasonic transducers and converted to electrical signals, followed by processing those electrical signals to form an image [11].

This combination of properties makes gold nanoparticles useful for applications within physics (plasmonics), biology (electron microscopy) and medicine (drug delivery and detection and treatment of diseases).

15.4 SI-BASED ELECTRODE SENSOR

Direct electrochemical detection has been explored for three decades for making stable and miniature sensors for biomedical applications. This sensing approach is adopted to replace enzyme-based sensors due to enzymes' intrinsic instability under physiological conditions [12]. To date, enzyme-free sensors have been made using noble metals, alloys of these metals, conducting polymers and carbon nanotubes. These sensors, while rendering device stability, still suffer from issues related to material selectivity and/or electrode poisoning due to adsorption of reaction intermediates. Presently, the enzyme, glucose oxidase (GOx), is used as the sensing element for glucose sensors.

Fig. 15.7 shows a high sensitivity sensor electrode using a thin film of silicon nanoparticles as the active material. Not only, nano silicon is one of the least toxic materials, but due to its highly efficient luminescence properties, it will enable dual amperometric/optical sensing [13].

The figure show the electrode being decorated by silicon nanoparticles. The preparation of electrode sensor starts from a metal or heavily doped

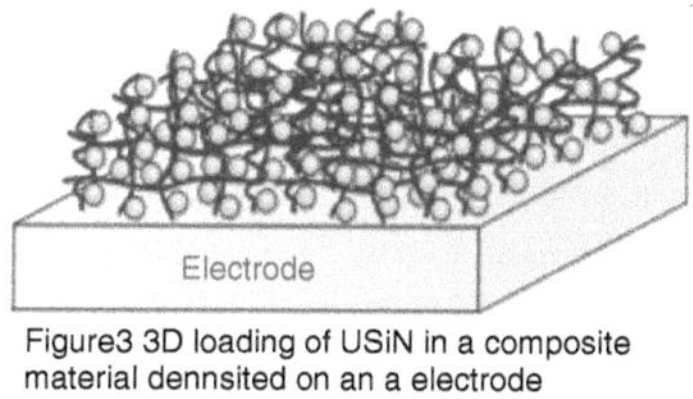

FIGURE 15.7 **High sensitivity sensor electrode for chemical substance using a thin film of silicon nanoparticles as the active material with dual amperometric and optical sensing.**

silicon electrode. The electodes are then loaded with silicon nanoparticles. Heavily doped (q < 0.005 Ù cm) n-type silicon wafers. The wafer with its surface containing the native oxide is cleaned with ethanol, isopropanol and de-ionized water. The wafer surface is then covered with a mask to achieve a working area of 1 mm × 1 mm. A drop of 0.1 ml of the Si29 colloid is spread on the wafer surface, and the sample is incubated for 10 h and then rinsed with de-ionized water. In conclusion, we demonstrated that highly doped silicon substrate or conducting substrates that are decorated with fluorescent silicon nanoparticles can be used as amperometric sensors for glucose. The device is more sensitive than GOx electrode with improved selectivity, stability, and ameanability to miniaturization and implantation for in vivo use. External light may be used to modulate the response, enabling phase sensitive detection.

15.5 METAL-BASED NANOPARTICLES

The use of iron oxide nanoparticles in biomedical applications is still a highly researched area. Moreover, Ag, Fe–Pt and TiO2, for possible use in biomedical applications. Ag nanoparticles show high electrical conductivity, high thermal conductivity, chemical stability, catalytic activity and enhanced Raman scattering and these properties have been used successfully in drug delivery and cancer therapy. Fe–Pt nanoparticles exhibit unique properties such as superparamagnetism, high coercivity, chemical stability, oxidation resistance and biocompatibility making them appealing materials for use in magnetic hyperthermia, MRI contrast agents, drug delivery and biosensors. Fe–Pt nanoparticles have been shown to lower *T2* relaxation times for MRI imaging and as successful drug carriers for targeted drug delivery. More research is needed for potential use in magnetic hyperthermia. TiO2 nanoparticles have many unique properties such as biocompatibility, chemical stability and photocatalytic properties which have led to an increase in the use of these nanoparticles in drug delivery, bio-imaging, photoablation therapy and biosensors. TiO2 nanoparticles were shown to be successful nanocarriers for drug delivery. The photocatalytic property of TiO2 nanoparticles has made it popular as photosensitiser agents for photodynamic therapy; however, more research is needed in this area.

REFERENCES

[1] Piccione B, Aspetti CO, Cho C-H, Agrawal R. Tailoring light–matter coupling in semiconductor and hybrid-plasmonic nanowires. Rep ProgPhys 2014;77:086401.

[2] (a) Nicholaides C, Nayfeh MH, and Clark CW, eds., Atoms in strong fields, Plenum (1989). (b) Glab W, Nayfeh MH. Stark induced resonances in the photoionization of hydrogen. Phys Rev. A 1985;31:530–2.

[3] Nedyalkov N, Imamova S, Atanasov P, Obara M. Gold nanoparticles as nanoheaters and nanolenses in the processing of different substrate surfaces. J. Phys. Conf. Ser. 2010;223:012035.

[4] Nanophotonics, http://www.hamamatsu.com/eu/en/technology/innovation/nanophotonics/index.html

[5] Kirui DK, et al. Targeted near-IR hybrid magnetic nanoparticles for in vivo cancer therapy and imaging, Nanomedicine: Nanotechnology. Exp. Biol. Med. 2013;9:702–11.

[6] Kang B, Mackey MA, El-Sayed MA. Nuclear Targeting of Gold Nanoparticles in Cancer Cells Induces DNA Damage, Causing Cytokinesis Arrest and Apoptosis. Journal of the American Chemical Society 2010;132:1517–9.

[7] Chen J, Keltner L, Christophersen J, Zheng F, Krouse M, et al. New technology for deep light distribution in tissue for phototherapy. Cancer J. 2002;8. 154-163.

[8] Kayal S, Ramanujan RV. Anti-cancer drug loaded iron-gold core-shell nanoparticles (Fe@Au) for magnetic drug targeting. Nanosci Nanotechnol J. Sep 2010;10(9):5527–39.

[9] High Field Magnet Laboratory (HFML) in Nijmegen and Leiden University: https://www.nwo-i.nl/en/fom-history/annual-reports/highlights/highlights2013/gigantic-diamagnetism-of-gold-nanorods-nr-78132/

[10] Maximenko Y, Gorin D, Ates H, Svenskaya Y, Elhalawany N, Yamani Z, Nayfeh MH. Layer-by-Layer Growth of Charged Polymers and Silicon Nanoparticles. BioNanoScience 2016;6(2):147–52. doi: 10.1007/s12668-015-0190-9.

[11] Yasun E, Kang H, Erdal H, Cansiz S, Ocsoy I, Huang Y-F, Tan W. Cancer cell sensing and therapy using affinity tag-conjugated gold nanorods. Interface Focus 2013;23. DOI:101098/rsfs.2013.0006, 2017, http://rsfs.royalsocietypublishing.org/content/royfocus/3/3/20130006.full.pdf.

[12] James Neal Craig. Fabrication and Investigation of an Enzyme-free, Nanoparticle-based Biosensor for Hydrogen Peroxide Determination. College of Engineering and Computer Science, University of Central Florida. Department of Material Science and Engineering. University of Central Florida; 2016.

[13] Wang G, Siu-Tung Y, Mantey K, Munir HN. Fluorescent Si nanoparticle-based electrode for sensing biomedical substances. Opt. Commun. 2008;281:1765.

Chapter 16

Nanoeffects in Ancient Technology and Art and in Space

16.1 METAL NANOPARTICLES (PLASMONICS) IN GLASS AND POTTERY

Glass is an amorphous form of silicon oxide; it is generally transparent, as well as hard and rigid at ordinary temperatures. In ancient times glass was made by melting a mixture of sand, potash, and lime.

16.1.1 Stained Glass

It was found that if metallic oxide powders are added during the melting process, the resulting glass would have specific colors depending on the metal powder used. Copper oxides impart green or bluish green color, cobalt imparts deep blue, and gold imparts wine red to the glass. Red glasses have been found in Italy that dated back to the late Bronze Age. In addition, gold nanoparticles were identified in some red tesserae of Roman times [1]. A vivid example is the well-known Roman Lycurgus Cup in glass dated from the 4th century CE and currently exhibited in the British Museum as shown in Fig. 16.1A [2], and the elaborate and complex multicolor stained glass window shown in Fig. 16.1B [3]. The phenomena stayed essentially unsolved till the last decade. It is now believed that the colors are attributed to white light excitation of plasmon surface modes of electron oscillations (see Chapter 8) in different metal nanoparticles incorporated during the manufacturing process [4,5].

16.1.2 Lusterware Pottery

In the Middle Ages, Abbasid and Fatimid Islamic potters were able to exploit the outstanding optical properties of metallic powders in more sophisticated ways. They were able as early as 836–883 CE to develop new chemical technology that allowed them to create complex nano metal-based multilayer structures. First they were incorporated in ceramics, which resulted in the emergence

Fundamentals and Applications of Nano Silicon in Plasmonics and Fullerines.
http://dx.doi.org/10.1016/B978-0-323-48057-4.00016-5

FIGURE 16.1 It is currently exhibited in the British Museum. Medieval stained glass windows (Courtesy: NanoBioNet and www.nano.gov) [2]

of lusterware, a special type of glazed ceramics, with striking optical effects. Second, the distribution of powders was programmed to include multilayers such that luster decoration color can change depending on the angle from which it is observed. An example of these types of ceramic decorations is given in Fig. 16.2 [6]. The color change is spectacular and brilliant. A variety of very intense colored metallic shine, including golden-yellow, blue, green, pink, etc., can be observed.

16.1.3 Gustav Mie Theory

Artists kept mixing silver and gold powders in glass or ceramics over many centuries to fabricate colorful glass or pottery; but it was not possible to produce a scientific explanation as to how these ingredients worked. The first step towards explaining the optical phenomenon had to wait until 1908 when Gustav Mie developed a theory of the optical properties of metallic particle colloids [7]. He showed that the color of a metal nanoparticle depends on its size, as well as on the optical properties of the precursor metal and the adjacent dielectric materials. But full understanding of stained glass and lusterware required even more time and 19–21th-century developments, including electromagnetic theory, electromagnetic wave nature, and interference/diffraction and scattering of light, solid state theory, plasma theory, nano scale phenomenon, advanced material characterization technologies, such as electron-based imaging and material analysis. In fact, it was only in the last two decades the scientific community has succeeded in solving the problem [8]. Those medieval artists or chemical industrialists were actually "nanotechnologists" synthesizing metal nanoparticles and

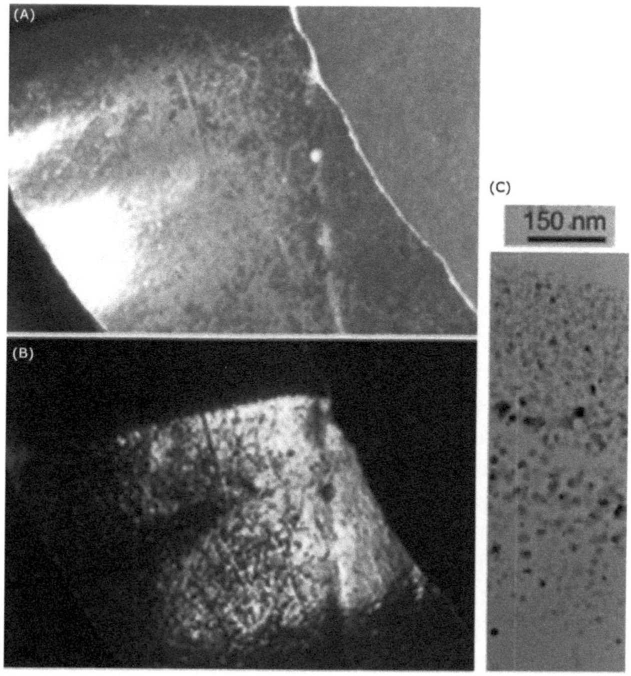

FIGURE 16.2 A 9th century CE lusterware from Mesopotamia (Susa). (A) Lusterware appears red. (B) It appears blue/green when viewed from different angle. (C) TEM image showing the presence of metal nanoparticle. (Micrographs, courtesy D. Chabanne).

harnessing what we today call plasmonics: a new field based on electron oscillations in metals called plasmon.

16.1.4 Red Stain

As was discussed in Section 7.3, the phenomenon can be easily understood in terms of color mixing effects. When white light shines only the G (green) component resonates with the electrons in the gold nanoparticles and is absorbed, hence removed. The B (blue) light is weakened by suffering scattering in all direction. The remaining R (red) component passes through. This is the reason that stained glass mixed with gold nanoparticles appear red to the naked eye.

16.1.5 Glaze Reflections

Studies conducted in 2005 concluded that the multi-layer features give rise to light interference phenomena and scattering through rough interfaces, which adds to the surface plasmon effect and strongly contributes to the observed color [9]. Moreover, several material analysis studies [10–12] confirmed that the color of the luster decorations come from metallic nanoparticles. For instance, the presence of metal particles was directly confirmed by conducting high resolution transmission electron microscopy (TEM) and imaging along with material analysis using electron energy loss spectroscopy (EELS). Generally, copper and silver clusters were incorporated by applying a mixture of a paint, which contained copper and silver salt powders onto a glazed ceramic. This was followed by annealing in a reducing atmosphere. It is now believed that the basic mechanism involves the two processes of ion exchange and crystallization (nucleation and crystal growth) of copper and silver metallic nanoparticles inside the glassy matrix [13].

The ingenious technology was way ahead of its time hence it was based on empirical chemical means, which made the technology vulnerable to extinction. In fact the technology has nowadays partially been lost. Vincent Reillon and collaborators [14] performed elaborate optical modeling, which confirmed the role of multilayer interference and interface scattering in the color of metallic shine reflection (specular direction). On the experimental side, modern artists [11] conducted experiments to re-create luster decorations. They used modern kilns, which allowed alternative oxidization (oxygen flux) and reduction (CO flux) phases during the firing [4]. The best recreations achieved exhibited partial multi-layer structure with a weaker organization.

Several material analysis studies [10-12] confirmed that the color of the luster decorations come from metallic nanoparticles. For instance, the presence of metal particles was directly confirmed by conducting high resolution TEM and imaging along with material analysis using EELS. Generally, copper and silver clusters were incorporated by applying a mixture of a paint, which contained copper and silver salt powders onto a glazed ceramic. This was followed by annealing in a reducing atmosphere. It is now believed that the basic mechanism involves the two processes of ion exchange and crystallization (nucleation and crystal growth) of copper and silver metallic nanoparticles inside the glassy matrix [13].

Fig. 16.3 shows a schematic of the light interference phenomenon as we understand it at this time. It is believed that the luster was made by creating a glaze by first applying copper and silver salts to clay pottery. The pottery is then baked at high temperatures in the presence of reducing chemical conditions. During the heating process metal ions from the salts migrate to the outer part of the glaze and are reduced (neutralized) to metals. At high temperature the metal atoms diffuse and agglomerate to produce nanoparticles. The overall process results in a double layer of metal nanoparticles with a small amount of glaze in between them as shown. When the cured pottery is exposed to light, it would penetrate and it will essentially suffer two light reflections. First, reflection occurs off the first layer. The light penetrating the first layer would reflect off the second layer of nanoparticles. Interference effects take place between the two reflected light components, creating a luster effect that results from both constructive and destructive interference.

The ingenious technology however was way ahead of its time, hence it was based on empirical chemical means, which made the technology vulnerable

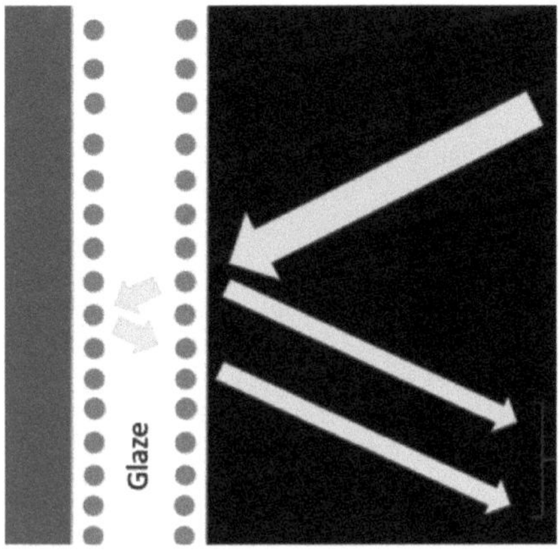

FIGURE 16.3 Glaze reflections from ancient pottery glaze: a schematic of the light interference phenomenon between a reflection from the front nanolayer of the glaze and from the back nanolayer of the glaze.

to extinction. In fact the technology has nowadays partially been lost. Vincent Reillon and collaborators [14] performed elaborate optical modeling, which confirmed the role of multilayer interference and interface scattering in the color of metallic shine reflection (specular direction). On the experimental side, modern artists [11] conducted experiments to re-create luster decorations. They used modern kilns, which allowed alternative oxidization (oxygen flux) and reduction (CO flux) phases during the firing [4]. The best recreations achieved exhibited partial multi-layer structure with a weaker organization.

16.2 CARBON NANOTUBE IN SWORDS

As seen earlier, the Abbasid and Fatimid era saw many fascinating advances in science and technology. The period witnessed an aggressive and stimulating advanced chemical and manufacturing technology of glass and pottery, exploiting advanced lighting, color, and optical effects of metal nano structures produced from mixed silver and gold powders. Not only the chemical industry was advanced, material, and metallurgical industry, including thermo-mechanical forging and annealing was sophisticated. Metal smiths subjected steel cakes, which were imported from ancient India to such protocols to refine the steel to its exceptional quality [15]. Domestic, as well as military products were produced. One of the products of the industry is the famous Damascus blades. Those were pivotal in the fight against the Crusaders. The blades were remarkably hard, sharp, and light weight [16].

The secret recipe was lost to time around 1750CE. The reasons are not entirely clear. In fact Peter Paufler and colleagues at Dresden's Technical University in Germany [17] claimed to unlock the secret (Fig. 16.4). They discovered carbon nanotubes in the microstructure of a 17th century Damascus sabre. In fact recent

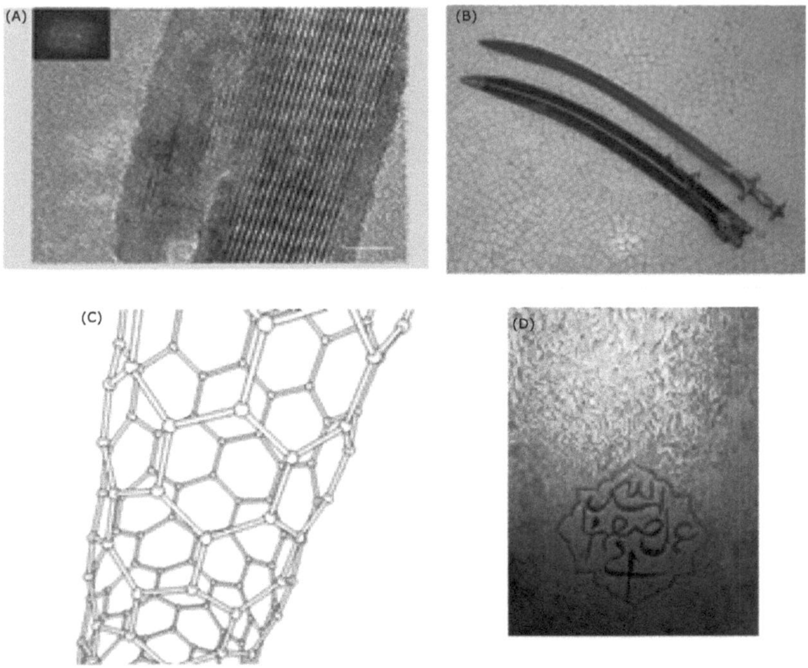

FIGURE 16.4 Carbon nanotubes in the microstructure of a 17th century Damascus sabre using high resolution transmission electron microscopy of samples of the blades.

high-resolution TEM of samples of the blades revealed the presence of carbon nanotube, which afforded the products exceptional material properties. The Damascus blades were forged from steel (*wootz*). TEM revealed carbon nanotube.

In modern times, scientists use an arc-evaporation procedure to make the nano tube [18]. A high current is passed between two graphite electrodes in an atmosphere of helium. This causes the graphite to vaporize, some of it condense on the cathode to form the carbon nanotubes. On another track iron is mixed with a carbon content between 0.02 % and 1.7 % by weight. The mixture is melted in a very hot furnace to produce an alloy. Carbon and other elements act as a hardening agent, preventing dislocations in the iron atom crystal lattice from sliding past one another. By varying the amount of alloying elements and their distribution in the iron one can control qualities, such as the hardness, elasticity, ductility, and tensile strength of the resulting steel [19].

16.3 NANOSILICON IN SPACE

16.3.1 Red and Blue Rectangle

Scientists have observed a mysterious red glow (and now also a blue one) emanating from the Milky Way and other galaxies that cannot be explained by any usual methods [20,21]. In fact an extended red emission (ERE) has been seen in a wide variety of dusty environments, ranging from the diffuse interstellar medium of our Galaxy, reflection nebulae, planetary nebulae, HII regions, to other galaxies. It has been generally attributed to photoluminescence (PL) by some component of interstellar grains, powered by ultraviolet (UV)/visible photons. According to NASA and the International Astronomy Union (IAU), the ERE feature is one of a number of spectral phenomena in the interstellar medium that still remains unexplained [20]. Interstellar spectra taken in the visible part of the spectrum over the last four decades have indeed shown the broad or extended emission features that cannot be explained by the sharp emission lines from atomic neutrals or ions.

While the broad structure less emission band typical of the ERE was thought to be limited to the wavelength range of 600–850 nm, recent observations by Vijh et al. [22] reported the discovery of blue luminescence; thus extending the range of the appearance of PL in astrophysical environments to the blue/UV [22]. The red and blue may exhibit different spatial distribution or they may originate from different objects. The red rectangle in which the red extended emission was first discovered has now been additionally identified with blue extended emission, which makes it more like pink as shown in Fig. 16.5.

Silicon in space has become a subject of great interest to the scientific community. In fact, a workshop was organized [23] on May 16–19, 2012 on "Silicon in Space" (see Fig. 16.6). It convened 30 experts in various fields (observation, theory, modeling, laboratory astrophysics, materials, and fundamental science) to address the state of the art of silicon-based (mostly nano- and micro-structured) materials (silicon, silica, silicates, etc.) in an astrophysical context.

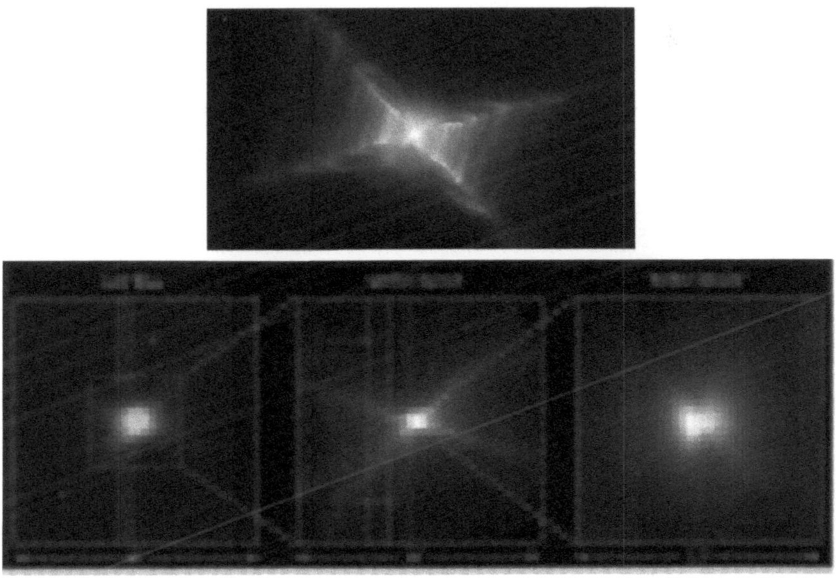

FIGURE 16.5 The red rectangle in which the red extended emission was first discovered. It has now been additionally identified with blue extended emission, which makes it more like pink

FIGURE 16.6 **Silicon in Space: workshop was organized on May 16–19, 2012 at the Villa Vigoni, (a lovely resort at the Lake Como, Italy) on "Silicon in Space." It was organized by the Laboratory Astrophysics and Cluster Physics Group of the Max Planck Institute for Astronomy at the Friedrich Schiller University Jena, Germany [23].**

16.3.2 Nanosilicon and Nanoaromatic

Observations of the polarized brightness in the coronal Fe XIII line at 1074.7 nm during the 2001 June 21 solar eclipse led to the suggestions that the observation could be interpreted as being due to the presence of silicon nano-particles near the sun [24]. The presence of the nano-particles is suggested to explain a tangentially polarized emission, extending radially from a low-temperature and low-density coronal hole with a different radial slope than the coronal background. A predominantly radial polarization direction is observed in the rest of the corona where the high-electron densities obscure the dust signal.

16.3.2.1 Silicon Nanoparticle Carriers

On the basis of experimental data, it was suggested that crystalline silicon nanoparticles with 1.5 nm–5 m diameters as the carrier, that is, cause the ERE, the broad emission band within the 0.54 and 0.95 μm spectral range that is seen in dust-rich objects, such as reflection nebulae as well as in the diffuse interstellar medium [25–27]. The silicon nano-particle hypothesis is also indirectly supported by observation of pre-solar diamond nano-particles in laboratory analysis of meteorites [28,29].

The data not only show that the nanoparticles could provide a close match to the observed ERE spectra, they also satisfy the quantum efficiency requirement. It was estimated that the nanoparticles account for 5% of the total interstellar dust mass and $\approx$6 Si per 10^6 H atoms. Modeling [26] the interstellar extinction curve taking SNPs as an interstellar dust component gave $\approx$18 Si per 10^6 H atoms, or $\sim$50% of solar Si abundance.

Recently, there has been extensive research both observation of interstellar medium and solar corona during eclipses in the visible spectrum, as well as in the infrared and laboratory measurements, as well as simulations to study these phenomena. The eventual goal is to identify a nano silicon-based carrier with

FIGURE 16.7 Silicon nanoparticles (A) prototype mode of 1 nm diameter can be assigned a molecular structure Si29H24. (B) Laboratory measurements showed the availability of a discreet set of clusters ranging from 1 to 3 nm in size. 1 nm clusters exhibit blue emission, while 3 nm particles exhibit red emission. (C) High-quality commercial quantities of the nanoparticles with repeatable structure and characteristics are produced in the laboratory.

a (i) plausible mechanism for formation that (ii) exhibits the optical characteristics of the background while (iii) maintaining brightness and withstanding harsh radiation without being destroyed. It was argued that nano silicon-based is a good candidate carrier of the mystery background that exhibits the required attributes. The carrier is a hydrogenated silicon nano cluster or super molecule.

Laboratory measurements showed the availability of a discreet set of clusters ranging from 1 to 3 nm in size. One nanometer clusters exhibit blue emission, while 3 nm particles exhibit red emission shown in Fig. 16.6. High-quality commercial quantities of the nanoparticles with repeatable structure and characteristics are produced in the laboratory (Fig. 16.7).

16.3.2.2 Spectra

Analysis of the spectra has shown that ultra-small silicon grains can account for the spectral content of the extended red and blue emission. The red luminescence of the 3-nm carrier with excitation at a wavelength 365 nm is given in Fig. 16.8A. The blue luminescence of the 1-nm carrier with excitation at a wavelength 365 nm is given in Fig. 16.8B.

Additional PL spectra with excitation wavelength from 250–350 nm were recorded. The spectra were recorded on a photon counting spectrofluorometer with a Xe arc-lamp light source and 4 nm bandpass excitation and emission monochromaters. Selected emission spectra from one of the samples are shown

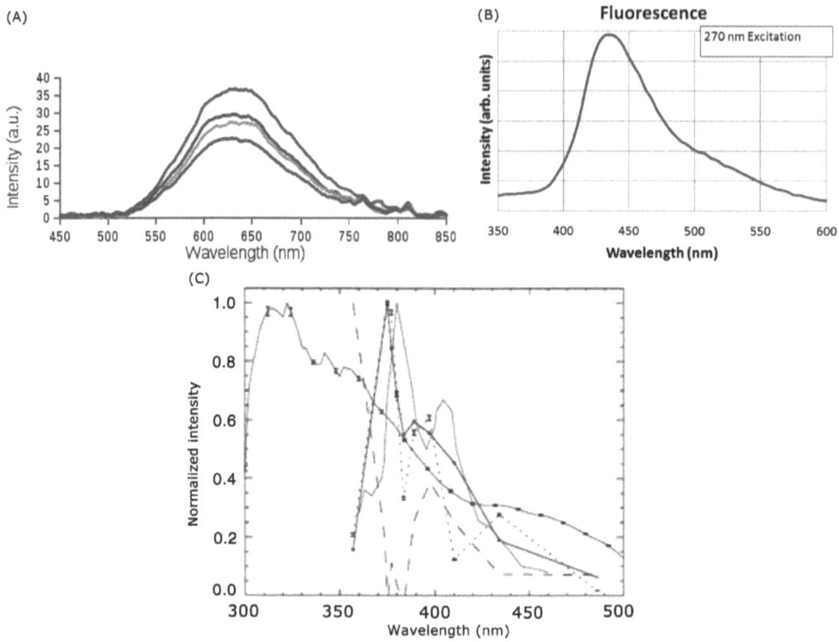

FIGURE 16.8 (A) The red luminescence of the 3-nm carrier with excitation at a wavelength 365 nm. (B) The blue luminescence of the 1-nm carrier with excitation at a wavelength 365 nm. (C) (Solid, Dot and Dash) Luminescence spectra taken by Vijh et al. [22] in three positions in the nebula south of the central star HD 44179 in the Red Rectangle. (*Blue*) the sum of photoluminescence response of 1 nm silicon nanoparticles under several excitation wavelengths covering the UV-visible range 250–440 nm. (*Red*) luminescence spectrum for anthracene.

in Fig. 16.8C, with each spectrum originating from a different excitation wavelength [30]. Curves a–g correspond to emission-line profile spectra for a typical sample with corresponding excitation wavelengths of 250, 260, 270, 280, 290, 310, 320, 335, 340, and 350 nm, respectively. Distinct emission bands are observed centered at 310, 360, 390, and 440 nm. The response of the silicon nanoparticles in the range 120–255 nm has not been measured because of propagation problems under ambient conditions. However, from the trend seen in Fig. 16.8C, it is plausible that the luminescence due to this portion of the spectrum lies mostly in the range below 310 nm, outside the domain of the reported observation of the blue luminescence. Assuming a flat spectrum for the illuminating star in the range 254–390 nm, the 10 selected responses in this range can then be summed. The result of the sum is shown as a blue curve in Fig. 16.8C [27].

16.3.2.3 Stability

It has also been established that these molecular like species are rugged and do not dim, and are not disintegrated by the harsh radiation or ionization

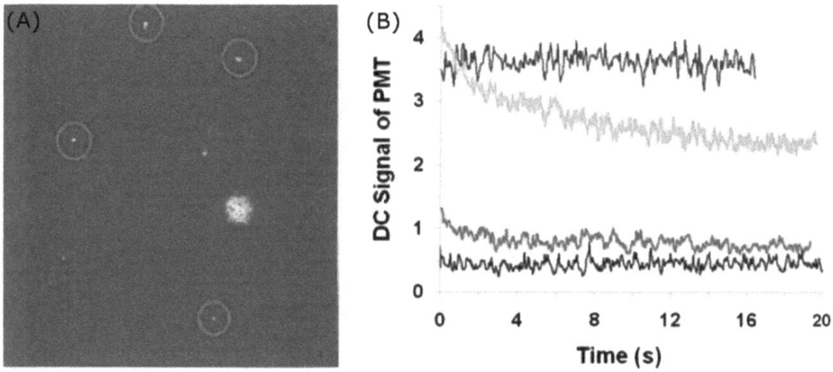

FIGURE 16.9 Stability: (A) 1 nm nanoparticles were immobilized in agarose gel at a very low density. (B) The emission (in a.u.) over the first 80 s of interaction. The laser beam was parked on stationary particles. The excitation power was 10 mW.

environment of the galaxies. Laboratory experiments were conducted [31]. First, 1 nm nanoparticles were immobilized in agarose gel at a very low density as shown in Fig. 16.9A. The laser beam was parked on stationary particles. The excitation power was 10 mW. The emission (in a.u.) over the first 80s of interaction is given in Fig. 16.9B. For later times, the emission (not shown here) continues to be at the level reached at the 80 s mark. Apart from for a brief period at the beginning of the interaction, the particle emission reaches a steady bright level. At higher incident intensities, the gel shows softening, and damage.

A small amount of the particles were also dispersed on a quartz substrate [31]. Fig. 16.10A shows a 10 × 10 μm^2 image of the substrate taken under femtosecond pulsed illumination at incident intensity of 30 mW (outside the microscope) as shown in Fig. 16.10A. The particle image shows luminescent

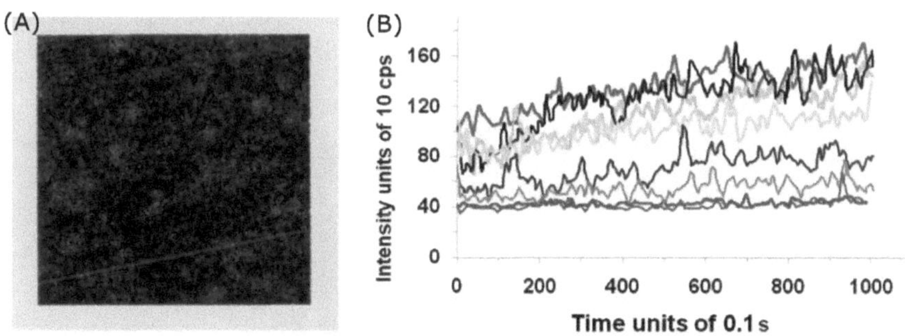

FIGURE 16.10 (A) A 10 × 10 μm^2 image of the substrate with a small number of particles dispersed on a quartz substrate, taken under femtosecond pulsed illumination at incident intensity of 30 mW (outside the microscope). The particle image shows luminescent particle sites. (B) The emission (in a.u.) over the first 100 s of interaction with the beam parked on the stationary particles at an excitation power of 10 and 50 mW, demonstrating the stability of the emission.

particle sites. The beam was parked on the stationary particles at an excitation power of 10 and 50 mW. The emission (in a.u.) over the first 100 s of interaction, given in Fig. 16.10B, demonstrates the stability of the emission. But, for a brief period near the beginning of the interaction, the emission rises slowly to a stable level. The excitation beam was mechanically blocked for a few second during the steady state region but upon reapplication of the beam, the emission was recovered at the steady state level.

16.3.2.4 Transport by Magnetic Field

Their heavy ion version is optically active, and may be tied to and locked into paths controlled by the coronal magnetic field. It has been known or assumed that, in charged clusters, the charge distribution is uniformly distributed, with the particles having no dipole moment. Atomistic first principle theoretical calculations shows that the charge is not uniformly distributed over the charged 1 nm particle, with the particle having a permanent dipole moment. The dipole moment is significant; being ~2.5 Debye with the center of electron charge being displaced from center of the positive charge by ~1.25 A. Fig. 16.11A–C shows the charge distribution in the neutral, charged particle, and the direction of the dipole moment, respectively. The equilibrium coordinates of the Si atoms inside the particle before and after ionization are computed. Very little change is found in the bond lengths, but bond angles change to cause asymmetry. The particle swells by ~0.08 A, making the particle non-spherical. The electric dipole direction is in the direction of swelling indicating the charge distribution correlates with the shape of the particle [21].

The electronic charge distribution in the (a) neutral particle and (b) stripped particle ($Si_{29}H_{24}^{+}$). (c) The electric dipole moment in the stripped particle.

16.3.2.5 Imprinting Spectra

Sharp transitions of heavy ions may be imprinted on the extended emission of the grains as shown in Fig. 16.12. This has been demonstrated by first incubating a bundle of fiber glass in a small volume of Si nanoparticle colloid [32].

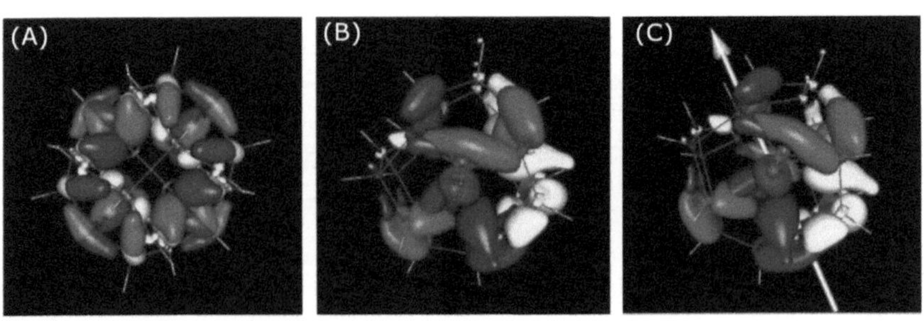

FIGURE 16.11 (A–C) The charge distribution in the neutral, charged 1-nm particle, and the direction of the dipole moment.

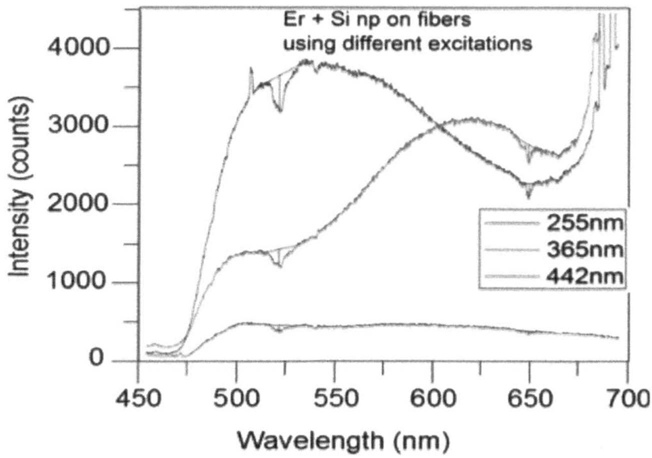

FIGURE 16.12 Imprinting sharp transitions of heavy ions on the ERE (extended red emission) using excitation at 255nm, 365 nm, and at 442 nm.

Upon drying under ambient conditions, the nanoparticles get drawn to the fibers, forming a thin nanofilm. The film is observable by the naked eye under UV irradiation with the luminescence spectrum showing the characteristic emission of Si nanoparticles. This step is followed by a similar incubation process of the coated fibers in an erbium chloride salt solution. The metal ions are drawn to the fibers by electrostatic attraction, forming a metal nanofilm coating over the nanoparticle film. Fig. 16.12 shows the luminescence spectrum taken with excitation using excitation at 255nm, 365 nm, and at 442 nm of the coated fibers along with the luminescent spectrum of the controlled sample of Si nanoparticles without Er. In addition to the red band characteristics of the nanoparticles, it exhibits absorption dips or holes burned at the absorption lines of erbium ions at 650 nm and 520 nm. Those sharp dips indicate resonant reabsorption.

16.3.2.6 Nanoparticles Under Pressure

Silicon nanoparticles grow brighter with extreme pressure and do not fragment (Fig. 16.13) [33]. 1-nm nanoparticles were dispersed in a PMMA matrix. A piston-cylinder-style diamond anvil cell to exert high-pressure on the nanoparticles. Pressures were determined using standard ruby fluorescence techniques. The absorption measurements were performed with a xenon arc or tungsten–halogen light source, Kratos monochromater, and a photomultiplier with photon counting. Fiber optics were used to direct the light from the monochromater to the pressure cell. Fig. 16.13A presents the absorption of the particles in PMMA as a function of the excitation photon energy in the range 370–286 nm or 3.35–4.34 eV under several values of pressure:atmosphere, 1.64, 3.36, 4.72, and 6.10 GPa. The absorption band is seen to slightly blue shift and widen at higher pressure. Fig. 16.12A also shows the luminescence spectrum under incident

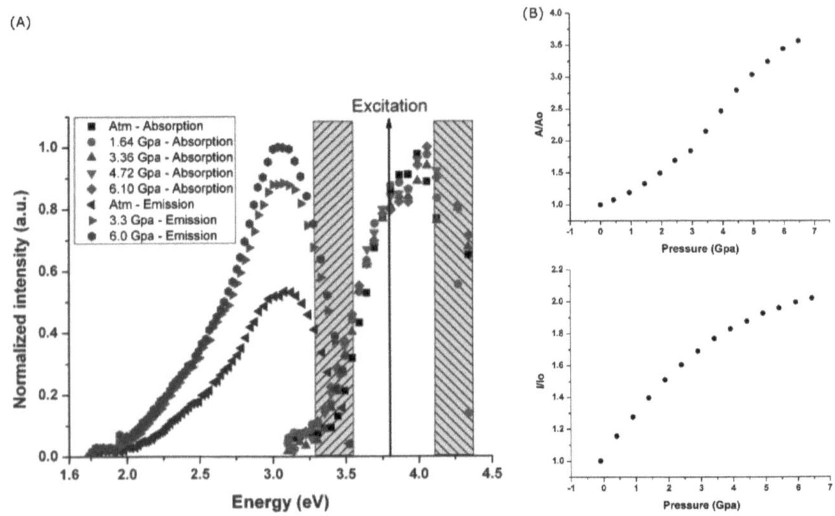

FIGURE 16.13 Si nanoparticles under extreme pressure. (A) The absorption and emission spectra of the particles in PMMA as a function of the excitation photon energy under several values of pressure: from atmosphere to 6.10 GPa. (B) Peak of absorption and emission bands as a function of pressure.

excitation of wavelength of 325 nm (3.8 eV), at the pressures:atmosphere, 3.36 and 6.10 GPa. The luminescence is dominated by a strong blue/ UV band. Note as disused in Chapter 9 that the broadening and the extended red tail in the emission are not homogeneous broadening resulting from a size distribution. Rather, it is inhomogeneous broadening due to the indirect nature of silicon.

Fig. 16.13B shows the absorption increasing gradually with pressure, increasing by a factor of ~3 over a pressure of 5 GPa. The peak of the luminescence intensity with pressure shown in Fig. 16.13B rises linearly by more than twofold, saturating at 6 GPa. As to the spectral distribution, we observe no significant change except for a slight blue shift of ~5 nm at the higher pressures. The observed gradual pressure dependence of the absorption and the luminescence is non-bulk like. Bulk-like absorption stays unchanged until the applied pressure reaches the threshold where sharp uptake takes place.

One can see that, ultrasmall 1-nm Si particles exhibit linear uptakes in the absorption and in the luminescence and with insignificant spectral change. The is unlike the behavior of bulk silicon which exhibits sharp threshold at ~11 GPa. The gradual absorption uptake and luminescence yield is consistent with silicon-metal transformation on the surface. The insignificant change in the spectral content of the luminescence points to surface stability for particles, which are smaller than the bulk fragmentation domain (20–50 nm). First-principles atomistic calculations yield absorption behavior that exhibits gradual uptake followed by sharp uptake at ~9–11 GPa. The results point to the conclusion that two-dimensional surface-like phase transformations are manifested by linear uptake in absorption and luminescence. In bulk single-crystal silicon, the semiconducting diamond-to-metallic ß-Sn phase transformation nucleates on defects, and is manifested by a sharp uptake in light absorption at a threshold of ~11 GPa, accompanied with the creation of nanosized (20–50 nm) fragmentation domains.

16.3.2.7 Production of Si Nanoparticles in Space

While significant advances has been made in this field, there are still more questions to be answered. One key question is to understand the mechanisms responsible for the production of the silicon grain carrier in the galaxies. In the laboratory, researchers use chemical etching of a silicon crystal in a mixture of hydrogen fluoride and hydrogen peroxide while being electrically charged. However, it remains doubtful whether this procedure could be mimicked in space. Interest has already moved on to look ahead at potential future revelations that utilizes a silicon rich hydrated silicate as precursor, which when subjected elevated temperature and pressure under certain conditions become favorable for the formation of those silicon rich grains [21].

16.3.2.8 Aromatic Complexes Carrier

It has been found that silicon nanoparticle is not the only candidate as a carrier for the extended emission: polycyclic aromatic hydrocarbons (PAHs) nano cluster or supermolecule also presents potential [20, 22]. It is a planar sheet, consisting of a small number of aromatic hexagon rings terminated with hydrogen. Stability is a major problem with molecular-like species when subjected to very harsh UV radiation, which is so strong that most known polyatomic interstellar molecules would be destroyed, turning into highly stripped atomic ions. It was found through simulations and experimentation that doubly ionizing PHAs makes them more rugged, and hence more viable. It was also noted that it is not straightforward to verify as they are a highly reactive charged species and are not readily accessible for laboratory study. The hypothesis is the enhanced likelihood for the existence of PAH molecules with less than 40 C atoms in regions not heavily exposed to UV radiation, such as in the case of the Red Rectangle nebula surrounding the star HD 44179.

16.3.2.9 Hydrogen in the Corona

Significant Lyman α, the finger print of hydrogen, is found to originate from the inner sun corona, implying the presence of a significant amount of hydrogen. Other lines of hydrogen are not observed, indicating that hydrogen in this region is not predominantly formed by recombination in electron–proton collision, which results in a cascade emission of multiple lines. As only a small fraction of hydrogen can survive in neutral form in the coronal temperature of millions of degrees, it is plausible that most of the neutral hydrogen or helium is due to direct entry or carried by nano grains from cooler interstellar regions, followed by resonance scattering of the sun's wide-band radiation at the $L\alpha$ by entering hydrogen. Neutrals can serve as local probes to elucidate the mechanisms for heating the solar atmosphere and accelerating the solar wind, and act as a proxy for protons in the solar wind, and may reflect the fate of cooler grains from solar, as well as interstellar origin. Resonance scattering of radiation at 121.6 nm, a well-known emission from hydrogen atoms, was very powerful in discovering the presence of significant amounts of neutral hydrogen in the inner corona and in the galaxies.

The synchrotron source at the Daresbury facility in the UK was used to record the response of mono-dispersed H-terminated 1-nm silicon nanoparticles to ionizing ultraviolet (VUV), with excitation energies covering the range of 4.5–22 eV. The measurements were taken by a collaboration including Shadia Habbal and Huw Morgan of the IFA at the University of Hawaii, University of Wales, Aberystwyth University (UK) and the Nayfeh's group at the University of Illinois. Thin films were formed on silicon wafers using simple drop evaporation starting from a suspension. The sample chamber has $<10^{-10}$ Torr at 10 K. The setup included a five meter normal incidence concave grating monochromator. Two interchangeable gratings cover the photon energy range of 5–35 eV (or, equivalently, 2500–350 A) with a best achievable resolution of 2 meV (0.05 A). A peak flux of 1.4×10^{12} photons/s has been measured at around 23 eV. Fig. 16.14A presents the intensity of the peak of blue band at 425 nm as a function of the energy of the incident photon (excitation spectrum). It is recorded at 10 K in the range 5–12.5 eV. It displays the trend in the strength of the luminescence with excitation energy. The spectrum has been normalized to the incident intensity. It shows that the luminescence intensity at 12.5 eV is sizable, being as high as ~18% of that for excitation at photon energy of 6.5 eV.The measurement shows minimum luminescence at 10.2 eV, the Lyman α wavelength. First principle atomistic theoretical simulations conducted using the time-dependent density functional theory (TDDFT)

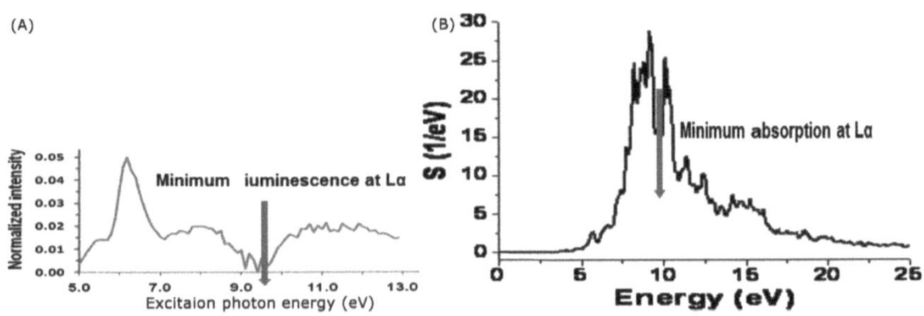

FIGURE 16.14 (A) The intensity of the peak of blue band at 425 nm as a function of the energy of the incident photon (excitation spectrum). It is recorded at 10 K. The measurement show minimum luminescence at 10.2 eV, the Lyman α wavelength. (B) First principle atomistic theoretical simulations of the absorption conducted using the time dependent density functional theory (TDDFT) show a minimum at 10.2 eV.

given in Fig. 16.14B show a minimum at 10.2 eV [34]. It is this resonance scattering by the hydrogen coating on the nano Si grains which strongly deflect s and hence protects the particles from the harsh Lyman α of the sun allowing safe entry and eventually release of hydrogen.

Observations of the polarized brightness in the coronal Fe XIII line at 1074.7 nm during the 2001 June 21 solar eclipse led to the suggestions that the observation could be interpreted as being due to the presence of silicon nano-particles near the sun [24]. The presence of the nano-particles is suggested to explain a tangentially polarized emission{2{extending radially from a low-temperature and low-density coronal hole with a different radial slope than the coronal background. A predominantly radial polarization direction is observed in the rest of the corona where the high electron densities obscure the dust signal.

The basis for this suggestion is provided by laboratory preparation of silicon nano-crystals and measurements of their optical properties [35]. Quantum confinement effects in the nano-particles cause a high photoluminescence (PL) quantum yield. Aside from PL in the visible regime where the ERE is observed, silicon nano-particles show PL in the infrared, namely for 3.5 nm sizes in the spectral band of the Fe XIII line. The fact that the emitted photons are unpolarized and that emission is restricted to a small spectral interval are major points of the silicon nano-particle suggestion to explain the 2001 eclipse observations. During the 1994 solar eclipse, in contrast, the observed radial slope of brightness in the spectral range around 1074.7 nm was comparable to the background [36]. Moreover, recent spectroscopic observations above a coronal hole region did not show any evidence of PL of silicon nano-particles at 1074.7 nm; that work lead the authors to conclude that silicon nano-particles are either absent or are present below the detection limit [37]. The hypothesis of survival of grains near the sun was analyzed by Mann et al. [38]. They examined the sublimation sequence of silicate dust in the inner solar system. It was shown that only metal oxides survive at small distances from the sun, while silicon sublimates at moderate temperatures. Thus it was concluded that it is not reasonable to suggest the presence of Si nano-particles in the vicinity of the sun.

16.4 NANOTECHNOLOGY, LIGHT (OPTICS), AND SPACE

Optics has been over many decades, centuries, and even millennia an important springboard in the development of science and astronomy, as well as the understanding of nature of matter. The advancement of science was also very pivotal in the development of better understanding of light and development of optics applications and most recently of photonics. In fact it has been a true two-way street between science and optics. The relationship between the emerging field of nano science and nano technology and optics is not an exception. Although nano structures afford novel optical, electronic, chemical, derivative bio properties, the relation to optics has had special importance.

16.4.1 Optics and Science Development

Optics especially visualization has been an important springboard in the development of science and our understanding of nature and the universe over many decades, centuries and even millennia [39]. In fact it has been a true two-way street. The advancement of science was also very pivotal in the development of better understanding of light and development of optics applications and most recently of photonics. As to applications, presently, light and optics, especially laser light play a vital role in daily life. Laser light is used in the making and playing compact disc (CD). Laser light beams read bar codes for prices at groceries. Printing images on paper uses lasers. Light is used in digital cameras. It allows computers and telephones to communicate with each other over fiber-optic cables, hence empowering the internet. There are many ways in which light is used in medicine, from producing medical images to performing eye surgery or treatment of arthritis. It is not to be forgotten that visualization and light have a major role in education. Optics at the same time is considered not only as a cognitive tool in science education but also as an epistemic object that can potentially support students to understand aspects of the nature of science [40].

As to science development, the role of optics is well documented in discerning the phenomenon such as blackbody radiation, sun spectra, stars, photoelectric effect, rainbow, eclipse, continuous spectra and discrete hydrogen and atomic spectra, special relativity [http://sciexplorer.blogspot.com/2012/10/atoms-part-3-atoms-together-can-emit.html]. In fact, the relationship between our understanding of nature (science) and optics has seen very important syndetic fruits and they propelled each other over many decades and centuries.

16.4.2 Optics and Nanoscience

The relationship between the emerging fields of nano science and technology and optics not an exception. Although nano structures afford novel optical, electronic, chemical, derivative bio properties, the relation to optics has a special importance. This can be understood by realizing that optics involves the study of the behavior and properties of light in vacuum, as well as the study of its interactions with matter in the gas, liquid, and solid states [39]. In fact, the connection between atomic physics *and* optical physics in modern times became apparent by the discovery of spectral lines and attempts to describe and explain the phenomenon including blackbody radiation, which resulted in the birth of the first quantum model of the atom (the Bohr atom model). Limitation of the Bohr model to Hydrogen spectra lead to an entirely new mathematical model of matter and light: quantum mechanics.

As one can see, optics encompasses the construction of instruments that use or detect light that may serve many other fields, over a wide range of the electromagnetic waves from UV to infrared light. Nanotechnology or nano science and technology, on the other hand aims at construction, understanding, and putting to use miniaturized matter [41–43]. Miniaturization of all types of matter, including dielectrics, metals, semiconductors, polymers, etc. affords interesting novel properties, especially optical properties. The novel size regime is intermediate between the largest molecules and 100 nm. In this regime, phenomenon may not be as predictable as those observed at larger scales. Using nano structures as building blocks to translate the novel properties into advanced miniaturized devices is at the heart of nanotechnology. Most notably, miniaturization of matter impacts the interaction and propagation of light in matter, especially and most interestingly in metal and semiconductors. In a nano metal medium, light can be guided, confined and focused far beyond what conventional dielectric technology offers. Whereas in a dielectric medium, focusing is wavelength-limited (diffraction-limited) to ~300 nm at best, in a nano metal medium it is independent of the incoming wavelength, with no limit as it is only limited by the size of the nanostructure, practically confining light into super intense "hot-spots." Moreover, nano metal shows strong color change; and become hot and highly interactive with the environment upon light illumination. On the other hand upon miniaturization, semiconductors provide strong size-dependent luminescence covering the full range of the RGB colors of light; becoming much brighter and may turn material from dull to glowing nature; finally synergetic optical functionalities have been observed upon integration of the functionalities of nano-metal and semiconductor. This affords unprecedented functionalities with potential to enable exciting applications in service of fields as diverse as electronics, opto- and photo-electronics, elementary particles, biomedicine, nanolithography and fabrication, and energy harvest and lighting.

16.4.3 Natural Nonman-Made Nanostructures

It is to be noted that naturally existing colloids, micelles, polymer molecules, PAHs, phase-separated regions in block copolymers, for example fall in this size regime. More recently, naturally unknown but interesting classes of nanostructures, such as carbon nanotubes [44], silicon nanoparticles [45,46], metal nanoparticles and nanorods [47–49], and compound semiconductor quantum dots [50] have been designed and fabricated. The application of such extremely small particles can find applications across all fields [43,51]. Miniaturization of silicon and metal down to the nanoscale regime is turning to be the most important areas from the point of view of fundamental science and application. They impact the behavior, properties, and interactions of light with matter, especially metals and semiconductors, and it enables novel advanced devices with application in electronics, photonics, energy and lighting, and biomedicine.

REFERENCES

[1] Colomban P, March G, Mazerolles L, Karmous T, Ayed N, Ennabli A, Slim H. Raman identification of materials used for jewelry and mosaics in Ifriqiya. J. Raman Spectrosc. 2003;34:205–13.

[2] Freestone I, Meeks N, Sax M, Higgitt C. The Lycurgus Cup – a Roman nanotechnology. Gold Bull. 2007;40:270–7.

[3] Medieval stained glass window, Courtesy: NanoBioNet and www.nano.gov

[4] Angelini I, Artioli G, Bellintani P, Diella V, Gemmi M, Polla A, Rossi A. Chemical analyses of bronze age glasses from Frattesina di Rovigo, northern Italy. J. Archaeol. Sci. 2004;31:1175–84.

[5] Artioli G, Angelini I, Polla A. Crystals and phase transitions in protohistoric glass materials. Phase Trans. 2008;81:233–52.

[6] Mirguet C, Roucau C, Sciau P. Transmission electron microscopy a powerful means to investigate the glazed coating of ancient ceramics. J. Nano Res. 2009;8:141–6.

[7] (a)Mie G. Beiträgezuroptiktrübermedien, speziellkolloidalermetallösungen. Ann. Phys. 1908;25:377–445. (b)Mie G. BeiträgezurOptiktrüberMedien, speziellkolloidalerMetallösungen. Ann. Phys. 1908;25:377–452.

[8] H.-P. Wagner, M. Kaveh-Baghbadorani, Plasmonics: Revolutionizing light-based technologies via electron oscillations in metals, 2015; http://phys.org/news/2015-06-plasmonics-revolutionizing-light-based-technologies-electron.html

[9] Chabanne D. Le décor de lustremétallique des céramiquesglaçurées (IXème-XVIIème siècles), Matériaux, couleurs et techniques. University Bordeaux; 2005. 3 PhD thesis.

[10] Bobin O, Schvoerer M, Miane JL, Fabre JF. Colored metallic shine associated to luster decoration of glazed ceramics: a theoretical analysis of the optical properties. J. Non-Cryst. Solids 2003;332:28–34.

[11] (a)Colomban P. The use of metal nanoparticles to produce yellow, red and iridescent color, from bronze age to present times in luster pottery and glass: solid state chemistry, spectroscopy and nanostructure. J. Nano Res. 2009;8:109–32. (b)Cizer S. History, Technique and Art of Luster. Narhdere Izmir: Dokuz Eylul University; 2010.

[12] Lafait J, Berthier S, Andraud C, Reillon V, Boulenguez J. Physical colors in cultural heritage: surface plasmons in glass. C. R. Phys. 2009;10:649–59.

[13] Roqué J, Molera J, Cepria G, Vendrell-Saz M, Perez-Arantegui J. Analytical study of the behaviour of some ingredients used in luster ceramic decorations following different recipes. Phase Trans. 2008;81:267–82.

[14] Reillon V, Berthier S. Modelization of the optical and colorimetric properties of lustered ceramics. Appl. Phys. A 2006;83:257–65.

[15] Sanderson S. Materials: carbon nanotubes in an ancient Damascus sabre, sharpest cut from nanotube sword. Nature 2006;444:286.

[16] Figiel L.S. On Damascus Steel. Atlantis Arts Press; 1991. pp. 10–11, ISBN 978-0-9628711-0-8.

[17] Reibold M, Paufler P, Levin AA, Kochmann W, Pätzke N, Meyer DC. Materials: carbon nanotubes in an ancient Damascus sabre. Nature 2006;444:286.

[18] Harris PJF. Carbon Nanotube Science: Synthesis, Properties and Applications. Cambridge, UK: Cambridge University Press; 2009.

[19] Kar KK. Carbon Nanotubes: Synthesis, Characterization and Applications. Research Publishing Service; 2011.

[20] Ruth Dasso Marlaire, NASA: https://www.nasa.gov/centers/ames/news/releases/2007/07_17AR.html

[21] Munir N., Total Eclipse: International Innovation, UK, 90, May 2012

[22] (a)Vijh UP, Witt AN, Gordon KD. Blue luminescence and extended red emission: possible connections to the diffuse interstellar bands. ApJL 2004;606:65. (b)Witt AN. Blue luminescence and extended red emission: possible connections to the diffuse interstellar bands. In: Cami J, Cox NLJ, editors. The Diffuse Interstellar Bands Proceedings IAU Symposium No. 297, 2013. International Astronomical Union; 2014. doi: 10.1017/S1743921313015810.

[23] T. Henning, C. Jäger, F. Huisken, Silicon in Space , Villa Vigoni, Workshop May 16–19, 2012.

[24] Habbal SR, Arndt MB, Nayfeh MH, Arnaud J, Johnson J, Hegwer S, Woo R, Ene A, Habbal F. On the detection of the signature of silicon nanoparticle dust grains in coronal holes. ApJL 2003;592:87.

[25] Witt AN, Gordon KD, Furton DG. Silicon nanoparticles: source of extended red emission? ApJL 1998;501:111.

[26] Zubko VG, Smith TL, Witt AN. Silicon nanoparticles and interstellar extinction. ApJL 1998;511:57.

[27] Nayfeh MH, Habbal SR, Rao S. Crystalline Si nanoparticles as carriers of the blue luminescence in the red rectangle nebula. Astrophys. J. 2005;621:L121–4.

[28] Dai Z-R, Bradley JP, Joswiak DJ, Brownlee DE, Hill HGM, Genge MJ. Possible *in situ* formation of meteoritic nanodiamonds in the early Solar System. Nature 2002;418:157.

[29] Fisenko AV, Verkhovsky AB, Semenova LF, Ivanov AV, Pillinger CT. The Kaidun meteorite: interstellar diamond in the CR and CI carbonaceous components. Geochem. Integr. 2000;38. S294 2000 http://www.maik.rssi.ru/cgi-bin/search.pl?type=abstract&name=geochems&number=3&year=0&page=294.

[30] Rao S, Sutin J, Clegg R, Gratton E, Nayfeh MH, Habbal S, Tsolakidis A, Martin RM. Excited states of tetrahedral single-core Si_{29} nanoparticles. Phys. Rev. B 2004;69:205319.

[31] Nayfeh M, Kumar A, Stephenson LD, Nelson AJ, Mantey K, Kwit M. Measurement of the photostability of Si nanoparticles under UVA and near infrared irradiation. J. Appl. Phys. 2010;107:064316.

[32] Hoang T, Stupca M, Maximenko Y, Elhalawany N, Carr C, Yu H, Nayfeh MH. Complex of heavy magnetic ions and luminescent silicon nanoparticles. J. Appl. Phys. 2013;114:164319.

[33] Mantey K, Zhu A, Boparai J, Nayfeh M, Marsh C, Al Chaar G. Observation of linear solid-solid phase transformation in silicon nanoparticles. Phys. Rev. B 2012;85:085417.

[34] Rao S, Sutin J, Clegg R, Gratton E, Habbal S, Nayfeh MH, Tsolakidis A, Martin R. Excited states of tetrahedral single-core Si29 nanoparticles. Phys. Rev. B 2004;69:205319.

[35] Nayfeh MH, Barry N, Therrien J, Akcakir O, Gratton E, Belomoin G. Stimulated blue emission in reconstituted films of ultrasmall silicon nanoparticles. Appl. Phys. Lett. 2001;78:1131.

[36] Kuhn JR, Penn MJ, Mann I. The near-infrared coronal spectrum. ApJL 1996;456:67.

[37] Singh J, Sakurai T, Ichimoto K, Hagino M, Yamamoto T. Existence of nanoparticle dust grains in the inner solar corona? ApJL 2004;608:69.

[38] Murad E. On the existence of silicon nanodust near the Suningrid Mann. Astrophys. J. 2005;624:125–8.

[39] McGraw-Hill Encyclopedia of Science and Technology (5th ed.), McGraw-Hill, 1993.

[40] Erduran ES, Mäntylä T. The role of visual representations in scientific practices: from conceptual understanding and knowledge generation to 'seeing' how science works. Int. J. STEM Educ. 2015;2:11.

[41] Drexler KE. Engines of Creation: The Coming Era of Nanotechnology. Doubleday; 1986.

[42] Richard Feynman, There's plenty of room at the bottom, http://www.nature.com/nnano/journal/v4/n12/full/nnano.2009.356.html

[43] What is nanotechnology? http://www.nano.gov/nanotech-101/what/definition; Benefits and Applications, Nanotechnology National Initiative: http://www.nano.gov/you/nanotechnology-benefits

[44] Ichihashi S, Iijima T. Single-shell carbon nanotubes of 1-nm diameter. Nature 1993;363:603–5.

[45] Nayfeh MH, Rogozhina E, Mitas L. Silicon nanoparticles: Next generation of ultrasensitive fluorescent markers. In: Isabelle MB, editor. Synthesis in Functionalization Surface Treatment of Nanoparticles. American Scientific Publishers; 2002.

[46] Nayfeh MH, Mitas L. Silicon nanoparticles: New photonic and electronic material at the transition between solid and molecule. In: Kumar V, editor. Nanosilicon. Elsevier; 2007.

[47] Lohse SE, Murphy CJ. The quest for shape control: a history of gold nanorod synthesis. Chem. Mater. 2013;25:1250–61.

[48] Babak BN, El-Sayed MA. Preparation and growth mechanism of gold nanorods (NRs) using seed-mediated growth method. Chem. Mater. 2003;15:1957–62.

[49] Nayfeh MH. Optics in nanotechnology. Optics in Our Time. New York: Springer; 2017.

[50] Murray CB, Kagan CR, Bawendi MG. Synthesis and characterization of monodisperse nanocrystals and close-packed nanocrystal assemblies. Annu. Rev. Mater. Res. 2000;30:545–610.

[51] Applications of nanotechnology, Foresight Institute; https://www.foresight.org/nano/applications.html

Chapter 17

Nanotechnology and Society: From Lab to Consumer

17.1 MASS PRODUCTION OF NANOMATERIAL

A tremendous number and variety of applications are contemplated for nanotechnology [1]. Commercialization of nanotechnology as of today, however suffers from several fundamental killing problems that may slow its realization as the platform technology of the 21st century [2]. Unlike atoms, which are naturally identical, particles are man-made and generally come out non-identical, that is, they are not mono dispersed with a wide size and shape distribution. Most advanced device applications, especially those that require a small number of nanoparticles, such as single electron transistors or memory cells, or single molecule imaging and control, etc. This could turn to be pivotal. In this regard, there is hope that natural etching/organization may help form objects in exact shapes without machining, thus saving manufacturing time and cost. Carbon and silicon tend to form specific nano structures due to the sp2 and sp3 symmetry; however synthesis is not without defects and imperfections.

Nanotechnology as of today suffers from several fundamental problems that slow large scale commercialization. In fact, the problem is not totally financial or managerial, rather of fundamental mass production and safety. One of the main challenges is the question of whether the World will be able to create nanomaterial:

- with precision, control, and repeatability,
- in due time,
- at a competitive cost and large enough volume,
- while satisfying consumer's demand and safety and environmental protection

It has been said that the Holy Grail in this business is to produce large quantities of nano components that are identical size and shape. In this regard there is hope or expectation that self-limiting interactions, reactions or etching may help us form objects in exact shapes without machining, thus saving manufacturing

Fundamentals and Applications of Nano Silicon in Plasmonics and Fullerines.
http://dx.doi.org/10.1016/B978-0-323-48057-4.00017-7

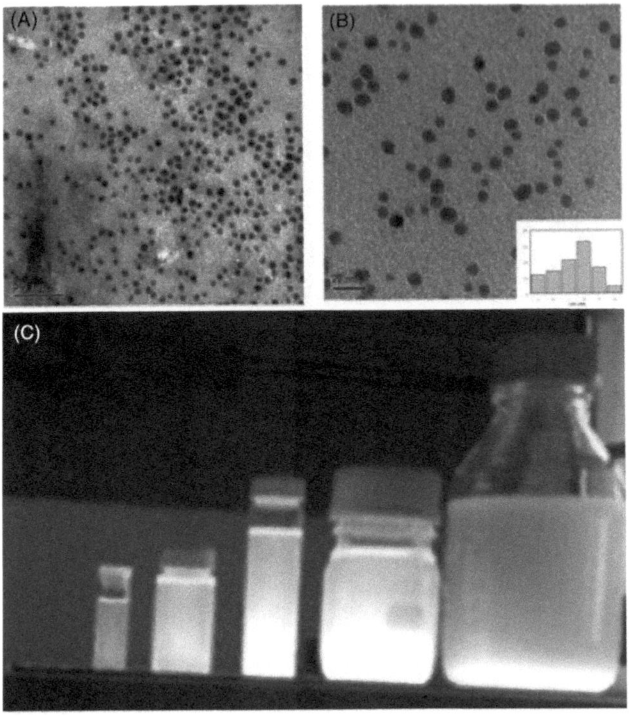

FIGURE 17.1 (A) An image of metal nanoparticles. (B) zoom on a section of the distribution. The image shows particles of different shapes and sizes. (C) Silicon nanoparticles can be synthesizes which are 85% of one size (2.9 nm) with the 15% consists of 1, 1.67, and 2.2 nm.

time and cost. Fig. 17.1A shows an image of metal nanoparticles. The image shows particles of different shapes and sizes. Two materials are amenable to such concepts. Carbon is one material; its sp2 coordination allows to make hollow magic fullerene structures such as spherical shells, cylinders or sheets or buck ball, nanotubes and graphene respectively. H-terminated silicon is another material; its sp3 coordination allows formation of magic filled fullerenes (solid particles), such as particles of 1 nm ($Si_{29}H_{24}$), 1.67 nm, 2.2 nm and 2.0 nm particles. However, both systems suffer from a weak amenability to defects both during or post formation. Fig. 17.1B shows that silicon nanoparticles can be synthesized, which are 85% of one size (2.0 nm) with the 15% consist of 1, 1.67, and 2.2 nm. Even if those conditions are met, there is still the major problem of delivering the nanoparticles to the right places and in a short order. If we have to move and handle one nano at a time, it will take infinitely long time. Concepts like self-assembly and natural organization if applicable would make the technology viable. But for particles to self-assemble they need to be identical. These are major challenges and obstacles that make or break any nano tech application.

17.2 PRODUCTS AND START-UPS AND MANAGEMENT

Although nanotech still has a science-fiction aura, there are actually consumer products already available which employ nanotech (see Fig. 17.2). Successful applications where nanotechnology has made a difference include burn and wound dressings; dental-bonding agent, water filtration devices; stain-free clothing and mattresses; forks, spoons, dishes, pots, and clothes that do not get dirty or wet because they are coated with nanomaterial are also on the market; industrial catalysts, protective and glare-reducing coatings for eyeglasses; coatings that allow for easier cleaning glass; sunscreens and cosmetics; step assists on vans, bumpers and catalytic converters on cars; longer-lasting tennis balls; light-weight, stronger tennis racquets. The chemical industry seems to be the first beneficiary of nanotechnology. Many companies or startups are benefiting from the sale of nanoparticle materials as a commodity with the market approaching so far 500 million dollars.

As of today, however, there is no real breakthrough yet in the area of advanced device applications, and solving disease remains an open question. Although the risk of spending the billions of dollars is high while challenges

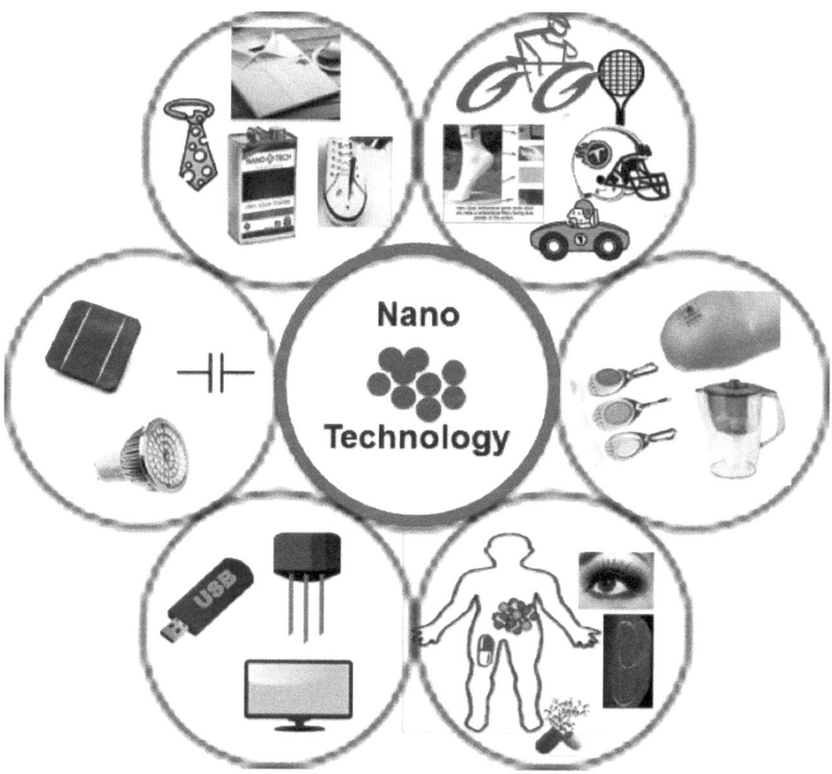

FIGURE 17.2 **Cartoon of application of nanotechnology.**

are not yet solved, the payoff could be astronomical, with room for discovery/innovation and inventership. This is a good time for developing countries with reasonable resources to participate and harvest the fruits of cutting edge of science and technology.

For illustration we mention here some start companies that were established in the early 2000. Many of them have disappeared. Studies showed that 9 out of 10 start-ups fail due to a variety of reasons, including bad management. In many cases, bad management is caused by the fact that there is an acute shortage of nanotech entrepreneur. This is a result of the fact that the nano field belongs to the high-tech advanced field and it is highly disciplinary requiring expertise in several advanced disciplines, such as physics, chemistry, optics, biology, and electrical engineering. As such there is a shortage or lack of nanotech entrepreneurs. Due to being highly disciplinary, requiring expertise in several advanced disciplines, it required the inventors or research professors to be involved in the companies at the managerial level which created conflicts between research and commercialization, which may have been the reason behind the failure of many start-ups.

The University at Albany is one of the pioneers to realize and address this problem. For instance its college of Nanoscale Science and Engineering (CNSE) and School of Business Partnered to create a first-ever joint degree program dedicated to producing highly skilled business leaders equipped to compete in today's global high-tech industry (Nano + MBA program to produce "Industry-Ready" Leaders (high tech management). Other examples include a *Master* in *Nanotechnology* has been jointly organized in Italy by CTVEN, an association founded by the University of Padova. In addition to courses in nanosciences, business administration, and project *management* are included in training; www.msm.cam.ac.uk/nanoenterprise Royal Holloway *College*, London, Physics, DK Three-year research *education* on the top of the *Master degree* devoted to Ph.D. students [3].

Many universities also instituted policies to encourage faculty to take their research results into the commercial domain. Academia in the west and east coasts alike has been active as early as the 1960s in promoting taking IP generated on their campuses to commercialization. This resulted in the Silicon Valley, North Carolina's Research Triangle Park, Kendall Square in Cambridge/Boston, and several others in recent years. In recent years, the rest of the country has realized the importance of this to their communities. Many universities established technology management initiatives on intellectual property (IP) and commercialization, including licensing, start-ups (business incubator). At the University of Illinois a State University, a position of vice president for commercialization, as well as a research park and incubator were created. In addition a center for Nanoscale Science and Technology was established for facilitating nanotechnology entrepreneurship. University of Wisconsin even provided on campus space and labs for start-ups. Even government Labs went through this transformation.

17.3 INTELLECTUAL PROPERTY AND FREEDOM OF OPERATION IN NANOTECHNOLOGY

It is easier to patent in the nano field but it is expected to be more difficult to have freedom of operation when patenting nano-material and products [4]. The general doctrine in patent law is that products of nature cannot be patented; however, atomically engineered elements can be patented but is an uphill effort. For instance, C Sixty Inc., a Canadian start-up company, filed several patents, five of which have been granted, for fullerenes (carbon nano particles and carbon nano tubes). This will place a hold on the use of this "molecule" in drug delivery. On the other hand, Nayfeh and collaborators of the University of Illinois and Nanosi Advanced Technologies, Inc. filed for 24 patents on silicon nano particle or molecule technology, 21 of which have already issued. Two of the issued patents are material composition patents to own the silicon molecules. The silicon "molecule" constitutes a wide platform technology, with applications in electronics, photonics, energy, lighting, biomedicine, cosmetics, paint, etc.

17.3.1 The Bayh-Dole Act

The federal government invests more than $75 billion a year in government sponsored R&D. With this funding the US government had accumulated 28,000 patents, with fewer than 5% of those patents were commercially licensed. The low commercial yield was due to the fact that the general policy on federal contracts and grants obligated inventors (wherever they worked) to assign inventions to the federal government; whereby the government retain titles and would license them only nonexclusively. In addition to ownership there was no uniform policy among federal agencies on patents or on the transfer of technology to the private sector, which caused a maze of rules and regulations.

In search of actions that may address the economic malaise of the 1970s, Bayh and Dole of the US CongressDecember 12, 1980 legislationPatent and Trademark Law The Bayh–Dole Act or Patent and Trademark Law Amendments Act). The key change was in ownership of inventions made with federal funding; it permits a university, small business, or non-profit institution to elect to pursue ownership of an invention in preference to the government.

One example of the pre-Bayh–Dahl era is the story of spinning off of a start-up company (ORTEC) from Oak Ridge National Laboratory (ORNL). ORNL is a government lab, which was managed at the time by the US Atomic Energy Commission. Silicon surface barrier detectors had been developed in the Physics Division of ORNL by John Walter and John Dabbs during the middle to late 1950s as radiation detectors. As the attributes and value of the new detectors became known, other researchers became interested. Having exhausted the realistic possibilities for making detectors at ORNL itself, a group of ORNL scientists led by Harold W. "Hal" Schmitt called for a commercial track outside of their ORNL responsibilities, namely a private company. The team sought the ORNL administration's approval to establish and own a private company without leaving ORNL employment. The process however was found to face several challenges: In addition to the above existing patent laws, some of the policies and rules had to be created; ORNL's concern about potential outside use of proprietary information and conflicts of interest have to be resolved; the ORNL employees must work with the company on a limited basis (as owners), and not as salaried employees. Because fabrication was complex, with each detector had to be tested, some expertise from ORNL needed to be involved, but their work must be carried completely apart from and subsidiary to their ORNL jobs and responsibilities and that testing may not be conducted using ORNL facilities. Despite legal difficulties encountered due to some testing on the compound ORTEC was the first successful technology transfer experiment took place.

17.4 OPEN INNOVATION: ACCELERATE AND LEAPFROG STI

The basic nature of nanotechnology is that it is knowledge intensive based on innovation, and synergetic effects among different sciences and disciplines, which necessities fast (light speed) collection and dissemination of information on breakthroughs, innovation, IP, and initiatives [5].

Science and technology and innovation (STI) is the engine of independent development; and is the driving force for sustainable economic growth; therefore, it is a major guarantee for poverty eradication and social prosperity. Countries worldwide have now come to realize the pivotal role of STI, and many have adopted aggressive STI national policies. However, processes proved to be inherently complex, incremental, slow, and costly. What needed to narrow the gap are strategies and mechanisms with potential to lead to significant strides and breakthroughs. Current research and innovation-based entrepreneurial activities in the emerging fields of biotechnology, ICT, advanced materials and nanotechnology to achieve green growth will be highlighted.

It has been proven that S&T&I advancement is the engine for improving the capability for independent development; and is the driving force for sustainable economic growth; hence a major guarantee for poverty eradication and social prosperity. Developing countries have now begun to realize the pivotal role of S&T, and many of them have adopted aggressive S&T national policies aiming to join the technology bandwagon. However, follow up processes have proven to be inherently complex, incremental, slow, and costly. Despite such attempts, the gap between the advanced and developing countries is widening instead of narrowing.

Open innovation (Fig. 17.3) is a new paradigm that assumes: (1) Two-way relation: when firms look to advance their technology they can and should use external ideas as well as internal ideas internal and external paths to market (customers, rival companies, and academic institutions). (2) Does not freeze IP assets: Internal inventions not being used in a firm's business should be licensed, offered to joint ventures or spin-offs. (3) New culture of doing things: It goes beyond using external sources: Major change in the use, management, and employment

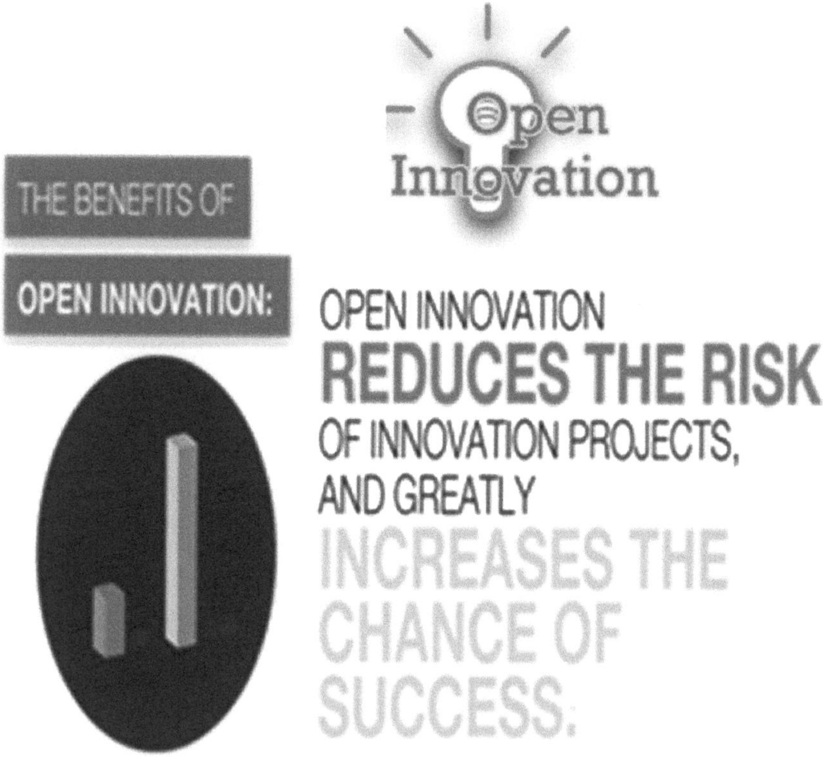

FIGURE 17.3 **Open innovation paradigm.**

of IP (as in technical research-driven generation of IP). (4) Secrecy and feedback companies should not keep their own innovations secret unto themselves. Rather it applies open platform operation, such that customers, employees and even competitors can build on one's innovation. Moreover it should employ a dynamic operation (Feedback). This results in the creation of an ongoing, evolving community of users, doers and creators which is key to success. (5) Crowdsourcing such that it obtain services, ideas, or content from group of people (online community) rather than from traditional employees or suppliers.

The concept Open innovation is feasible and in fact it makes sense for several reasons, including (i) boundaries between a firm and its environment have become more permeable; innovations can easily transfer inward and outward; (ii) knowledge has become widely distributed, (iii) it has become more economical to buy or license processes or inventions (i.e., patents) from other companies than relying entirely on one's own research; and (iv) innovating with partners provides sharing risk and sharing reward.

The model of open innovation is generally good in practice. But issues differ depending on the status of the two sides (private industry, private academia, state academia, federal government, international entity; all colors of legality, the independence of the legal system, language). The risk of open Innovation include theft, infringement, and variations; as well as issues with regard to legal authority of the contract (in which country the agreement will be legally executed), as well as mismatches (size, etc.). For instance multinational corporations cannot afford to stay away from China. But to remain competitive they must develop mechanisms that allow them to minimize the risk of losing critical know-how. IP protection is the No. 1 challenge for multinational corporations operating in China. According to the US government, China accounted for nearly 80% of all IP thefts from US-headquartered organizations in 2013, amounting to an estimated $300 billion in lost business. Among European manufacturers, the loss of IP in China reduced potential profits by 20%. The most important way to avoid problems when defending IP rights in Korea is to be prepared. Make sure that you can anticipate any potential issues.

An example that can be mentioned is an open innovation attempt between NanoSi and Wintek. After signing a confidentiality agreement (NDA) and Exchanging visits the two sides drafted am agreement of joining efforts to develop a device. The two sides selected the background patents to be utilized (from those owned by UIUC/Nanosi and from those owned by Wintek) and agreed on sharing new IP generated from the cooperation. An agreement was negotiated and signed.

The "fast follower" strategies involving innovation tools, such as international cooperation as well as exchange of IP, complemented by an integrated, aggressive process of transfer, digestion, and re-innovation to accelerate and leapfrog STI, as well as commercialization with mutual benefits and win–win results.

17.4.1 "Fast Follower" versus "First Mover" Business Model

It is not sufficient to adopted aggressive STI national policies as processes proved to be inherently complex, incremental, slow, and costly. What needed to narrow the gap are strategies and mechanisms with potential to lead to significant strides and breakthroughs. First, one has to be an ***Innovator*** by being first to develop or patent an idea, or being a ***Product Pioneer*** by being First to have a working model. It is also important to select a model of development: "First Mover" versus "Fast Follower" model (Fig. 17.4).

FIGURE 17.4 "First Mover" versus. "Fast Follower" model.

In fact, a 1993 paper by Peter N. Golder and Gerard J. Tellis had a description of what happens to startup companies entering new markets. In their analysis, they found almost half of the market pioneers who are first to sell the product (First Movers) in their sample of 500 brands in 50 product categories failed. Even worse, the survivors' mean market share was lower than thought. Further, their study shows that those who did not enter the market first but at an average of thirteen years later than the pioneers (Fast Followers) have much greater long-term success.

Therefore, we believe that the Fast Follower model strategy is more suited for developing countries for several reasons. First, when targeting a world market it is much more difficult to understand the business model, customer's problems and the features needed to solve those problems. In this case there is not much advantage of being "First Mover." First Movers tend to launch basically in the dark, without really fully understanding customer problems or the product features that solve those problems. They are very susceptible to burning through the cash. Fast followers win more often because they learn from the first-mover problems.

17.4.2 Open Innovation With Developing Countries

Another interesting feature of nanotechnology is the fact that some aspect or phase of nanotechnology is low cost, and of highly interdisciplinary nature. In this regard, there is ample opportunity to pool resources: human, financial, and technical.

International cooperation in conjunction with firm adoption of combined process of importation, digestion, and re-innovation, on the other hand may effectively accelerate and leapfrog STI in developing countries. International collaborations allows sharing and exchanges of IP, as well as could produce a wealth of new IP with ownership shared by both sides that is capable of catalyzing and leveraging additional funding and commercialization, resulting in true transfers of know-how and technology.

Although STI has always been of world-wide importance, increasing globalization challenges such as intensified economic globalization, the rise of new global players and the provision of global public goods, makes international cooperation between industrial countries and developing countries necessary if not imperative. Even in Europe critical mass and large-scale infrastructure requirements call not only for national partnerships but also for strong international partnerships. One must now seek to learn and benefit from good development practices elsewhere in the world. For instance, European and US industries are not only investing in industrialized economies, but are rapidly investing in emerging economies, such as in China and in India, with researchers and students looking beyond training opportunities at home into world-class centers anywhere in the World. By involving their scientific and technological expertise, countries can have significant opportunities to be players at the

World scene, allowing them to meet political, social, economic, and humanitarian commitments to sustainable development fields ranging from global climate change and biodiversity to fulfilling the Millennium Development Goals. In addition, participating countries in partnerships may also have ample opportunities play a role in the implementation of international agreements, such as on biodiversity and climate change.

The new approach to international cooperation must aim to rise to these challenges by way of innovative mechanisms for promoting an all-encompassing international research collaboration. It must aim to address three interdependent objectives:

- supporting scientific and economic development through strategic partnerships in selected fields of science and by engaging forged between industrial and developing countries;
- facilitating contacts with partners with the aim of providing better access to research carried out elsewhere in the world;
- addressing specific problems that not only face the industrial countries but also face developing countries, that is, have a global character (e.g., by contributing towards Millennium Development Goals, addressing global climate change, combating biodiversity loss, water and energy scarcity).

As to specific mechanisms that have the potential to lead to significant strides and jumps with mutual benefits and win–win results for all sides, we outline here the following describing how the new approach to international cooperation is intended to work with particular focus on implementation.

(i) establishing agreements, joint research projects, and training program with well-established science parks, as well as with universities, research institutes, and enterprises; establishing joint labs or research centers; stationing institutes and engineering centers in the park;
(ii) establishing a service center for start-ups, industrial incubators, and a start-ups park for returned overseas students;
(iii) homing on emerging fields of advanced material and nano-enabled advanced material and devices
(iv) creating a culture of invention, IP and prototyping;
(v) importing high caliber talent; creating an environment such that the park becomes a rendezvous for domestic and overseas R&D institutes, high tech businesses, and high caliber personnel

The above mechanisms augmented by strong international cooperation, as well as exchange of IP, complemented by an integrated, aggressive process of transfer, digestion, and re-innovation would certainly accelerate and leapfrog STI, as well as commercialization with mutual benefits and win–win results.

But there are acute mismatches in carrying out open innovation with developed countries. Collaborations, especially international collaborations with developing countries are not without problems. In carrying out collaborations, acute mismatches in training, resources and reward, interest and needs, and cultural sensitivities may pose as barriers. Cultural differences could be somewhat alleviated if countries involve and harness expatriates in their Diaspora when setting up international collaborations. Lack of resources may be offset since scientists operating in an environment which lacks resources tend to be more creative and inventive. The more serious problem however is with regard to finding common interest. In fact, it is not straightforward to home in on projects that could be of interest to all parties involved. Collaborative areas and projects must be screened and selected by all such that both industrial and developing countries find mutual benefits and scientists on both sides could participate on the same footing at the cutting edge of science and technology. The area of advanced material and nanotechnology is an area, which would be ideal for such collaborations because it is highly interdisciplinary allowing scientists of different backgrounds to participate at the cutting forefront of research with ample opportunity for discovery and invention while many aspects of this field is low cost. In addition, this field promises the development of novel advanced devices including low cost devices with applications in public health, water collection and purification, renewable energy and biofuels. Finally, the payoff could be significant since such collaborations are expected to generate a wealth of new IP with ownership shared by both sides.

17.4.2.1 International Open Cooperation is Win–Win

The Research Park and the University of Illinois emphasize and encourage international cooperation in terms of partnerships. Examples of which include partnerships with Singapore, France, Brazil, Sweden, Vietnam, and Korea. A partnership involving researchers from Illinois, Egypt, Turkey, Pakistan, Abu Dhabi, and Jordan is being setup with sponsorship from US national science foundation (NSF).

Recently the University of Illinois and SHTP in Vietnam executed a collaborative agreement, with initial activities focusing on the use of advanced material and nanotechnology for the development of devices for electronics, photonics and renewable energy, biomedical, and water and environment. We recently proposed to establish a partnership mechanisms to enable and facilitate communication among researchers from the USA and Asia involving the University of Illinois, University of Chicago, and Cleveland State University on the US side and researchers from Tokyo University for Agriculture and Technology, National Institute for Material Science, University of Tsukuba in Japan, The Vietnam National University, and the Saigon High Tech Park University in Vietnam to examine the feasibility of initiating and catalyzing a coherent, complimentary partnership among the institutions to conduct joint studies to engineer novel hybrid sensitizer super molecules that integrate the benefit of dye molecule sensitizer and silicon nanoparticle sensitizer to alleviate problems limiting the efficiency and stability hence the use of organic photovoltaics (OPVs), drawbacks that must be overcome before OPVs can be seriously considered as energy sources. The initiative is expected to make significant contribution to the solutions of the present energy crisis, and to yield transformational energy technologies, which will get renewable energy to a point where it is economically competitive with current technology.

The activities of the project include:

- Planning visits to the research sites in order to:
  - Survey and collect data on human and technical resources at the participating sites
  - Design, establish and survey intra communication networks
- Hold a workshop/working meeting of the expert scientists from these institutions in order to assist designing multi instructional research projects
- Facilitate and ensure collaborative work to prepare proposals for presentation to US NS, US AID and World Bank funding agencies

Finally, among some of the other international multidisciplinary activities the University of Illinois is involved in include (see Appendix 1 for details): Advanced Digital Sciences Center (ADSC); I^2CNER; ADM Institute for the Prevention of Post-Harvest Loss; Computation-based Science and Technology Research Centre (CaSToRC); Center for Agricultural, Biomedical, and Pharmaceutical Nanotechnology (CABPN); Emergent Behaviors of Integrated Cellular Systems Center (EBICS); [Reference: University of Illinois College of Engineering, International Engagement Committee, 2012].

In conclusion, we argued that the "fast follower" strategy is more appropriate than First Mover strategy provided it involves innovation tools, such as international cooperation as well as exchange of IP, complemented by an integrated, aggressive process of transfer, digestion, and re-innovation to accelerate and leapfrog, STI as well as commercialization with mutual benefits and win-win results for all sides.

17.4.3 Nanotechnology Catalyst for Global Cooperation

Another interesting feature of nanotechnology is the fact that some aspect or phase of nanotechnology is low cost, and of highly interdisciplinary field. In this regard, there is ample opportunity to pool resources: human, financial and technical. It is also considered as one of the most effective catalyst for substantial university–industry–government partnerships. It is even suited for real international cooperation where developing countries can participate and contribute at the cutting edge.

Nano-structured material-based research is a field in which both industrialized and developing nations can participate on the frontier and secondly advances in material impact many branches of science and technology which implies that the central investment in a facility can directly benefit several diverse areas of research. Thus the "*critical mass*" of scientists necessary for the success of a central research facility need not be made up of specialists in just one field. Even though the principal techniques and ultimate goals of, e.g., physics and medical research may be very different, a common focus on nano-based research can provide an opportunity for pooling the scientific expertise of many individuals. Moreover, nano structuring provides impetus to shifting toward more high value added materials

Nano-technology and advanced materials are emerging as cornerstones for establishing knowledge-based industries. Nanotechnology has advantageous features for research, industry and commercialization for several reasons. First, the relatively low capital cost of some aspects of nanotechnology compared to other advanced scientific equipment makes nano-based research a field in which developing nations can participate on the frontier. Secondly, the domain of nanotechnology is in highly interdisciplinary research and development, thus involving many branches of science and technology, enabling pooling of human resources. In this regard, an investment in a central nano facility can directly benefit several diverse areas of research, reducing the cost to an affordable level to non-industrially advanced countries. In knowledge-based technology, nanotechnology stimulates enormous growth at the "cutting edge" of science, and has advantageous features for industry and commercialization. It also stimulates interest in learning at all levels of the society. As an exciting domain it captures the imagination of the young, especially the K-12 kids, as well as the layman. Finally, nanotechnology and advanced material technology are ideal for international collaboration, as has been recognized by the NSF program: World Materials Network (WMN).

17.4.4 Nanodivide – Widening and Closing the Nanodivide

The world is already divided along the lines of information and knowledge. In information-based society and knowledge-based society there is IT divide (internet usage may be an indicator) which correlates with inequality in the distribution of wealth. The gap can be huge. By incorporating, for example silicon nano particles in transistors and memory cells, nanotechnology improve the computing power of transistors and improve the storage capability of memory elements (see last week's free scale announcement). Information acquisition, retrieval, and storage become faster, more abundant, and cheaper [6].

The World stands to be divided even further with the advent and adoption of nanotechnology. This gap is likely to be exacerbated forming the so-called: nano-divide. Nano-based or nano-enabled advanced devices in other sectors will also contribute to the widening of the gap.

If the industrial world were to coordinate and collaborate with the developing world, empowerment of developing countries may be realized in nano technology, and the nano-divide or other divides may ease or close.

- Establish global fund
- Access to innovations and IP
- Sharing risks on human health and environment
- Establishing partnerships
- Adopting joint global policies
- Developing low cost technologies
- Building trust and sharing

But there are acute deficiencies in infrastructure and resources facing reaching this goal. High nanotech costs for acquisition of IP rights, nanotech infrastructure; lack of human and policy capacity; trade barriers, the political instability in the countries constitute further barriers, although these are not unique to nanotechnology.

Low average incomes and low government spending on R&D and even healthcare, in many countries worsens the scenario for nanotechnology to even be considered in less-developed countries, Lack of a comprehensive standards infrastructure, database or regulatory framework including protocols for risk and life cycle assessment for nanotechnologies.

Lack of tools to assure quality and evaluate and mitigate hazards and risks.

17.5 PARTNERSHIPS: ACADEMIA–INDUSTRY–GOVERNMENT

Arenas and environments for bridging academic innovation and commercialization in modern economies [7].

It has been recognized that research at academic and government laboratories constitutes the foundation of development, especially industry in the Western World (Fig. 17.5). It has not however been straightforward to transfer this research to industry, namely going from *lab to consumer with mass production,* which rendered a large volume of generated knowledge, patents or IP with little direct impact; depriving society from their benefits. Speedy commercialization processes have been hampered by all sides of the equation alike: the researcher, institution, industry, and the government. Some of the impeding factors involve cultural barriers while others involve lack of organization, constraint of institutional missions and public policy. The last decade however witnessed several novel developments, including (i) considerable cultural change, with all sides showing interest in a paradigm change; (ii) the emergence of nano, information, advanced material and bio technologies, which have proven to be an effective exciting bridge between science and commercialization; (iii) the rise of modern high tech science parks, and (iv) the emergence of the digital revolution which allowed fast storage and retrieval of information.

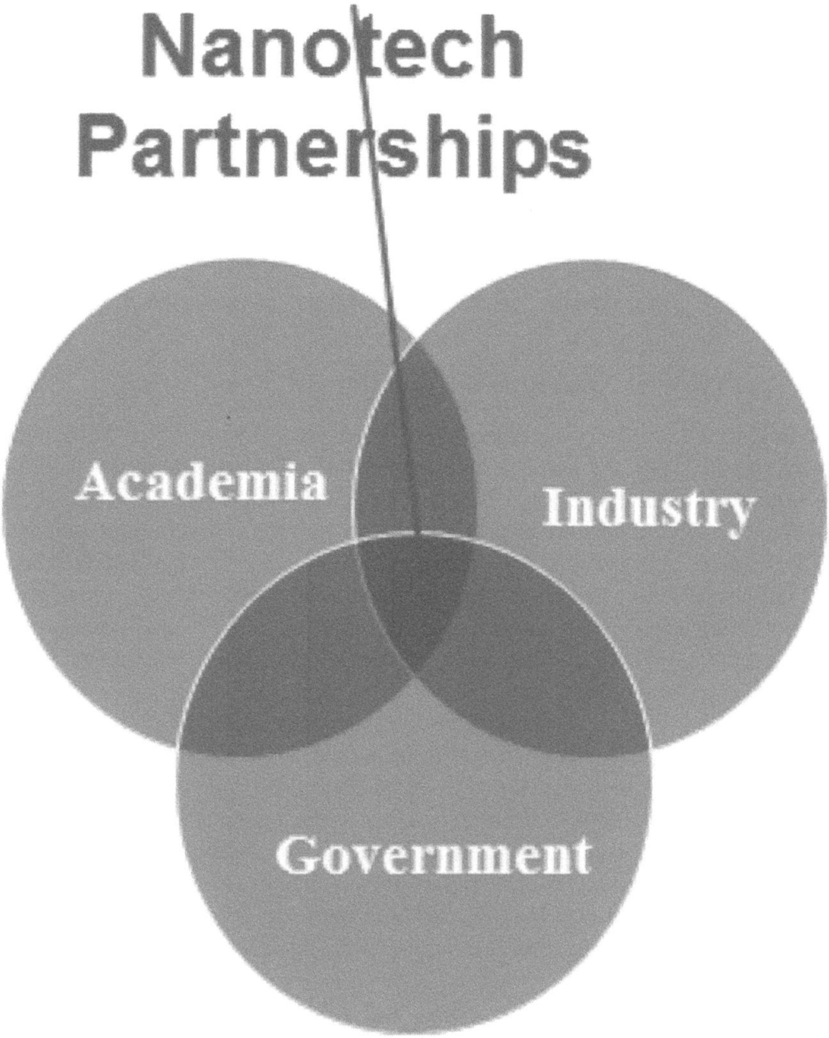

FIGURE 17.5 **Confluence of the partnerships among academia, government and industry.**

These developments helped catalyze inventive means to bridge innovative research and commercialization. During this agonizing era of paradigm transformation, the University changed from being agnostic or outright hostile to full supporter of commercialization, as it saw economic value. Operational steps were adapted to fund and facilitate the process by the University, including building science parks, involving private business and local city governments as well as the state and federal governments. Toward this goal, the Office of Technology Management (OTM) of the University of Illinois, for example, acquired the services of Deloitte and Touche and employed *student assistants to* review a large backlog of IP. Concurrently, Illinois Ventures was established as a Consortium of Business community, City & State Government, and the University. The University encouraged faculty to found startups (NanoSi Advanced Technologies, Inc. was among the initial few) and assisted with writing the business plan and providing legal framework, management, as well as identifying investors and international contacts, and arranging for interaction with industries. Operating in the field of nanotechnology made the process easier because it is one of the most effective catalyst for substantial university–industry–government partnerships. It is even suited for real international cooperation where developing countries can participate and contribute at the cutting edge.

17.5.1 Priorities and Role of Government

Due to limited resources, activities must be governed by priorities. Those must be set according to the features and demands of all countries involved. Priority areas for S&T cooperation bearing great socio-economic significance for all must be identified and implemented step by step. Government guidance (Fig. 17.6) is very instrumental in achieving this goal by establishing related policies and providing funds from governments. This shall play the role in guidance of encouraging and supporting enterprises, universities and colleges and research institutes to participate in the cooperation with financial contributions in order to forge synergies.

With a focus on S&T capacity building, priority areas of common interest will be selected out for cooperation in basic and applied research, technological demonstration, and transfer of research findings. Priority is given to areas concerning the people's well-being, such as water resources, food, health, energy, environment, etc., and others that would greatly facilitate socio-economic development. The cooperation shall be focused on upgrading African countries' capacity in S&T policies making and management, promoting the development of S&T industries and improving people's livelihood through science and technology.

One method of enforcing priorities and guiding or orienting the development is by allocating research funding in the areas. Moreover, Federal agencies adopted a scheme requiring the participation of industry or government labs along with academia.

17.5.2 Academia–Industry Partnership

The role of academia–industry partnership in science and technology and economic development cannot be over emphasized. To promote such endeavors a paradigm-shifting legislation known as Bayh-Dole Act of 1980 was passed by the US Congress allowing small businesses, and non-profits such as universities to own inventions and patents on federally-funded research projects. Among the indicators are *patents awarded*, active patents, and licenses granted. *To innovate*

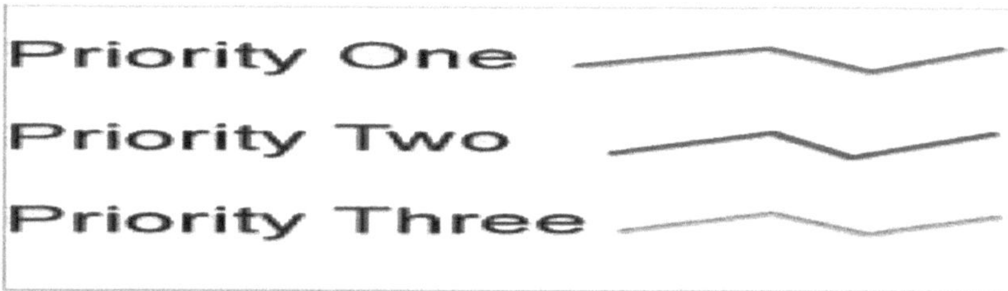

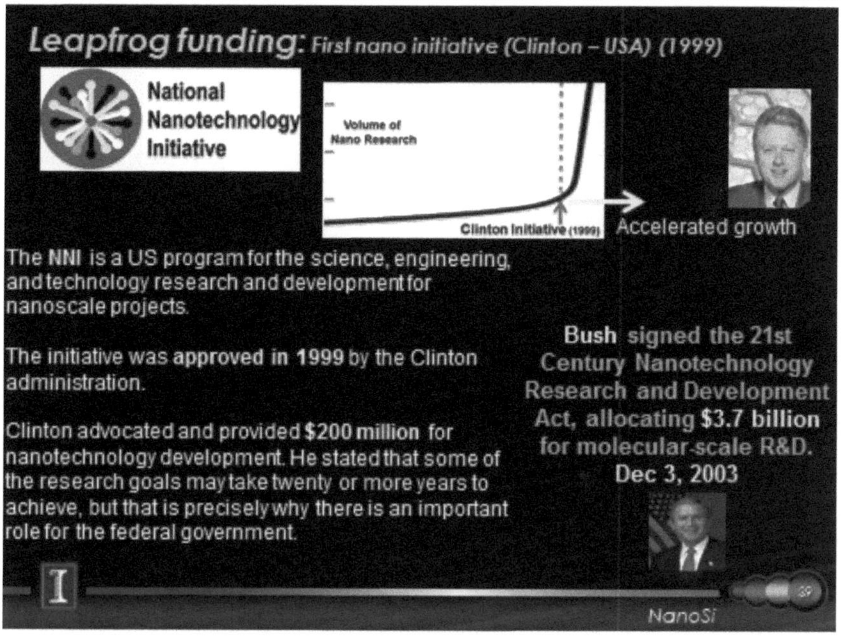

FIGURE 17.6 **The Clinton initiative and nanotechnology national initiative (NNI).**

is to translate knowledge and ideas into economic growth and social wellbeing. The University of Illinois is a leader in conducting groundbreaking research that forms the foundation for meaningful innovation. Discoveries and developments at the University of Illinois, such as LEDS, Plasma Displays, MRI, Worldwide Web Browser, PLATO, and Flexible Electronics have led to establishment of multi-billion dollar industries, products and services that have had a profound and broad impact on human lives. The Association of University Technology Managers (AUTM) estimated that in 1998 alone, *$33 billion of US economic activity was traced to academic licensing* from the US universities (AUTM, 1999).

Some like Rifkin have characterized the dawning of the Third Industrial Revolution, heralding the convergence of communication technology and renewable energy technologies, whereby individual homes, offices, and factories will produce their own energy, and will share/sell the surplus over the Internet aka Energy Internet (EI or ENet). This is bound to have broad societal impact in how we interact with each other, how we educate our children, and how we live and interact with the environment. This is where academic startups/companies like NanoSi Advanced Technologies have a pivotal role to play.

I provide some specific examples from activities both at the Illinois Research Park and the University Campus. The Research Park was established as an arm of the University of Illinois to serve faculty and students to develop their IP. It was founded 12 years ago and is only 2 miles from the campus center of the University. It has been designed to be a small but effective park. It provides an environment where technology-based businesses can work with the research faculty and students on collaborative research and access UI services. The Research Park has 90 companies and employs more than 1200 people in high-tech jobs. It houses sites for 9 Fortune 500 companies including Yahoo, Caterpillar, State Farm, ADM, Sony, Raytheon, SAIC, Abbott, and others. The Park also includes an incubator facility called Enterprise Works – a startup incubator, which houses at any time about 38–40 companies. Since its inception the incubator has seen 55 startups graduating. In addition, there are 53 startup/small businesses in the Research Park. The good fraction of the companies addresses IT, advanced material and nanotechnology, biomedical and renewable energy and lighting including harvest and storage. NanoSi Advanced Technologies, a company which I started, graduated from the incubator and it allocated its operations in the research Park. It focuses on development of silicon nanomaterial and uses them to develop advanced prototype device, such as nano silicon solar cells, organic-inorganic thin film solar cells, biofuel cells, super capacitors, nanophosphor for White light LED lighting. It presently collaborates with Wintek on improving displays for hand-held devices and with PolyBrite International on development of nanoLED based white lighting.

Startup Cafe is another example of the activities of the research park. It is a luncheon series that highlights successful entrepreneurs with connections to the Research Park as guest speakers. The forum gives a chance for entrepreneurs to share their story and provide their insights on their success, things they wish they would have known before getting started, and lessons from their entrepreneurial journey.

17.5.3 Commercialization Initiative (Obama's Whitehouse) (Fig. 17.7)

Since the announcement of the national nanotechnology initiative (NNI), and over $20 billion invested in R&D by the US Federal government alone we find fifteen years later that consumers have not seen the promised dream. Nanotechnology has delivered some mundane products but it yet to deliver advanced products, such as smart anticancer therapeutics that could destroy tumors, and lighter, thinner body armor that could save the lives of soldiers. This has dampened enthusiasm. Indeed commercial targets have lagged behind.

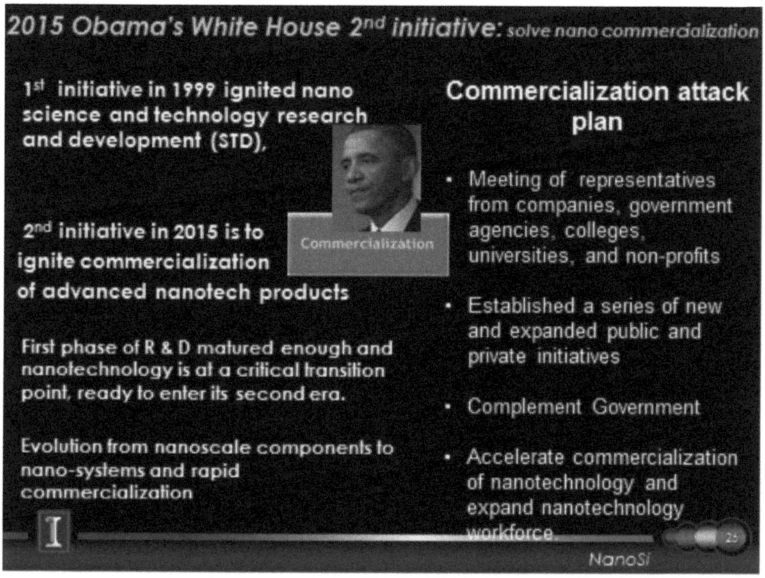

FIGURE 17.7 **Commercialization initiative (Obama's Whitehouse).**

The problem is not financial or managerial, rather more of fundamental mass production and safety. Nanotechnology as of today suffers from several fundamental problems that slow large scale commercialization. The challenges include the ability to:

- create nano material with precision, control, and repeatability,
- in due time,
- at a competitive cost and large enough volume,
- while satisfying consumer's demand and safety and environmental protection

The White House is of the belief that it is not time to buckle and call it quit. Just like its first initiative in 1999, which ignited nano-science and technology research and development, the White House is weighing in again; but this time to ignite the second phase, namely commercialization of advanced nanotech products. It has recently realized that the first phase of R and D has matured enough and that nanotechnology is at a critical transition point, ready to enter its second era. This next technological generation will see the evolution from nanoscale components to nano-systems and rapid commercialization. Thus in recognition of the urgent need it facilitated a meeting of representatives from companies, government agencies, colleges and universities, and non-profits in which they established a series of new and expanded public and private initiatives that complement the Government effort to accelerate the commercialization of nanotechnology and expand the nanotechnology workforce.

FIGURE 17.8 Pehr Anderson took his class project to the real world and was acquired for $90 million [7].

17.6 TECH ANGELS, VENTURES, MAVERICKS, AND RISK

Pehr Anderson went from N Dakota to MIT to enroll into a master degree (Fig. 17.8). There was a class project where he described Ethernet telephone prototype. Before he finished the degree he dropped out to take his class project to the real world. He got a B grade, but 3 years after he dropped out he turned the project into a corporation NBX that took the telephone into a 90-person business and was later acquired by 3Com Corporation for $90 million [7].

I personally know Pehr and his wife Teresa as I invited both of them to an NSF sponsored meeting in Amman, Jordan.

The Venture Cafe (Fig. 17.9)

FIGURE 17.9 Teresa Esser presents in her book Venture café vivid secrets, strategies, and stories from America's High-Tech.

How do you take a start-up to $100 million [8]. In this book, Teresa Esser presents vivid secrets, strategies, and stories from America's High-Tech Written for anyone interested in cutting edge entrepreneurship, this work offers an inside account of what business and technology mavericks are really talking about. Teresa Esser is an active member of the MIT entrepreneurial community and lives in Cambridge, Massachusetts. Venture Café offers examples of inside account of business and technology mavericks. It teaches lessons that no one will or can tell you. This book provides the next best thing to being there.

It has been said that *risk takers* rule the world but also can go to Jail. The truism "nothing ventured, nothing gained" is an often heard phrase. But risk-taking must be balanced with prudence. Karen Firestone, chairman, CEO and co-founder of Aureus Asset Management, knows how to strike that balance. Risk-taking is a factor in everybody's life, all through your day, every single day, and we may be aware of it or not. Being sensible about it is another challenge, and I think that what makes it even harder today is that when people are under increased levels of stress – which we all are – we tend to act more impulsively, and we don't think things through in a logical or sensible way. Given that I spend most of my work day, and a lot of my time, thinking about risk-taking professionally, it's sort of a natural outgrowth that I think about risk more generally – the way that we deal with it in complicated fashions and ways to be more sensible.

It seems like, especially since the recession, we are in a period where the concept of understanding risk-taking and actually taking that risk has been heightened.

How can we possibly, for example, make an investment? How can we possibly put money into the stock market when we see all of this happening around us? Yet what we talk about here at our company lately is that the market seems to be one of the few, in a sense, calm locations for thought and reasonable consideration of what to value.

Some universities such as MIT are creating activities to train student on risk taking and out of norm. The Eugene McDermott Award in the Arts at MIT recognizes rising, innovative talents and offers the recipient a $100,000 prize and a campus residency. Established in 1974 by the Council for the Art at MIT, the Eugene McDermott Award in the Arts at MIT is bestowed upon individuals whose artistic trajectory and body of work indicate that they will achieve the highest distinction as leaders in their fields. One of the most generous arts honors in the United States, the Award reflects MIT's commitment to risk-taking, problem solving and to the idea of connecting creative minds across disciplines. The Award is considered an investment in the recipient's future creative work, rather than a prize for a particular project or lifetime of achievement. A distinctive feature of the Award is a campus residency, which includes a celebratory event at which the Award is presented, a public presentation of the artist's work and significant interactions with students, faculty and staff. The goal of the residency is to provide the recipient unparalleled access to the creative energy and cutting-edge research found in the MIT community and to have the recipient connect with departments, laboratories and research centers throughout the Institute in ways that will be mutually enlightening.

On July 21, 2016 we announced the creation of a $250K cash prize award for disobedience. This idea came after a realization that there's a widespread frustration from people trying to figure out how can we effectively harness responsible, ethical disobedience aimed at challenging our norms, rules, or laws to benefit society.

The award will go to a person or group engaged in what we believe is extraordinary disobedience for the benefit of society. Specifically, we'd like to call out action that seeks to change society in positive ways and is consistent with a set of key principles. These principles include non-violence, creativity, courage, and taking responsibility for one's actions. We're seeking both expected and unexpected nominees. This could include – but isn't limited to – those engaged in scientific research, civil rights, freedom of speech, human rights, and the freedom to innovate.

"*Scared money never wins*" is an old gambler's adage from the 1940s. It is as timely today as it was when it was coined. Only now, the phrase should be top-of-mind for CEOs and their management teams as they strive to define and implement successful strategies in today's uncertain business environment. It is time to be bold. Unless your company is literally in the gambling industry (and if you happen to run a real casino, you know the truth about that "scared money" adage), you understand that business is not a game of chance. You're not "betting" on the future – you're planning, executing and managing for growth, even though that future is fraught with uncertainties. Every age in history has its challenges and uncertainties. And every age in history has leaders who figure out not only how to survive, but how to thrive and grow amid those challenges. Leaders must have the forethought, diligence and nimbleness to take advantage of opportunities. They must lead the development of realistic and actionable growth strategies that incorporate business, product and operational improvements to capture market share. And, all the while, having the courage to make the necessary changes and upgrade talent at all functional levels to meet the evolving needs of the business. Bold leadership is never easy, and its results are never guaranteed, but fear not investing in success is always worthwhile.

17.7 NANOCULTURE AND SOCIETAL IMPLICATIONS

There is a vague understanding of what nanotechnology, although a lot of people are aware of the term. There is great shortage if not absence of S and T based education, as well as nano culture at academic and research institutions, decision makers and public at large. Specifically, until recently there has been very little and vague activity in nano technology in formal academic programs, educational material, nanotechnology knowledge in existing curricula, and outreach programs. The problems may be alleviated through dissimilation of information through schools and society at large, exhibitions, mass media, and Internet [9].

17.7.1 Integration of Research, Education, and Training

Nanotechnology is a far-reaching science; effectively impact almost every part of our life. The "next industrial revolution" requires several agents including creation of a knowledge base and the infrastructure to the scientific, technological and societal impacts and implications; motivate and attract the best and brightest students into the field of nanoscale engineering for high-achieving careers in academia, industry and government; Development of curriculum at both the graduate and undergraduate levels, that may also involve industry; outreach activities to facilitate dissemination of both the research and educational program to industry and to other universities and pre-college students, and the public at large.

17.7.1.1 Where are the Youth in Nanotechnology?

Nanotechnology is ideal for training projects since it captures the imagination of the young and the old, and serves as a rally for all: scientists, students, funding sources, decision makers, politicians, and the general public.

Nanotechnology is a far-reaching science, and is expected to effectively penetrate almost every part of life in the society in the very near future. While great deal remains to be discovered and understood about these ultra-small materials, the scientific community is already calling it the "next industrial revolution." This revolution can only be possible by creating the knowledge base and the infrastructure to evaluate its scientific, technological and societal impacts and implications. The proposed center is indeed the essential step towards accomplishing the goal. It will develop the knowledge base and disseminate it to other educational institutions at every level, industry, and community at large. It will develop the infrastructure to explore the unprecedented opportunities nanoscience and nanotechnology has to offer. Further, it will educate and train the new generation of scientists and qualified and well-trained work force to realize the overarching goals.

Among the educational activities are the development of curriculum, courses, and workshops, on nano-science and nanotechnology including but not limited to nanoscale manufacturing processes, nano-biotechnology, and nanoengineering.

The unit will be responsible for establishing programs in the following areas

- Minor in nanotechnology: principles of nanotechnology, hands on training, applications, IP & commercialization
- Research experience for undergraduates (REUs)
- General education: K-12 schools (REHS)
- Teaching the teacher (research experience for teachers: RETs)
- Outreach program: Training professionals and industrialists, public dissemination, and communications

17.7.1.2 Training Initiative Teaching/Training Core

To realize the training mission of the proposed center, a teaching/training core will be established. This core will primarily serve undergraduate and graduate students as well as technicians from the industry. It will be open to all universities, research organizations, general school education, industry and the community at large. The funds will be used to upgrade general Chem-Phys-Optics-Electro lab, construct a nanomaterial lab with general purpose characterization, and demo modules as well as to construct class modules in nanotechnology. With the proposed core, we hope to create highly trained local human resources in advanced technology that may increase the competitiveness of the Kazakh economy consistent with existing knowledge-based global economy.

17.7.1.3 Undergraduate Student Start-Ups

In analogy with the so-called March madness basketball sport event, a Student Startup Madness (SSM) event has been established to encourage competition and interest in commercialization of nanotechnology. The activity is a nationwide tournament-style competition for college student digital media startups, culminating with the "Entrepreneurial Eight" national finals at South by Southwest (SXSW) Interactive. SSM is open to college student digital media startups from colleges across the USA. Each round of the nationwide tournament builds momentum, awareness and social media buzz, drawing attention to colleges and universities as sources for innovation, entrepreneurship and talent while showcasing innovative university entrepreneurship programs and encouraging college students to start businesses. From all online applications, 64 teams will be chosen by the SSM Selection Committee for the first round of competition

From the 64 first-round competitors, half (32 teams) will advance to the semi-final regionals round. From the regionals will emerge the "Entrepreneurial Eight," the top eight student teams that will be invited to pitch to a celebrity panel of well-known entrepreneurs, venture capitalists and investors at the SSM national finals at South by Southwest in Austin, Texas, in March 2017.

The University of Illinois, for example established a Student Startup Initiative. It is a program to train students in business. The Enterprise Works hosts also a number of Student Startups. Enterprise Works partners with the UI Illinois Launch program to host student startups, allowing students to have free space in the incubator. Examples of Student Teams at Enterprise Workshave included projects such as preservation of milk for up to 3 days without refrigeration to extend shelf in developing nations, low-cost prosthetic arms for developing nations, gear shifting system, makes pushing a wheelchair easier, creating rapid solar cells at lower manufacturing costs using new technology from the College of Engineering, software to reduce printing costs by formatting efficiently, selling specialty teas with functional properties, Merge FM: allowing fans to interact with their favorite bands during the creative process. Students may submit proposals to be considered for access to the co-working space and shared lab facilities in Enterprise Works. Proposals are approved on a competitive basis.

FIGURE 17.10 **Startup Cafe is a speaker/luncheon series at the research park of the University of Illinois.**

Occasionally, shared space that is dedicated to student startups may be available but is not guaranteed.

Another example of the activity at the research park of the University of Illinois is the startup café (Fig. 17.10). Startup Cafe is a speaker/luncheon series intended to highlight successful entrepreneurs. The guest entrepreneur speakers have connections to the Research Park. The forum is designed to allow entrepreneurs to share their story and experiences, both successes and failures, that is, entrepreneurial journey with the residents. By giving their insights on their success or failure, the program is expected to be highly motivating.

Drexler the Undergrad Who Moved the World

Kim Eric Drexler was born on April 25, 1955, in Oakland, California. As a freshman at MIT in 1974, Drexler began to wonder whether it was possible to reproduce mechanically what cells manage biologically and to manipulate atoms in complex patterns to make essentially anything that's physically possible. Drexler's ideas were respectable enough that in 1981, the prestigious *Proceedings of the National Academy of Sciences* published his first major paper, "Molecular Engineering: An Approach to the Development of General Capabilities for Molecular Manipulation." His first book, *Engines of Creation*, published in 1986. Drexler earned his PhD in 1991 in molecular nanotechnology. Drexler presented his theories to Congress in 1992. He testified before the Senate Subcommittee on Science, Technology, and Space during a hearing about "new technologies for a sustainable world." Subcommittee chair Al Gore declared his enthusiasm and vowed to fund exploratory research. The same year, Drexler published a 550-page blueprint for a molecular manufacturing system titled *Nanosystems: Molecular Machinery, Manufacturing, and Computation*.

At present time, many units are springing out around the country targeting the young. The units are entrusted with the responsibilities of establishing programs in the following areas

- Hands on training, applications, IP and commercialization
- REUs
- Teaching the teacher (RETs)
- Outreach program: Training professionals and industrialists, public dissemination, and communications.

Some units focus on undergraduate training and senior thesis research. They initiate educational and awareness programs within society, focusing on the need for nanotechnology, future prospects, and knowledge based society. In fact, such units may turn out to be pivotal in producing a contingency of trained students in the field of nano-technology and creation of a nano-culture on campus to go forward for establishing an applied unit that would focus on some applications that are relevant to the society.

17.7.1.4 Lab on a Van Energizer

Building Opportunity Out of Science and Technology (BOOST): Energizing Young Scientists.

This is a proposal to build a "lab on a van" energizer. The van is essentially a comprehensive, self-sufficient advanced training unit. The van will serve young scientists, including undergraduate and graduate students, professionals, industry, general school education, and the community at large. The initial grand tour involves "docking" at institutions in several cities of engagements, and leveraging local technical, human and financial resources, as well as communication facilities (Fig. 17.11).

The van carries facilities for synthesis or/and handling nano and advanced material, with general purpose characterization, demo modules, simulation and computation, and application modules, digital platforms, as well as a variety of instruction modules. Nanotechnology is ideal for training projects since it captures the imagination of the young and the old, and serves as a rally for all: scientists, students, funding sources, decision makers, politicians, and the general public. It attracted billions upon billions of research funding dollars, and thousands upon thousands of scientists and engineers. It affords us to dream and be futuristic in hope of solving the major problems facing the human race, such as acute disease, the energy, lighting, and food crunch. It is interdisciplinary, which enables pooling of human and funding resources. Finally, many aspects of the technology are low cost, allowing developing countries to participate and collaborate in state-of-the art R&D at the same footing with industrial countries.

With a "lab on a van" concept, we hope to catalyze science and technology excellence and to connect them with the global scientific community, catalyze the creation of large number of highly trained human resources and leaders in

FIGURE 17.11 **Lab on a Van for energizing the young.**

advanced technology that may increase the competitiveness of economy and improves quality of life consistent with knowledge-based global economies.

The program includes training modules in the following areas

- hands on or/and web-based virtual technical training and instruction (synthesis/fabrication, characterization, demonstration, and simulation of nano and advanced material, such as silicon nanoparticles and nanowires, graphene, carbon nanotubes, phosphor, polymer, and other metal and oxide nanostructures and composites.); low-cost applications in renewable energy and lighting, water, environment, electronics, and biomedicine.
- competency and sustainability: science writing, grant writing, accessing and sharing data, communication and outreach; scientific rigor, transparency, and ethics.
- Commercialization: IP, generation and protection, licensing, partnerships with industry, startups, high tech management, angel and venture capital
- Group work, collaborations, and partnerships, networking.
- safety and ethical use: toxicity of chem, bio, and nano material; biocompatibility, code of conduct and ethical use of S&T, and toxic materials in the military.

17.7.1.5 Nanotechnology for Children

Many of us are familiar of the statement: "Who says you can't explain theoretical physics to kids"? Nanotechnology is not an exception. Nanotechnology, coined the technology of this century and the meeting ground of all fields attracted billions of research funding dollars, and thousands of scientists and engineers. It is simply exciting, novel and ventures into areas of science and technology beyond nature where we never been. It captures the imagination of the young and the old, the scientist and the layman, and affords us to dream and be futuristic in hope of solving the problems facing the human race, such as acute disease, energy, lighting and food crunch. But how do you explain it to children especially since it relies on advanced quantum concepts of material characteristics. There have been many *efforts to* simplifying science in general to children, with concerted effort to reach out to children to explain nanotechnology and incite them to get interested in science and technology via several agents. Nanotechnology points to the future; it aims to have fun with this fascinating revolution and inspire readers to learn more. Especially as children and young adults become more technology-fluent, the demand grows each day for easy to understand books explaining how nanotechnology will be used in buildings, cars, clothing, entertainment, education, fashion, food, furniture, homes, media, medicine, personal hygiene, space exploration, sports, telecommunication, etc.

17.7.1.6 Entertainment and Humor

Some nanotechnology books for children and young adults are designed to use entertainment and humor to educate and entertain readers of all ages about the strengths and advantages of nanotechnology. Author, Mark Tomassoni saw a need for light-hearted books to educate and entertain readers of all ages about the benefits of nanotechnology. Each book presents an insightful and engaging look at how nanotechnology is evolving and what the future portends. For example he uses the nanotechnology characters, Nano and Nana to present nanotechnology. He allows Nano and Nana to encounter a full range of strange and powerful forces in remote areas of the nano-world. They apply their vast powers of super computing, genomics, biomechanics, and artificial intelligence to overcome illnesses, pollution, starvation, and intergalactic communication; apply unique nanotechnology skills to overcome many of today's ills and make the world a better place.

17.7.1.7 Story-Telling, Science Fiction

World- renowned British physicist and mathematician Stephen Hawking teams up with his daughter Lucy, journalist and novelist to use science story-telling, for simplification of science for children. It is possible through story-telling to present complex science material in an exciting, simplified fashion, touching the present as well as the future. They use science fiction and adventure fantasy wrapped and interwoven with science concepts and facts to make the topic of science interesting and accessible to middle school children. The book tells the story of a middle school child called George and an easy going astro physicist called Eric who embodies the person of professor Hawking himself, whereby the professor uses his personal speaker, a powerful table top supercomputer called "Cosmos," which is able to open a window and propel them into the universe, as well as capable of manipulating space, time and matter. The voyage takes them to the far reaches of space and back. Along the way, the adventure teaches little George and the reader a lot about science and how the universe works, including Stephen Hawking's latest theories about black holes and quantum mechanics. The end result of the work is an informative and entertaining book, explaining planets, stars, comets, black holes and more to readers of all ages. I should also emphasize that the book transcends science. I believe it constitutes a good idea of Middle School drama.

17.7.1.8 Dr. Nano Fiction Character

The author Munir Nayfeh is another example of efforts to simplifying nanotechnology to children (Fig. 17.12). Under the trade mark of Dr. Nano, he wrote short stories in which science and science fiction are interwoven. These include the chocolate cake; the gold ring (the monks). Moreover, he was inspired by Steven Hawking writing to write a short story "the cosmos and the nanos," which utilizes Stephen Hawking's "Cosmos" supercomputer concept to present nanotechnology to children and adults alike. Drafts of some of my short stories on nanotechnology were tried at Dr. Howard Elementary here in Champaign. Some of these stories were translated to other languages. In collaboration with

FIGURE 17.12 Dr. Nano fiction character: Munir Nayfeh, under the trade mark Dr. Nano, simplifies nanotechnology to (top) children (at Dr Howard Elementary School in Champaign, Illinois); and to school teachers (Virginia Biotechnology Institute) (bottom) to students at Iskendrun University in Turkey.

Ebtesam AlOlayan of King Abdullah Institute for Nanotechnology (KAIN), some were produced in Arabic.

17.7.1.9 Shrinking the World Metaphor

Nanotech is on the cusp of applied sciences and involves manipulating matter on the molecular and atomic levels. "I would call it 'Honey, I Shrunk the Doctor,'" says Irfan Ahmad, the associate director of the Center for Nanoscience and Technology, of the research. Ahmad says that the technologies tiny sensors could be used to quickly diagnose injuries and illnesses, which could be crucial for soldiers on the battlefield. This metaphor was used to present in an article, written with Brian Cunningham for the IEEE nanotechnology Magazine, a brief snapshot of some of the agricultural, biological, electronics, medical, and manufacturing applications currently under active research and development at the Center for Nanoscale Silence and Technology (CNST) collaborator at the University of Illinois at Urbana-Champaign. The trend has been to prefix the word nano to any application (such as nanoagriculture, nanoelectronics, nanomedicine, etc.) that leverages features at the nanoscale to design and develop materials, processes, sensors, devices, and systems.

On 22 December 2009, Ignitee Digital Solutions Pvt Ltd launched a global digital media handbook called *Honey! I Shrunk the World*. The book includes digital basics, internet and mobile statistics of countries, list of websites, trends and predictions, a news roundup, case studies of brands who have won awards, tables and a glossary James May and Joe Dunton stared the 29 minutes episode of Honey I shrunk the World. In the episode which was aired in July 2007, May concludes that the major breakthroughs in transportation over the last 100 years shrunk the world not by allowing us to see more of it, but – thanks to television, computers, and fiber optic cable – by increasingly bringing it to us. In The Opinion Pages under the title "Foreign Affairs; Honey, I Shrunk the World" Thomas Friedman wrote in September 12, 1999 that the internet has been adopted in Australia with such zeal for a simple reason: Australia is a million miles from everywhere, and sparsely populated across a vast continent. The Internet can shrink that distance between Australia and the world, and between Aussies and Aussies, for almost no cost. "For years we have been praying for faster airplanes – we thought that would be our salvation," says Annie Hayes, an Australian conference organizer. "With the Internet, we can be so much more in touch with the world now, without actually leaving here."

17.7.1.10 Nanotech Pioneers: From Scientists, to Futuristic, to Politicians: Where are They Taking Us? [10]

In 2005, Steven Edward, an insider and experienced science writer in the area of nanotechnology, presented a book in which he gave a vivid look at nanotechnology (Fig. 17.13). Steven Edwards's goal was to bring nanotechnology closer to the science-interested general public, as well as scientists, students, organizations, journalists, politicians, and entrepreneurs. Edwards believes that

FIGURE 17.13 **The cover of a book written in 2005 by Steven Edward, an insider and experienced science writer in the area of nanotechnology.**

the nanometer scale is where physicists, chemists, biologists, materials scientists and engineers met to create a new technology – nanotechnology. He often laments weather nanotechnology is hype, hope, or horror?

Edwards says that although nanotech still has a science-fiction aura, there are actually consumer products already available which employ nanotech – sunscreen, paint, tennis balls, non-staining fabrics, and some recently approved drugs. The variety of new products and technologies that will spin out of nanoscience is limited only by the imagination of the scientists, engineers and entrepreneurs drawn to this new field. He presents enough facts and details to separate the hype from the real nanotech that is just around the next corner:

- superior materials – stronger, lighter, harder, better
- smart drugs that specifically target infected or malicious cells
- life-like prosthetics, such as limb, joint or organ replacements, even artificial eyes
- brain/computer interfaces
- new computer architectures with immense capacities and abilities
- new kinds of power sources that lessen our fossil fuel dependence

The book presents a realistic view of how this new field of technology will affect people in the near future. The author often laments weather nanotechnology is hype, hope, or horror? But he presents no military application, or wild fantasies of horror nano-robot predators and other science fiction stuff.

The book is in close contact with many pioneers in nanotechnology, and includes their backgrounds to allow readers, especially college students considering a career in the field, to better imagine themselves in such positions. However, technology does not develop in a vacuum, and this book also looks at the social, political and economic changes attendant upon the development of nanotechnology. This book is for the science-interested general public, as well as chemists, students, lecturers, chemical organizations, materials scientists, journalists, politicians, industry, physicists, and biologists.

Among the most famous pioneers he mentions: Isaac Assiminove, Isaac, Newton, Schrödinger, Van Der Waals, Darwin, Feynman, Neil Bohr David Awschalom, Gerhard Binning, De Broglie, Steffen Hawkins, and Edison. He also includes Shakespeare, James Bond, and Bill Gates. Among the politicians he mentions several US presidents and vice presidents: George Washington, Jimmy Carter, Ronald Reagan, Bill Clinton, Al Gore, and George Bush. It is also interesting to note that the University of Illinois at Urbana-Champaign has been recognized as the only single institution which has three pioneers: Ken Suslick, Michael Strano, and Munir Nayfeh. Finally, the document mentions only two of Arab origin as pioneers: Mongi Bawendi (Tunisian origin for his pioneering work on CdS nanoparticles) and Munir Nayfeh (for his work on silicon nanoparticles).

17.7.1.11 Student's Contributions to the IT Revolution

The University of Illinois at Urbana-Champaign (UIUC) has bred some of the most remarkable tech visionaries in history (Fig. 17.14). They built companies that essentially changed tech history as we know it. Those include mosaic, Netscape, PayPal and YouTube. First, mosaic was invented at the National

FIGURE 17.14 Student pioneers and tech visionaries from the University of Illinois in the information technology neo revolution: Marc Andreessen, Eric Bina, Chad Hurley, Jawad Karim, Steven Shih, Max Levchyn. They had roles in Netscape, Mosaic, paypal, and YouTube.

Center for Supercomputing Applications (NCSA) at the University of Illinois Urbana-Champaign beginning in late 1992. It has been credited with popularizing the World Wide Web. NCSA released the browser in 1993, and officially discontinued development and support four years later. However, it can still be downloaded from NCSA. Second, Netscape was also developed at University of Illinois by Marc Andreessen of NCSA and Silicon Graphics' Jim Clark. Eric Bina, a graduate of UIUC co-created Mosaic and was among the first employees of Netscape. Netscape Navigator was later developed by Netscape, which employed many of the original Mosaic authors. Netscape Navigator's code descendant is Mozilla Firefox. Microsoft licensed Mosaic to create Internet Explorer in 1995. Netscape had monstrous impact as one of the first and most important startups to shape the internet. As to PayPal, Chad Hurley, Jawad Karim, Steven Shih, Max Levchyn of UIUC students founded it in 1998. In 2005, Chad Hurley, Jawed Karim and Chen of UIUC founded YouTube, with Chen having the position of chief technology officer.

17.7.2 Environment, Fears and Concerns, Safety, Military and Social Issues

There is a danger in overestimating nanotechnology as being a solution for all: and also a danger in viewing nanotechnology as a "solution" to everything of the challenges. Some may look at it as a diversion because in some cases its application may undermine local alternatives and more appropriate approaches to dealing with the problems at hand as well as siphoning resources while causing health hazard. In countries, which rely on single commodities it might hinder commodity economy. Advances in science and technology may have profound impacts on commodity production (especially oil) and trade. There are concerns that nanotechnology will change the commodity markets, disrupt trade and eliminate jobs. Others believe that it can be used by enemies as a genetic weapon [11,12].

Nanotechnologies may have unintended negative side effects or consequences for health and environment because nanomaterial are too small to be seen (hard to see, follow, or track); toxic material and methodology; heavy metal poisoning; carbon poisoning. On the other hand it may have intended (deliberate) effects, including polluting and contaminating water and air resources as well as can be used as Military ammunition and weapons. In fact some may go as far as believing that it can be used for Nano-based genetic discrimination against an individual, family or race; or for nano-based enhancement of the performance of the human body (voluntary or involuntary).

When working in nanotechnology one needs to consider safety issues including effects on genes, heredity and society. The technology in large part uses toxic chemical methodology which has potentially harmful effects. Also in the case of heavy metal nanoparticles, such as cadmium, lead or tungsten, there is evidence that such particles may not flush and stay as a poisoning centers in one's body. Being small it is hard to track and it is impossible to find or remove.

The nano poisoning is similar to the dime (dense inert metal explosives) weapon, which was most recently used against Gaza civilians. Dime bombs produce an unusually powerful blast within a relatively small area, spraying super-heated micro-scale shrapnel of powdered heavy metal tungsten alloy (HMTA), which makes forced entry into organs and cells. It would not be surprising if the shrapnel included submicron particles ~100–200 nm in size. Scientific studies have found that HMTA is chemically toxic, damages the immune system, rapidly causes cancer, and attacks DNA (genotoxic). Very little testing of its adverse effects if any is available. Tungsten nano powder with an average particle size of 30–120 nm has recently become available commercially. Means of thermal heating, as well as of induced force entry, such as ballistic entry into cells must be developed to conduct real testing.

In fact Europe & Japan: are already debating banning the use of heavy metals nanos in house-hold applications, such as commercial displays. US FDA control is expected to follow suit. We should note that silicon-based nanotechnology is least toxic and it is the most biocompatible and compatible with the environment. It does not require further processing or shielding and is expected to dissolve in the body into silicic acid, an acid found in the body. Ebtesam Al Olayan from King Saud University is presently testing the toxicity of silicon nanoparticles, dissolved in cosmetic products, using in vitro, as well as animal testing. We are reminded of how much silicon we deal with directly or indirectly by just mentioning Si-based body implants and computers.

Others foresee the doomsday scenario of a world ruined by nanotechnology as its assembler robots run amok, beyond all human control, reducing everything to a homogenous dust as they replicate themselves without check.

Writing about technology will not be complete without addressing the question of technology versus military. In particular it is interesting to reflect on the origins, nature, and effects of technology on ancient, as well as on modern and postmodern world. We believe that this is a subject that warrants an article of its own. But briefly here we can say that there has been a raging debate on who drives who and how did the interrelationship between them develop or change with time. At this time there is a belief that the military, especially modern military is the driver of technology development to a large extent. For instance a whole lot of technology development in the US today is driven by mission oriented projects funded by the Army, Air Force, Navy, and Defense-related agencies.

The majority of people however believe that, to a large extent, the military just exploits technology advances that are driven by business and consumer demand. Certainly, the use of technology for hunting, food collection and survival is as old as the human race. In the Iron Age, iron was used prominently as tools and weapons for hunting. Iron-working for the military was enhanced by the carburization process, a process by which carbon is introduced into a metal to make the surface harder and more abrasion resistant. Metal, primarily iron and brass, began to be used for arms and armor. Wood, earth and stone were used for construction of army facilities. The decisive use of steel in military action as crowned by the development of the Damascus sword was thought to be a deliberate development.

Paul Virilio is one of the most aggressive critics of technology, especially military technology. He believes that technology is transforming the world and even the human species, with a highly destructive impact. This is countered by some who believe that this is a flawed conception of technology that is excessively negative and one-sided. It misses the empowering aspects of new technologies.

17.7.2.1 Nanobullets?

When working in nanotechnology one needs to consider safety issues including effects on genes, heredity and society. The technology in large part uses toxic chemical methodology, which has potentially harmful effects. Also in the case of heavy metal nanoparticles such as cadmium, lead, or tungsten, there is evidence that such particles may not flush and stay as a poisoning centers in one's body. Being small it is hard to track and it is impossible to find or remove. The nano-poisoning is similar to the dime (dense inert metal explosives) weapon, which was recently in the news. Dime bombs produce an unusually powerful blast within a relatively small area, spraying super-heated micro-scale shrapnel of powdered HMTA, which makes forced entry into organs and cells. It would not be surprising if the shrapnel included submicron particles ~100–200 nm in size. Scientific studies have found that HMTA is chemically toxic, damages the immune system, rapidly causes cancer, and attacks DNA (genotoxic). Very little testing of its adverse effects if any is available.

In a paper Niels Harrit group, Chemistry Department, University of Copenhagen, Denmark discussed the nano-thermite weapon in the aftermath of the September 11 attack. They noted that reactions of material become faster when the material is reduced to the nano regime. Nano aluminum, rust-powder, and other additives are reactive and can explode producing intense heat as high as 2500°C, which can melt iron. Such explosions produce more energy than dynamite and may be used as rocket fuel.

17.7.2.2 Regulations of Nanomaterials in USA and European Union (EU): REACHnano

In the US, there is no federal or state legislation specific to nanomaterials. Nanomaterials are managed by current regulatory framework for chemicals, pesticides, food, cosmetics, and drugs through the EPA and FDA agencies. Nanomaterials are considered as chemical substances under the Toxic Substance Control Act (TSCA). In the process, manufacturers must submit premanufacture notices of new nanomaterial to the Agency prior to manufacturing or importing them. TSCA identifies nanoscale substances by their molecular identity rather than physical characteristics such as particle size. For example, the spectrum of multi-walled carbon nanotubes are considered new substances/nanomaterials. According to the 2015, updates of nanomaterial regulations

(TSCA section 8(a)) in USA, certain chemical substances that are manufactured or processed as nanoscale materials, reporting, and recordkeeping would are required. Moreover production volume, methods of manufacture and processing, exposure and release information, and available health and safety data must be filed.

In the EU, producers and importers, as well as down-stream users must submit registration dossiers in which they identify, characterize, and evaluate their substances, mixtures, and products. REACH is the primary EU regulation on the health, safety, and environmental risks of chemical substances (Fig. 17.15).

FIGURE 17.15 **Nanotechnology industries associations (NIA), and the "REACHnano" project, developed to help the industry and stakeholders to conduct risk assessment of nanomaterials.**

In May 2014, Annexes amendment were proposed in 2014 to ensure efficacy of nanomaterial registration. This position paper discusses these proposed amendments.

Development of a web-based REACH Toolkit to support the chemical safety assessment of nanomaterials/LIFE11 ENV/ES/000549 easy-to-use tools to support the risk assessment of nanomaterials.

The "REACHnano" project was developed to help the industry and stakeholders to conduct risk assessment of nanomaterials. For example, "REACHnano" presents standard testing models for risk characterization for nanomaterials, as well as provides a set of exposure scenarios and makes available accumulated information about physicochemical, toxicological, and ecotoxicological properties of nanomaterials and related exposure-, use-, and risk-management measures.

According to the Center of International Environmental law, the important revision of reach annexes for nanomaterial are shown in Fig.17.16. The nanotechnology industries associations (NIA) has established a service to assist its members manage their nanomaterials. In addition to the general REACH-nano information the NIA provides and offers its members individual consultations.

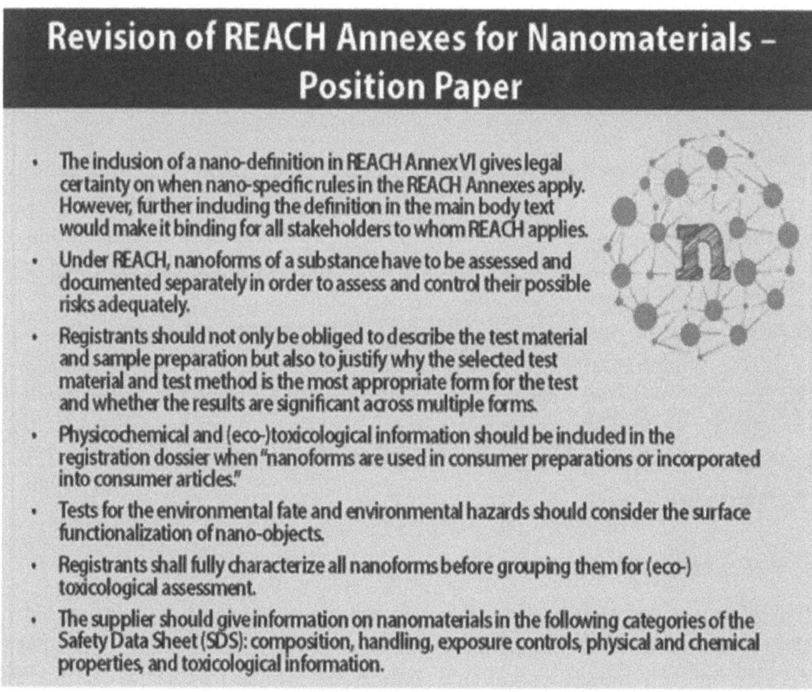

FIGURE 17.16 Important revision of reach annexes for nanomaterial according to the Center of International Environmental law. http://www.ciel.org/research-publications/.

17.7.3 Nanotechnology Strategy for the Poor: The Grand Challenge Initiative for Developing Countries

Peter Singer has been talking about a grand challenge initiative of nanotechnology. This has been in response to a proposal by Canadian Prime Minister Martin. He proposed devoting 5% of the country's research and development budget to technologies that would help the developing world. "If every industrialized country pitched in, we could see $50 billion. Peter Singer came up of the Top Ten application of nanotechnology that would be relevant in this regards of course in addition to defense and Data storage or communication. Singer list is follows

1. Energy storage, production, conversion: semiconductor quantum dots or nanoparticles boost the efficiency of solar cells); carbon nanotubes, which are stronger than steel, help store hydrogen
2. Agricultural productivity enhancement: nanoporous minerals Zeolites can control release of nutrients
3. Water treatment and remediation: Nano filters that can desalinate water and purify water of bacteria and viruses much more efficiently and with much less volume. Nanomagnets that can be sprayed onto oil spills and separate the oil from the water, cleaning up pollution
4. Disease screening and diagnosis: labs-on-a-chip uses just a drop of a patient's blood to conduct a sophisticated battery of clinical tests.
5. Drug-delivery: intelligent drug-delivery nanosystems can release drugs at the right time in a single dose and last a long time for treating tuberculosis or HIV
6. Food processing and storage: nanofilms can do away with the need for refrigerators
7. Air-pollution clean-up: titanium-dioxide nanoparticles in sidewalks or house paint that, in the presence of sunlight, catalyze the breakdown of several pollutants.
8. Construction materials for cheaper, stronger housing (super tents with great insulation and soundproofing)
9. Health-monitoring: They could help patients monitor key signs, such as glucose or cholesterol levels,
10. Pest detection and control, including improved pesticides and insect repellents.

17.7.4 Ethics in Nanotechnology

UNESCO has been advocating the adoption of a code for ethics in nanotechnology. It organized expert meetings on Nanotechnologies and Ethics in several Regions. The meetings gathered experts in the field of nanotechnologies, ethics, and law from the region and outside the region. During the two-day meetings, experts exchanged information and opinions on how to address ethical issues of nanotechnologies. At the closure of the meetings participants adopted a recommendation which urges UNESCO to start the process of elaboration of Declaration on Ethics of Nanotechnologies. Among the issues raised is that (i) military research and applications should be limited to defense and security systems, not for political purposes or aggression; (ii) prohibit the use of the technology to produce genetic-targeted weapons that selectively favor or disfavor a race; (iii) users must require training and awareness, such that participants teams should have experts who have sufficient training and awareness in ecology and public safety; and (iv) scientists and their organizations must also be held accountable for the willful, fraudulent or irresponsible misuse of the science. The code also advocates (i) an open source mode in which dissemination of results to the public is encouraged as much as possible, especially, non-military government-funded research; (ii) and adoption of a General research ethics in which Published research should adhere to the standard scientific method used in any other science with regard to accuracy, crediting collaborators, contributing sources, mutual trust of motives, etc.; (iii) finally commercialization and business must adhere to clarity and accuracy of labeling of products, avoiding conflict of interest with regard to promotion of services, incorporation of long-term efficient use of resources, recycling of toxic materials, adequate compensation for workers and other fair labor practices, contributing to public education and supporting reasonable legislation to deal with legal and social issues.

17.8 GLOBAL STATUS OF NANOTECHNOLOGY

According to the *UNESCO Science Report: towards 2030*, the interest in nanotechnology is spreading across the globe as more and more countries enter the race for leadership in the new technological front [13]. The report states that in many countries, nanoparticles have become part of daily life. They are also found in clothing, tires, fertilizers and cosmetics, for instance. Given the breadth of potential applications, nanotechnology has become a research priority for many countries. We outline here some of the main indicators.

First, in industrially advanced countries, Thomson Reuters' data concludes the scientific, as well as innovation output in the field of nanotechnology is still going strong with The Scientific nanotechnology articles per million population in 2013 (Thomson Reuters' data cited by the *UNESCO Science Report* and analyzed by Statnano).

Switzerland	198
Republic of Korea	150
Germany	93
France	79
USA	69
Japan	56

The number of patents per 100 articles on nanotechnology (according to the *UNESCO Science Report* and analyzed by Statnano) ranks as follows:

USA	44
Japan	30
R. of Korea	27
Germany	22
Switzerland	17
France	15

Second, the interest in nanotechnology has picked up in BRICS countries (Brazil, Russian Federation, India, China and South Africa). Recent activities have indicated that all of these countries have been striving to become nanotechnology hubs. Their contribution to this field nevertheless remains relatively modest, with nanotechnology articles per million population: China: 25; Russian Federation 23; Brazil 9; South Africa 9' India 6. Although the academic output in nanotechnology of these countries is growing, related patents and products are not always progressing at the same pace: in the ratio of nanotechnology patents to 100 articles on nanotechnology (2015): South Africa 2.47; China 2.28; Brazil 1.67; India 1.61; Russian Federation 0.72. This is compared to Italy 4.46; UK 8.39; and Canada 10.08.

Third, nanotechnology has taken advanced manufacturing to another level. It has become interwoven with nanotechnology, with nanotechnology constituting a key element. It is being pursued by a widening circle of industrial countries that include Australia, Canada, China, France, Germany, Japan, the Republic of Korea and USA leading the effort. Through advanced manufacturing, governments hope to enhance national competitiveness and create jobs. As an example we mention here the Advanced Manufacturing Partnership launched by president Obama in 2013. Steered by a committee drawn from the industrial, labor and academic sectors, this partnership benefits from an investment of US$ 2.9 billion under the Revitalize American Manufacturing Act (2014). These funds, which are to be matched by private and non-federal partners, are being used to create an initial network of up to 15 institutes, including several with a focus on additive manufacturing, such as three-dimensional (3D) printing, digital manufacturing and design, lightweight manufacturing, wide band semiconductors, flexible hybrid electronics, integrated photonics, clean energy and revolutionary fibers and textiles. Another example is China. Advanced manufacturing has become the focus of one of China's 16 mega-engineering programmers to 2020, by which time the country plans to be "innovation-driven." In 2014, advanced manufacturing was incorporated in Canada's revised research strategy, *Seizing Canada's Moment: Moving Forward in Science, Technology and Innovation*, with a focus on automation, including robotics, lightweight materials and technologies, additive manufacturing, quantum materials and nanotechnology. It is clear that the impetus of the interest in nanotechnology in this regard is the belief that through advanced manufacturing, governments hope to enhance national competitiveness and create jobs.

17.8.1 Commercialization-Science Lag

In 2014, it was realized that the first phase of R and D, which was initiated in 1999 has matured enough and that nanotechnology is at a critical transition point, ready to enter its second era, namely commercialization. In other words, only 15 years after the onset of interest in nanotechnology that interest and support shifted to commercialization issues. In the Russian Federation, on the other hand, commercialization of nanoproduct was emphasized right from the onset and was ahead of fundamental research. Although the production of nanomaterials has grown considerably, academic articles on nanotechnology do not seem to be progressing as quickly as in some other economies. Nor does Russian scientific activity appear to have translated, as yet, into a significant amount of patented inventions. There were just 0.72 patents per 100 articles on nanotechnology in 2015, down from 1.18 five years earlier. Over 500 companies were engaged in manufacturing nanotech products in 2013, according to the state corporation Rusnano. In 2010, the *UNESCO Science Report* had stressed the significance of the Russian *Strategy for Nano-industry Development* (2007). By 2013, sales of nanotechnology-related products exceeded RUB 416 billion (more than US$ 15 billion). This is 11% over the target fixed in 2007 and means that the industry has grown 2.6 times since 2011.

17.8.2 Other Developing Countries

Several developing countries, particularly Iran and Malaysia are venturing in nanoworld. For instance, the number of articles on nanotechnology between 2009 and 2013 showed strong growth in both of these countries, with Iran now is in seventh rank while Malaysia is in the 22nd rank. As to the number of articles per million population, it quadrupled in Malaysia to 41, and tripled in Iran to 59.Thus these countries overtook Japan in these categories. This progress has not extended, however to the patent category. For example in 2015, the ratio of nanotechnology patents to articles in Iran was 0.41 per 100 articles while it was 0.73 in Malaysia. Moreover, on the commercialization side 143 nanotech companies have been established in eight industries over the last 10 years. In Saudi Arabia, there has been a concerted effort to venture in nanotechnology, with King Abdulaziz City for Science and Technology (KACST) acting as a technology incubator. It fosters ties between research universities and between the public and private sectors in a wide range of areas, including nanotechnology and advanced materials. Moreover, patenting is highly encouraged with KACST acting as the national patent office.

17.9 THE INTERNET OF THINGS (IOT)

In the emerging field of Internet of Things (IoT), objects, animals or people are provided with unique identifiers, such as a radio frequency ID (RFID)[14] (Fig. 17.17). Using this technology and appropriate sensors data over a network

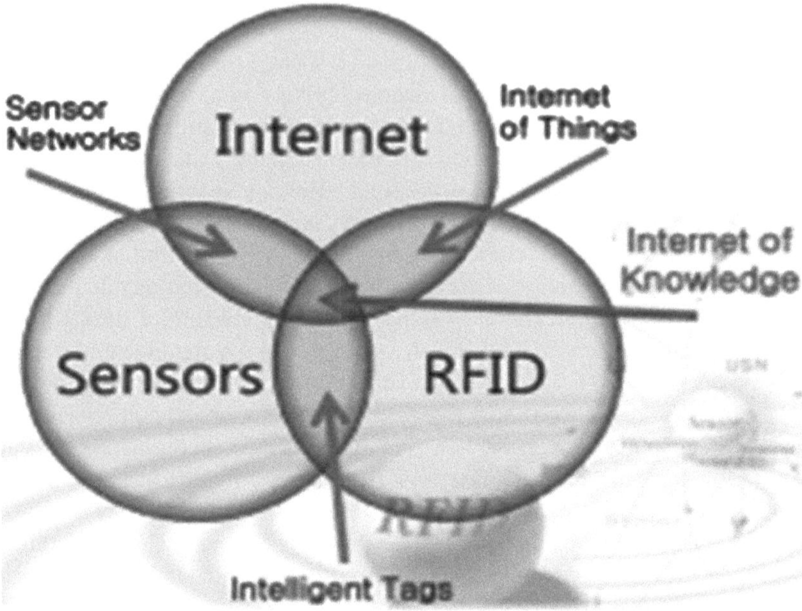

FIGURE 17.17 **Internet of Things (IoT).** *(From http://www.bit-tech.net/news/internet-of-things-government-cash/1/)*

(the internet for example) is transferred to a central facility for analysis without human-to-human or human-to-computer interaction. IoT evolved from the convergence of wireless technologies, micro-electro mechanical systems (*MEMS*) and the networks Internet.

Rank	Country	Devices	Rank	Country	Device
1	South Korea	37.9	13	Canada	11.6
2	Denmark	32.7	14	Italy	10.2
3	Switzerland	29.0	15	Brazil	9.2
4	United States	24.9	16	Japan	8.2
5	Netherlands	24.7	17	Australia	7.9
6	Germany	22.4	18	Mexico	6.8
7	Sweden	21.9	19	Poland	6.3
8	Spain	19.9	20	China	6.2
9	France	17.6	21	Colombia	6.1
10	Portugal	16.2	22	Russia	4.9
11	Belgium	15.6	23	Turkey	2.3
12	United Kingdom	13.0	24	India	0.6

FIGURE 17.18 **List of countries by IoT devices online per 100 inhabitants.**

The technology is expanding very rapidly. In fact according to the World Economic Forum, *nanosensors and the Internet of Nano things* has made it to the top of the list of the *technological innovations for the year 2016.* The market is expected to more than double over 4 years (2016–2020) from $4.26 to $9.69 billion.

IoT products can be classified broadly into five different categories: smart wearable, smart home, smart city, smart environment, and smart enterprise. A list of countries by IoT devices online per 100 inhabitants (published in 2015) (Fig. 17.18). The IoT, is rapidly expanding the online universe from computers and mobile gadgets to ordinary pieces of the physical world: thermostats, cars, door locks, even pet trackers. It is expected to up to 30 billion of them to be online by 2020. It appears that an explosion of the number if connected items has begun. Smart functions can be remotely controlled by artificial intelligence, such as a front door designed to unlock only when the owner arrives. Another sensor which could be a lifesaver is an implanted heart monitor. It can be programed to call the doctor if the heart shows signs of failing.

The integration of IoT and nanotechnology has been coined the Internet of Nanothings (IoNT) (Fig. 17.19). There is also the internet of bio-nanothings (IoBNT), but it is still in theoretical research stages. IoNT infrastructure uses nano devices and sensors, IoT, networks, cloud computing, big data technologies, etc. In-body networks, for example could monitor real-time blood, sickness, breath tests, etc.

17.10 NANOTECH AND ENERGY SECURITY

It is believed that nanotechnology can help realize energy security. Energy security has become one of the leading issues in the world today. It requires

- uninterrupted physical availability
- price which is affordable
- respecting environment concerns

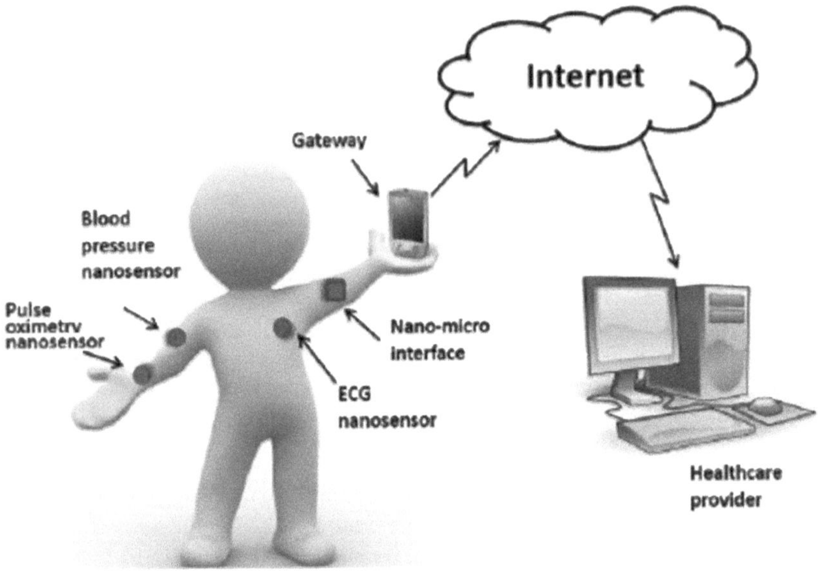

FIGURE 17.19 **Integration of IoT and nanotechnology.**

The energy resources presently available include:

- Fossil (oils, coal, natural gas). In fact, oil has become what oxygen is to humans but for the economy
- Renewable (solar, biomass, hydro (water), geothermal, wind
- Nuclear

Nanotechnology can help the oil and gas industry find new fields and extract more from existing ones. Numerous research initiatives, corporate R&D programs, and start-up companies are seeking nanotech solutions to oil supply troubles and distresses in terms of better equipment, smarter fluids, and new sensors for exploration. Energy applications of nanotechnology can be realized through creation of "new" materials, affect thermal and electrical conductivity, allow hydrogen production and storage, and more efficient solar power, batteries, fuel cells, etc. More efficient bulbs for lighting, enhancing oil recovery, and discovery.

There is no single solution. Nuclear is not the solution, but a pivotal component. Diversification of supply is a vital component of energy security. A package may include:

- Oil + natural gas + coal
- Nuclear: provides steady source
- Renewable: fill in the gaps, Intermittent
- Conservation
- Technology and innovation (nanotechnology): enhanced recovery, discovery, uptake of renewables, conservation

The new resources will become more useful as the price of exporting and importing oil increases due to the increase of demand.

17.11 THE PARIS AGREEMENT AND THE ENVIRONMENT AND NANOTECHNOLOGY

The US contributes about 15% of global emissions of carbon. But it is also a significant source of finance and technology for developing countries in their efforts to fight rising temperatures [15]. Nevertheless, President Trump's withdrawal from the agreement will

- Deprive the Globe from a major contributor to the reduction effort
- Inability of the reaming countries to pick up the US share will make it more difficult for the world to reach the goals that it set for itself in the Paris agreement (keeping global temperature rises well under 2C).
- The absence of the moral leadership of the US in this arena may prompt other countries to slack or lose interest and pull out.
- Drop in US investment in renewable energy, hence slow down or reverse tumbling price of renewables. Such drop is encouraging emerging economies to leapfrog to greener sources. In recent auctions in India, the price of solar energy was 18% lower than the average price for electricity generated by coal-fired plants.
- Danger of loss of support of corporate America. Leaders of companies, such as Google, Apple and hundreds of other including major fossil fuel producers, such as Exxon Mobil, who pledged support to the agreement may reverse direction.
- Drop in US funding in the field of nanotechnology as renewable energy is one of the major drivers of research and development of the technology. It is continuing to search for renewable energies for sustainable development and new technologies for carbon capture and environmental protection. In fact clean energy and environmental applications often demand the development of novel nanomaterials that can provide shortest reaction pathways for the enhancement of reaction kinetics. Understanding the physicochemical, structural, microstructural, and surface properties of nanomaterials is vital for achieving the required efficiency, cycle life, and sustainability in various technological applications. In general nanotechnology has been impacting many sectors of energy and environment including clean and sustainable technology; solar energy collection and conversion; hydrogen and fuels cells; energy storage and novel generation; nanotech for oil and gas; environment protection and remediation; water technologies; smart grid evolution.
- Clean water shortages as treatment and purification benefits tremendously from membrane processes, nanosized materials and the combination from both offer a wide range of possibilities and ways for water treatment by means of filtration and catalytic processes.

There is still a silver lining. Despite the withdrawal from the accord, US carbon will continue to drop. The projections are that they will fall about half as much as had been planned by President Obama. This is because there are already technologies procured and operation. Also may cities, and companies pledged to adhere to it. Moreover, if the trend in price drop continue, market consumer and market will save the day. There is yet another factor. US energy production is now powered more by gas than by coal. The fracking revolution has seen a huge jump in the production and a huge drop in the price of natural gas. Energy producers like gas because it is flexible and integrates better with renewable sources which are also growing rapidly.

17.12 BIG DATA, NANOTECHNOLOGY, AND ELECTIONS

It has been said that the age of big data has arrived, as we are witnessing explosions in data and applications [16]. The term big data has been coined to describe the situation for which there is (Fig. 17.20):

- Large or complex data such that standard data processing methodologies and software become inadequate.
- Large velocities of accumulation and processing of data;
- Large variety, variability, and complexity of data.

From technology point of view, big data is accompanied with innovative techniques for analyzing data, development of sophisticated technologies, such as business intelligence, cloud computing and databases; as well as advanced data visualization. In fact, one can also see the imprint of nanotechnology in big data technologies, but big data impacts it as well.

Other signs of the arrival of big data is the significant demand for data-literate managers, trained professionals in deep analysis and data scientists as well as large growth of Web traffic, social networks as well as software and sensors. A vivid example is monitoring shipments, suppliers and customers. Another example of data abundance is derived from politics. Politics is becoming increasingly data-intensive, involving computer-automated analysis of voters and trends in elections, blog postings, congressional speeches and press releases, and news articles, looking for insights into how political ideas spread and how people vote. Similar examples may be derived from science and sports, advertising and public health, academia, business, and government. Another example is the IoT. This is enabled by growth of two technologies, one of which is digital communicating sensors in industrial equipment, automobiles, electrical meters, and shipping crates, etc. Those sensors can measure and communicate location, movement, vibration, temperature, humidity, even chemical changes in the air. This capability of communicating sensors along with computing intelligence gave rise to the *IoT* or the *Industrial Internet* (discussed in 17.9).

Studies show that big data is growing at 50 % a year, or near doubling every two years. This is not unlike Moore's law known for electronics. A marriage

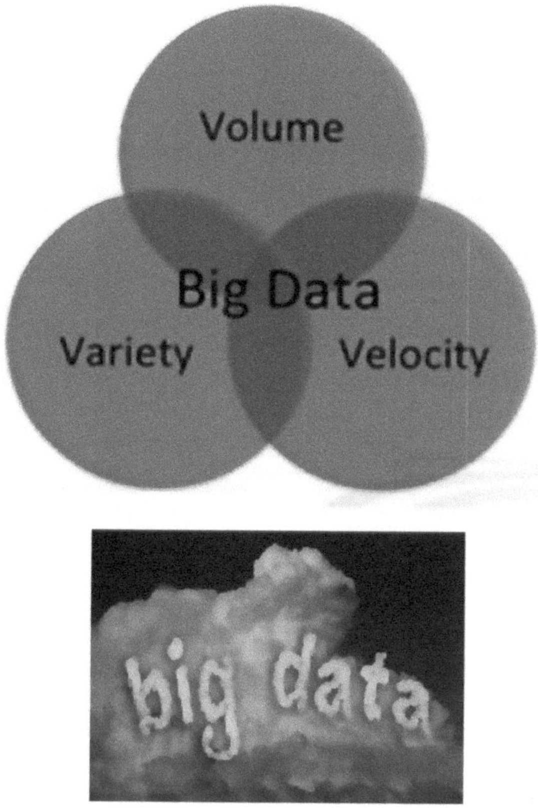

FIGURE 17.20 **Big data (definition).**

between electronics, nanotechnology and Big Data promises to propel the next industrial revolution. For example miniaturization of material (nanoparticles), sensor technology, energy and storage devices and development of smart materials enable collection of massive amount of data as it enables embedding the nano-sensors in concrete, engines, human body, etc. The wealth of new data, in turn, accelerates advances in computing and nanotechnology a virtuous circle of Big Data. This includes pushing the limits of chip technology towards produced the 5-nm node chops and less needed to meet the emerging demands of cloud computing and Big Data systems.

In the health sector, nanotechnology is making pharma become Big Pharma (Fig. 17.21). Nanoparticles is moving strongly into health products, such as sunscreens, fabric protectors, plastic food liners, and other products. A big splash is around the corner as plans are underway to incorporate nanoparticles in pills (tablets, capsules and vial caps) for making smart drug delivery. Through the nanoparticles, data will be stored in the pill, including product information (strength and expiration date), manufacturing information (location date, batch

FIGURE 17.21 **Big data and its impact on the pharmaceutical industry.**

and lot number) and distribution information (country, distributor, wholesaler and chain). Moreover, such information can be read by medical technicians who could then track what drugs one took in the past. It is a lot like a nano barcode incorporated directly.

As to relevance to politics, data scientists made pioneering innovations on applying big data to politics during the 2012 elections. Big data-driven decision making, optimization of information, activities and fund raising. Hundreds of thousands of computer simulations were ran going over how the elections might pan out. Technology allowed sending personalized messages to each voter; even family members received different campaign messages. Cloud computing played a big role in the campaign, where data was run using Amazon web services and used open source software to write programs.

But big data in politics a double edge sword. In the 2016, presidential elections it may have made it possible to hack data bases, e-mails accounts to reach voters en mass with personalized message (Fig. 17.22). With huge data sets and complex operations there is increased risk of false findings or implantation of false information and foul play. These disadvantages and more, however are not going to hold back progress in Big Data. On the contrary, a new initiative by the United Nations wants to utilize Big Data to develop digital early-warning signals for economical, health and even political issues, such as jobs, spending trends, or disease outbreaks. The idea here is that much of the data needed in the analysis can be derived from messages in social networks and text messages.

17.13 NANOTECHNOLOGY AND CONSUMER PRODUCTS

We end this chapter with some deliverables for the consumer. Nano has already found its way into so many types of consumer products that one may use every day. Fig. 17.23 shows images of some products which are presently available on the market. They are actually assigned part and ordering numbers. They include charcoal-based and silica-based nano tooth brushes; paints for cars and buildings, coating for shoes; screen covers for I-phones etc.

There are more sophisticated products. Those include aerogel wall insulator, one third the thickness of conventional insulation; nanoparticles-based

FIGURE 17.22 **Big data and its impact on politics.**

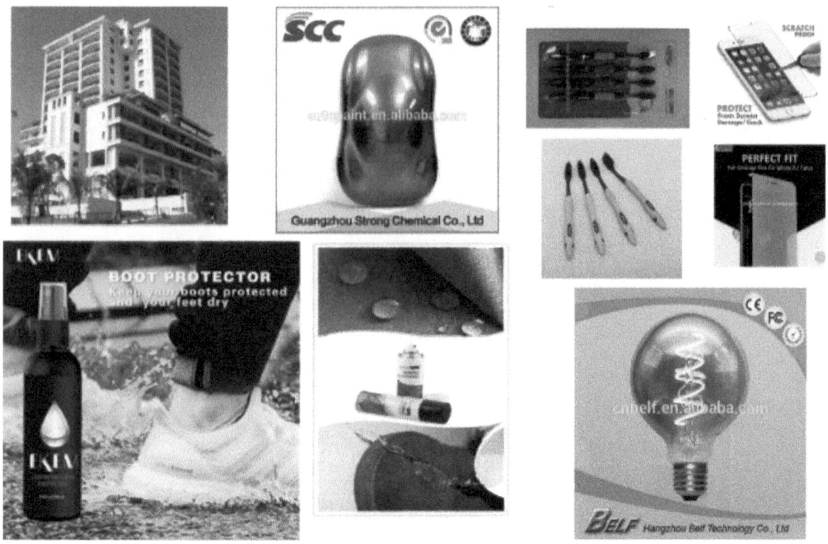

FIGURE 17.23 **Some of nanotechnology consumer products.**

sunscreens or coatings that block UV or kill bacteria on surfaces, or provide better soap (cleaning products); highly stretchable clothing (made using protein-based fabric) that stretch as much as 1500 percent; sports goods with enhanced performance (tennis balls, rackets, bicycles, fishing rods, etc.); enhanced energy and fire products, such as nanoparticle-based solar cells to charge cell phone and other hand-held devices; carbon nanofibers-based flame retardant for furniture.

In electronics, nanotechnology enabled the replacement of arcade size video games with very sophisticated home Playstations, X-Boxes, and Game Cubes. Other electronic applications include faster and more powerful computers, palm pilots (blackberries), flash drives, digital cameras and displays, cell phones, LCDs, LEDs, MP3's, electronic ink displays, thin film batteries, and flexible electronics.

In the biotech field, numerous useful products are becoming available, including bandages embedded with silver nanoparticles, drug delivery via a patch, time release thin films for implantations into the human body (for example screws, joints, and stents), respiration monitors utilizing nano-materials.

REFERENCES

[1] Nanotechnology: Concepts, Methodologies, Tools, and Applications: Concepts, Methodologies, Tools, and Applications, Management Association, Information Resources, IGI., Global, Feb 28, 2014; M. Karkare, Nanotechnology: Fundamentals and Applications, I.K. International Pvt Ltd, Oct 23, 2008; M.H. Fulekar, Nanotechnology: Importance and Applications, I.K. International Pvt Ltd, 2010; V. Sharma, Korada, N. Hisham, Hamid, Engineering Applications of Nanotechnology: From Energy to Drug Delivery, Springer, Jan 9, 2017.

[2] Escoffier L, Helwegen W. Nanotechnology Commercialization for Managers and Scientists. Boca Raton, FL: CRC Press; 2012.
[3] Education in Nanoscience and Engineering: Materials Research Society, vol. 931, 2006, April 17-21, 2006, San Francisco, California, USA.
[4] The Bayh-Dole Act: Selected Issues in Patent Policy and the Commercialization of Technology, Wendy H. Schacht, Congressional Research Service, Library of Congress, 201; Nanotechnology Intellectual Property Rights: Research, Design, and Commercialization, Prabuddha Ganguli, Siddharth Jabade, CRC Press, Jun 25, 2012.
[5] Open Innovation: The New Imperative for Creating and Profiting from Technology, Henry William Chesbrough, Harvard Business Press, 2006; Nanovate: Commercializing Disruptive Nanotechnologies, Mohab Anis, Ghada AlTaher, Wesam Sarhan, Mona Elsemary, Springer, Dec 8, 2016.
[6] Cozzens SE, Wetmore J. Nanotechnology and the Challenges of Equity, Equality and Development. Springer Science & Business Media; 2010.
[7] Block FL, Keller MR. State of Innovation: The U.S. Government's Role in Technology Development. Abingdon, UK: Routledge; 2015.
[8] The Venture Cafe: Secrets, Strategies, and Stories from America's High-Tech Entrepreneurs, Teresa Esser, Grand Central Publishing, Mar 14, 2002; Nanotechnology: "Risk, Ethics and Law" Geoffrey Hunt, Michael Mehta, Taylor & Francis, Jun 17, 2013.
[9] Light JF, Aznar MF. Nanoscience Education, Workforce Training, and K-12 Resources. Boca Raton, FL: CRC Press; 2010.
[10] Edwards SA. The Nanotech Pioneers: Where Are They Taking Us? Hoboken, NJ: John Wiley & Sons; 2008.
[11] Nanotechnology: Ethics and Society Deb Bennett-Woods CRC Press, Apr 29, 2008; Nanotechnology & Society: Current and Emerging Ethical Issues, Fritz Allhoff, Patrick Lin, Springer Science & Business Media, Apr 3, 2008.
[12] Asmatulu R. Nanotechnology Safety, Newnes. Technology & Engineering. Elsevier: Amsterdam; 2013.
[13] https://en.unesco.org/unesco_science_report.
[14] Buyya R, Vahid AD. Internet of Things: Principles and Paradigms. Internet of Things: Science Fiction or Business Fact? (PDF). Amsterdam: Elsevier; 2014. May 11; Harvard Business Review, Boston, November 2014. Retrieved 23 October 2016.
[15] The Paris Agreement: Climate Change, Solidarity, and Human Rights, Judith Blau, Springer, Mar 21, 2017; Paris Climate Agreement: Beacon of Hope, Ross J. Salawitch, Timothy P. Canty, Austin P. Hope, Walter R. Tribett, Brian F. Bennett, Springer International Publishing, Jan 12, 2017.
[16] Mayer-Schönberger V, Cukier K. Big Data: A Revolution That Will Transform How We Live, Work, and Think. Boston, MA: Houghton Mifflin Harcourt; 2013.

Chapter 18

Final Remarks

18.1 ENHANCED INTEREST AND SUPPORT OF SCIENCE

The full technical impact of nanotechnology remains to be seen but there is already a great deal of impact on the support for science: We mention here the following points.

- Interest in nanotechnology had synergetic effects on science development in general.
- Federal funding of nanotechnology initiatives has allowed for expansion into diverse fields and the impetus for expanding the scope of research of several fields, especially biomedicine.
- Some funding awarded to nanotechnology initiative research may have made it to science development in general.
- Nanotechnology's tools acquired have found applications and users by other programs in diverse fields, from biology to device physics. In fact, early successes led to a significant influx of resources and interest in science and nanotechnology in particular that led to the establishment of centralized research programs and facilities. Further government initiatives (at federal, state, and local levels) have firmly cemented these disciplines as "big science," with efforts increasingly concentrated at select laboratories and centers.
- In many respects, these trends have triggered or stimulated certain changes in academic science over the past 20 years, with a greater emphasis on applied science and research that can be more directly utilized for commercial applications
- Increased linkage between academia and industry for physics and materials science research, and a greater tendency towards more collaborative and centralized projects

18.2 ENHANCED INTEREST AND SUPPORT OF EDUCATION

Nanotechnology had a deep influence on the quality of education weather in the K-12 domain or college domain. This has been through additional federal or state funding or through the institution of programs for collaboration between colleges and industry and schools and in some part by the interest

Fundamentals and Applications of Nano Silicon in Plasmonics and Fullerines.
http://dx.doi.org/10.1016/B978-0-323-48057-4.00018-9

in nanotechnology and programs through additional funding to schools, training teachers, and opportunities for students. The interest in nanotechnology has ignited and rekindled interest of professional researchers at University level of education. It had prompted people to re-realize that education has long been recognized as an important factor for growing the fields of nanoscience and nanotechnology and solidifying and expanding their roles in the global economy. This has not been limited to the United States, but there is growing interest in many countries in developing educational programs across the full spectrum of educational levels from K-12 to postgraduate studies. Various formal and informal educational practices are being designed and tested that promote general awareness of ST in general and nanoscience and nanotechnology in particular, as well as provide advanced learning and skills development, including through group learning and peer assessment. Interest in innovative learning models that are being applied at the undergraduate level in order to train future leaders at the interface of engineering and management.

In the United States, many diverse programs funded by the NSF at the national are designed just to do that. Those include STEM, REU, iREU, LEF, RET, International Winter School, Nanooze, Equipment raining, National Outreach Activates. At the local level, programs can act as a test bed for new activities. Alternatively, local programs can leverage resources of the local site and community to offer programs which would not be possible on a national scale. Examples include Nanoexpress, Camps, Workshops, Tours, Local outreach activities. Many US universities are taking advantage of these programs. The University of Illinois Center for Nanoscale Science and Technology (CNST), for example putting together an innovative, fully integrated with the research to create and sustain a continuous, robust, and diverse pipeline in science, technology, engineering, and mathematics (STEM) education from elementary school through to graduate studies for workforce development. Affiliated undergraduates, graduate students, post-docs, researchers, professors, and staff are integrated in accomplishing this educational mission.

18.3 DEVELOPMENT OF ENHANCED MATERIAL AND COMPOSITES

On the fundamental level, nanotechnology succeeded in producing material and composites with novel and enhanced characteristics. Miniaturization produces nanoparticles. Nanoparticles are innovative because of several features. Solid nanoparticles, that is, nanoparticles of semiconductor, metal, oxides, ceramic, polymer, and alloys do not in general exist in nature. Second, in those the surface is large which makes them chemical reaction catalyzed on surface. Moreover, being small allows bringing that are absorbed on their surface brings adsorbed species closer allowing reactions to proceed at lower temperatures. Also their surfaces are exotic as it is less surface structure less rigid than bulk surface. There are other fundamental material characteristics. Confinement of electrons or positive charge changes the material characteristics (mass of electron, dielectric constant, conductivity, impurity level, mechanical vibration, which may drastically impact/change the optical, chemical, mechanical, thermal and electronic properties, etc.). Sub layering reveals mechanical strength (graphite is soft, but single layers are very strong-tubes, sheets). We can chemically modify surfaces by attaching additional species to design new surfaces (super molecules) and make them smart and interactive with biological processes.

One striking example is how miniaturization impacts the properties of matter, one of which is light interaction and propagation in metal and semiconductors. Unlike semiconductors, coupling of light to oscillating free electrons in metals allows metal nanostructures to concentrate or focus light to spots limited only by their size, much beyond the wavelength of incoming light, which enables novel concepts of metal-based lenses. Moreover, strong concentration, absorption and scattering induce drastic color change, as well as temperature rise and interactivity with the surrounding environment. On the other hand, miniaturization of semiconductors affords the material strong size-dependent luminescence and color, to the degree that it can take indirect material, such as silicon from being the dullest material to a glowing material. Integration of plasmonic and semiconductor effects in hybrid architectures is promising synergetic applications.

As a UNESCO report finds, in many countries, nanoparticles have become part of daily life. They are also found in clothing, tires, fertilizers and cosmetics, for instance. Given the breadth of potential applications, nanotechnology has become a research priority for many countries. Nanoscience and nanotechnology are a priority field for today's innovation leaders.

Nanotechnology had some success in developing combinations of materials (hybrid nano composites). New functional organic–inorganic nanomaterial composites enabled novel, non-traditional electronic, optical, magnetic and biomedical applications. For example, hybrid nanocomposites provide an attractive, versatile, technological platform for emerging high-added-value applications, such as photovoltaic cells and light-emitting devices, lithium-ion batteries, supercapacitors, and biosensors.

18.4 ENHANCED MANUFACTURING

Another success of nanotechnology is advanced manufacturing. Nanotechnology has taken manufacturing to another level, where it has become interwoven with manufacturing, even becoming a key element in the process [1]. In fact is being pursued big time by a widening circle of industrial countries. At this time Australia, Canada, China, France, Germany, Japan, the Republic of Korea, and USA are leading the effort. The impetus of the integration of nanotechnology with manufacturing is the belief that advanced manufacturing will help countries enhance national competitiveness and create jobs. As an example we mention here the advanced manufacturing partnership that was launched by president Obama in 2013. The partnership was propelled by an investment of US $ 2.9 billion. It was funded under the "Revitalize American Manufacturing Act (2014)." Additional funds are expected to come from private and non-federal US partners as matching funds. The partnership was overseen by a committee drawn from the industrial, labor, and academic sectors. The partnership created a network of 15 institutes. Several of the institutions focus on additive manufacturing, such as three-dimensional (3D) printing, digital manufacturing and design, lightweight manufacturing, wide band semiconductors, flexible hybrid electronics, integrated photonics, clean energy, and revolutionary fibers and textiles. China is another example in this advanced manufacturing emerging field. This program and others are intended to transform China into an "innovation-driven" country. In 2014, Canada revised its research strategy, which is entitled *Seizing Canada's Moment: Moving Forward in Science, Technology and Innovation* by incorporating advanced manufacturing. The focus of the initiative is automation, including robotics, lightweight materials and technologies, additive manufacturing, quantum materials, and nanotechnology.

18.5 NANOTECHNOLOGY CONSUMER PRODUCTS

Nanotechnology has already ventured into so many types of consumer products that one may use every day ranging from clothing to tennis racquets to charger of cell phones. Some of those were covered in 17.2 and 17.13. In fact the consumer world is exploding with "nanotechnology enhanced" products.

Currently there are numerous products on the market that are the result of nanotechnology. We briefly mention some examples. These products all use nanostructured materials to give them enhanced performance. In the sporting industry there is tennis balls that last longer, tennis rackets that are stronger, golf balls that fly straighter, nano ski wax that is easier to apply and more effective than standard wax, and bowling balls that are harder. In the paint applications, there is car wax that fills cracks more effectively and gives shinier looks. For cleaning, nano products are available to keep eyewear and other optical devices cleaner and more durable. In the clothing industry, there are pants that repel water, and stain shirts and shoe inserts that keep cool in the summer and warm in the winter, and non-stinking socks (nano sliver particle additive). Picture papers with nano-ceramic coatings that afford sharper and higher quality printed photos. The world of electronics has benefited from nanotechnology by providing faster and more powerful electronic devices, including computers, palm pilots (blackberries), flash drives, digital cameras and displays, cell phones, electronic ink displays, thin film batteries, and flexible electronics. Products were also introduced in the biotech world including bandages embedded with silver nanoparticles, patch-based drug delivery.

18.6 CHALLENGES

Nanotechnology (particle-by-particle) as of today, however suffers from several fundamental problems that may slow its realization as the platform technology of the 21st century. The problem is not totally financial or managerial, rather more of mass production, quality control and safety. Unlike atoms, which are naturally identical, particles are man-made and generally come out non-identical, that is, are not mono dispersed with a wide size and shape distribution. Most advanced device applications, especially those that require a small number of nanoparticles, such as single electron transistors or memory cells, or single molecule imaging and control, etc. this could turn to be pivotal. In this regard, there is hope that natural etching/organization may help form objects in exact shapes without machining, thus saving manufacturing time and cost. Carbon and silicon tend to form specific nano structures due to the sp2 and sp3 symmetry; however synthesis is not without defects and imperfections.

Another challenge is the processing time or speed. We have seen the development of the concept of nanofabrication develop from atom-by-atom to particle-by-particle and to self-organization or assembly. The evolution has been commencement with the desire to speed up the process so it may become commercially viable. The excitement about particle-by-particle is not only due to the fact that it speeds the process because a silicon particle of 10 nm diameter for example has ~30,000 atoms, which cuts down the time needed by this factor compared to atom-by-atom procedure, as well as it is easier to observe and handle such larger particles. Moreover, the good news particles may not offer properties not found in single atoms nor they are found in bulk material, namely offer novel effects. Self-assembly and organization may help speed up formation of macroscopic systems or super lattices of nanoparticles; but self-assembly requires identical objects to proceed.

These novel and unprecedented nanoscale functionalities enable a variety of exciting applications including a variety of futuristic applications. But there are novel exciting applications which have been demonstrated; those in general utilize a large volume but do not require high quality mono dispersed nanoparticles; others may be around the corner for the consumer use; those tend to require reasonably narrow size distribution such as hyperthermia treatment of cancer, monitoring cell death through molecular changes, substance identification and detection, photovoltaic thin film solar cells, solid state white lighting, photodetectors, subwavelength waveguides and antennas for commercial, military and elementary particles applications, nanofabrication, integration of optics and electronics, art of stained glass and lusterware, etc.

In any case it has been said that the *Holy Grail* in this business is to produce large quantities of nano components that are identical in size and shape with precision, control and repeatability, in due time, competitive cost, large enough volume that satisfy consumer demand and meet safety requirements. Thus the question that begs an answer is, When? When, if ever, do we see the predicted

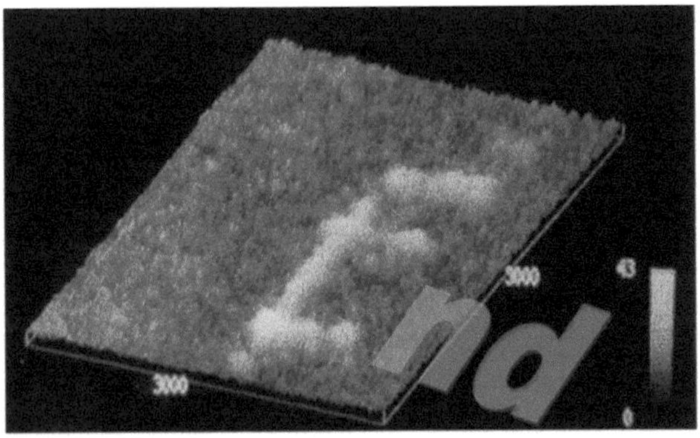

FIGURE 18.1 Letter "E" written with aluminum atoms on a silicon surface using a scanning tunneling microscope tip and organometallic aluminum gas as ink. The method was described in Section 5.5.3.

benefits? In one year? Ten? A hundred? This book presented a picture of where nanotechnology is now and highlighted the *state of the art* best guesses of those currently involved in industry and research establishments at the likely time scale of its development to the status of a mature technology in everyday use.

This book has shown that a lot has been accomplished but much more remains to be realized. Thus the pivotal question: Is nanotechnology still hot? Nearly steady state sustainable funding; new commercialization unitive in advanced countries; developing countries joining; increased number of patents, companies, scientific publications, conferences and books. It appears the march is not reversible.

I end this book with a nanotechnology image of Fig. 18.1, with the letter "E" for "End."

REFERENCES

[1] David R. Nanotechnology for Sustainable Manufacturing. Boca Raton, FL: CRC Press; 2014.

Index

M

N

O

P

Q

R

S

T

U

V

X

FSC
www.fsc.org
MIX
Papier | Fördert gute Waldnutzung
FSC® C083411

Kolibri 360 GmbH
Ferdinand-Jühlke-Straße 7
99095 Erfurt, Deutschland
produktsicherheit@kolibri360.de